Environmental Analysis of Organic Pollutants

Environmental Analysis of Organic Pollutants

Editor

Wenbin Liu

MDPI • Basel • Beijing • Wuhan • Barcelona • Belgrade • Manchester • Tokyo • Cluj • Tianjin

Editor
Wenbin Liu
Chinese Academy of Sciences
China

Editorial Office
MDPI
St. Alban-Anlage 66
4052 Basel, Switzerland

This is a reprint of articles from the Special Issue published online in the open access journal *Molecules* (ISSN 1420-3049) (available at: https://www.mdpi.com/journal/molecules/special_issues/analysisorganic_pollutant).

For citation purposes, cite each article independently as indicated on the article page online and as indicated below:

LastName, A.A.; LastName, B.B.; LastName, C.C. Article Title. *Journal Name* **Year**, *Volume Number*, Page Range.

ISBN 978-3-0365-5315-3 (Hbk)
ISBN 978-3-0365-5316-0 (PDF)

© 2022 by the authors. Articles in this book are Open Access and distributed under the Creative Commons Attribution (CC BY) license, which allows users to download, copy and build upon published articles, as long as the author and publisher are properly credited, which ensures maximum dissemination and a wider impact of our publications.

The book as a whole is distributed by MDPI under the terms and conditions of the Creative Commons license CC BY-NC-ND.

Contents

About the Editor ix

Ben-Zhan Zhu, Miao Tang, Chun-Hua Huang and Li Mao
Detecting and Quantifying Polyhaloaromatic Environmental Pollutants by Chemiluminescence-Based Analytical Method
Reprinted from: *Molecules* 2021, 26, 3365, doi:10.3390/ molecules26113365 1

Hongmin Yin, Jiayi Ma, Zhidong Li, Yonghong Li, Tong Meng and Zhenwu Tang
Polybrominated Diphenyl Ethers and Heavy Metals in a Regulated E-Waste Recycling Site, Eastern China: Implications for Risk Management
Reprinted from: *Molecules* 2021, 26, 2169, doi:10.3390/molecules26082169 19

Eirini Chrysochou, Panagiotis Georgios Kanellopoulos, Konstantinos G. Koukoulakis, Aikaterini Sakellari, Sotirios Karavoltsos, Minas Minaidis and Evangelos Bakeas
Heart Failure and PAHs, OHPAHs, and Trace Elements Levels in Human Serum: Results from a Preliminary Pilot Study in Greek Population and the Possible Impact of Air Pollution
Reprinted from: *Molecules* 2021, 26, 3207, doi:10.3390/molecules26113207 37

Ying Han, Yumeng Fei, Mingxin Wang, Yingang Xue, Hui Chen and Yuxuan Liu
Study on the Joint Toxicity of BPZ, BPS, BPC and BPF to Zebrafish
Reprinted from: *Molecules* 2021, 26, 4180, doi:10.3390/molecules26144180 67

Shijian Xiong, Fanjie Shang, Ken Chen, Shengyong Lu, Shaofu Tang, Xiaodong Li and Kefa Cen
Stable and Effective Online Monitoring and Feedback Control of PCDD/F during Municipal Waste Incineration
Reprinted from: *Molecules* 2021, 26, 4290, doi:10.3390/molecules26144290 81

Liping Fang, Linyan Huang, Gang Yang, Yang Jiang, Haiping Liu, Bingwen Lu, Yaxian Zhao and Wen Tian
Development of a Water Matrix Certified Reference Material for Volatile Organic Compounds Analysis in Water
Reprinted from: *Molecules* 2021, 26, 4370, doi:10.3390/molecules26144370 97

Wei Liu, Yuxi Zhang, Shui Wang, Lisen Bai, Yanhui Deng and Jingzhong Tao
Effect of Pore Size Distribution and Amination on Adsorption Capacities of Polymeric Adsorbents
Reprinted from: *Molecules* 2021, 26, 5267, doi:10.3390/molecules26175267 111

Zhitong Liu, Ke Xiao, Jingjing Wu, Tianqi Jia, Rongrong Lei and Wenbin Liu
Distributions of Polychlorinated Naphthalenes in Sediments of the Yangtze River, China
Reprinted from: *Molecules* 2021, 26, 5298, doi:10.3390/molecules26175298 123

Rongrong Lei, Yamei Sun, Shuai Zhu, Tianqi Jia, Yunchen He, Jinglin Deng and Wenbin Liu
Investigation on Distribution and Risk Assessment of Volatile Organic Compounds in Surface Water, Sediment, and Soil in a Chemical Industrial Park and Adjacent Area
Reprinted from: *Molecules* 2021, 26, 5988, doi:10.3390/molecules26195988 133

Xiaohong Xue, Yaoming Su, Hailei Su, Dongping Fan, Hongliang Jia, Xiaoting Chu, Xiaoyang Song, Yuxian Liu, Feilong Li, Jingchuan Xue and Wenbin Liu
Occurrence of Phthalates in Bottled Drinks in the Chinese Market and Its Implications for Dietary Exposure
Reprinted from: *Molecules* 2021, 26, 6054, doi:10.3390/molecules26196054 145

Jiawei Zhang, Mengtao Zhang, Huanyu Tao, Guanjing Qi, Wei Guo, Hui Ge and Jianghong Shi
A QSAR–ICE–SSD Model Prediction of the PNECs for Per- and Polyfluoroalkyl Substances and Their Ecological Risks in an Area of Electroplating Factories
Reprinted from: *Molecules* **2021**, *26*, 6574, doi:10.3390/molecules26216574 . . . . **161**

Wei Guo, Junhui Yue, Qian Zhao, Jun Li, Xiangyi Yu and Yan Mao
A 110 Year Sediment Record of Polycyclic Aromatic Hydrocarbons Related to Economic Development and Energy Consumption in Dongping Lake, North China
Reprinted from: *Molecules* **2021**, *26*, 6828, doi:10.3390/molecules26226828 . . . . **173**

Tianao Mao, Haoyang Wang, Zheng Peng, Taotao Ni, Tianqi Jia, Rongrong Lei and Wenbin Liu
Determination of Hexabromocyclododecane in Expanded Polystyrene and Extruded Polystyrene Foam by Gas Chromatography-Mass Spectrometry
Reprinted from: *Molecules* **2021**, *26*, 7143, doi:10.3390/molecules26237143 . . . . **187**

Mengqi Dong, Yuanyuan Li, Min Zhu, Jinbo Li and Zhanfen Qin
Tetrabromobisphenol A Disturbs Brain Development in Both Thyroid Hormone-Dependent and -Independent Manners in *Xenopus laevis*
Reprinted from: *Molecules* **2022**, *27*, 249, doi:10.3390/molecules27010249 . . . . **195**

Qianqian Ma, Yanli Li, Jianming Xue, Dengmiao Cheng and Zhaojun Li
Effects of Turning Frequency on Ammonia Emission during the Composting of Chicken Manure and Soybean Straw
Reprinted from: *Molecules* **2022**, *27*, 472, doi:10.3390/molecules27020472 . . . . **207**

Jinbo Li, Yuanyuan Li, Min Zhu, Shilin Song and Zhanfen Qin
A Multiwell-Based Assay for Screening Thyroid Hormone Signaling Disruptors Using *thibz* Expression as a Sensitive Endpoint in *Xenopus laevis*
Reprinted from: *Molecules* **2022**, *27*, 798, doi:10.3390/molecules27030798 . . . . **229**

Enrico Paris, Monica Carnevale, Beatrice Vincenti, Adriano Palma, Ettore Guerriero, Domenico Borello and Francesco Gallucci
Evaluation of VOCs Emitted from Biomass Combustion in a Small CHP Plant: Difference between Dry and Wet Poplar Woodchips
Reprinted from: *Molecules* **2022**, *27*, 955, doi:10.3390/molecules27030955 . . . . **239**

Ying Han, Yumeng Fei, Mingxin Wang, Yingang Xue and Yuxuan Liu
Effects of BPZ and BPC on Oxidative Stress of Zebrafish under Different pH Conditions
Reprinted from: *Molecules* **2022**, *27*, 1568, doi:10.3390/molecules27051568 . . . . **251**

Ji-Fang-Tong Li, Xing-Hong Li, Yao-Yuan Wan, Yuan-Yuan Li and Zhan-Fen Qin
Comparison of Dechlorane Plus Concentrations in Sequential Blood Samples of Pregnant Women in Taizhou, China
Reprinted from: *Molecules* **2022**, *27*, 2242, doi:10.3390/molecules27072242 . . . . **265**

Joaquín Hernández Fernández, Yoleima Guerra and Heidi Cano
Detection of Bisphenol A and Four Analogues in Atmospheric Emissions in Petrochemical Complexes Producing Polypropylene in South America
Reprinted from: *Molecules* **2022**, *27*, 4832, doi:10.3390/molecules27154832 . . . . **279**

Dawid Kucharski, Robert Stasiuk, Przemysław Drzewicz, Artur Skowronek, Agnieszka Strzelecka, Kamila Mianowicz and Joanna Giebułtowicz
Fit-for-Purpose Assessment of QuEChERS LC-MS/MS Methods for Environmental Monitoring of Organotin Compounds in the Bottom Sediments of the Odra River Estuary
Reprinted from: *Molecules* **2022**, *27*, 4847, doi:10.3390/molecules27154847 . . . . **289**

About the Editor

Wenbin Liu

Prof. Dr. Wenbin Liu. After graduating from the Research Center for Eco-Environmental Sciences, Chinese Academy of Sciences in 2005, Dr. Liu worked in the same institute, first as an assistant professor, and then as an associate professor and professor. In 2022, he transferred to the University of the Chinese Academy of Sciences (UCAS) and now serves as a professor and Vice Dean of College of Resources and Environment, UCAS. His research focuses on environment policy, environmental processes and control technology regarding persistent organic pollutants (POPs). He has published more than 110 SCI papers and has won a series of awards, including the Second prize of National Science and Technology Progress Award and Outstanding Science and Technology Achievement Prize, Chinese Academy of Sciences.

Review

Detecting and Quantifying Polyhaloaromatic Environmental Pollutants by Chemiluminescence-Based Analytical Method

Ben-Zhan Zhu [1,2,*], Miao Tang [1,2], Chun-Hua Huang [1,2] and Li Mao [1,2,*]

[1] State Key Laboratory of Environmental Chemistry and Ecotoxicology, Research Center for Eco-Environmental Sciences, Chinese Academy of Sciences, Beijing 100085, China; miaotang_st@rcees.ac.cn (M.T.); chhuang@rcees.ac.cn (C.-H.H.)
[2] University of Chinese Academy of Sciences, Beijing 100049, China
* Correspondence: bzhu@rcees.ac.cn (B.-Z.Z.); limao@rcees.ac.cn (L.M.); Tel.: +86-10-62849030 (B.-Z.Z.)

Abstract: Polyhaloaromatic compounds (XAr) are ubiquitous and recalcitrant in the environment. They are potentially carcinogenic to organisms and may induce serious risks to the ecosystem, raising increasing public concern. Therefore, it is important to detect and quantify these ubiquitous XAr in the environment, and to monitor their degradation kinetics during the treatment of these recalcitrant pollutants. We have previously found that unprecedented intrinsic chemiluminescence (CL) can be produced by a haloquinones/H_2O_2 system, a newly-found $^\bullet$OH-generating system different from the classic Fenton system. Recently, we found that the degradation of priority pollutant pentachlorophenol by the classic Fe(II)-Fenton system could produce intrinsic CL, which was mainly dependent on the generation of chloroquinone intermediates. Analogous effects were observed for all nineteen chlorophenols, other halophenols and several classes of XAr, and a novel, rapid and sensitive CL-based analytical method was developed to detect these XAr and monitor their degradation kinetics. Interestingly, for those XAr with halohydroxyl quinoid structure, a Co(II)-mediated Fenton-like system could induce a stronger CL emission and higher degradation, probably due to site-specific generation of highly-effective $^\bullet$OH. These findings may have broad chemical and environmental implications for future studies, which would be helpful for developing new analytical methods and technologies to investigate those ubiquitous XAr.

Keywords: polyhaloaromatic compounds; chemiluminescence; analytical method; Fenton system; hydroxyl radicals

1. Introduction

1.1. Polyhaloaromatics (XAr) and Their Toxicity

Polyhaloaromatic compounds (XAr) have been found world-wide in pesticides, pharmaceuticals, flame retardants and personal care products [1–4]. Most of these compounds are persistent and widely existing in the environment because of their recalcitrant properties in the soil and water. More importantly, not only the oxidative DNA damage, but also protein and DNA adducts may be induced by these XAr compounds in vitro and in vivo systems [5–9], which possibly makes them carcinogenic to mammalian organisms [10,11]. One typical group of XAr are polyhalophenols, some of which, such as 2,4,6-trichlorophenol and pentachlorophenol (PCP, the widely-used wood preservative) have been classified as priority pollutants by the U.S. Environmental Protection Agency (US-EPA) [12]. PCP was also classified as a group I human carcinogen by the International Agency for Research on Cancer (IARC) [13]. PCP is potentially carcinogenic to mammals. Hepatocellular carcinomas and hemangiosarcomas were observed from B6C3F1 mice under exposure to PCP [14]. In individuals with occupational exposure to PCP, malignant lymphoma and leukemia in humans were also found to relate to PCP [15].

1.2. The Detection of XAr

The widespread distribution and highly-toxic nature, together with the recalcitrant and carcinogenic characteristics of these XAr, have raised public concerns about their potential risks to human health and ecosystems [4,16–21]. Therefore, detecting and quantifying these widespread polyhaloaromatic pollutants or pharmaceutics in the environment is crucial. The traditional analytical methods used to detect XAr, such as UV−Vis spectrophotometry, high-performance liquid chromatography (HPLC) and gas chromatography (GC) [22,23], usually have many shortcomings: such as low sensitivity, time-consuming, requiring sample pretreatment, expensive apparatus and complicated operation. Therefore, a sensitive, simple, low-cost and effective analytical method to detect and quantify the ubiquitous XAr is urgently needed.

Chemiluminescence (CL) is well regarded as a kind of light emission from complicated chemical reactions, during which high-energy excited-states can be generated and energy is released [24–27]. Since the CL intensity depends on the rate of the chemical reactions, it can be used to detect and quantify the specific compounds that determine the generation of CL emission. For the CL-based analytical method, which exhibits the excellent properties of relatively simple, rapid, sensitive, without complicated pretreatment [28], has been widely used as the analytical method in environmental analysis, clinical diagnosis and food safety monitoring [29–31]. So if the CL analytical method can be successfully applied to detect and quantify the highly-toxic XAr, it will be significant for the degradation and pollution control of these persistent and recalcitrant substances.

1.3. The Degradation and Treatment of XAr

A variety of methods and technologies have been used to degrade and treat recalcitrant XAr in the environment, including enzymatic biodegradation, physical adsorption and chemically advanced oxidation. Among them, advanced oxidation processes (AOPs) have been considered to be the most widely-used means for degrading and treating XAr, mainly because they are highly-effective and environmentally green [2,32,33]. Several alternative AOPs, such as Fenton and Fenton-like oxidation [17,34], UV-photolysis [35], and ozonation [36], have also been developed to effectively degrade and treat the recalcitrant XAr. In these green AOP systems, the most reactive intermediate for degradation is the strong oxidative radical species hydroxyl radical (•OH) [32].

1.4. Unprecedented •OH Generation and CL Emission Can Be Produced from H_2O_2 and Polyhaloquinones, the Carcinogenic Metabolites of XAr

The most well-known pathway for •OH generation is through the classic Fenton or Fenton-like reactions mediated by reactive transition metal ions [37,38]. We recently found an unprecedented metal-independent •OH-generating system: polyhaloquinones and H_2O_2, and the molecular mechanism of typical nucleophilic substitution coupling with homolytic decomposition for •OH generation was proposed [3,39–47]. More interestingly, an unexpected intrinsic CL emission can also be produced in this novel •OH-generating system, which was found to be specifically dependent on •OH production [44,48–51]. Taking the reaction of tetrachloro-1,4-benzoquinone (TCBQ, a carcinogenic quinone metabolite of PCP) with H_2O_2 as an example, a typical two-step CL emission can be clearly observed, which is directly dependent on the two-step generating processes of •OH [44]. Moreover, not only for TCBQ, but also for other polyhaloquinones, such as other chloroquinones, fluoroquinones, bromoquinones and halonaphthoquinones, similar intrinsic •OH-dependent CL was produced. These results revealed an unprecedented •OH-generating and CL-producing system: polyhaloquinones (XQs) and H_2O_2.

1.5. The Goal of This Paper

It has been previously known that a variety of XAr, such as PCP, can be chemically-degraded into haloquinones during the AOPs with the generation and involvement of •OH [17–19,52–54]. Then we wonder whether •OH-dependent CL emission can also be generated in the degradation of PCP mediated by these AOPs. In addition, as one of the important final products of TCBQ after the interaction with H_2O_2, 2,5-dichloro-3,6-dihydroxyl-1,4-benzoquinone (DDBQ, an inert halohydroxyl quinoid compound) would not react with H_2O_2 to generate •OH and produce CL [44], but it is not clear whether the addition of extra •OH will induce the production of CL emission. Therefore, in order to answer the above questions, the following issues were addressed in a series of our recent studies: (1) Can intrinsic CL be generated from the degradation of precursors XAr (such as PCP and other chlorophenols) mediated by AOPs involving •OH-generation? (2) If so, what is the potential molecular mechanism for the CL emission, and is there any potential structure–activity relationship (SAR) between CL emission and the structures of XAr? (3) Do •OH-generating systems induce the inert halohydroxyl quinoid compounds to generate CL? (4) If so, what is the role of the typical bidentate coordination sites on the structures of halohydroxyl quinoid compounds, and is the CL generated from them different to that from other XAr? (5) Can we develop a sensitive and selective CL-based analytical method for the detection and quantification of XAr? (6) What are the potential chemical and environmental implications?

2. Chemiluminescence-Based Analytical Methods Induced by Fe(II)-Fenton System for the Detection of XAr

2.1. Intrinsic •OH-Dependent CL Emission Can Be Generated from the Degradation of the Priority Pollutant PCP in Fe(II)-Fenton System

We have previously known that the reaction between TCBQ and H_2O_2 could unexpectedly produce highly-reactive •OH and specific •OH-dependent CL [44]. Moreover, PCP could be chemically degraded and converted to TCBQ in the classic •OH-generating Fe(II)-Fenton system [52,54]. Then, we wanted to know whether CL could be generated in the PCP degradation process mediated by AOPs composed of the classic Fe(II)-Fenton system. As expected, neither •OH nor CL emission was detected when incubating PCP with H_2O_2, whereas a remarkable CL emission (510–580 nm) was generated when extra •OH was introduced by adding Fe(II)-EDTA (Fe(II)-ethylenediamine tetraacetic acid, a classic Fenton reagent) (Figure 1A), indicating that the degradation of PCP mediated by •OH-generating Fe(II)-Fenton system indeed produce intrinsic CL emissions [48].

Interestingly, similar to the CL generated from TCBQ/H_2O_2, the CL derived from PCP/Fe(II)-Fenton system was also directly dependent on •OH generation, as shown by the following line of evidence [48]: (1) The CL emission from the PCP/Fe(II)-Fenton system was significantly inhibited by dimethyl sulfoxide (DMSO), a typical •OH scavenger (Figure 1B); (2) both the yields of •OH and the intensity of CL emission increased with the increasing dosage of Fenton reagents; (3) CL was also produced from the other widely known •OH-generating Fenton agent Fe(II)-NTA (nitrilotriacetic acid) [55], and a good correlation was observed between the CL emission and the kinetics of •OH formation.

Previous studies have reported that the degradation of chlorophenol could produce chloroquinone as intermediates [52,54]. Considering that obvious CL could be generated from the interaction of TCBQ (or other haloquinones) with H_2O_2, we hypothesized that the critical species initiating the CL emission from PCP/Fe(II)-Fenton system might be such chloroquinone intermediates. As expected, five transient chloroquinone intermediates were identified (Figure 1C,D), they were TCBQ, tetrachloro-1,4-hydroquinone (TCHQ), tetrachloro-1,2-hydroquinone (tetrachlorocatechol, TCC), trichlorohydroxyl-1,4-benzoquine (TrCBQ-OH) and 2,5-dichloro-3,6-dihydroxyl-1,4-benzoquinone (DDBQ) [48]. It should be noted that besides DDBQ, the other four chloroquinones could undergo interactions with H_2O_2 to generate CL, and the addition of an extra •OH-generating Fenton reagent can markedly enhance CL emission. These results verify that the CL production from PCP/Fe(II)-Fenton systems is originated from the generation of chloroquinones.

On the basis of our previously discovered mechanism for CL generation from the TCBQ/H_2O_2 system [44], and together with the above findings, we proposed the underlying molecular mechanism for •OH-dependent CL emission from the degradation of PCP mediated by the classic •OH-generating Fe(II)-Fenton system (Scheme 1) [48]:

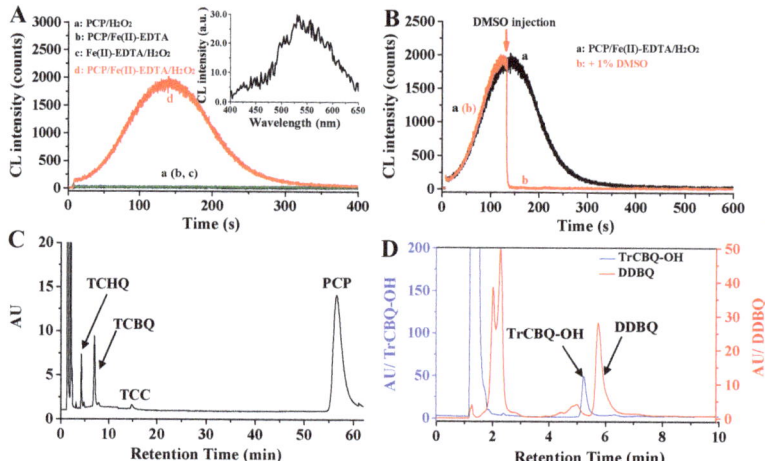

Figure 1. Intrinsic •OH-dependent CL was generated from PCP degradation mediated by Fe(II)-Fenton system, during which chloroquinones were formed as the critical intermediates [48]. (**A**) Intrinsic CL emission was produced by PCP/Fe(II)-Fenton system; (**B**) •OH scavenger DMSO can markedly inhibit CL emission; (**C,D**) several chloroquinone intermediates were generated from PCP during the CL emission. The CL measurement (**A,B**) was conducted through an ultraweak CL analyzer, with the CL signal recorded at a time interval of 0.1 s. Both the instant CL intensity and the intensity of total CL emission were recorded. The separation and identification of chloroquinone intermediates (**C,D**) generated from PCP degradation was conducted by HPLC analysis. For detailed experimental procedure, please refer to our previous study [48]. Fe(II)-EDTA, 1mM; H_2O_2, 100 mM. For (**A,B**) PCP, 20 µM; for (**C,D**) PCP, 1 mM. Copyright © 2015, American Chemical Society.

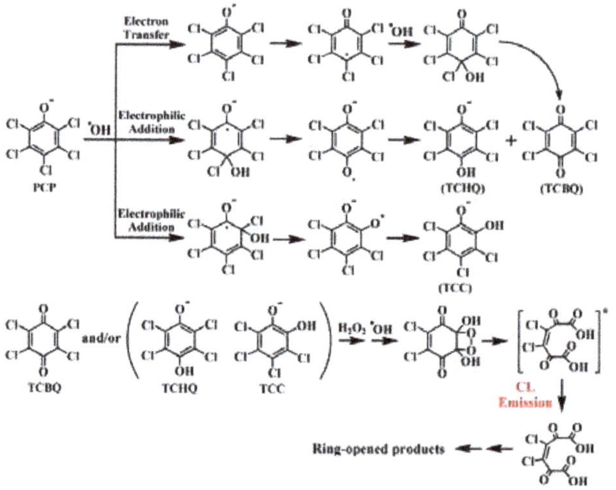

Scheme 1. The molecular mechanism for the unexpected •OH-dependent CL emission from the degradation of PCP mediated by Fe(II)-Fenton system [48]. The classic Fenton system could produce large amounts of reactive •OH, which further attacks PCP via electrophilic addition and/or electron transfer pathways, forming pentachlorophenoxyl radical and tetrachlorosemiquinone radicals. The latter radicals then convert to tetrachloroquinones, which further react with H_2O_2 to produce high-energy quinone-1,2-dioxetanes, and finally emit the intrinsic •OH-dependent CL as reported before. Copyright © 2015, American Chemical Society.

Since remarkable CL emission can be unambiguously generated from PCP/Fe(II)-Fenton system and increasing the concentration of Fenton reagents can enhance CL, these suggest that it is possible to develop an undiscovered novel CL-based analytical method for the detection and quantification of PCP. Further studies proved it was indeed the case [48]: the unique CL-generating property of PCP was used to develop a novel analytical method for detecting and measuring trace amounts of PCP, and it was found that the LOD (limit of detection) value was 1.8 ppb and the linear range (LR) was 2.6–18,620 ppb for PCP as detected by this CL method. Both the LOD and LR values are lower than the concentration of PCP (40 ppb) in the body fluids of people under non-occupational exposure, and much lower than PCP (19,580 ppb) concentration under occupational exposure [3]. Interestingly, the kinetics of CL emission was found to correlate well with the kinetics of PCP degradation: when PCP degradation achieved the maximum, CL emission was no longer observed (Figure 2). These results indicate that this novel CL-based analytical method could also be used to monitor the degradation kinetics of PCP.

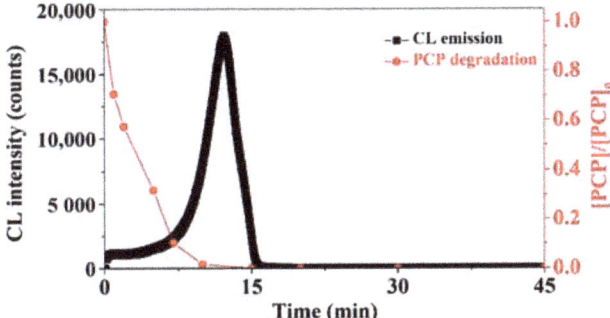

Figure 2. The CL emission from PCP/Fenton system correlated well with PCP degradation [48]. The CL emission was measured by an ultraweak CL analyzer, with the CL signal recorded at a time interval of 0.1s. The kinetics of PCP degradation was monitored by HPLC analysis. For detailed experimental procedure, please refer to our previous study [48]. PCP, 1 mM; Fe(II)−EDTA, 3 mM; H_2O_2, 300 mM. Copyright © 2015, American Chemical Society.

2.2. Analogous •OH-Dependent CL Emission from the Degradation of All 19 Chlorophenols and the Underlying Structure−Activity Relationship

The above findings that remarkable CL emission can be produced from the PCP/Fe(II)-Fenton system suggest that CL may also be generated from the interactions of other chlorophenols (CPs) with the classic Fe(II)-Fenton system. If so, there might be a close relationship between the CP structures and their abilities to generate CL. As expected, similar •OH-dependent intrinsic CL could also be generated from the other 18 CPs (Figure 3). The intensity of CL emission induced by CPs was strongly dependent on both chlorination level and chlorine substitution position. An obvious SAR between CPs structures and CL emission was observed [48–50]: (1) In general, as the chlorination level increases, the intensity of CL emission increases; (2) for CPs congeners, the CL increased in the order of para- < ortho- < meta-chlorine substitution with respect to the −OH group of CPs. For example, 2,5-dichlorophenol (DCP), 2,3,5-trichlorophenol (TCP) and 2,3,5,6-tetrachlorophenol (TeCP) generated the strongest CL emission among all the DCPs, TCPs and TeCPs congeners, respectively.

Actually, the properties that CL could be generated from all nineteen CPs in Fe(II)-Fenton system were also used to detect and quantify these ubiquitous CPs. As shown in Table 1, for those CPs that could produce an obvious CL emission, the LOD value could reach as low as 0.007 μM by this highly-sensitive CL analytical method.

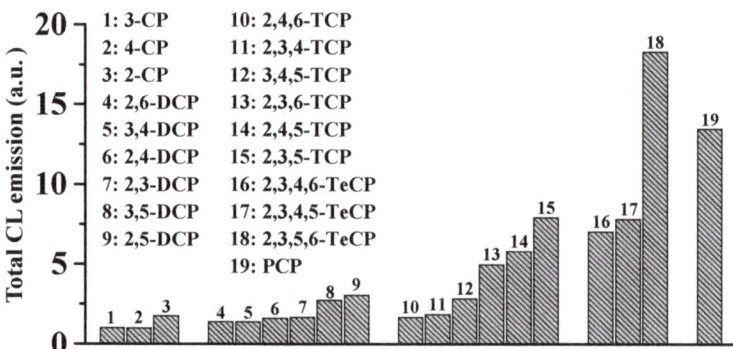

Figure 3. Intrinsic •OH-dependent CL were also generated from all 19 chlorophenols in AOPs mediated by Fe(II)-Fenton system [48]. The CL emission was measured by an ultraweak CL analyzer, and the intensity of total CL emission was recorded. CPs, 10 µM; Fe(II)–EDTA, 1 mM; H_2O_2, 100 mM. Copyright © 2015, American Chemical Society.

Table 1. The limit of detection (LOD) and linear range (LR) for the quantification of CPs by the novel CL-based analytical method mediated by Fe(II)-Fenton system [49,50]. Copyright © 2017, American Chemical Society. Copyright © 2016, with permission from Springer Nature.

CPs	LOD (µM)	LR (µM)
2,3,4-TCP	0.3	0.3~100
2,4,6-TCP	0.3	0.3~100
3,4,5-TCP	0.07	0.1~100
2,4,5-TCP	0.01	0.03~100
2,3,6-TCP	0.07	0.07~100
2,3,5-TCP	0.003	0.007~100
2,3,4,6-TeCP	0.01	0.03~100
2,3,4,5-TeCP	0.01	0.03~100
2,3,5,6-TeCP	0.007	0.01~100
PCP	0.007	0.01~100

Similar to the degradation of PCP, the degradation of other CPs by Fenton system-mediated AOPs also generated chloroquinones as the major intermediates, they were chloro-1,4-benzoquinones (CBQs), chloro-1,4-hydroquinones (CHQs) and chloro-1,2-hydroquinones (also called chlorocatechols, CCs) [49,50]. Moreover, we have previously known that, all the above chloroquinones could react with H_2O_2 to produce CL, and the addition of Fenton reagent can markedly enhance CL emission [48]. In the studies on CL emission from all the nineteen CPs, CL emission of CPs was found to primarily depend on the yields and types of the corresponding chloroquinone intermediates generated from CPs [49,50]: a good relationship was observed between the CL intensity of CPs and the total yields of corresponding CBQs/CHQs and TCC/3,4,6-TCC (tetrachlorocatechol and 3,4,6-trichlorocatechol, the two CCs which emit stronger CL) (Figure 4A,B). More interestingly, not only chloroquinone intermediates, but chlorosemiquinone radicals (CSQs•) were also produced during the CL emission of CPs in Fenton-like systems, and the types and yields of which were also in good agreement with the emission of CL and the generation of chloroquinone intermediates (Figure 4).

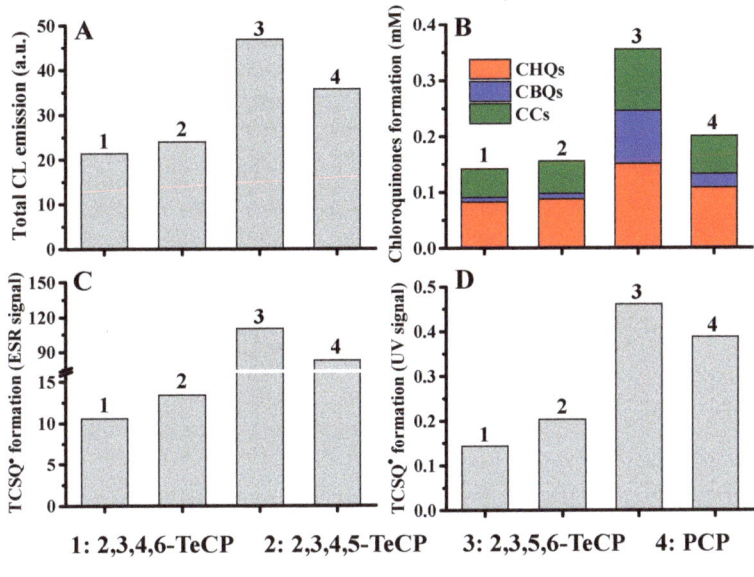

CCs: TCC and/or 3,4,6-TrCC

Figure 4. The intensity of CL emission from CPs correlated well with the total yields of the formation of corresponding chloroquinone intermediates and chlorosemiquinone radicals (taking PCP and three TeCPs for example) [50]. (**A**) The intensity of CL emission; (**B**) the total yields of CHQs/CBQs and CCs; (**C**,**D**) the yields of chlorosemiquinones measured by ESR (**C**) and UV−Vis method (**D**). The CL emission (**A**) was measured by an ultraweak CL analyzer, and the intensity of total CL emission was recorded. The maximum yields of chloroquinone intermediates (**B**) were measured by HPLC analysis. The maximum yields of CSQs• were measured by both monitoring the ESR signal of CSQs• through ESR analysis (**D**) and observing the typical UV−visible spectra of CSQs• through UV−visible analysis. For detailed experimental procedure, please refer to our previous study [50]. For (**A**), CPs, 30 µM; H_2O_2, 100 mM; Fe(II)-EDTA, 1 mM. For (**B**), CPs, 1 mM; H_2O_2, 1 mM; Fe(II)-EDTA, 3 mM. For (**C**,**D**), CPs, 0.5 mM; H_2O_2, 0.5 mM; Fe(II)-EDTA, 1.5 mM. Copyright © 2016, with permission from Springer Nature. In the tests of acute toxicity to *Photobacterium phosphoreum*, a good relationship was observed between the chemical structures of 19 CPs and their acute toxicity [49,50]: (1) The higher the level of chlorine substitution, the stronger the toxicity of CPs; (2) for CPs congeners, their toxicity increased in the order of non- < mono- < di-ortho position-chlorophenols. Moreover, the rules for the relationship between the CPs structures and their degradation rates during •OH-generating AOPs have been reported [56–58], which were listed as follows: (1) The higher the level of chlorine substitution, the slower the degradation rate of CPs; (2) for CPs congeners, the degradation rate decreased in the order of 3-/5- > 2-/4-/6-chlorine substitution CPs.

In summary, based on the above results, together with the previously reported studies on SAR [56,57,59,60], we found good correlations between the CP structures and their chemical activities (CL emission, toxicity and degradation rate) as following listed [49,50]: (1) The higher the level of chlorine substitution for CPs, stronger CL emission, higher toxicity and slower degradation could be observed; (2) for CPs congeners, CPs with position-3 or -5 chlorine substitution show stronger CL emission, higher toxicity and faster degradation; while those CPs congeners with position-6 chlorine substitution show much weak CL emission, lower toxicity, and slower degradation. These findings may suggest that, utilizing the distinct CL-generating property of CPs induced by classic Fe(II)-Fenton system, a novel CL-based method can be developed, to not only detect and quantify trace amounts of CPs in pure or real samples, but also provide valuable information for evaluating the toxicity or degradation rate of CPs.

2.3. Similar to Chlorophenols, Other Classes of XAr Could Also Generate •OH-Dependent Intrinsic CL Emission in the Degradation Mediated by Fe(II)-Fenton System

It should be noted that during the AOPs mediated by the •OH-generating Fe(II)-Fenton system, besides PCP and the other eighteen chlorophenols, similar •OH-dependent intrinsic CL was also generated from other halophenols and several other classes of XAr. These compounds include (Figure 5) [48]: other halophenols such as pentafluorophenol (PFP), 2,4,6-tribromophenol (2,4,6-TBP), the flame retardant 3,3′,5,5′-tetrabromobisphenol A (TBBPA), and the broad-spectrum antibacterial agent triclosan (TCS); halogenated naphthoquinone pesticides such as 2,3-dichloro-1,4-naphthoquinone (2,3-DCNQ); chlorophenoxyacetic acid herbicides such as the notorious 2,4,5-trichlorophenoxyacetic acid (2,4,5-T), one important component of Agent Orange; halogenated benzene biocides such as pentachlorobenzene (PCB); iodinated pharmaceuticals such as triiodothyronine (T3); These results indicate that most or even all XAr can generate •OH-dependent CL in the degradation mediated by the Fe(II)-Fenton system. Moreover, similar CL spectra were also observed from these XAr, which were found to be attributed to the analogous molecular mechanism and similarity in structures of light-emitting species.

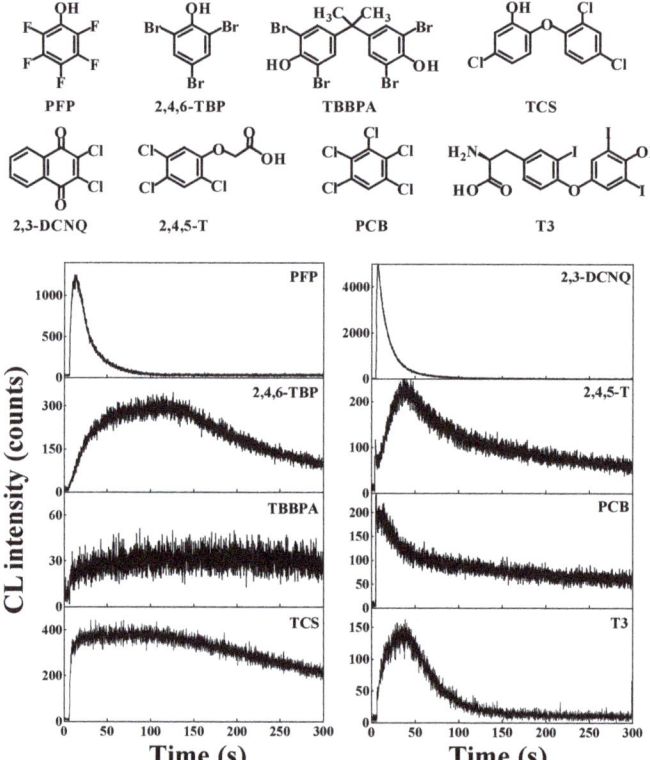

Figure 5. Similar •OH-dependent intrinsic CL were also produced from the degradation of several typical XAr in classic Fe(II)-Fenton system [48]. The CL emission was measured by an ultraweak CL analyzer, with the CL signal recorded at a time interval of 0.1 s. PFP, 2,4,6-TBP, TBBPA, TCS, 2,4,5-T and PCB, 30 µM; 2,3-DCNQ, 30 µM; T3, 3 µM; Fe(II)–EDTA, 1 mM; H_2O_2, 100 mM. Copyright © 2015, American Chemical Society.

Similarly, based on the CL emission properties of XAr, we developed a novel and sensitive CL analytical method to detect and quantify these ubiquitous XAr. As anticipated, we successfully detected and quantified traces of several typical XAr, including PFP, 2,4,6-

TBP, TBBPA, TCS, 2,3-DCNQ, 2,4,5-T, PCB and T3. For example, using this novel CL analytical method, we can directly detect concentrations as low as 0.03 μM for TCS, 0.07 μM for TBBPA, and 0.03 μM for T3 (Table 2) [48].

Table 2. The limit of detection (LOD) and linear range (LR) for the quantification of CPs by the novel CL-based analytical method mediated by Fe(II)-Fenton system [48]. Copyright © 2015, American Chemical Society.

XAr	LOD (μM)	LR (μM)
PCP	0.01~70	0.007
TCS	0.07~30	0.03
TBBPA	0.1~10	0.07
2,3-DCNQ	0.1~100	0.07
PCB	0.1~10	0.07
PCB	0.1~10	0.07
T3	0.03~1	0.03

More importantly, this novel CL analytical method based on Fe(II)-Fenton system has been utilized to evaluate and detect whether XAr is contained in an actual environmental sample, the discharge from a paper mill [48]. As anticipated, obvious CL emission could be generated from the discharge of paper mill in the presence of Fenton agent, and further analysis suggested that the discharge contained 2,4-dichlorophenol and 4-chlorophenol. Furthermore, this newly-developed CL-based analytical method can also be used for monitoring the degradation kinetics of XAr in their treatment mediated by AOPs. In the PCP/Fe(II)-Fenton system, the kinetics of CL emission correlated well with the kinetics of PCP degradation [48]: the profiles of CL emission coincided with the kinetic curves of PCP degradation, and no further CL emission could be generated when PCP degradation finished.

3. Chemiluminescence-Based Analytical Methods Induced by Co(II)-Fenton-Like System for the Detection of XAr

3.1. Distinct Intrinsic CL Emission in the Degradation of Halohydroxyl Quinoid Compounds by Co(II)-Fenton-Like System: Markedly Different from the CL Produced by Classic Fe(II)-Fenton System

As mentioned in the Introduction, unexpected •OH-dependent intrinsic CL could be generated from TCBQ/H_2O_2, with DDBQ formed as an important final product, whereas neither •OH nor CL could be generated by DDBQ and H_2O_2 [44]. However, it was unexpectedly discovered that a remarkable CL emission could be induced by adding some redox-active transition metal ions like Fe(II) and cobalt(II) (Co(II)), particularly for Co(II), which induced a much stronger CL emission than Fe(II) (Figure 6A,B) [51]. These results suggest that not only the key reactive oxygen species (ROS) intermediates for CL emission, but also the underlying molecular mechanism of the unexpected strong CL emission from DDBQ in Co(II)-Fenton-like system might be different from the CL in the classic Fe(II)-Fenton system.

It should be noted that, in the Co(II)-Fenton-like system, similar CL emissions were also observed when substituting DDBQ with other halohydroxyl quinones such as 2,5-dibromo-3,6-dihydroxyl-1,4-benzoquinone, chlorocatechols such as 3-chlorocatechol, 3,4-dichlorocatechol, 3,4,6-trichlorocatechol, 3,4,5-trichlorocatechol and tetrachlorocatechol (the latter two are typical effluents from bleached kraft pulp mills [61]), and other halocatechols such as tetrabromocatechol and tetrafluorocatechol (Figure 7A,B) [51].

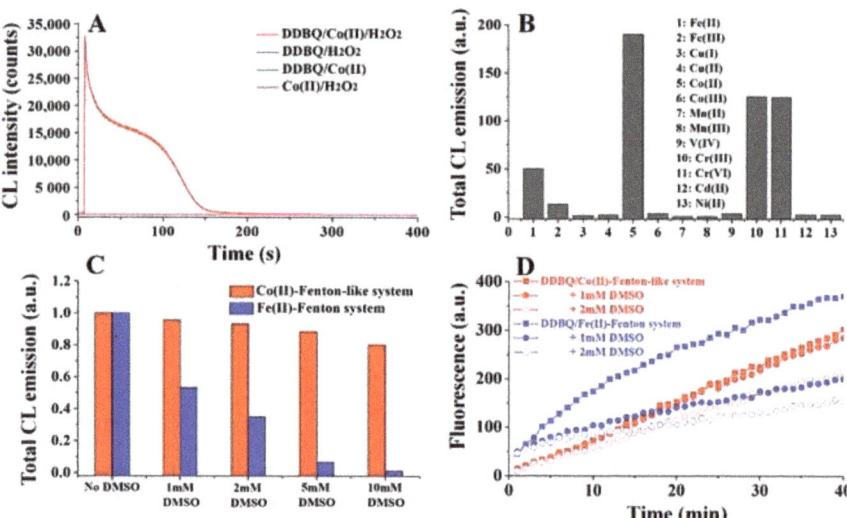

Figure 6. Distinct CL emission was generated from DDBQ in Co(II)-Fenton-like system, dependent on site specifically produced •OH [51]. (**A**) The CL emission generated from DDBQ/Co(II)-Fenton-like system; (**B**) the intensity of CL emission generated by Co(II) was higher than other active transition metal ions; (**C**) DMSO only slightly inhibited CL emission from DDBQ in Co(II)-Fenton-like system, but significantly inhibited CL in Fe(II)-Fenton system; (**D**) the effect of DMSO on the generation of •OH in DDBQ/Co(II)-Fenton-like system and DDBQ/Fe(II)-Fenton system. The CL emission (**A**–**C**) was measured by an ultraweak CL analyzer. Both the instant CL intensity and the intensity of total CL emission were recorded. (**D**) The generation of •OH from Fenton or Fenton-like systems was measured through fluorescence analysis with terephthalic as the •OH probe. For detailed experimental procedure, please refer to our previous study [51]. DDBQ, 10 μM; transition metal ions, 1 mM; H_2O_2, 100 mM. Copyright 2020, with permission from Elsevier.

3.2. Site Specifically Produced •OH, but Not Free •OH Is Responsible for the CL Production of Halohydroxyl Quinoid Compounds Induced by Co(II)-Fenton-Like System

We have previously known that the generation of free •OH was critical for CL generation by TCBQ (the precursor of DDBQ) and H_2O_2, and adding •OH scavenger could markedly inhibit CL emission [44]. However, we found it is not the case for the CL induced by DDBQ in Co(II)-Fenton-like system: CL emission was slightly inhibited by adding DMSO (Figure 6C). This strongly indicates that free •OH is not responsible for the CL emission induced by DDBQ in the Co(II)-Fenton like system.

However, although free •OH may not be involved in CL emission from the DDBQ/Co(II)-Fenton-like system, •OH generation was also detected and confirmed [51]. A relatively weak ESR signal of DMPO/•OH was observed from the CL reaction of DDBQ with the Co(II)-Fenton-like system, and adding DMSO could diminish the ESR signal of DMPO/•OH with the concurrent generation of the secondary radical DMPO/•CH₃, indicating •OH was indeed produced. It should be noted that the signal of secondary radical DMPO/•CH₃ was still relative weak even after the added DMSO completely diminished the signal of DMPO/•OH. However, for •OH production in the CL reaction of DDBQ with the classic free •OH-generating Fe(II)-Fenton system, the DMPO/•OH signal could be markedly diminished by adding DMSO, with the concurrent generation of the strong secondary radical DMPO/•CH₃. In addition, for the kinetics of •OH formation, adding DMSO only partly inhibited •OH generation from the CL reaction of DDBQ in the Co(II)-Fenton-like system, whereas it markedly inhibited •OH generation in the classic Fe(II)-Fenton system (Figure 6C,D). These results suggest that for the CL reaction of DDBQ with the Co(II)-Fenton-like system, •OH might be generated as the site specifically produced •OH, but not the free •OH.

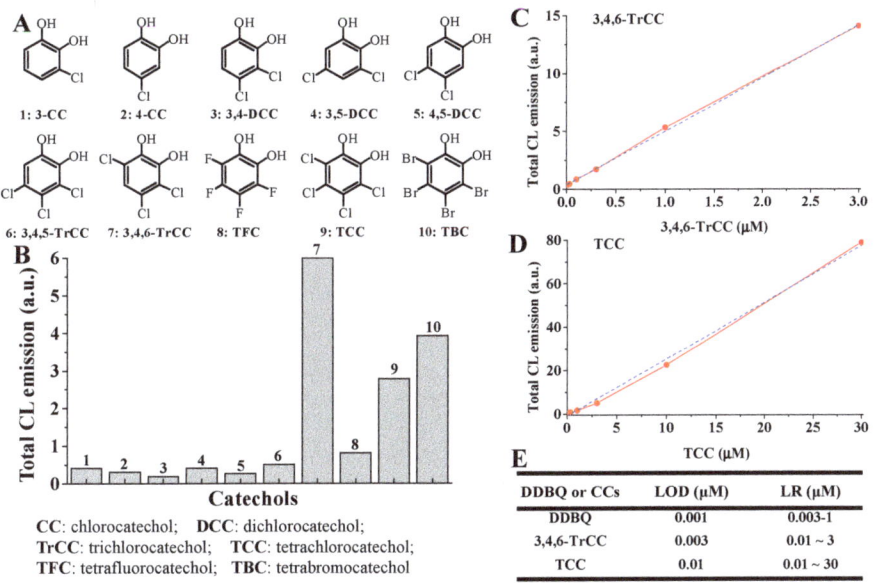

Figure 7. Analogous •OH-dependent CL emission were also generated from halocatechols with the typical Co(II)-Fenton-like system, which can be utilized to quantify trace amount of halocatechols and DDBQ [51]. (**A**) The structure of several halocatechols; (**B**) the total CL intensity for these halocatechols; (**C,D**) CL emission of 3,4,6-TrCC and TCC; (**E**) the LOD and LR for the quantification of DDBQ, 3,4,6-TrCC and TCC by the distinct CL analytical method mediated by Co(II)-Fenton-like system. The CL emission was measured by an ultraweak CL analyzer, and both the instant CL intensity and the intensity of total CL emission were recorded. For (**B**), Halocatechols, 10 μM; Co(II), 0.05 mM; H_2O_2, 100 mM; For (**C,D**), Co(II), 0.5 mM; H_2O_2, 100 mM. Copyright 2020, with permission from Elsevier.

Interestingly, besides the site specifically produced •OH, other ROS, such as $O_2^{•-}$ and $^1O_2$, were also generated from the CL reaction of DDBQ with the Co(II)-Fenton-like system [51]. The ROS intermediates from this distinct CL reaction of DDBQ with the Co(II)-Fenton-like system was significantly distinct from the CL reaction by TCBQ/H_2O_2 and the CL reaction induced by the classic Fe(II)-Fenton system mentioned above. Based on these results, we further varied the molar ratio of DDBQ:Co(II) to investigate the correlations between ROS production and CL emission, and to confirm which ROS is crucial for the generation of CL emission. As expected, not only the CL emission, but also the types and yields of ROS were significantly affected by changing the ratio of DDBQ:Co(II).

For the role of $O_2^{•-}$, it seems that $O_2^{•-}$ is not responsible for the CL generation from DDBQ in Co(II)-Fenton-like system [51]. $O_2^{•-}$ generation was hardly affected by changing the ratio of DDBQ:Co(II), suggesting that $O_2^{•-}$ is not the crucial ROS responsible for CL emission. Moreover, we also found that $^1O_2$ is not critical for initiating CL emission [51]. Although $^1O_2$ production could be enhanced by adding DDBQ, and large quantity of $^1O_2$ can be produced, most $^1O_2$ was generated only in the later stage. No matter how the ratio of DDBQ:Co(II) was varied, almost no $^1O_2$ was generated in the early stage, whereas CL emission had been generated in this stage.

However, for the role of •OH, a close relationship between •OH generation and DDBQ degradation was clearly observed: the higher the concentration of DDBQ, the lower the yields of •OH generation. Moreover, the degradation of DDBQ also correlated well with CL emission: the process of generating CL emission was accompanied by the degradation of DDBQ, and when DDBQ degradation achieved the maximum, no further CL emission can be observed. These results indicate that CL emission is closely related to and probably dependent on the generation of site specifically produced •OH. More interestingly, different from the results observed in the classic Fe(II)-Fenton system, the inhibitory effect of DMSO on the kinetics of both CL emission and •OH generation were relatively weak in Co(II)-Fenton-like system (Figure 6C,D). These results further confirmed that CL generation from the DDBQ/Co(II)-Fenton-like system is indeed dependent on site specifically produced •OH, but not free •OH, $O_2^{•-}$ and $^1O_2$ [51]. In other words, the site specifically produced •OH might be the initiating ROS for CL emission.

3.3. The Molecular Mechanism for the Site-Specific •OH-Dependent CL Emission of Halohydroxyl Quinoid Compounds in Co(II)-Fenton-Like System

Then the questions are how to generate site-specific •OH and how to induce the intrinsic CL emission of halohydroxyl quinoid compounds such as DDBQ? The further investigation indicated that during the CL reaction of DDBQ with Co(II)-Fenton-like system, Co(II) could combine with DDBQ to form a Co(II)-DDBQ complex through the active bidentate coordinate sites -C(O)C(OH) [62,63], and then the attack by H_2O_2 may result in the generation of site-specific •OH at the coordinates sites, which then induce the degradation of DDBQ and the concurrent CL emission [51].

In order to better clarify the molecular mechanism for the distinct CL emission from DDBQ in the Co(II)-Fenton like system, the conversion of Co(II) was investigated in detail. As anticipated, we further found that Co(II) was transformed to Co(III) on the basis of UV–Vis analysis on the production of the Co(III)-complex [64] and the XPS analysis [65,66] on the binding energy of Co for the samples before or after reaction. On the basis of all the above findings and our previous results, we proposed the underlying molecular mechanism for the distinct site-specific •OH-dependent CL emission from the degradation of DDBQ mediated by Co(II)-Fenton-like system (Scheme 2) [51]:

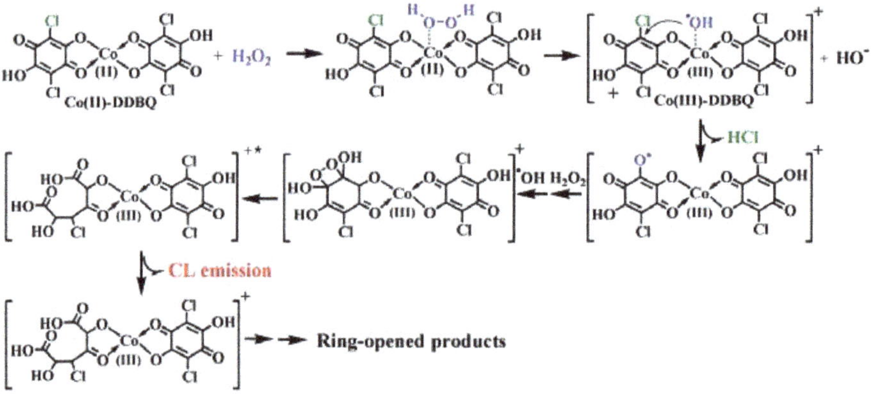

Scheme 2. The potential molecular mechanism for the unexpectedly strong CL emission from DDBQ/Co(II)-Fenton-like system mainly depends on site specifically produced •OH [51]. Copyright 2020, with permission from Elsevier.

3.4. Highly-Sensitive CL-Based Analytical Method for the Detection of Halohydroxyl Quinoid Compounds on the Basis of Co(II)-Fenton-Like System

As mentioned above, the CL intensity of DDBQ in Co(II)-Fenton-like system is markedly stronger than that in classic Fe(II)-Fenton system. Therefore, this distinct CL-generating property of DDBQ induced by Co(II)-Fenton-like system might be utilized to measure and quantify trace amounts of DDBQ, and to monitor the degradation of DDBQ as well. Indeed, by using this novel CL-based analytical method, the LR for the quantitative detection of DDBQ is 3–1000 nM, and the LOD value is as low as 1 nM [51], which is much lower than the LOD value (0.5 µM) of DDBQ by using the reported traditional HPLC method. Compared with the traditional analytical methods, such as HPLC [67] on the detection of DDBQ, this novel CL-based analytical method exhibits a series of advantages, such as being relatively simple, rapid, sensitive, and without a complicated pretreatment process. Moreover, the kinetics of CL emission was in good agreement with the kinetics of DDBQ degradation: CL can be produced during DDBQ degradation, and when DDBQ was completely degraded, CL emission was no longer observed (Figure 8B). Interestingly, we found that the efficiency for DDBQ degradation mediated by the Co(II)-Fenton-like system was much higher than the classic Fe(II)-Fenton system (Figure 8A) [51], possibly due to the site-specific •OH, which can be effectively used to degrade adjacent DDBQ, that was produced in the former system, whereas free •OH was produced in the latter system. These results indicate that in the treatment or degradation of those pollutants with structures containing bidentate coordination sites that can typically bind with transition metal ions, the technology consisting of AOPs mediated by the Co(II)-Fenton-like system may degrade target pollutants more effectively, because the degradation occurring site specifically exhibits a higher efficiency.

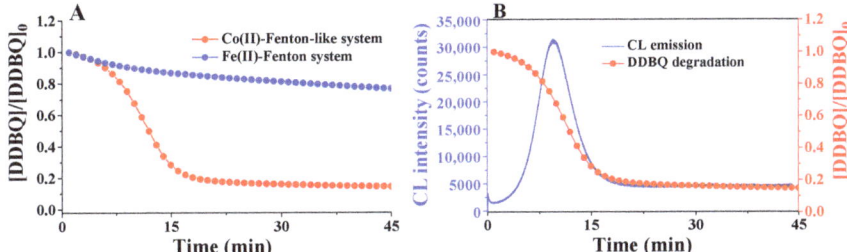

Figure 8. (**A**) The degradation of DDBQ in Co(II)-Fenton-like system was more effective than that in the classic Fe(II)-Fenton system, and (**B**) good correlation can be observed between CL emission and DDBQ degradation in Co(II)-Fenton-like system [51]. The kinetics of DDBQ degradation (**A**,**B**) was monitored by HPLC analysis. The CL emission (**B**) was measured by an ultraweak CL analyzer, with the CL signal recorded at a time interval of 0.1s. DDBQ, 3 mM; transition metal ions, 3 mM; H_2O_2, 300 mM. Copyright 2020, with permission from Elsevier.

More importantly, as mentioned in Section 3.1., in the Co(II)-Fenton-like system, typically strong CL emission can be produced from halocatechols. All the above halocatechols have a similar bidentate coordinated site(-C(OH)C(OH)), which can well coordinate with transition metal ions such as Co(II). Similarly, compared with Fe(II)-Fenton system, these halocatechols in the Co(II)-Fenton-like system produced significantly stronger CL emissions. Interestingly, the distinct CL analytical method based on the Co(II)-Fenton-like system can also be applied to detect and quantify halocatechols. By utilizing this unique CL-based method, the LOD values for 3,4,6-trichlorocatechol and tetrachlorocatechol can reach as low as 3 nM and 10 nM, respectively, [51] (Figure 7C–E). These results, together with above findings on the measurement of DDBQ, indicate that this novel CL analytical method based on the Co(II)-Fenton-like system can be applied to detect and quantify DDBQ (or its analogues), halocatechols and other halohydroxyl quinoid compounds.

4. The Advantages and Challenges of the Typical CL-Based Analytic Methods for the Detection of XAr in their Environmental Applications

Compared with the traditional methods for the detection and quantification of XAr, the novel CL-based analytical method mediated by the Fe(II)-Fenton or Co(II)-Fenton-like system displayed a series of advantages: (1) It is relatively selective. Obvious CL emission can be produced from samples containing XAr, whereas no CL can be observed from samples that do not contain XAr. (2) It is extremely sensitive. Both the LOD and LR values for the quantification of XAr by this CL-based method are lower than the values obtained from traditional analytical methods. (3) It is simple, fast and low-cost. This CL-based method is easy to operate, without expensive apparatus and complicated pretreatment.

However, this CL-based analytical method for the detection of XAr also faces potential challenges during its environmental applications. It can only evaluate whether a sample contains XAr, but cannot accurately identify the specific kind of XAr. In order to unambiguously identify the XAr contained in the sample, the CL-based method needs to be combined with other qualitatively analytical methods. Moreover, the organic pollutants that can be detected by this novel CL analytical method are limited: it can detect and measure only those halocompounds with aromatic structures, but not the non-halogenated aromatic compounds and haloaliphatic compounds. Therefore, it is necessary to develop a more efficient and sensitive CL-producing system based on this typical CL analytical method of XAr, to detect and quantify more environment pollutants. These would be addressed in our future studies.

5. Conclusions

On the basis of the above series of studies, we have made important progress in the research of CL emission generated from the degradation of ubiquitous XAr mediated by AOPs, which have broad chemical and environmental implications.

We found that the degradation of highly-toxic PCP in AOPs mediated by the classic Fe(II)-Fenton system can unprecedentedly produce intrinsic CL emission, specifically dependent on the generation of free •OH. Interestingly, besides PCP, all nineteen chlorophenols can be induced to generate •OH-dependent intrinsic CL by the classic Fe(II)-Fenton system, and the underlying SAR for CL of these chlorophenols was revealed: (1) In general, the CL emission increased with the increase of chlorination level; (2) for CP congeners, the CL emission decreased in the following order of 3-/5- > 2-/4-/6-chlorine substitution CPs; (3) the CL intensity for each CP was determined by the types and the total yields of corresponding chloroquinone intermediates and semiquinone radicals. Additionally, several kinds of XAr, including the broad-spectrum antibacterial agent triclosan, the flame retardant TBBPA, the widely used herbicides 2,4,5-T, and iodinated pharmaceuticals T3, were capable of generating similar •OH-dependent CL. Based on these results, a novel and sensitive CL analytical method was developed, which can not only detect and quantify these ubiquitous XAr, but also monitor their degradation kinetics, and provide useful information for evaluating their toxicity to organisms and degradation rate.

However, for those XAr with structures containing the group of bidentate coordination sites, such as halohydroxyl quinone and halocatechol, the degradation of which in AOPs mediated by the distinct Co(II)-Fenton-like system would generate significant CL emission, much stronger than that generated in the classic Fe(II)-Fenton system, which was found to be attributed to the site-specific generation of reactive •OH. Consequently, a more sensitive CL analytical method via the Co(II)-Fenton-like system was developed to detect and quantify halohydroxyl quinoid compounds or those compounds with similar structures, and monitor their degradation kinetics. Moreover, the AOPs consisting of the typical Co(II)-Fenton-like system can be effectively used to selectively degrade and treat halohydroxyl quinoid compounds, and this is because these compounds can bind with Co(II) to site-specifically produce the highly-reactive •OH, which can be effectively used to degrade the target pollutants.

Author Contributions: Review and editing, B.-Z.Z.; writing—original draft, M.T.; review and editing, C.-H.H.; writing—review and editing, L.M. All authors have read and agreed to the published version of the manuscript.

Funding: The work in this paper was supported by NSFC Grants (21577149, 22021003, 21621064 and 21836005); Strategic Priority Research Program of CAS Grant [No. XDB14030101]; "From 0 to 1" Original Innovation Project, the Basic Frontier Scientific Research Program, CAS (ZDBS-LY-SLH027).

Institutional Review Board Statement: Not applicable.

Informed Consent Statement: Not applicable.

Data Availability Statement: Not applicable.

Conflicts of Interest: The authors declare no conflict of interest.

References

1. Olaniran, A.O.; Igbinosa, E.O. Chlorophenols and other related derivatives of environmental concern: Properties, distribution and microbial degradation processes. *Chemosphere* **2011**, *83*, 1297–1306. [CrossRef]
2. Pera-Titus, M.; García-Molina, V.; Baños, M.A.; Giménez, J.; Esplugas, S. Degradation of chlorophenols by means of advanced oxidation processes: A general review. *Appl. Catal. B-Environ.* **2004**, *47*, 219–256. [CrossRef]
3. Zhu, B.-Z.; Shan, G.-Q. Potential mechanism for pentachlorophenol-Induced carcinogenicity: A novel mechanism for metal-independent production of hydroxyl radicals. *Chem. Res. Toxicol.* **2009**, *22*, 969–977. [CrossRef]
4. Dann, A.B.; Hontela, A. Triclosan: Environmental exposure, toxicity and mechanisms of action. *J. Appl. Toxicol.* **2010**, *31*, 285–311. [CrossRef]
5. Chignell, C.F.; Han, S.K.; Mouithys-Mickalad, A.; Sik, R.H.; Stadler, K.; Kadiiska, M.B. EPR studies of in vivo radical production by 3,3′,5,5′-tetrabromobisphenol A (TBBPA) in the Sprague–Dawley rat. *Toxicol. Appl. Pharmacol.* **2008**, *230*, 17–22. [CrossRef]
6. Song, Y.; Wagner, B.A.; Witmer, J.R.; Lehmler, H.-J.; Buettner, G.R. Nonenzymatic displacement of chlorine and formation of free radicals upon the reaction of glutathione with PCB quinones. *Proc. Natl. Acad. Sci. USA* **2009**, *106*, 9725–9730. [CrossRef]
7. Michałowicz, J.; Majsterek, I. Chlorophenols, chlorocatechols and chloroguaiacols induce DNA base oxidation in human lymphocytes (in vitro). *Toxicology* **2010**, *268*, 171–175. [CrossRef]
8. Carstens, C.P.; Blum, J.K.; Witte, I. The role of hydroxyl radicals in tetrachlorohydroquinone induced DNA strand break formation in PM2 DNA and human fibroblasts. *Chem. Biol. Interact.* **1990**, *74*, 305–314. [CrossRef]
9. Bukowska, B. 2,4,5-T and 2,4,5-TCP induce oxidative damage in human erythrocytes: The role of glutathione. *Cell Biol. Int.* **2004**, *28*, 557–563. [CrossRef]
10. Wang, Y.-J.; Ho, Y.-S.; Jeng, J.-H.; Su, H.-J.; Lee, C.-C. Different cell death mechanisms and gene expression in human cells induced by pentachlorophenol and its major metabolite, tetrachlorohydroquinone. *Chem. Biol. Interact.* **2000**, *128*, 173–188. [CrossRef]
11. Wang, Y.-J.; Lee, C.-C.; Chang, W.-C.; Liou, H.-B.; Ho, Y.-S. Oxidative stress and liver toxicity in rats and human hepatoma cell line induced by pentachlorophenol and its major metabolite tetrachlorohydroquinone. *Toxicol. Lett.* **2001**, *122*, 157–169. [CrossRef]
12. Igbinosa, E.O.; Odjadjare, E.E.; Chigor, V.N.; Igbinosa, I.H.; Emoghene, A.O.; Ekhaise, F.O.; Igiehon, N.O.; Idemudia, O.G. Toxicological profile of chlorophenols and their derivatives in the environment: The public health perspective. *Sci. World J.* **2013**, *2013*, 1–11. [CrossRef]
13. Guyton, K.Z.; Loomis, D.; Grosse, Y.; El Ghissassi, F.; Bouvard, V.; Benbrahim-Tallaa, L.; Guha, N.; Mattock, H.; Straif, K. Carcinogenicity of pentachlorophenol and some related compounds. *Lancet Oncol.* **2016**, *17*, 1637–1638. [CrossRef]
14. McConnell, E.E.; Huff, J.E.; Hejtmancik, M.; Peters, A.C.; Persing, R. Toxicology and carcinogenesis studies of two grades of pentachlorophenol in B6C3F1 mice. *Fundam. Appl. Toxicol.* **1991**, *17*, 519–532. [CrossRef]
15. Seiler, J.P. Pentachlorophenol. *Mutat. Res. Genet. Toxicol.* **1991**, *257*, 27–47. [CrossRef]
16. Covaci, A.; Harrad, S.; Abdallah, M.A.-E.; Ali, N.; Law, R.J.; Herzke, D.; De Wit, C.A. Novel brominated flame retardants: A review of their analysis, environmental fate and behaviour. *Environ. Int.* **2011**, *37*, 532–556. [CrossRef]
17. Zimbron, J.A.; Reardon, K.F. Fenton's oxidation of pentachlorophenol. *Water Res.* **2009**, *43*, 1831–1840. [CrossRef]
18. Gupta, S.S.; Stadler, M.; Noser, C.A.; Ghosh, A.; Steinhoff, B.; Lenoir, D.; Horwitz, C.P.; Schramm, K.-W.; Collins, T. Rapid total destruction of chlorophenols by activated hydrogen peroxide. *Science* **2002**, *296*, 326–328. [CrossRef]
19. Sorokin, A.; Seris, J.L.; Meunier, B. Efficient oxidative dechlorination and aromatic ring cleavage of chlorinated phenols catalyzed by iron sulfophthalocyanine. *Science* **1995**, *268*, 1163–1166. [CrossRef] [PubMed]
20. Peller, J.; Wiest, O.; Kamat, P.V. Mechanism of hydroxyl radical-induced breakdown of the herbicide 2,4-dichlorophenoxyacetic acid (2,4-D). *Chem. Eur. J.* **2003**, *9*, 5379–5387. [CrossRef]
21. Zhang, H.C.; Huang, C.H. Oxidative transformation of triclosan and chlorophene by manganese oxides. *Environ. Sci. Technol.* **2003**, *37*, 2421–2430. [CrossRef] [PubMed]
22. Eriksson, J.; Rahm, S.; Green, N.; Bergman, Å.; Jakobsson, E. Photochemical transformations of tetrabromobisphenol A and related phenols in water. *Chemosphere* **2004**, *54*, 117–126. [CrossRef]
23. Peiró, A.M.; Ayllón, J.A.; Peral, J.; Doménech, X. TiO$_2$-photocatalyzed degradation of phenol and ortho-substituted phenolic compounds. *Appl. Catal. B-Environ.* **2001**, *30*, 359–373. [CrossRef]
24. Schuster, G.B. Chemiluminescence of organic peroxides. Conversion of ground-state reactants to excited-state products by the chemically initiated electron-exchange luminescence mechanism. *Acc. Chem. Res.* **1979**, *12*, 366–373. [CrossRef]
25. Matsumoto, M. Advanced chemistry of dioxetane-based chemiluminescent substrates originating from bioluminescence. *J. Photochem. Photobiol. C* **2004**, *5*, 27–53. [CrossRef]
26. Widder, E.A. Bioluminescence in the ocean: Origins of biological, chemical, and ecological diversity. *Science* **2010**, *328*, 704–708. [CrossRef]
27. McCapra, F. Chemical generation of excited states: The basis of chemiluminescence and bioluminescence. *Methods Enzymol.* **2000**, *305*, 3–47. [CrossRef]
28. Grayeski, M.L. Chemiluminescence analysis. *Anal. Chem.* **1987**, *59*, 1243A–1256A. [CrossRef]
29. Wang, X.; Lin, J.-M.; Liu, M.-L.; Cheng, X.-L. Flow-based luminescence-sensing methods for environmental water analysis. *Trends Anal. Chem.* **2009**, *28*, 75–87. [CrossRef]
30. Marquette, C.A.; Blum, L.J. Chemiluminescent enzyme immunoassays: A review of bioanalytical applications. *Bioanalysis* **2009**, *1*, 1259–1269. [CrossRef]

31. Dodeigne, C.; Thunus, L.; Lejeune, R. Chemiluminescence as diagnostic tool. A review. *Talanta* **2000**, *51*, 415–439. [CrossRef]
32. Von Sonntag, C. Advanced oxidation processes: Mechanistic aspects. *Water Sci. Technol.* **2008**, *58*, 1015–1021. [CrossRef]
33. Wang, J.-L.; Xu, L.-J. Advanced oxidation processes for wastewater treatment: Formation of hydroxyl radical and application. *Crit. Rev. Environ. Sci. Technol.* **2012**, *42*, 251–325. [CrossRef]
34. Rastogi, A.; Al-Abed, S.R.; Dionysiou, D.D. Effect of inorganic, synthetic and naturally occurring chelating agents on Fe(II) mediated advanced oxidation of chlorophenols. *Water Res.* **2009**, *43*, 684–694. [CrossRef]
35. Lente, G.; Espenson, J.H. Photoaccelerated oxidation of chlorinated phenols. *Chem. Commun.* **2003**, 1162–1163. [CrossRef]
36. Hong, P.K.; Zeng, Y. Degradation of pentachlorophenol by ozonation and biodegradability of intermediates. *Water Res.* **2002**, *36*, 4243–4254. [CrossRef]
37. Halliwell, B.; Gutteridge, J.M.C. *Free Radicals in Biology and Medicine*, 5th ed.; Oxford University Press: Oxford, UK, 2015; p. 944.
38. Edwards, J.O.; Curci, R. Fenton type activation and chemistry of hydroxyl radical. In *Catalytic Oxidations with Hydrogen Peroxide as Oxidant*; Strukul, G., Ed.; Springer: Dordrecht, The Netherlands, 1992; pp. 97–151.
39. Qin, H.; Huang, C.-H.; Mao, L.; Xia, H.-Y.; Kalyanaraman, B.; Shao, J.; Shan, G.-Q.; Zhu, B.-Z. Molecular mechanism of metal-independent decomposition of lipid hydroperoxide 13-HPODE by halogenated quinoid carcinogens. *Free Radic. Biol. Med.* **2013**, *63*, 459–466. [CrossRef]
40. Zhu, B.-Z.; Zhu, J.-G.; Mao, L.; Kalyanaraman, B.; Shan, G.-Q. Detoxifying carcinogenic polyhalogenated quinones by hydroxamic acids via an unusual double Lossen rearrangement mechanism. *Proc. Natl. Acad. Sci. USA* **2010**, *107*, 20686–20690. [CrossRef] [PubMed]
41. Huang, C.-H.; Shan, G.-Q.; Mao, L.; Kalyanaraman, B.; Qin, H.; Ren, F.-R.; Zhu, B.-Z. The first purification and unequivocal characterization of the radical form of the carbon-centered quinone ketoxy radical adduct. *Chem. Commun.* **2013**, *49*, 6436–6438. [CrossRef] [PubMed]
42. Huang, C.-H.; Ren, F.-R.; Shan, G.-Q.; Qin, H.; Mao, L.; Zhu, B.-Z. Molecular mechanism of metal-independent decomposition of organic hydroperoxides by halogenated quinoid carcinogens and the potential biological implications. *Chem. Res. Toxicol.* **2015**, *28*, 831–837. [CrossRef]
43. Shao, J.; Huang, C.-H.; Kalyanaraman, B.; Zhu, B.-Z. Potent methyl oxidation of 5-methyl-2′-deoxycytidine by halogenated quinoid carcinogens and hydrogen peroxide via a metal-independent mechanism. *Free Radic. Biol. Med.* **2013**, *60*, 177–182. [CrossRef] [PubMed]
44. Zhu, B.-Z.; Mao, L.; Huang, C.-H.; Qin, H.; Fan, R.-M.; Kalyanaraman, B.; Zhu, J.-G. Unprecedented hydroxyl radical-dependent two-step chemiluminescence production by polyhalogenated quinoid carcinogens and H_2O_2. *Proc. Natl. Acad. Sci. USA* **2012**, *109*, 16046–16051. [CrossRef]
45. Zhu, B.-Z.; Zhao, H.-T.; Kalyanaraman, B.; Liu, J.; Shan, G.-Q.; Du, Y.-G.; Frei, B. Mechanism of metal-independent decomposition of organic hydroperoxides and formation of alkoxyl radicals by halogenated quinones. *Proc. Natl. Acad. Sci. USA* **2007**, *104*, 3698–3702. [CrossRef]
46. Zhu, B.-Z.; Shan, G.-Q.; Huang, C.-H.; Kalyanaraman, B.; Mao, L.; Du, Y.-G. Metal-independent decomposition of hydroperoxides by halogenated quinones: Detection and identification of a quinone ketoxy radical. *Proc. Natl. Acad. Sci. USA* **2009**, *106*, 11466–11471. [CrossRef] [PubMed]
47. Zhu, B.-Z.; Zhao, H.-T.; Kalyanaraman, B.; Frei, B. Metal-independent production of hydroxyl radicals by halogenated quinones and hydrogen peroxide: An ESR spin trapping study. *Free Radic. Biol. Med.* **2002**, *32*, 465–473. [CrossRef]
48. Mao, L.; Liu, Y.-X.; Huang, C.-H.; Gao, H.-Y.; Kalyanaraman, B.; Zhu, B.-Z. Intrinsic chemiluminescence generation during advanced oxidation of persistent halogenated aromatic carcinogens. *Environ. Sci. Technol.* **2015**, *49*, 7940–7947. [CrossRef]
49. Gao, H.-Y.; Mao, L.; Li, F.; Xie, L.-N.; Huang, C.-H.; Shao, J.; Shao, B.; Kalyanaraman, B.; Zhu, B.-Z. Mechanism of intrinsic chemiluminescence production from the degradation of persistent chlorinated phenols by the Fenton system: A structure–activity relationship study and the critical role of quinoid and semiquinone radical intermediates. *Environ. Sci. Technol.* **2017**, *51*, 2934–2943. [CrossRef]
50. Gao, H.-Y.; Mao, L.; Shao, B.; Huang, C.-H.; Zhu, B.-Z. Why does 2,3,5,6-tetrachlorophenol generate the strongest intrinsic chemiluminescence among all nineteen chlorophenolic persistent organic pollutants during environmentally-friendly advanced oxidation process? *Sci. Rep.* **2016**, *6*, 33159. [CrossRef]
51. Mao, L.; Gao, H.-Y.; Huang, C.-H.; Qin, L.; Huang, R.; Shao, B.; Shao, J.; Zhu, B.-Z. Unprecedented strong intrinsic chemiluminescence generation from degradation of halogenated hydroxy-quinoid pollutants by Co(II)-mediated advanced oxidation processes: The critical role of site-specific production of hydroxyl radicals. *Chem. Eng. J.* **2020**, *394*. [CrossRef]
52. Fang, X.W.; Schuchmann, H.-P.; Von Sonntag, C. The reaction of the •OH radical with pentafluoro-, pentachloro-, pentabromo- and 2,4,6-triiodophenol in water: Electron transfer vs. addition to the ring. *J. Chem. Soc. Perkin Trans. 2* **2000**, 1391–1398. [CrossRef]
53. Mvula, E.; Von Sonntag, C. Ozonolysis of phenols in aqueous solution. *Org. Biomol. Chem.* **2003**, *1*, 1749–1756. [CrossRef] [PubMed]
54. Czaplicka, M. Photo-degradation of chlorophenols in the aqueous solution. *J. Hazard. Mater.* **2006**, *134*, 45–59. [CrossRef] [PubMed]
55. Goldstein, S.; Meyerstein, D.; Czapski, G. The Fenton reagents. *Free Radic. Biol. Med.* **1993**, *15*, 435–445. [CrossRef]

56. Tang, W.Z.; Huang, C. Effect of chlorine content of chlorinated phenols on their oxidation kinetics by Fenton's reagent. *Chemosphere* **1996**, *33*, 1621–1635. [CrossRef]
57. Tang, W.Z.; Huang, C. The effect of chlorine position of chlorinated phenols on their dechlorination kinetics by Fenton's reagent. *Waste Manag.* **1995**, *15*, 615–622. [CrossRef]
58. Oturan, N.; Panizza, M.; Oturan, M.A. Cold incineration of chlorophenols in aqueous solution by advanced electrochemical process electro-Fenton. Effect of number and position of chlorine atoms on the degradation kinetics. *J. Phys. Chem. A* **2009**, *113*, 10988–10993. [CrossRef]
59. Smith, S.; Furay, V.; Layiwola, P.; Filho, J.M. Evaluation of the toxicity and quantitative structure—Activity relationships (QSAR) of chlorophenols to the copepodid stage of a marine copepod (*Tisbe battagliai*) and two species of benthic flatfish, the flounder (*Platichthys flesus*) and sole (*Solea solea*). *Chemosphere* **1994**, *28*, 825–836. [CrossRef]
60. Padmanabhan, J.; Parthasarathi, R.; Subramanian, V.; Chattaraj, P.K. Group philicity and electrophilicity as possible descriptors for modeling ecotoxicity applied to chlorophenols. *Chem. Res. Toxicol.* **2006**, *19*, 356–364. [CrossRef]
61. McKague, A.B. Some toxic constituents of chlorination-stage effluents from bleached kraft pulp mills. *Can. J. Fish. Aquat. Sci.* **1981**, *38*, 739–743. [CrossRef]
62. Molčanov, K.; Jurić, M.; Kojić-Prodić, B. Stacking of metal chelating rings with π-systems in mononuclear complexes of copper(II) with 3,6-dichloro-2,5-dihydroxy-1,4-benzoquinone (chloranilic acid) and 2,2′-bipyridine ligands. *Dalton Trans.* **2013**, *42*, 15756–15765. [CrossRef]
63. Verdaguer, M.; Michalowicz, A.; Girerd, J.J.; Berding, N.A.; Kahn, O. EXAFS study and magnetic properties of copper(II) chloranilato and bromanilato chains: A new example of orbital reversal. *Inorg. Chem.* **1980**, *19*, 3271–3279. [CrossRef]
64. Tanaka, H.; Fukuoka, T.; Okamoto, K. Cobalt(III) oxidation pretreatment of organic phosphorus compounds for a spectrophotometric determination of phosphorus in water. *Anal. Sci.* **1994**, *10*, 769–774. [CrossRef]
65. McIntyre, N.S.; Cook, M.G. X-ray photoelectron studies on some oxides and hydroxides of cobalt, nickel, and copper. *Anal. Chem.* **1975**, *47*, 2208–2213. [CrossRef]
66. Fu, L.; Liu, Z.; Liu, Y.; Han, B.; Hu, P.; Cao, L.; Zhu, D. Beaded cobalt oxide nanoparticles along carbon nanotubes: Towards more highly integrated electronic devices. *Adv. Mater.* **2005**, *17*, 217–221. [CrossRef]
67. Sarr, D.H.; Kazunga, C.; Charles, M.J.; Pavlovich, J.G.; Aitken, M.D. Decomposition of tetrachloro-1,4-benzoquinone (*p*-chloranil) in aqueous solution. *Environ. Sci. Technol.* **1995**, *29*, 2735–2740. [CrossRef]

Article

Polybrominated Diphenyl Ethers and Heavy Metals in a Regulated E-Waste Recycling Site, Eastern China: Implications for Risk Management

Hongmin Yin [1], Jiayi Ma [2], Zhidong Li [3], Yonghong Li [2], Tong Meng [1] and Zhenwu Tang [1,2,*]

1. College of Environmental Science and Engineering, North China Electric Power University, Beijing 102206, China; hongminyin@ncepu.edu.cn (H.Y.); mengtong@ncepu.edu.cn (T.M.)
2. College of Life and Environmental Sciences, Minzu University of China, Beijing 100081, China; 17010176@muc.edu.cn (J.M.); 20040254@muc.edu.cn (Y.L.)
3. Cangzhou Ecology and Environment Bureau, Cangzhou 061000, China; lzdhdm@163.com
* Correspondence: zwtang@muc.edu.cn; Tel.: +86-10-68936927

Abstract: Serious pollution of multiple chemicals in irregulated e-waste recycling sites (IR-sites) were extensively investigated. However, little is known about the pollution in regulated sites. This study investigated the occurrence of 21 polybrominated diphenyl ethers (PBDEs) and 10 metals in a regulated site, in Eastern China. The concentrations of PBDEs and Cd, Cu, Pb, Sb, and Zn in soils and sediments were 1–4 and 1–3 orders of magnitude lower than those reported in the IR-sites, respectively. However, these were generally comparable to those in the urban and industrial areas. In general, a moderate pollution of PBDEs and metals was present in the vegetables in this area. A health risk assessment model was used to calculate human exposure to metals in soils. The summed non-carcinogenic risks of metals and PBDEs in the investigated soils were 1.59–3.27 and 0.25–0.51 for children and adults, respectively. Arsenic contributed to 47% of the total risks and As risks in 71.4% of the total soil samples exceeded the acceptable level. These results suggested that the pollution from e-waste recycling could be substantially decreased by the regulated activities, relative to poorly controlled operations, but arsenic pollution from the regulated cycling should be further controlled.

Keywords: regulated e-waste recycling; polybrominated diphenyl ethers (PBDEs); heavy metals; environmental media; vegetable; risks

1. Introduction

The amount of waste of electronic and electrical equipment is continually increasing with the widespread use of electronic products. The Global E-waste Monitor reported that the world generation of electronic waste (e-waste) reached 41.8, 44.7, and 53.6 million Mt in 2014, 2016, and 2019, respectively; and it is expected to grow to 74.7 million tons by 2030 [1–3]. Correspondingly, the e-waste from Asia, Americas, Europe, and Africa in 2019 were at 24.9, 13.1, 12.0, and 2.90 Mt, respectively [3]. However, only 15.6% to 20.0% of e-waste was documented to be collected and properly recycled in 2014, 2016, and 2019 around the world [1–3]. Therefore, the safety of the recycling of e-waste is of increasing concern.

E-waste usually contains or is contaminated with the multiclass hazardous chemicals. Typically, high concentrations of lead, cadmium, and mercury are often observed in e-wastes due to these metals used widely in battery [4]. Some lead compounds are also used as a stabilizer or plasticizer [5,6]. Chromium was used in data tapes, floppy disks, and are usually used as pigments in many electrical and electronic products [5,7]. Copper is frequently used in printed circuit boards [8,9]. Antimony trioxide is used as flame retardants [10,11]. In addition, many organic chemicals are also used widely in electrical and electronic products as flame retardants. For example, polybrominated diphenyl ethers

(PBDEs) and dechlorane plus (DPs) are mainly used in cables. DPs are usually used in connector coating and electric wires, and PBDEs are usually found in printed circuit boards [12,13]. Additionally, polychlorinated biphenyls (PCBs) are used in transformers and capacitors [14]. These chemicals are easily released into the environment during the dismantling or treatment of e-waste and can cause adverse effects on the human body [15–18]. Meanwhile, many kinds of toxic pollutants such as polycyclic aromatic hydrocarbons (PAHs) and polychlorinated dibenzo-*p*-dioxins and dibenzofurans could be produced in the process of e-waste incineration [19].

In the past decades, large amounts of e-waste were transported to developing countries for recycling. Typically, about 70% of the world's e-waste were processed in China in 2012 [20]. Due to the lack of related technology and cost savings, e-waste was recycled in many countries by the primary method, inducing manual dismantling, crude processing of circuit board, acid treatment to recover metals, open burning, and dumping of residues [21,22]. This led to many crude e-waste recycling sites formed in many low-income countries, such as India, China, Ghana, and Nigeria [15,23–25]. In China, the irregulated e-waste recycling sites were mainly concentrated in family workshops in Shantou, Qingyuan, and Taizhou. Mechanical recovery dominated in these irregulated process, which was mainly to extract precious metals. Many previous surveys found that serious environmental pollution occurs in these sites. Labunska et al. reported that the PBDE concentrations reached 390 $\mu g \cdot g^{-1}$ in soils, from an e-waste open-burning area, and reached 220 $\mu g \cdot g^{-1}$ in sediments surrounding a circuit board shredding in the Guiyu town, Southern China [26]. Moeckel et al. also reported that PAH concentrations caused by e-waste dumping activities were up to 16.0 $\mu g \cdot g^{-1}$ in soils from Agbogbloshie, Ghana [27]. Meanwhile, Wu et al. found that the mean concentrations ($\mu g \cdot g^{-1}$) of Sb, Sn, and Ag reached 5,003, 2,527, and 47.9, respectively, in surface soils, at an e-waste burnt site in Guiyu, China [28]. The serious environmental pollutions could directly cause high levels of human internal exposure to these chemicals, likely resulting in high risks to human health [29,30].

In recent years, the irregulated e-waste recycling was gradually banned and relevant policies and technical regulations were developed and implemented in many countries [3,31]. The e-waste recycling operations is usually concentrated to industrial parks and many pollution-control measures are applied to recycling processes [32]. Currently, many venous industrial parks are established in China to concentrate the recycling activities of e-waste [21]. Many pollution-control measures, including safe storage, closely mechanical dismantling, and negative pressure to control dismantling dust, are performed in the recycling processes [21]. These measures are generally considered to be effective in reducing chemical emissions. However, there are few systematic investigations of the environment contamination surrounding the regulated recycling sites and the associated human health risks remain unknown.

In this study, we used a regulated e-waste recycling site in Eastern China, as a case study. Our main aim was to characterize the concentrations and risks of PBDEs and metals, two characteristic pollutants in the e-waste recycling process, in the soils, sediments and the vegetables surrounding the site. Then, the contamination levels and the risks posed by these chemicals in this site and in other areas, and in particular in irregulated e-waste recycling sites, were compared, in order to understand if the current pollution-control measures in the regulated e-waste recycling are effective and safe. Our results will provide important insight into these chemicals in regulated e-waste recycling area, and might be useful in informing the development of risk management measures.

2. Results and Discussion

2.1. Contamination of PBDEs

The concentrations of PBDEs in soil, sediment, and vegetable samples are summarized in Table 1. In the soil samples, the detection frequency of BDE-209 was highest (100%), followed by BDE-154 (33%). However, the detection frequencies of most other congeners were less than 20%. The concentrations of BDE-209 and Σ_{21}PBDEs were 3.05–1331 and 3.05–1366 ng·g^{-1}, respectively. Higher concentrations (ng·g^{-1}) of Σ_{21}PBDEs were observed in the sites S$_{19}$ (1366) and S$_{20}$ (593) (Figure S1 from Supplementary Materials). The e-waste was piled up at the two sampling sites. BDE-209 accounted for 94.7% of the concentration of Σ_{21}PBDEs. The octa-BDEs (including BDE-196, -197, -201, -202, -203, and BDE-205) and the nona-BDEs (including BDE-206, -207, and BDE-208) represented 1.86% and 2.16% of the concentrations of Σ_{21}PBDEs, respectively. The results reflected that the e-waste recycled in the site might originate mainly from China. Many previous studies confirmed that deca-BDE predominated in the e-waste from China [33–35].

Table 1. Summary of polybrominated diphenyl ether (PBDE) concentrations (ng·g^{-1} dry weight), with basic statistical parameters, for the soils, sediments, and vegetables from the regulated electronic waste recycling site.

	Soils		Sediments		Vegetables	
	Min–Max	Median	Min–Max	Median	Min–Max	Median
BDE-28	<0.002	<0.002	<0.002–0.043	<0.002	<0.003–0.18	0.05
BDE-47	<0.003–0.11	<0.003	<0.003–0.81	<0.003	0.005–1.27	0.26
BDE-66	<0.006–0.03	<0.006	<0.006–0.62	<0.006	<0.007–0.19	0.05
BDE-85	<0.013	<0.013	<0.013	<0.013	<0.013–0.18	<0.013
BDE-99	<0.007–0.22	<0.007	<0.007–0.31	<0.007	<0.007–0.30	0.110
BDE-100	<0.010	<0.010	<0.010–1.26	<0.010	<0.010–0.36	0.02
BDE-138	<0.005	<0.005	<0.005	<0.005	<0.005–0.06	<0.005
BDE-153	<0.008–0.33	<0.008	<0.008–0.23	<0.008	<0.008–0.20	<0.008
BDE-154	<0.022–14.0	<0.022	<0.022–0.93	0.12	<0.022–0.76	0.20
BDE-183	<0.018–0.81	<0.018	<0.018–0.13	<0.018	<0.019–0.30	0.019
BDE-190	<0.035	<0.035	<0.035	<0.035	<0.035	<0.035
BDE-196	<0.054–1.09	<0.054	<0.054	<0.054	<0.054–0.44	<0.054
BDE-197	<0.073–1.04	<0.073	<0.073–0.70	<0.073	<0.073–0.37	<0.073
BDE-201	<0.002–0.74	<0.002	<0.002–0.73	<0.002	<0.003–0.46	<0.003
BDE-202	<0.002–0.55	<0.002	<0.002–0.54	<0.002	<0.003–0.37	<0.003
BDE-203	<0.105–1.56	<0.105	<0.105	<0.105	<0.105–0.37	<0.105
BDE-205	<0.079	<0.079	<0.079	<0.079	<0.079	<0.079
BDE-206	<0.045–8.70	<0.045	<0.045–1.74	<0.045	0.34–1.76	0.89
BDE-207	<0.081–3.96	<0.081	<0.081–4.12	<0.081	<0.082–1.12	0.61
BDE-208	<0.183–1.99	<0.183	<0.183–1.81	<0.183	<0.184–1.24	0.54
BDE-209	3.05–1,331	8.09	4.27–314	10.8	90.2–419	201
Σ_{21}PBDEs	3.05–1,336	9.23	4.31–327	11.6	93.9–425	204

In the sediment samples, the detection frequencies of BDE-209 and BDE-154 were 100% and 57.1%, respectively. Low detection frequencies of the other congeners were also observed. The concentrations of BDE-209 and Σ_{21}PBDEs in sediments were in the range of 4.27–314 and 4.31–327 ng·g^{-1}, respectively. Higher concentrations of Σ_{21}PBDEs were found in the sites S$_{24}$ and S$_{25}$. BDE-209 represented 93.2% of the Σ_{21}PBDE concentrations. In general, the contributions of other individual congeners were less than 1%.

The detection frequencies of individual congeners in the vegetable samples were different from those of the soil and sediment samples. BDE-28, -47, -66, -99, -100, -154, -183, -206, -207, -208, and BDE-209 were detectable in more than 50% of vegetable samples, likely reflecting the stronger bioaccumulation of these chemicals. The median concentrations of BDE-209 and Σ_{21}PBDEs in vegetable samples were 201 and 204 ng·g^{-1}, respectively. The highest Σ_{21}PBDE concentration was found in the baby's breath samples at site V$_7$ (425 ng·g^{-1} dry weight), a site near the soil sampling site S$_{20}$. Interestingly, the congener profiles of PBDEs in vegetables were similar to those in soils and sediments, although the detection frequencies of congeners differed, suggesting the vegetable PBDEs mainly originated from the emission from the recycling operations in the site.

The PBDE concentrations found in this study were compared to those observed from the IR-sites, UI-areas, and the AR-areas (Tables S3–S5 from Supplementary Materials). The median concentrations of PBDEs of soils and sediments in our study were 1–4 orders of magnitude lower than those separately reported in some IR-sites in China (Figure 1). In particular, the maximum concentration of total PBDEs in the soils from an IR-site in Guiyu was one order of magnitude higher than that in this area [36]. Huang et al. [37] reported that the maximum concentration of total PBDEs in the sediment in an IR-site reached 1030 µg·g^{-1}, four orders of magnitude higher than our result. For the vegetables, however, the concentrations of PBDEs found in this study were generally higher than those reported in the IR-sites. In this study, most vegetable samples were collected within the recycling plant. In most IR-sites selected, almost all vegetable samples were collected far from the recycling plant [38,39]. In addition, the differences of plant species and their growth period among different studies might also be responsible for the comparison results.

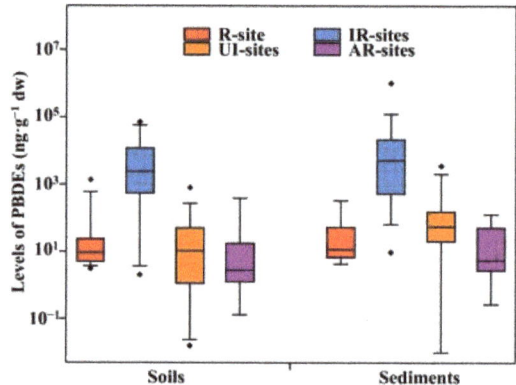

Figure 1. Comparison of the PBDE concentrations (ng·g^{-1} dry weight) in soils and sediments in the study area with the other areas. R-site, regulated e-waste recycling site (this study); IR-sites, irregulated e-waste recycling sites; UI-areas, urban and industrial areas; AR-areas, agricultural and rural areas (data from the Tables S3 and S4 from Supplementary Materials) The maximum, minimum, and median (or mean) of PBDE concentrations in each literature were selected as a total sample. The horizontal lines represent the 10th, 50th, and 90th percentiles, and the boxes represent the 25th and 75th percentiles; the asterisk below or above indicates outliers.

The contamination levels of PBDEs in this site were compared with those in the UI-areas. Median concentrations of PBDEs of soils and sediments in this site were comparable to or lower than those in the UI-areas (Figure 1). Typically, the median concentrations of soil PBDEs from an industrial area in Tianjin and an urban area in Shenyang-Fushun, China, were similar to those reported in this study [40,41]. The concentrations of sediment PBDEs in this study were generally lower than those reported in UI-areas. Even, the concentrations of total PBDEs in some urban rivers in the Yangtze River Delta and some urban reaches of the Pearl River were one order of magnitude higher than those observed in this area [42,43]. The concentrations of total PBDEs in vegetable samples observed in this study were lower than those reported in pine needle from Shanghai and camphor bark from the Jiangsu Province, China [44,45]. However, our results were one to two orders of magnitude higher than those observed in the UI-areas in Inner Mongolia and in the Zhejiang Province, China [46,47].

The PBDE concentrations observed in this study were also compared with those reported in some agricultural regions in China (Figure 1). In general, the median PBDE concentrations in soils and sediments from our study were 6.36 and 2.12 times higher than those in the AR-areas (Tables S3 and S4 from Supplementary Materials). However, the median PBDE concentrations of soils in Shanghai and sediments in the Chaohu Lake were one order of magnitude higher than those in our study [48,49]. This was likely due to the fact that these samples from the AR-areas were collected from surrounding industrial or commercial areas. The PBDE concentrations in vegetables in the AR-areas were generally lower than our results [46,50].

Further, the profiles of PBDEs in soils, sediments, and vegetables between this study and other areas were compared. Based on our collected literature, BDE-209 contributed a median of 84%, 45%, and 40% of total concentrations of PBDEs in the soils, sediments, and vegetables, respectively, in the IR-sites (Tables S3–S5 from Supplementary Materials). However, the penta-BDEs and octa-BDEs usually contributed a lot to total PBDE concentrations. Typically, the sum of BDE-47 and BDE-99 accounted for 27%, 63%, and 38% of the total PBDE concentrations in the soil, sediments, and vegetables in these IR-sites, respectively [36,51,52]. Historically, a large amount of e-waste mainly imported from European and American countries were irregularly recycled in China. These e-wastes usually contained penta-BDEs and octa-BDEs. In the UI-areas, BDE-209 contributed a median of 95%, 80%, and 96% of the total PBDE concentrations in soils, sediments, and vegetables, respectively, which was similar to our results [45,48,53]. In addition, the median contributions of BDE-209 ranged from 85% to 91% in the AR-areas. These results further indicated the difference in the sources of e-waste between the IR-sites and this site.

In China, the technical specifications of pollution control in e-waste recycling were developed and implemented in 2010. In the specifications, many measures including controlling dust emission, forbidding crud waste recycling, and harmless disposal of residues were required. In general, the observed PBDE concentrations in the investigated site was much lower than those in IR-sites, and comparable to those in UI-areas. Additionally, media concentration of PBDEs in sediments across China reached 15.8 ng·g^{-1}, which was slightly higher than that in this study [54]. This indicated that the pollution control measures taken by this recycling plant might be effective.

2.2. Contamination of Heavy Metals

The descriptive statistics of the heavy metals in soils, sediments, and vegetables in this study are summarized in Table 2. With the exception of zinc, the mean concentrations of metals in this site were higher than their background values of soils [55]. Especially, the mean concentrations of Mn, Sb, and As were 2.86, 2.34, and 2.16 times as much as their background values. In general, based on China's standard for soil risk control (GB 36600-2018 and GB 15618-2018), the concentrations of soil metals (excluding As) in the factory and in the surrounding farmland were lower than the risk screening values in industrial and agricultural areas, respectively. The concentrations of As, Cr, and Ni in this site were 1.65, 1.52, and 1.31 times as much as those in Chinese soils, while the concentrations of other metals were comparable to or lower than those across China [56]. Further, the pollutions of soil metals were assessed using the geoaccumulation index (I_{geo}) values and the results are shown in Figure S2 from Supplementary Materials. In our study, the metals of Mn, As, Sb, Hg, and Cr ranged from unpolluted to moderately polluted, with the mean I_{geo} values of 0.87, 0.47, 0.45, 0.35, and 0.04, respectively. The I_{geo} values for Mn and Sb were 1.53 and 1.52 at the 95th percentile, belonging to moderately polluted levels. The mean I_{geo} values showed that the other metals in soils belonged to the unpolluted levels ($I_{geo} \leq 0$).

The concentrations of heavy metals in sediments were higher than the soil background values (Table 2). In particular, the concentrations of Hg and Sb were 39.9 and 23.0 times as much as the corresponding background values, respectively. According to the numerical sediment quality guidelines, however, the mean concentrations of metals (excluding Cr) were between the threshold effect concentration (TEC) and probable effect concentration (PEC), indicating that the pollution might have low adverse effects [57]. In addition, the mean I_{geo} values showed that heavy metals in the investigated sediments were generally at unpolluted to moderately polluted levels (Figure S2). It should be noted that the maximum I_{geo} values of Cd, Hg, Sb, and Zn in S_{24} with 3.60, 7.44, 6.51, and 3.49, respectively, belonged to heavily and extremely polluted, likely due to the rainfall, wastewater discharge, and poor water mobility in this sampling site.

Table 2. Descriptive statistics of metal concentrations (μg·g^{-1}) in soils, sediments, and vegetables collected from a regulated electronic waste recycling site in China.

Sampling Sites	Statistics	As	Cd	Cr	Cu	Hg	Mn	Ni	Pb	Sb	Zn
Soils (n = 21)	Range	11.2–31.3	0.08–0.22	73.2–159	14.9–101	0.02–0.09	1077–2991	21.8–52.4	23.7–63.3	0.96–8.07	32.2–170
	Mean	20.0	0.12	104	27.9	0.04	1839	38.9	33.4	2.11	54.0
Background values [1]		9.30	0.08	66.0	24.0	0.019	644	25.8	25.8	0.90	63.5
Risk screening values		60 [2]	65 [2]		18,000 [2]	38 [2]	–	900 [2]	800 [2]	–	–
		25 [3]	0.6 [3]	250 [3]	100 [3]	3.4 [3]	–	190 [3]	170 [3]	–	300 [3]
Sediments (n = 7)	Range	7.41–18.0	0.11–1.53	98.1–294	17.5–132	0.03–4.95	780–1684	27.3–61.8	26.3–226	1.52–123	52.9–1071
	Mean	12.0	0.42	144	41.6	0.76	1145	40.7	62.3	20.7	226
TEC [4]		9.79	0.99	43.4	31.6	0.18		22.7	35.8		121
PEC [4]		33.0	4.98	111	149	1.06		48.6	128		459
Vegetables (n = 8)	Range	0.49–1.72	0.03–0.31	3.28–13.4	4.00–16.5	0.05–0.37	86.7–284	1.30–6.14	1.09–9.93	0.06–2.76	21.8–73.7
	Mean	0.95	0.14	5.49	9.99	0.17	177	3.39	3.94	0.62	42.2

[1] Background values (μg·g^{-1}) of metals in soils in Shandong Province, China [55]. [2] Limits in the Chinese standard (GB 36600-2018). [3] Limits in the Chinese standard (GB 15618-2018). [4] TEC, the threshold effect concentration; PEC, the probable effect concentration [57].

The concentrations of metals in the study vegetables differed (Table 2). The highest concentrations (ng·g^{-1}) of Cd (311) and Hg (370) were observed in the pine needle and wheat samples, respectively. As for the other metals, relatively high concentrations were found in the baby's breath samples. In particular, the Mn concentration reached 284 μg·g^{-1}. This phenomenon suggested that the vegetables had species-specific absorption capacities [58].

The metal concentrations in the soils, sediments, and vegetables in this study were compared with those in some IR-sites (Tables S6–S8 from Supplementary Materials). Concentrations of Cd, Cu, Pb, Sb, and Zn in the soils in this study were 1–3 orders of magnitude lower than those from the IR-sites [28,59]. Wu et al. [60] reported mean concentrations of As and Ni in an e-waste burning site in Guiyu, which were 2.52 and 6.58 times higher than our results, respectively. However, Mn concentrations in soils found in this study were comparable to those from the e-waste recycling areas (Figure 2a). The reported concentrations of metals in the sediments were generally 1–2 orders of magnitude lower than those found in the IR-sites (Figure 2b). The Cr concentrations in the investigated sediments was 1–2 orders of magnitude higher than those in the IR-sites [61,62], suggesting that the release of heavy metals during different e-waste recycling activities differed [5]. Concentrations of As, Cu, and Zn in vegetables from this regulated site were lower than those in the IR-sites [50]. The contamination levels of Cd, Hg, Mn, Ni, and Pb were comparable to those in IR-sites [36,50,63]. Compared to the IR-sites, the concentrations of Cd, Cu, Pb, Sb, and Zn in this site were generally at lower levels, especially in soils and sediments.

The concentrations of metals found in our study were also compared to those observed in the UI-areas (Figure 2). The concentrations of heavy metals in soils in this regulated site were generally comparable to those observed in urban and industrial soils. However, the concentrations of Hg and Zn in soils in this study were 1–3 orders of magnitude lower than those from UI-areas [64,65], while concentrations of As and Ni were 2.02 to 1.52 times that in the UI-areas [64,66,67]. The concentrations of most metals, except for As and Cd, in the sediments in our study were higher than those in the UI-areas (Figure 2b). Yang et al. [68] reported that the concentrations of As and Cd in the Wuhan section of the Yangtze River were 1.23 and 1.62 times as much as those in our study, respectively. The contamination levels of Cu and Ni in the sediments were comparable to those in the UI-areas [68]. For the vegetables, the concentrations of Cd, Cu, Pb, and Zn found in this study were 1–2 orders of magnitude lower than those from UI-areas [69,70] (Table S8 from Supplementary Materials). However, Bi et al. [71] reported that the mean concentration of As in the leafy vegetables from Shanghai was two orders of magnitude lower than our study. Generally, the concentrations of metals (excluding As) in soils and vegetables in our study were comparable to or lower than those in the UI-areas, while the concentrations of sediment metals were slightly higher than those in the UI-areas.

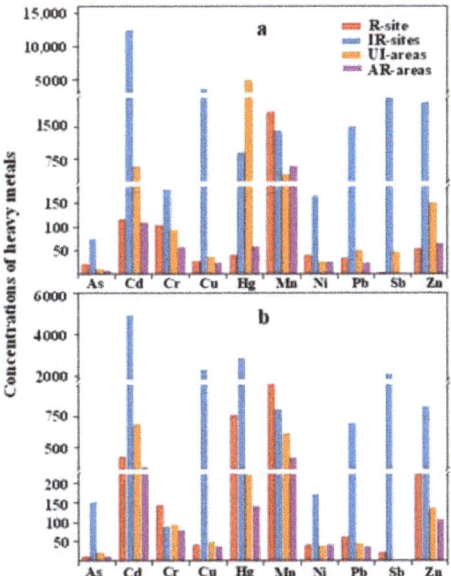

Figure 2. Comparison of heavy metal concentrations in soils (**a**) and sediments (**b**) in the study area with those in other areas. The units of Cd and Hg were ng·g^{-1}, and those of other metals were μg·g^{-1}. R-site, regulated e-waste recycling site (this study); IR-sites, irregulated e-waste recycling sites; UI-areas, urban and industrial areas; AR-areas, agricultural and rural areas. The mean concentrations of heavy metals in each study were selected. Data are from Tables S6 and S7 from Supplementary Materials.

Compared to the AR-areas, the concentrations of heavy metals in soils in this study were relatively high (Figure 2). The mean concentrations of soil heavy metals (excluding Hg and Zn) in our study were 1.06 to 3.21 times as much as those in the AR-areas (Figure 2a). In the sediments, the concentrations of all metals were generally comparable to or higher than those in the AR-areas. The concentrations of Cu, Ni, Pb, and Zn in the vegetable samples were 1–2 orders of magnitude higher than those from the AR-areas, and the contamination of Cd and Cr in the AR-areas were comparable to those in this regulated site [72–74]. This indicated that the regulated e-waste recycling operations could reduce the release of metals to some extent, although the metal concentrations were still higher than those in the AR-areas.

2.3. Health Risk Assessment

The non-carcinogenic risks from exposure to PBDEs and metals in the soils in this site were calculated to further understand the potential adverse effects on human health. Figure 3 depicts the non-carcinogenic risks for children and adults. In this study, only four congeners, i.e., BDE-47, BDE-99, BDE-153, and BDE-209, were selected to calculate the hazard quotient (HQs), due to the lack of RfDs of the other congeners. The calculated results of individual PBDEs are depicted in Table S9 from Supplementary Materials. For adults, the HIs of PBDEs in soils ranged from 7.15×10^{-6}–2.98×10^{-3}, and the HIs of metals ranged from 0.25–0.51, indicating that the risks from PBDEs and metals in the soils in this area were acceptable. For children, the HIs of PBDEs in soils ranged from 3.89×10^{-5}–1.62×10^{-2}; however, the HIs of soil metals ranged from 1.59–3.27, indicating the unacceptable non-carcinogenic risks to children. As, Cr, and Mn were the main contributors to the total risks, with a median of 47%, 21%, and 10% of the HIs. In particular, the HQs of As in 71.4% of the total soil samples were higher than 1.0, which exceeded the acceptable level.

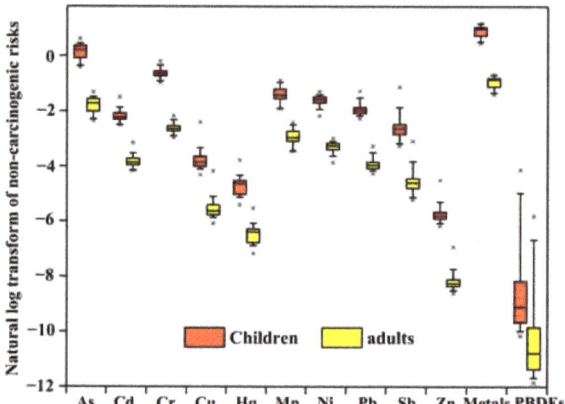

Figure 3. Non-carcinogenic risks to children and adults from exposure to metals and PBDEs in soils. Horizontal lines represent the 5th, 50th, and 95th percentiles, and the boxes represent the lower quartile and upper quartile; the asterisk below or above indicates outliers.

Many studies reported that the non-carcinogenic risks from these chemicals in IR-sites were much higher than the risk levels in our study. Typically, Ge et al. [75] reported that the non-carcinogenic risks of PBDEs for children in an e-waste dismantling site in South China, at the 95th percentile, reached 0.486, which was two orders of magnitude higher than our result (0.007). Zhang et al. [76] also found that the HQs for Pb, Cd, Hg, and Cu in an abandoned e-waste recycling plant in Taizhou were higher than 10. These results suggested that the non-carcinogenic risks from the chemicals released from the regulated e-waste recycling were substantially reduced, but further control of the risks from some specific chemicals are imperative and necessary.

3. Materials and Methods

3.1. Study Area

The regulated e-waste recycling site, surrounded by farmland, is located in a small town on the outskirt of Qingdao, Eastern China, and covers an area of about 43 hectares. The area has a warm and humid climate, and annual average temperature, precipitation, and wind speed of 11.3 °C, 732 mm, and 2–3 m·s^{-1}, respectively [77]. E-waste recycling activities in this park lasted for 13 years. By the end of 2015, the annual treatment capacity of e-waste in this site reached 500,000 tons. In the site, e-waste was dismantled physically with mechanization. Many pollution-control measures, including classified storing, classified dismantling, closed crushing, negative pressure in workshop, were implemented in the recycling site, based on China's technical specifications of pollution control for this recycling. Typically, the dust containing metals and organic pollutants was removed by adsorption in the negative pressure workshop. The fragments of printed circuit board were transported to other sites for subsequent treatment. The operations carried out in the factory followed the OHSAS 18001 guidelines.

3.2. Sample Collection

In June 2015, soil, sediment, and vegetable samples from this regulated e-waste recycling site were collected. The sampling sites are depicted in Figure S1 from Supplementary Materials. Based on the technical guidelines on soil sampling recommended by China [78], a total of 21 composite soil samples (0–10 cm deep) were collected in the factory (S_1–S_7 and

S$_{19}$–S$_{21}$) and the surrounding farmland (S$_8$–S$_{18}$). The farmland was about 300 to 2000 m from the e-waste dismantling workshop. Each soil sample consisted of four sub-samples. According to the Chinese specification [79], seven composite sediment samples (ca. 1 kg) were collected randomly from the surrounding ditches (S$_{22}$–S$_{28}$). Each sediment sample consisted of four sub-samples. Additionally, the wheat leaf (*Triticum aestivum* L.) (V$_1$ and V$_4$) was collected from the surrounding farmland. The metasequoia leaf (*Metasequoia glyptostroboides*) (V$_2$), pine needle (*Pinus massoniana* Lamb) (V$_3$), rattan grass (*Vulpia myuros*) (V$_5$), aspen leaf (*Populus alba* L.) (V$_6$ and V$_8$), and a whole plant of baby's breath (*Gypsophila paniculata* L.) (V$_7$) were collected within the site. Four parallel samples were randomly selected for each vegetable. The soil and vegetable samples were wrapped in sealed aluminum foil bags (washed with acetone before being used). The sediment sample was placed in a washed aluminum box. All samples were freeze-dried and then sealed before being ground, homogenized, sieved (a 150-mesh, approximately 100-μm, stainless sieve), and stored at −20 °C, until analysis within two months.

3.3. Sample Analysis

In the study, twenty-one PBDE congeners and ten metals in the soil, sediment, and vegetable samples were measured. The extraction and cleanup procedures for the detection of PBDEs were conducted on the basis of the methods described by Tang et al. [80], which are described in the Section 1 in Supplementary Materials. The selected PBDE congeners were analyzed by a gas chromatography/mass spectrometry (Agilent 7890A/5975C, Agilent Technologies, Santa Clara, CA, USA), and the instrumental conditions were described in the Section 2 in Supplementary Materials. A spiked blank, procedural blank, and matrix-spiked sample was measured after every 8–10 samples, to monitor the performance of the method and matrix effects. All samples were determined twice. The recoveries of PBDEs in spiked soil/sediment and vegetable samples were 81.5–127% and 81.0–127%, and their relative standard deviations (RSDs) were 4.52–16.6% and 4.52–16.5%, respectively. The limit of quantification (LOQ) for PBDEs in the soil/sediment and vegetable samples were in the range of 0.002–0.250 and 0.003–0.252 ng·g^{-1} dry weight, respectively, based on a signal-to-noise ratio of 10:1.

For metal measure, the soil and sediment samples were digested by the HNO$_3$: HF: HClO$_4$ (*v:v:v* = 5:10:2) method, which was described by Tang et al. [81]. The vegetable samples were digested with a microwave digestion system, based on the methods described by Cheng et al. [58]. The sample preparation is described in the Section 1 in Supplementary Materials. The concentrations of As and Hg were analyzed using an atomic fluorescence spectrometer (XGY-1011A, IGGE, Langfang, China), and the concentrations of the other eight metals were determined with an inductively coupled plasma mass spectrometer (Agilent 7500a, Agilent Technologies, Santa Clara, CA, USA). Reagent blanks and procedural blanks were analyzed to check for interference with the analysis. The duplicate samples were analyzed to evaluate the precision of the analysis (produced values within ±10%). Geochemical standard reference soil samples (GSS-17 and GSS-25) and vegetable samples (GSB-2, GSB-4, and GSB-5) were analyzed and the relative standard deviations (RSDs) of the results were generally better than 10%. All samples were determined twice.

3.4. Statistical Analysis and Risk Assessment

Statistical analyses of PBDE and metal concentrations in soils, sediments, and vegetables were conducted using SPSS 18.0 (IBM SPSS, Armonk, NY, USA). Origin 8.0 software was used to draw the figures (Origin Lab, Northampton, MA, USA). For concentrations below LOQ, the value was set to half of LOQ during data analysis. The Kolmogorov–Smirnov test was used to test the normality of data. In this study, the observed concentrations of PBDEs and metals were compared with those reported in some previous investigation in irregulated e-waste recycling sites (IR-sites), urban and industrial areas (UI-areas), and agricultural and rural areas (AR-areas). For the IR-sites, only the literature reported for the mechanical recycling site was selected. The information about the pollution from the PBDE manufacturing and metal smelting was excluded in the data screening for the UI-areas. In addition, the PBDEs and metal concentrations reported for agricultural areas close to cities or remote areas were also excluded from the literature collection for AR-areas. After screening and sorting, the PBDE concentrations reported in 37 references and metal concentrations from 44 references were collected, and the information was depicted in Tables S3–S8 from Supplementary Materials, respectively.

The I_{geo} was used to evaluate the metal pollution in the soil and the sediment samples. The calculation method is described in Section 3 in Supplementary Materials. According to Muller et al. [82], the corresponding relationships between I_{geo} and the levels of pollution were defined as follows—unpolluted ($I_{geo} \leq 0$), unpolluted to moderately polluted ($0 < I_{geo} \leq 1$), moderately polluted ($1 < I_{geo} \leq 2$), moderately to heavily polluted ($2 < I_{geo} \leq 3$), heavily polluted ($3 < I_{geo} \leq 4$), heavily to extremely polluted ($4 < I_{geo} \leq 5$), and extremely polluted ($I_{geo} > 5$). In addition, the non-carcinogenic risks from exposure to metals and PBDEs in soils were also estimated. According to the Exposure Factors Handbook [83], the average daily doses (ADDs) (mg·kg^{-1}·day^{-1}) through ingestion, dermal contact, and inhalation for both adults and children were estimated using Equations (1)–(3), as follows:

$$ADD_{ingest} = C_{soil} \times IngR \times EF \times ED / BW / AT \times 10^{-6} \quad (1)$$

$$ADD_{dermal} = C_{soil} \times SA \times AF \times ABS \times EF \times ED / BW / AT \times 10^{-6} \quad (2)$$

$$ADD_{inhale} = C_{soil} \times InhR \times EF \times ED / PEF / BW / AT \quad (3)$$

where ADD_{ingest}, ADD_{dermal}, and ADD_{inhale} are the daily exposure to metals through soil ingestion, dermal contact, and inhalation absorption, respectively; C_{soil} is the concentration of metal in soil; $IngR$ and $InhR$ are the ingestion and inhalation rates of soil, respectively; EF is the exposure frequency; ED is the exposure duration; BW is the body weight of the exposed individual; AT is the time period over which the dose is averaged; SA is the exposed skin surface area; AF is the adherence factor; and PEF is the particle emission factor. The value for each factor was selected according to the literature, which is shown in Table S1 from Supplementary Materials. ABS is the dermal absorption factor; given in Table S2 from Supplementary Materials. The non-carcinogenic risks from metals could be assessed using the hazard quotient (HQ), which is the ratio of the ADD to the reference dose (RfD, mg·kg^{-1}·day^{-1}) of a metal, for the same exposure pathway. The value of RfD for each metal is given in Table S2 from Supplementary Materials. In this study, the hazard index (HI) was defined as the sum of the hazard quotient (HQ) values. If the hazard index HI ≤ 1, the non-carcinogenic risks were acceptable. The potential non-carcinogenic health effects occurred when HI > 1 [84].

4. Conclusions

This study reports the concentrations of PBDEs and metals in the soils, sediments, and vegetables surrounding regulated e-waste recycling sites in China. The results indicated that the concentrations of these typical chemicals in this site were much lower than those in irregulated e-waste recycling sites, and are generally comparable to those in the urban and industrial areas. This study suggested that the regulated operations can substantially reduce the pollution of PBDEs and metals from the e-waste recycling. The non-carcinogenic risks from these chemicals were also reduced in comparison to those in the irregulated e-waste recycling areas. However, this study found that the concentrations of some specific chemicals in this site were still relatively high. In particular, arsenic pollution in the recycling site were relatively serious and the non-carcinogenic risks from soil arsenic to local children exceeded the acceptable level. This suggested that further pollution control in regulated e-waste recycling is needed.

It should be noted that this is a pilot study with a small set of samples, and some limitations should be noted. The reported concentrations of chemicals depended on many factors, including the types of e-waste, the content of chemicals in the waste, the amount of recycling waste, the method of recycling, and the time of plant operation, as well as the local climate conditions. There might also have been some uncertainty associated with the selection of other studies, for comparison in this study. For example, the sample size and sampling times in each study varied. The difference of plant species and the growth period of plants might cause uncertainty in the evaluation of vegetable pollution in this site. In addition, the nature of the risk assessment also renders the findings uncertain. In particular, the bioavailability of the chemicals in soils were not considered in this study, which likely resulted in an overestimation of the potential risks. More importantly, the occurrence and the associated risks from many other hazardous chemicals released from the e-waste recycling were not investigated. More research is need to investigate the risks from multiclass pollutants surrounding the site and the effectiveness of the current pollution control in the e-waste recycling operation.

Supplementary Materials: The following are available online, Sample preparation, Instrumental Analysis, Geoaccumulation index, Table S1: Exposure factors and values used in the non-carcinogenic risk estimation, Table S2: Reference dose of different exposure routes and skin absorption factors, Table S3: Polybrominated diphenyl ether concentrations in soils from the regulated e-waste recycling area and from other areas, Table S4: Polybrominated diphenyl ether concentrations in sediments from the regulated e-waste recycling area and from other areas, Table S5: Polybrominated diphenyl ether concentrations in vegetables from the regulated e-waste recycling area and from other areas, Table S6, Mean metal concentrations in soils from the regulated e-waste recycling area and from other areas, Table S7: Mean metal concentrations in sediments from the regulated e-waste recycling area and from other areas, Table S8: Mean metal concentrations in vegetables from the regulated e-waste recycling area and from other area, Table S9: Non-carcinogenic risks for individual polybrominated diphenyl ethers in soils, Figure S1: Map of the soil, sediment, vegetable sampling sites, Figure S2: Box plots of the geoaccumulation index for heavy metals in soils and sediments. References [28,36–53,56,58–74,80–82,84–135] are cited in the Supplementary Materials.

Author Contributions: Conceptualization, Z.T.; Formal analysis, H.Y.; Validation, H.Y.; Data curation, H.Y.; Investigation, H.Y., J.M., Z.L., Y.L., T.M., and Z.T.; Methodology, Z.T.; Project administration, Z.T.; Funding acquisition, Z.T.; Resources, H.Y.; Writing—original draft, H.Y.; Writing—review and editing, Z.T. All authors have read and agreed to the published version of the manuscript.

Funding: This research was supported by the National Key R&D Program of China (grant no. 2018YFC1900104) and the Protection Public Welfare Science and Technology Research Program of China (201309023).

Institutional Review Board Statement: Not applicable.

Informed Consent Statement: Not applicable.

Data Availability Statement: Not applicable.

Acknowledgments: We thank Lu Jin for her help for sample analyses. The authors would like to acknowledge Yang Li for his help for sample collection.

Conflicts of Interest: The authors declare no conflict of interest.

Sample Availability: Samples of soils, sediments, and vegetables are not available from the authors.

References

1. Baldé, C.P.; Wang, F.; Kuehr, R.; Huisman, J. *The Global E-Waste Monitor 2014: Quantities, Flows and Resources*; United Nations University, IAS–SCYCLE: Bonn, Germany, 2015; Available online: http://ewastemonitor.info/download/global-e-waste-monitor/ (accessed on 14 December 2020).
2. Baldé, C.P.; Forti, V.; Gray, V.; Kuehr, R.; Stegmann, P. The Global E-Waste Monitor 2017: Quantities, Flows and Resources. 2017. Available online: http://ewastemonitor.info/home-new-with-more-news/ (accessed on 14 December 2020).
3. Forti, V.; Baldé, C.P.; Kuehr, R.; Bel, G. *The Global E-Waste Monitor 2020: Quantities, Flows and the Circular Economy Potential*; 2020; Available online: http://ewastemonitor.info/ (accessed on 14 December 2020).
4. Ackah, M. Informal e-waste recycling in developing countries: Review of metal(loid)s pollution, environmental impacts and transport pathways. *Environ. Sci. Pollut. Res.* **2017**, *24*, 24092–24101. [CrossRef] [PubMed]
5. Han, Y.; Tang, Z.W.; Sun, J.Z.; Xing, X.Y.; Zhang, M.N.; Cheng, J.L. Heavy metals in soil contaminated through e-waste processing activities in a recycling area: Implications for risk management. *Process. Saf. Environ. Prot.* **2019**, *125*, 189–196. [CrossRef]

6. Wang, Q.; Shaheen, S.M.; Jiang, Y.; Li, R.; Slaný, M.; Abdelrahman, H.; Kwon, E.; Bolan, N.; Rinklebe, J.; Zhang, Z. Fe/Mn- and P-modified drinking water treatment residuals reduced Cu and Pb phytoavailability and uptake in a mining soil. *J. Hazard. Mater.* **2021**, *403*, 123628. [CrossRef] [PubMed]
7. Song, Q.; Li, J. A systematic review of the human body burden of e-waste exposure in China. *Environ. Int.* **2014**, *68*, 82–93. [CrossRef]
8. Adie, G.U.; Sun, L.; Zeng, X.; Zheng, L.; Osibanjo, O.; Li, J. Examining the evolution of metals utilized in printed circuit boards. *Environ. Technol.* **2017**, *38*, 1696–1701. [CrossRef]
9. Zhang, W.H.; Wu, Y.X.; Simonnot, M.O. Soil contamination due to e-waste disposal and recycling activities: A review with special focus on China. *Pedosphere* **2012**, *22*, 434–455. [CrossRef]
10. Ernst, T.; Popp, R.; Wolf, M.; van Eldik, R. Analysis of eco-relevant elements and noble metals in printed wiring boards using AAS, ICP-AES and EDXRF. *Anal. Bioanal. Chem.* **2003**, *375*, 805–814. [CrossRef]
11. Robinson, B.H. E-waste: An assessment of global production and environmental impacts. *Sci. Total Environ.* **2009**, *408*, 183–191. [CrossRef]
12. Qiu, X.; Marvin, C.H.; Hites, R.A. Dechlorane Plus and Other Flame Retardants in a Sediment Core from Lake Ontario. *Environ. Sci. Technol.* **2007**, *41*, 6014–6019. [CrossRef]
13. Liyanage, D.; Walpita, J. Organic pollutants from E-waste and their electrokinetic remediation. *Handb. Electron. Waste Manag.* **2020**, 171–189. [CrossRef]
14. Khalid, F.; Hashmi, M.Z.; Jamil, N.; Qadir, A.; Ali, M.I. Microbial and enzymatic degradation of PCBs from e-waste-contaminated sites: A review. *Environ. Sci. Pollut. Res.* **2021**, *28*, 10474–10487. [CrossRef]
15. Awasthi, A.K.; Zeng, X.; Li, J. Environmental pollution of electronic waste recycling in India: A critical review. *Environ. Pollut.* **2016**, *211*, 259–270. [CrossRef]
16. Morf, L.S.; Tremp, J.; Gloor, R.; Schuppisser, F.; Stengele, M.; Taverna, R. Metals, non-metals and PCB in electrical and electronic waste-actual levels in Switzerland. *Waste Manag.* **2007**, *27*, 1306–1316. [CrossRef]
17. Stenvall, E.; Tostar, S.; Boldizar, A.; Foreman, M.R.; Möller, K. An analysis of the composition and metal contamination of plastics from waste electrical and electronic equipment (WEEE). *Waste Manag.* **2013**, *33*, 915–922. [CrossRef]
18. Zhang, Q.; Hu, M.; Wu, H.; Niu, Q.; Lu, X.; He, J.; Huang, F. Plasma polybrominated diphenyl ethers, urinary heavy metals and the risk of thyroid cancer: A case-control study in China. *Environ. Pollut.* **2021**, *269*, 116162. [CrossRef]
19. Chung, T.L.; Liao, C.J.; Chang-Chien, G.P. Distribution of polycyclic aromatic hydrocarbons and polychlorinated dibenzo-*p*-dioxins/dibenzofurans in ash from different units in a municipal solid waste incinerator. *Waste Manag. Res.* **2010**, *28*, 789–799. [CrossRef]
20. Zhang, K.; Schnoor, J.L.; Zeng, E.Y. E-waste recycling: Where does it go from here? *Environ. Sci. Technol.* **2012**, *46*, 10861–10867. [CrossRef]
21. Chi, X.; Streicher-Porte, M.; Wang, M.Y.; Reuter, M.A. Informal electronic waste recycling: A sector review with special focus on China. *Waste Manag.* **2011**, *31*, 731–742. [CrossRef]
22. Yu, J.L.; Williams, E.; Ju, M.T.; Shao, C.F. Managing e-waste in China: Policies, pilot projects and alternative approaches. *Resour. Conserv. Recycl.* **2010**, *54*, 991–999. [CrossRef]
23. Adeyi, A.A.; Oyeleke, P. Heavy metals and polycyclic aromatic hydrocarbons in soil from e-waste dumpsites in Lagos and Ibadan, Nigeria. *J. Health Pollut.* **2017**, *7*, 71–84. [CrossRef]
24. Kaifie, A.; Schettgen, T.; Bertram, J.; Löhndorf, K.; Waldschmidt, S.; Felten, M.K.; Kraus, T.; Fobil, J.N.; Küpper, T. Informal e-waste recycling and plasma levels of non-dioxin-like polychlorinated biphenyls (NDL-PCBs)—A cross-sectional study at Agbogbloshie, Ghana. *Sci. Total Environ.* **2020**, *723*, 138073. [CrossRef]
25. Kim, S.S.; Xu, X.; Zhang, Y.; Zheng, X.; Liu, R.; Dietrich, K.N.; Reponen, T.; Xie, C.; Sucharew, H.; Huo, X.; et al. Birth outcomes associated with maternal exposure to metals from informal electronic waste recycling in Guiyu, China. *Environ. Int.* **2020**, *137*, 105580. [CrossRef]
26. Labunska, I.; Harrad, S.; Santillo, D.; Johnston, P.; Brigden, K. Levels and distribution of polybrominated diphenyl ethers in soil, sediment and dust samples collected from various electronic waste recycling sites within Guiyu town, southern China. *Environ. Sci. Proc. Imp.* **2013**, *15*, 503–511. [CrossRef]
27. Moeckel, C.; Breivik, K.; Nøst, T.H.; Sankoh, A.; Jones, K.C.; Sweetman, A. Soil pollution at a major West African e-waste recycling site: Contamination pathways and implications for potential mitigation strategies. *Environ. Int.* **2020**, *137*, 105563. [CrossRef]
28. Wu, Q.; Leung, J.; Du, Y.; Kong, D.; Shi, Y.; Wang, Y.; Xiao, T. Trace metals in e-waste lead to serious health risk through consumption of rice growing near an abandoned e-waste recycling site: Comparisons with PBDEs and AHFRs. *Environ. Pollut.* **2019**, *247*, 46–54. [CrossRef]
29. Jose, A.; Ray, J.G. Toxic heavy metals in human blood in relation to certain food and environmental samples in Kerala, South India. *Environ. Sci. Pollut. Res.* **2018**, *25*, 7946–7953. [CrossRef]
30. Meng, T.; Cheng, J.; Tang, Z.; Yin, H.; Zhang, M. Global distribution and trends of polybrominated diphenyl ethers in human blood and breast milk: A quantitative meta-analysis of studies published in the period 2000–2019. *J. Environ. Manag.* **2020**, *280*, 111696. [CrossRef]
31. Ye, J.; Kayaga, S.; Smout, I. Regulating for e-waste in China: Progress and challenges. *Munic. Eng.* **2009**, *162*, 79–85. [CrossRef]

32. He, M.; Shen, H.; Li, Z.; Wang, L.; Wang, F.; Zhao, K.; Liu, X.; Wendroth, O.; Xu, J. Ten-year regional monitoring of soil-rice grain contamination by heavy metals with implications for target remediation and food safety. *Environ. Pollut.* **2019**, *244*, 431–439. [CrossRef]
33. Huang, F.; Wen, S.; Li, J.; Zhong, Y.; Zhao, Y.; Wu, Y. The human body burden of polybrominated diphenyl ethers and their relationships with thyroid hormones in the general population in Northern China. *Sci. Total Environ.* **2014**, *466–467*, 609–615. [CrossRef]
34. Ji, X.; Ding, J.; Xie, X.; Cheng, Y.; Huang, Y.; Qin, L.; Han, C. Pollution status and human exposure of decabromodiphenyl ether (BDE-209) in China. *ACS Omega* **2017**, *2*, 3333–3348. [CrossRef] [PubMed]
35. Li, Y.; Duan, Y.P.; Huang, F.; Yang, J.; Xiang, N.; Meng, X.Z.; Chen, L. Polybrominated diphenyl ethers in e-waste: Level and transfer in a typical e-waste recycling site in Shanghai, Eastern China. *Waste Manag.* **2014**, *34*, 1059–1065. [CrossRef] [PubMed]
36. Alabi, O.A.; Bakare, A.A.; Xu, X.; Li, B.; Zhang, Y.; Huo, X. Comparative evaluation of environmental contamination and DNA damage induced by electronic-waste in Nigeria and China. *Sci. Total Environ.* **2012**, *423*, 62–72. [CrossRef] [PubMed]
37. Huang, C.; Zeng, Y.; Luo, X.; Ren, Z.; Tang, B.; Lu, Q.; Gao, S.; Wang, S.; Mai, B. In situ microbial degradation of PBDEs in sediments from an e-waste site as revealed by positive matrix factorization and compound-specific stable carbon isotope analysis. *Environ. Sci. Technol.* **2019**, *53*, 1928–1936. [CrossRef] [PubMed]
38. Wang, S.; Wang, Y.; Luo, C.; Li, J.; Yin, H.; Zhang, G. Plant selective uptake of halogenated flame retardants at an e-waste recycling site in southern China. *Environ. Pollut.* **2016**, *214*, 705–712. [CrossRef] [PubMed]
39. Wen, S.; Yang, F.; Li, J.G.; Gong, Y.; Zhang, X.L.; Hui, Y.; Wu, Y.N.; Zhao, Y.F.; Xu, Y. Polychlorinated dibenzo-p-dioxin and dibenzofurans (PCDD/Fs), polybrominated diphenyl ethers (PBDEs), and polychlorinated biphenyls (PCBs) monitored by tree bark in an E-waste recycling area. *Chemosphere* **2009**, *74*, 981–987. [CrossRef] [PubMed]
40. Han, Y.; Cheng, J.; He, L.; Zhang, M.; Ren, S.; Sun, J.; Xing, X.; Tang, Z. Polybrominated diphenyl ethers in soils from Tianjin, North China: Distribution, health risk, and temporal trends. *Environ. Geochem. Health* **2021**, *43*, 1177–1191. [CrossRef] [PubMed]
41. Xiang, X.X.; Lu, Y.T.; Ruan, Q.Y.; Lai, C.; Sun, S.B.; Yao, H.; Zhang, Z.S. Occurrence, distribution, source, and health risk assessment of polybrominated diphenyl ethers in surface soil from the Shen-Fu Region, Northeast China. *Environ. Sci.* **2020**, *41*, 368–376. (In Chinse) [CrossRef]
42. Mai, B.; Chen, S.; Luo, X.; Chen, L.; Yang, Q.; Sheng, G.; Peng, P.; Fu, J.; Zeng, E.Y. Distribution of polybrominated diphenyl ethers in sediments of the Pearl River Delta and adjacent South China Sea. *Environ. Sci. Technol.* **2005**, *39*, 3521–3527. [CrossRef]
43. Zhang, T.; Yang, W.L.; Chen, S.J.; Shi, D.L.; Zhao, H.; Ding, Y.; Huang, Y.R.; Li, N.; Ren, Y.; Mai, B.X. Occurrence, sources, and ecological risks of PBDEs, PCBs, OCPs, and PAHs in surface sediments of the Yangtze River Delta city cluster, China. *Environ. Monit. Assess.* **2014**, *186*, 5285–5295. [CrossRef]
44. Jia, H.H.; Wang, X.T.; Cheng, H.X.; Zhou, Y.; Fu, R. Pine needles as biomonitors of polybrominated diphenyl ethers and emerging flame retardants in the atmosphere of Shanghai, China: Occurrence, spatial distributions, and possible sources. *Environ. Sci. Pollut. Res.* **2019**, *26*, 12171–12180. [CrossRef]
45. Shi, S.X.; Zeng, L.Z.; Zhou, L.; Zhang, L.F.; Zhang, T.; Dong, L.; Huang, Y.R. Polybrominated diphenyl ethers in camphor bark from speedy developing urban in Jiangsu Province. *Environ. Sci.* **2011**, *32*, 2654–2660. (In Chinese)
46. Chen, Y.; Zhang, A.; Li, H.; Peng, Y.; Lou, X.; Liu, M.; Hu, J.; Liu, C.; Wei, B.; Jin, J. Concentrations and distributions of polybrominated diphenyl ethers (PBDEs) in surface soils and tree bark in Inner Mongolia, northern China, and the risks posed to humans. *Chemosphere* **2020**, *247*, 125950. [CrossRef]
47. Wang, J.; Lin, Z.; Lin, K.; Wang, C.; Zhang, W.; Cui, C.; Lin, J.; Dong, Q.; Huang, C. Polybrominated diphenyl ethers in water, sediment, soil, and biological samples from different industrial areas in Zhejiang, China. *J. Hazard. Mater.* **2011**, *197*, 211–219. [CrossRef]
48. Li, F.; Zhang, H.; Meng, X.; Chen, L.; Yin, D. Contamination by persistent toxic substances in surface sediment of urban rivers in Chaohu City, China. *J. Environ. Sci.* **2012**, *24*, 1934–1941. [CrossRef]
49. Wu, M.H.; Pei, J.C.; Zheng, M.; Tang, L.; Bao, Y.Y.; Xu, B.T.; Sun, R.; Sun, Y.F.; Xu, G.; Lei, J.Q. Polybrominated diphenyl ethers (PBDEs) in soil and outdoor dust from a multi-functional area of Shanghai: Levels, compositional profiles and interrelationships. *Chemosphere* **2015**, *118*, 87–95. [CrossRef]
50. Wang, J.; Liu, L.; Wang, J.; Pan, B.; Fu, X.; Zhang, G.; Zhang, L.; Lin, K. Distribution of metals and brominated flame retardants (BFRs) in sediments, soils and plants from an informal e-waste dismantling site, South China. *Environ. Sci. Pollut. Res.* **2015**, *22*, 1020–1033. [CrossRef]
51. Luo, Q.; Cai, Z.W.; Wong, M.H. Polybrominated diphenyl ethers in fish and sediment from river polluted by electronic waste. *Sci. Total Environ.* **2007**, *383*, 115–127. [CrossRef]
52. Ma, J.; Addink, R.; Yun, S.; Cheng, J.; Wang, W.; Kannan, K. Polybrominated dibenzo-p-dioxins/dibenzofurans and polybrominated diphenyl ethers in soil, vegetation, workshop-floor dust, and electronic shredder residue from an electronic waste recycling facility and in soils from a chemical industrial complex in eastern China. *Environ. Sci. Technol.* **2009**, *43*, 7350–7356. [CrossRef]
53. Li, W.L.; Ma, W.L.; Jia, H.L.; Hong, W.J.; Moon, H.B.; Nakata, H.; Minh, N.H.; Sinha, R.K.; Chi, K.H.; Kannan, K.; et al. Polybrominated diphenyl ethers (PBDEs) in surface soils across five Asian countries: Levels, spatial distribution, and source contribution. *Environ. Sci. Technol.* **2016**, *50*, 12779–12788. [CrossRef]
54. Yin, H.; Tang, Z.; Meng, T.; Zhang, M. Concentration profile, spatial distributions and temporal trends of polybrominated diphenyl ethers in sediments across China: Implications for risk assessment. *Ecotoxicol. Environ. Saf.* **2020**, *206*, 111205. [CrossRef]

55. China National Environmental Monitoring Center (CNEMC). *The Soil Background Value in China*; China Environmental Science Press: Beijing, China, 1990. (In Chinese)
56. Chen, H.; Teng, Y.; Lu, S.; Wang, Y.; Wang, J. Contamination features and health risk of soil heavy metals in China. *Sci. Total Environ.* **2015**, *512–513*, 143–153. [CrossRef]
57. MacDonald, D.D.; Ingersoll, C.G.; Berger, T.A. Development and evaluation of consensus-based sediment quality guidelines for freshwater ecosystems. *Arch. Environ. Contam. Toxicol.* **2000**, *39*, 20–31. [CrossRef]
58. Cheng, J.; Zhang, X.; Tang, Z.; Yang, Y.; Nie, Z.; Huang, Q. Concentrations and human health implications of heavy metals in market foods from a Chinese coal-mining city. *Environ. Toxicol. Pharmacol.* **2017**, *50*, 37–44. [CrossRef]
59. Luo, C.; Liu, C.; Wang, Y.; Liu, X.; Li, F.; Zhang, G.; Li, X. Heavy metal contamination in soils and vegetables near an e-waste processing site, South China. *J. Hazard. Mater.* **2011**, *186*, 481–490. [CrossRef]
60. Wu, Q.; Du, Y.; Huang, Z.; Gu, J.; Leung, J.; Mai, B.; Xiao, T.; Liu, W.; Fu, J. Vertical profile of soil/sediment pollution and microbial community change by e-waste recycling operation. *Sci. Total Environ.* **2019**, *669*, 1001–1010. [CrossRef]
61. Luo, Y.; Luo, X.; Yang, Z.; Yu, X.; Yuan, J.; Chen, S.; Mai, B. Studies on heavy metal contamination by improper handling of e-waste and its environmental risk evaluation. IV. Heavy metal contamination of sediments in a small scale valley impacted by e-waste treating activities. *Asian J. Ecotoxicol.* **2008**, *3*, 343–349. Available online: http://www.cnki.com.cn/Article/CJFDTOTAL-STDL200804007.htm (accessed on 8 April 2021). (In Chinese).
62. Wang, F.; Leung, A.O.; Wu, S.C.; Yang, M.S.; Wong, M.H. Chemical and ecotoxicological analyses of sediments and elutriates of contaminated rivers due to e-waste recycling activities using a diverse battery of bioassays. *Environ. Pollut.* **2009**, *157*, 2082–2090. [CrossRef]
63. Zheng, J.; Chen, K.H.; Yan, X.; Chen, S.J.; Hu, G.C.; Peng, X.W.; Yuan, J.G.; Mai, B.X.; Yang, Z.Y. Heavy metals in food, house dust, and water from an e-waste recycling area in South China and the potential risk to human health. *Ecotoxicol. Environ. Saf.* **2013**, *96*, 205–212. [CrossRef]
64. Luo, X.S.; Yu, S.; Zhu, Y.G.; Li, X.D. Trace metal contamination in urban soils of China. *Sci. Total Environ.* **2012**, *421–422*, 17–30. [CrossRef]
65. Wu, Z.; Zhang, L.; Xia, T.; Jia, X.; Wang, S. Heavy metal pollution and human health risk assessment at mercury smelting sites in Wanshan district of Guizhou Province, China. *RSC Adv.* **2020**, *10*, 23066–23079. [CrossRef]
66. He, A.; Li, X.; Ai, Y.; Li, X.; Li, X.; Zhang, Y.; Gao, Y.; Liu, B.; Zhang, X.; Zhang, M.; et al. Potentially toxic metals and the risk to children's health in a coal mining city: An investigation of soil and dust levels, bioaccessibility and blood lead levels. *Environ. Int.* **2020**, *141*, 105788. [CrossRef] [PubMed]
67. Wang, S.; Cai, L.M.; Wen, H.H.; Luo, J.; Wang, Q.S.; Liu, X. Spatial distribution and source apportionment of heavy metals in soil from a typical county-level city of Guangdong Province, China. *Sci. Total Environ.* **2019**, *655*, 92–101. [CrossRef] [PubMed]
68. Yang, Z.; Wang, Y.; Shen, Z.; Niu, J.; Tang, Z. Distribution and speciation of heavy metals in sediments from the mainstream, tributaries, and lakes of the Yangtze River catchment of Wuhan, China. *J. Hazard. Mater.* **2009**, *166*, 1186–1194. [CrossRef] [PubMed]
69. Liang, J.; Fang, H.L.; Zhang, T.L.; Wang, X.X.; Liu, Y.D. Heavy metal in leaves of twelve plant species from seven different areas in Shanghai, China. *Urban. For. Urban. Gree.* **2017**, *27*, 390–398. [CrossRef]
70. Zhang, Q.; Yu, R.; Fu, S.; Wu, Z.; Chen, H.; Liu, H. Spatial heterogeneity of heavy metal contamination in soils and plants in Hefei, China. *Sci. Rep.* **2019**, *9*, 1049. [CrossRef]
71. Bi, C.; Zhou, Y.; Chen, Z.; Jia, J.; Bao, X. Heavy metals and lead isotopes in soils, road dust and leafy vegetables and health risks via vegetable consumption in the industrial areas of Shanghai, China. *Sci. Total Environ.* **2018**, *619–620*, 1349–1357. [CrossRef]
72. Gao, Z.Q.; Fu, W.J.; Zhang, M.J.; Zhao, K.L.; Tunney, H.; Guan, Y.D. Potentially hazardous metals contamination in soil-rice system and it's spatial variation in Shengzhou City, China. *J. Geochem. Explor.* **2016**, *167*, 62–69. [CrossRef]
73. Luo, Y.; Zhao, X.; Xu, T.; Liu, H.; Li, X.; Johnson, D.; Huang, Y. Bioaccumulation of heavy metals in the lotus root of rural ponds in the middle reaches of the Yangtze River. *J. Soils Sediments* **2017**, *17*, 2557–2565. [CrossRef]
74. Zeng, L.; Zhou, F.; Zhang, X.; Qin, J.; Li, H. Distribution of heavy metals in soils and vegetables and health risk assessment in the vicinity of three contaminated sites in Guangdong Province, China. *Hum. Ecol. Risk Assess.* **2018**, *24*, 1901–1915. [CrossRef]
75. Ge, X.; Ma, S.; Zhang, X.; Yang, Y.; Li, G.; Yu, Y. Halogenated and organophosphorous flame retardants in surface soils from an e-waste dismantling park and its surrounding area: Distributions, sources, and human health risks. *Environ. Int.* **2020**, *139*, 105741. [CrossRef]
76. Zhang, Q.; Ye, J.; Chen, J.; Xu, H.; Wang, C.; Zhao, M. Risk assessment of polychlorinated biphenyls and heavy metals in soils of an abandoned e-waste site in China. *Environ. Pollut.* **2014**, *185*, 258–265. [CrossRef]
77. National Bureau of Statistics (NBS). *China County Statistics Yearbook-2018 (Township Volume)*; China Statistics Press: Beijing, China, 2019; p. 268. (In Chinese)
78. Chinese Ministry of Environmental Protection (CMEP). Technical Guidelines for Investigation on Soil Contamination of Land for Construction. Available online: http://kjs.mee.gov.cn/hjbhbz/bzwb/jcffbz/201912/t20191209_748223.shtml (accessed on 25 December 2020). (In Chinese)
79. Ministry of Land and Resources of the People's Republic of China (MLRC). Specification on Geochemical Reconnaissance Survey (1:50000) (DZ/T 0011–2015). 2015. Available online: http://www.jianbiaoku.com/webarbs/book/82739/2260214.shtml (accessed on 25 December 2020). (In Chinese).

80. Tang, Z.; Huang, Q.; Cheng, J.; Yang, Y.; Yang, J.; Guo, W.; Nie, Z.; Zeng, N.; Jin, L. Polybrominated diphenyl ethers in soils, sediments, and human hair in a plastic waste recycling area: A neglected heavily polluted area. *Environ. Sci. Technol.* **2014**, *48*, 1508–1516. [CrossRef]
81. Tang, Z.; Huang, Q.; Yang, Y.; Nie, Z.; Cheng, J.; Yang, J.; Wang, Y.; Chai, M. Polybrominated diphenyl ethers (PBDEs) and heavy metals in road dusts from a plastic waste recycling area in north China: Implications for human health. *Environ. Sci. Pollut. Res.* **2016**, *23*, 625–637. [CrossRef]
82. Muller, G. Index of Geo-Accumulation in Sediments of the Rhine River. *GeoJournal* **1969**, *2*, 108–118.
83. United States Environmental Protection Agency (USEPA). *Child-Specific Exposure Factors Handbook, EPA-600-P-00-002B*; National Center for Environmental Assessment: Washington, DC, USA, 2002. Available online: https://cfpub.epa.gov/si/si_public_record_report.cfm?Lab=NCEA&dirEntryId=55145 (accessed on 31 January 2021).
84. United States Environmental Protection Agency (USEPA). *Risk Assessment Guidance for Superfund (EPA/540/1-89/002). Human Health Evaluation Manual (Part. A)*; Office of Emergency and Remedial Response: Washington, DC, USA, 1989; Volume 1. Available online: https://www.epa.gov/sites/production/files/2015-09/documents/rags_a.pdf (accessed on 31 January 2021).
85. United States Environmental Protection Agency (USEPA). *Baseline Human Health Risk Assessment Vasquez Boulevard and I-70 superfund site Denver*; Denver (Co); 2001. Available online: https://www.epa.gov/sites/production/files/documents/hhra_vbi70-ou1.pdf (accessed on 31 January 2021).
86. Beijing Municipal Bureau of Quality and Technical Supervision (BMBQTS). *Environmental Site Assessment Guideline (DB11/T 656–2009)*; BMBQTS: Beijing, China. Available online: https://www.baidu.com/link?url=V_DFtVvt6IPOY1rE8yRzPcRze-thX5lehk2IHdQwQaJjH1eiAyXEoLMjpbms94iS-1Izhtj_Fcjk7wOIORT9VKt9wtu3WtMYWfv7tuTO5I-hj3xHk_NcnwSI2FjD2KVHYRC2I6meKP9RGI_3-Bks4VyEYqqI1xLRQaFeCxldTau7o6h4rX2prhW98uGe0X3rygS9Wqp_yBUUmFAAStpgwLkZlkudFA7gP0KMR-2x-yw0D-MOziCKHN0yiXGFmSGB&wd=&eqid=af84f9d00001ea5d00000003606f0895 (accessed on 8 April 2021). (In Chinese).
87. Smith, R.L. Use of Monte Carlo Simulation for Human Exposure Assessment at a Superfund Site. *Risk Anal.* **1994**, *14*, 433–439. [CrossRef]
88. United States Environmental Protection Agency (USEPA). *Risk Assessment Guidance for Superfund: Volume III-Part A, Process for Conducting Probabilistic Risk Assessment*; EPA: Washington, DC, USA, 2001. Available online: https://www.epa.gov/sites/production/files/2015-09/documents/rags3adt_complete.pdf (accessed on 31 January 2021).
89. Health Canada (HC). *Federal Contaminated Site Risk Assessment in Canada-Part II: Health Canada Toxicological Reference Values (TRVs) and Chemical-Specific Factors*; Government of Canada: Ottawa, ON, Canada, 2004. Available online: http://publications.gc.ca/site/eng/387683/publication.html (accessed on 8 April 2021).
90. Integrated Risk Information System (USEPA IRIS). *2,2',4,4'-Tetrabromodiphenyl ether (BDE-47) (CASRN 5436-43-1)*; US Environmental Protection Agency: Washington, DC, USA, 2008. Available online: https://cfpub.epa.gov/ncea/iris/iris_documents/documents/toxreviews/1010tr.pdf (accessed on 31 January 2021).
91. Integrated Risk Information System (USEPA IRIS). *2,2',4,4',5-Pentabromodiphenyl ether (BDE-99) (CASRN 60348-60-9)*; US Environmental Protection Agency: Washington, DC, USA, 2008. Available online: https://cfpub.epa.gov/ncea/iris/iris_documents/documents/subst/1008_summary.pdf (accessed on 31 January 2021).
92. Integrated Risk Information System (USEPA IRIS). *2,2',4,4',5,5'-Hexabromodiphenyl ether (BDE-153) (CASRN 68631-49-2)*; US Environmental Protection Agency: Washington, DC, USA, 2008. Available online: https://cfpub.epa.gov/ncea/iris/iris_documents/documents/subst/1009_summary.pdf (accessed on 31 January 2021).
93. Integrated Risk Information System (USEPA IRIS). *2,2',3,3',4,4',5,5',6,6'-Decabromodiphenyl ether (BDE-209) (CASRN 1163-19-5)*; US Environmental Protection Agency: Washington, DC, USA, 2008. Available online: https://cfpub.epa.gov/ncea/iris/iris_documents/documents/subst/0035_summary.pdf (accessed on 31 January 2021).
94. Leung, A.O.W.; Luksemburg, W.J.; Wong, A.S.; Wong, M.H. Spatial Distribution of Polybrominated Diphenyl Ethers and Polychlorinated Dibenzo-*p*-dioxins and Dibenzofurans in Soil and Combusted Residue at Guiyu, an Electronic Waste Recycling Site in Southeast China. *Environ. Sci. Technol.* **2007**, *41*, 2730–2737. [CrossRef]
95. Wu, X.F.; Tang, Z.W.; Shen, H.Y.; Huang, Q.F.; Tao, Y. Characterization of polybrominated diphenyl ethers (PBDEs) in dismantling and burning sites in electronic waste polluted area, south China. *Environ. Sci. Technol.* **2013**, *36*, 84–89. (In Chinese) [CrossRef]
96. Wang, H.M.; Yu, Y.J.; Han, M.; Yang, S.W.; Li, Q.; Yang, Y. Estimated PBDE and PBB Congeners in Soil from an Electronics Waste Disposal Site. *Bull. Environ. Contam. Toxicol.* **2009**, *83*, 789–793. [CrossRef]
97. Jiang, Y.; Wang, X.; Zhu, K.; Wu, M.; Sheng, G.; Fu, J. Occurrence, compositional profiles and possible sources of polybrominated diphenyl ethers in urban soils of Shanghai, China. *Chemosphere* **2010**, *80*, 131–136. [CrossRef]
98. Li, K.; Fu, S.; Yang, Z.Z.; Xu, X.B. Composition, Distribution and Characterization of Polybrominated Diphenyl Ethers (PBDEs) in the Soil in Taiyuan, China. *Bull. Environ. Contam. Toxicol.* **2008**, *81*, 588–593. [CrossRef] [PubMed]
99. Jiang, Y.; Wang, X.; Zhu, K.; Wu, M.; Sheng, G.; Fu, J. Occurrence, compositional patterns, and possible sources of polybrominated diphenyl ethers in agricultural soil of Shanghai, China. *Chemosphere* **2012**, *89*, 936–943. [CrossRef] [PubMed]
100. Sun, J.; Pan, L.; Zhan, Y.; Lu, H.; Tsang, D.C.; Liu, W.; Wang, X.; Li, X.; Zhu, L. Contamination of phthalate esters, organochlorine pesticides and polybrominated diphenyl ethers in agricultural soils from the Yangtze River Delta of China. *Sci. Total. Environ.* **2016**, *544*, 670–676. [CrossRef] [PubMed]
101. Huang, Y.; Zhang, D.; Yang, Y.; Zeng, X.; Ran, Y. Distribution and partitioning of polybrominated diphenyl ethers in sediments from the Pearl River Delta and Guiyu, South China. *Environ. Pollut.* **2018**, *235*, 104–112. [CrossRef]

102. Fu, J. Study on the accumulation of dechorane plus: Compared with polybrominated diphenyl ethers in sediment from the Yongning River with an e-waste dismantling plant in Taizhou, Southeastern China. Master's Dissertation, Zhejiang University of Technology, Hangzhou, China, 2015. Available online: http://cdmd.cnki.com.cn/Article/CDMD-10337-1016036270.htm (accessed on 8 April 2021). (In Chinese).
103. Xiong, J.; An, T.; Zhang, C.; Li, G. Pollution profiles and risk assessment of PBDEs and phenolic brominated flame retardants in water environments within a typical electronic waste dismantling region. *Environ. Geochem. Heal.* **2015**, *37*, 457–473. [CrossRef]
104. Yuan, X.; Wang, Y.; Tang, L.; Zhou, H.; Han, N.; Zhu, H.; Uchimiya, M. Spatial distribution, source analysis, and ecological risk assessment of PBDEs in river sediment around Taihu Lake, China. *Environ. Monit. Assess.* **2020**, *192*, 1–13. [CrossRef]
105. Da, C.; Wang, R.; Huang, Q.; Mao, J.; Xie, L.; Xue, C.; Zhang, L. Sediment Records of Polybrominated Diphenyl Ethers (PBDEs) from the Anhui Province Section of Yangtze River, China. *Bull. Environ. Contam. Toxicol.* **2021**, *106*, 334–341. [CrossRef]
106. Da, C.; Wang, R.; Ye, J.; Yang, S. Sediment records of polybrominated diphenyl ethers (PBDEs) in Huaihe River, China: Implications for historical production and household usage of PBDE-containing products. *Environ. Pollut.* **2019**, *254*, 112955. [CrossRef]
107. Sun, J.L.; Chen, Z.X.; Ni, H.G.; Zeng, H. PBDEs as indicator chemicals of urbanization along an urban/rural gradient in South China. *Chemosphere* **2013**, *92*, 471–476. [CrossRef]
108. Zhao, Y.X.; Qin, X.F.; Li, Y.; Liu, P.Y.; Tian, M.; Yan, S.S.; Qin, Z.F.; Xu, X.B.; Yang, Y.J. Diffusion of polybrominated diphenyl ether (PBDE) from an e-waste recycling area to the surrounding regions in Southeast China. *Chemosphere* **2009**, *76*, 1470–1476. [CrossRef]
109. Tian, M.; Chen, S.J.; Wang, J.; Luo, Y.; Luo, X.J.; Mai, B.X. Plant Uptake of Atmospheric Brominated Flame Retardants at an E-Waste Site in Southern China. *Environ. Sci. Technol.* **2012**, *46*, 2708–2714. [CrossRef]
110. Yang, Z.Z.; Zhao, X.R.; Zhao, Q.; Qin, Z.F.; Qin, X.F.; Xu, X.B.; Jin, Z.X.; Xu, C.X. Polybrominated Diphenyl Ethers in Leaves and Soil from Typical Electronic Waste Polluted Area in South China. *Bull. Environ. Contam. Toxicol.* **2008**, *80*, 340–344. [CrossRef]
111. Huang, H.; Zhang, S.; Christie, P. Plant uptake and dissipation of PBDEs in the soils of electronic waste recycling sites. *Environ. Pollut.* **2011**, *159*, 238–243. [CrossRef]
112. Luo, Y.; Yu, X.; Yang, Z.; Yuan, J.; Mai, B. Studies on heavy metal contamination by improper handling of e-waste and its environmental risk evaluation I. Heavy metal contamination in e-waste open burning sites. *Asian J. Ecotoxicol.* **2008**, *3*, 34–41. Available online: https://en.cnki.com.cn/Article_en/CJFDTOTAL-STDL200801008.htm (accessed on 8 April 2021). (In Chinese).
113. Luo, Y.; Luo, X.; Yang, Z.; Yu, X.; Yuan, J.; Chen, S.; Mai, B. Studies on heavy metal contamination by improper handling of e-waste and its environmental risk evaluation. II. Heavy metal contamination in surface soils of e-waste disassembling workshops within villages and the adjacent agricultural soils. *Asian J. Ecotoxicol.* **2008**, *3*, 123–129. Available online: https://xueshu.baidu.com/usercenter/paper/show?paperid=97bb4a53b576730d0e9eec5d88acfda0&site=xueshu_se (accessed on 8 April 2021). (In Chinese).
114. Tang, X.; Shen, C.; Shi, D.; Cheema, S.A.; Khan, M.I.; Zhang, C.; Chen, Y. Heavy metal and persistent organic compound contamination in soil from Wenling: An emerging e-waste recycling city in Taizhou area, China. *J. Hazard. Mater.* **2010**, *173*, 653–660. [CrossRef]
115. Qing, X.; Yutong, Z.; Shenggao, L. Assessment of heavy metal pollution and human health risk in urban soils of steel industrial city (Anshan), Liaoning, Northeast China. *Ecotoxicol. Environ. Saf.* **2015**, *120*, 377–385. [CrossRef]
116. Zheng, Y.M.; Chen, T.B.; He, J.Z. Multivariate geostatistical analysis of heavy metals in topsoils from Beijing, China. *J. Soils Sediments* **2007**, *8*, 51–58. [CrossRef]
117. Zhang, M.K.; Ke, Z.X. Heavy metals, phosphorus and some other elements in urban soils of Hangzhou city, China. *Pedosphere* **2004**, *14*, 177–185. [CrossRef]
118. Shi, G.; Chen, Z.; Xu, S.; Zhang, J.; Wang, L.; Bi, C.; Teng, J. Potentially toxic metal contamination of urban soils and roadside dust in Shanghai, China. *Environ. Pollut.* **2008**, *156*, 251–260. [CrossRef]
119. Xia, X.; Yang, Z.; Cui, Y.; Li, Y.; Hou, Q.; Yu, T. Soil heavy metal concentrations and their typical input and output fluxes on the southern Song-nen Plain, Heilongjiang Province, China. *J. Geochem. Explor.* **2014**, *139*, 85–96. [CrossRef]
120. Yang, P.; Mao, R.; Shao, H.; Gao, Y. An investigation on the distribution of eight hazardous heavy metals in the suburban farmland of China. *J. Hazard. Mater.* **2009**, *167*, 1246–1251. [CrossRef]
121. Zheng, Y.P. On Heavy metal pollution status and its assessment for the farmland soil of Henan Province. *J. Environ. Manage. Coll. Chin.* **2007**, *3*, 44–46. (In Chinese) [CrossRef]
122. Geng, J.M.; Wang, W.B.; Wen, C.P.; Yi, Z.Y.; Tang, S.M. Concentrations and distributions of selenium and heavy metals in Hainan paddy soil and assessment of ecological security. *Acta Ecol. Sin.* **2012**, *32*, 3477–3486. [CrossRef]
123. Zhao, Y.; Wang, Z.; Sun, W.; Huang, B.; Shi, X.; Ji, J. Spatial interrelations and multi-scale sources of soil heavy metal variability in a typical urban–rural transition area in Yangtze River Delta region of China. *Geoderma* **2010**, *156*, 216–227. [CrossRef]
124. Ni, D.; Fang, H.G.; Wang, L.J. Present situation of soil environmental quality and protection countermeasures in Ji County. *Till. Cultivation* **2012**, *2*, 36–37. (In Chinese)
125. Chai, Y.; Guo, J.; Chai, S.; Cai, J.; Xue, L.; Zhang, Q. Source identification of eight heavy metals in grassland soils by multivariate analysis from the Baicheng–Songyuan area, Jilin Province, Northeast China. *Chemosphere* **2015**, *134*, 67–75. [CrossRef] [PubMed]
126. Niu, L.; Yang, F.; Xu, C.; Yang, H.; Liu, W. Status of metal accumulation in farmland soils across China: From distribution to risk assessment. *Environ. Pollut.* **2013**, *176*, 55–62. [CrossRef] [PubMed]

127. Quan, S.X.; Yan, B.; Lei, C.; Yang, F.; Li, N.; Xiao, X.M.; Fu, J.M. Distribution of heavy metal pollution in sediments from an acid leaching site of e-waste. *Sci. Total. Environ.* **2014**, *499*, 349–355. [CrossRef] [PubMed]
128. Chen, L.; Yu, C.; Shen, C.; Zhang, C.; Liu, L.; Shen, K.; Tang, X.; Chen, Y. Study on adverse impact of e-waste disassembly on surface sediment in East China by chemical analysis and bioassays. *J. Soils Sediments* **2010**, *10*, 359–367. [CrossRef]
129. Du, Y.; Wu, Q.; Kong, D.; Shi, Y.; Huang, X.; Luo, D.; Chen, Z.; Xiao, T.; Leung, J.Y. Accumulation and translocation of heavy metals in water hyacinth: Maximising the use of green resources to remediate sites impacted by e-waste recycling activities. *Ecol. Indic.* **2020**, *115*, 106384. [CrossRef]
130. Lin, C.; He, M.; Li, Y.; Liu, S. Content, enrichment, and regional geochemical baseline of antimony in the estuarine sediment of the Daliao river system in China. *Geochem.* **2012**, *72*, 23–28. [CrossRef]
131. Guan, Q.; Cai, A.; Wang, F.; Wang, L.; Wu, T.; Pan, B.; Song, N.; Li, F.; Lu, M. Heavy metals in the riverbed surface sediment of the Yellow River, China. *Environ. Sci. Pollut. Res.* **2016**, *23*, 24768–24780. [CrossRef]
132. Mao, L.; Liu, L.; Yan, N.; Li, F.; Tao, H.; Ye, H.; Wen, H. Factors controlling the accumulation and ecological risk of trace metal(loid)s in river sediments in agricultural field. *Chemosphere* **2020**, *243*, 125359. [CrossRef]
133. Bo, L.; Wang, D.; Li, T.; Li, Y.; Zhang, G.; Wang, C.; Zhang, S. Accumulation and risk assessment of heavy metals in water, sediments, and aquatic organisms in rural rivers in the Taihu Lake region, China. *Environ. Sci. Pollut. Res.* **2014**, *22*, 6721–6731. [CrossRef]
134. Zhang, P.; Qin, C.; Hong, X.; Kang, G.; Qin, M.; Yang, D.; Pang, B.; Li, Y.; He, J.; Dick, R.P. Risk assessment and source analysis of soil heavy metal pollution from lower reaches of Yellow River irrigation in China. *Sci. Total. Environ.* **2018**, *633*, 1136–1147. [CrossRef]
135. Xia, W.; Wang, R.; Zhu, B.; Rudstam, L.G.; Liu, Y.; Xu, Y.; Xin, W.; Chen, Y. Heavy metal gradients from rural to urban lakes in central China. *Ecol. Process.* **2020**, *9*, 1–11. [CrossRef]

Article

Heart Failure and PAHs, OHPAHs, and Trace Elements Levels in Human Serum: Results from a Preliminary Pilot Study in Greek Population and the Possible Impact of Air Pollution

Eirini Chrysochou [1], Panagiotis Georgios Kanellopoulos [1], Konstantinos G. Koukoulakis [1], Aikaterini Sakellari [2], Sotirios Karavoltsos [2], Minas Minaidis [3] and Evangelos Bakeas [1,*]

[1] Laboratory of Analytical Chemistry, Department of Chemistry, National and Kapodistrian University of Athens, Panepistimiopolis, 15784 Athens, Greece; eirinichr@chem.uoa.gr (E.C.); gpkan@chem.uoa.gr (P.G.K.); kkoukoulakis@chem.uoa.gr (K.G.K.)

[2] Laboratory of Environmental Chemistry, Department of Chemistry, National and Kapodistrian University of Athens, Panepistimiopolis, 15784 Athens, Greece; esakel@chem.uoa.gr (A.S.); skarav@chem.uoa.gr (S.K.)

[3] General Hospital "LAIKO", 11527 Athens, Greece; minaidis64@hotmail.com

* Correspondence: bakeas@chem.uoa.gr

Abstract: Cardiovascular diseases (CVDs) have been associated with environmental pollutants. The scope of this study is to assess any potential relation of polycyclic aromatic hydrocarbons (PAHs), their hydroxylated derivatives, and trace elements with heart failure via their direct determination in human serum of Greek citizens residing in different areas. Therefore, we analyzed 131 samples including cases (heart failure patients) and controls (healthy donors), and the respective demographic data were collected. Significantly higher concentrations ($p < 0.05$) were observed in cases' serum regarding most of the examined PAHs and their derivatives with phenanthrene, fluorene, and fluoranthene being the most abundant (median of >50 µg L^{-1}). Among the examined trace elements, As, Cd, Cu, Hg, Ni, and Pb were measured at statistically higher concentrations ($p < 0.05$) in cases' samples, with only Cr being significantly higher in controls. The potential impact of environmental factors such as smoking and area of residence has been evaluated. Specific PAHs and trace elements could be possibly related with heart failure development. Atmospheric degradation and smoking habit appeared to have a significant impact on the analytes' serum concentrations. PCA–logistic regression analysis could possibly reveal common mechanisms among the analytes enhancing the hypothesis that they may pose a significant risk for CVD development.

Keywords: cardiovascular diseases (CVDs); heart failure; polycyclic aromatic hydrocarbons (PAHs); trace elements; serum

1. Introduction

Air pollution is a major public health problem with a plethora of consequences on humans and other living beings [1]. An estimated measure of more than 4.2 million annual global premature deaths related to air pollution has been reported [2]. Outdoor air pollution is considered the fifth greatest risk factor for all-cause mortality, which is higher than the acknowledged risk factors, including poor diet and low exercise, and the first among environmental risk factors [3]. Air pollution consists of plenty of diverse pollutants partitioned in the gas phase, such as volatile organic compounds (VOCs), nitrogen oxides (NO$_x$), sulfur dioxide (SO$_2$), carbon monoxide (CO), etc. [4–6] and in the particle phase including polycyclic aromatic hydrocarbons (PAHs), trace elements, polychlorinated biphenyls (PCBs), etc. [7,8]. As a result of the presence of different harmful substances, the International Agency for Research on Cancer (IARC) has classified air pollution as a human carcinogen [9]. Globally, over 90% of individuals live in areas where air pollution levels exceed the World Health Organization (WHO) guidelines [2]. The well-known "Harvard Six Cities Study" was the origin of the link of air pollution to mortality from lung cancer and cardiopulmonary disease [10]. In this light, many studies addressed geographic differences of exposures, for example PAHs and particulates exposures in urban and rural areas of Czech Republic [11], or temporal sources of pollution, such as the New York World Trade Center disaster, which was a transient event of PAHs, dioxins, and inorganic dusts exposure of a well-defined population [12], with air pollution being associated with mortality [13–16].

Although the lung is the main receptor of air contaminants, air pollution is significantly associated with cardiovascular diseases (CVDs) [17,18]. In particular, exposure to atmospheric particulate matter (PM) has been correlated with increased arrhythmia incidents, carotid intima-media thickness, which is a marker of subclinical atherosclerosis, with the progression of inflammation and hypertension [19–21], as well as with reduced heart rate variability (HRV) [22] with particles with diameter <0.3 µm being the most crucial PM fraction to the reduction of the cardiac autonomic function [23]. Moreover, a 10–30% increase of the death risk from ischemic heart disease per 10 µg m^{-3} increase of PM$_{2.5}$ (particles with aerodynamic diameter <2.5 µm) has been estimated [24]. Generally, air pollution's implications on CVDs may lead to higher mortality rates than those caused by the air pollution impact on respiratory diseases [3]. In Europe, about 3.9 million deaths annually have been attributed to CVDs making them the leading cause of mortality, with more than 85 million European citizens living with CVDs during 2015 [25].

PM-bound PAHs are associated with CVDs, with studies involving humans and other mammals underlining the link of tricyclic PAHs, and especially phenanthrene with arrhythmias, the aggravation of heart failure, heart attacks, and other complications involving atherosclerosis and ischemia [17,26]. Occupational exposure to PAHs has been also linked with alterations in cardiac autonomic function, as implied by the decreased HRV [27]. The cardiovascular toxicity of PAHs involves aryl hydrocarbon receptor (AhR), reactive oxygen species (ROS), and/or reactive electrophilic metabolites, with the cardiotoxic effects not being limited to benzo[a]pyrene (BaP) but also to other PAHs including pyrene (PYR), phenanthrene (PHE), and benzo[e]pyrene (BeP) [28]. PAHs mixtures and especially PHE cause cardiotoxicity independent of the (AhR) pathway with various toxicant and cellular pathways involving atherosclerosis, cardiac arrhythmias, and cardiac hypertrophy [26]. For instance, the DNA methylation, caused by exposure to PHE, may induce cardiac hypertrophy with a mechanism that involves the reduction of the miR-133a expression [29]. PAHs constitute a group of compounds formed by the incomplete combustion of carbonaceous material, which can be emitted into the atmosphere from both natural and anthropogenic sources, including vehicular emissions, domestic heating, power plants, tobacco smoke, and solid waste incineration [30,31]. In developing countries, organic wastes burning for domestic needs such as cooking [32] is another source of PAHs. After exposure, PAHs bound to the smallest sized particles in PM$_{2.5}$ can enter the systemic circulation un-metabolized and reach various organs [26].

Apart from PAHs, environmental trace elements are a noteworthy but overlooked source of CVDs risk [33]. Various studies implied that trace elements including As, Cd, Hg, and Pb may constitute an important factor to CVDs development [34–37]. Although other trace elements such as Co, Cu, Fe, Se and Zn are essential for the human organism [38–41], high exposure to them is also associated with the risk of CVD development [24]. Airborne particle-bound trace elements have both natural and anthropogenic origin, including natural dust emissions [42], coal [43] and oil combustion [44], the production of iron, steel, cast iron, etc. [45]. However, metal exposure is usually neglected by the agencies that produce guidelines about cardiovascular prevention [33].

There is adequate data based on measurements of urinary metabolites of PAHs trying to link PAHs levels with CVDs [46–48] and other diseases, including rheumatoid arthritis [49] and diabetes [50]. However, data of PAHs levels in plasma and serum related to CVDs are limited [26]. The determination of PAHs in different human matrices provides different information based on the selected matrix. For example, urine samples are more closely related with metabolites derived from biotransformation procedures; hair samples are considered ideal for long-term exposure studies; whereas blood, serum, saliva, or exhaled breath samples are more associated with unmetabolized PAHs recent exposure [51]. Moreover, blood-borne PAHs compared to their respective metabolites or adducts are less susceptible to variability from inter-subject differences in metabolism and excretion [52]. In addition, serum is a more familiar and efficient way for the monitoring of PAHs compared to other tissues [53]. PAHs have been measured in serum samples not only in an effort to investigate any potential association with different types of cancer such as leukemia [54] and bladder cancer [55] but also for the estimation of background burden values [56,57]. Moreover, PAHs have been measured at the maternal serum in order to investigate the possible transplacental transfer from the mother to the fetus [58], while others have proposed an inversed trend of maternal serum PAHs and a decreased birth weight [59].

Trace elements are measured in various biological samples, including blood, serum, erythrocytes, and urine, with the whole blood being the preferred matrix used for the biomonitoring of the toxic Pb, Cd, and Hg as they are concentrated in the erythrocytes. The determination of serum trace elements is also prone to errors due to hemolysis, which may lead to possible errors in the results [60]. However, the elemental composition of serum is widely studied, as it provides important insights for the state of a human organism [61] and because it is the most exchangeable blood compartment [60]. Serum trace elements' levels have been associated with various CVDs, including coronary artery disease (CAD) [62], carotid artery atherosclerosis in maintenance hemodialysis patients [63], and other diseases, such as asthma, various allergic diseases [64], and different types of leukemia [65,66].

Heart failure is a clinical syndrome provoked by multiple causes [67], with the CAD and arterial hypertension being the leading causes and with heart dysfunction, valvular disease, tachyarrhythmias, diabetes mellitus, myocarditis, and infiltrative disorders being some of the other causes [68].

In this perspective, the aim of this work is the determination and comparison of PAHs' and OHPAHs' trace elements' concentrations in human serum of heart failure patients and healthy donors, all residing in different areas of Greece, and the investigation of the potential impact of different environmental factors including smoking habit and area of residence.

2. Results and Discussion

2.1. Cases and Controls

The total concentrations of PAHs, OHPAHs, and trace elements for cases and controls are presented in Figures 1–3. More detailed information about the mean, median, and concentration ranges of the analytes is shown in Tables S1–S3.

2.1.1. PAHs

Almost the 70% of the studied PAHs presented detection frequencies over 65% (Table S1). In particular, naphthalene (NAP), acenaphthene (ACE), fluorene (FL), PHE, fluoranthene (FLT) and PYR were detected in 100% of both cases' and controls' samples. Anthracene (ANT), benzo[a]anthracene (BaA), benzo[b,k]fluoranthenes (BFA), and dibenzo[a,h]anthracene (DBA) were found at detection rates varying from 65.7% (DBA-control samples) to 93.7% (BFA-cases samples). Lower detection rates were found for chrysene (CHR) (35.4% in cases and 28.6 in controls), while the rest of the studied PAHs, including BaP, acenaphthylene (ACY), indeno[1,2,3 cd]pyrene (IPY), and benzo[ghi]perylene (BPE) were detected at ≤25% of the samples. As shown in Figure 1, PHE was the dominant PAH in both cases' and controls' serum followed by FL, with the concentrations in cases' serum being approximately four and three times higher, respectively ($p < 0.05$) (Table S1). Significantly higher concentrations in cases' samples were also observed for the following PAHs with a concentration descending sequence of ACE, FLT, DBA, PYR, NAP, ANT, BaA, and BFA with FLT, PYR, and DBA median values being almost 4-fold higher than those observed in controls' serum. CHR, BaP, ACY, and IPY did not differ significantly ($p > 0.05$), since in many cases and controls samples, their concentration was below the limits of detection (LoD). The relative profile of PAHs, regarding the number of rings in each molecule, in cases' serum is displayed in Figure 4. The dominance of 3-ring-PAHs is highlighted as they accounted for the 73% of the total measured PAHs. As a result, the contribution of low molecular weight PAHs (LMW PAHs) was almost three times higher than that of high molecular weight PAHs (HMW PAHs). Consequently, the significantly higher concentrations of the majority of PAHs and especially LMW PAHS, detected in this work, may play a pivotal role in the development of heart failure.

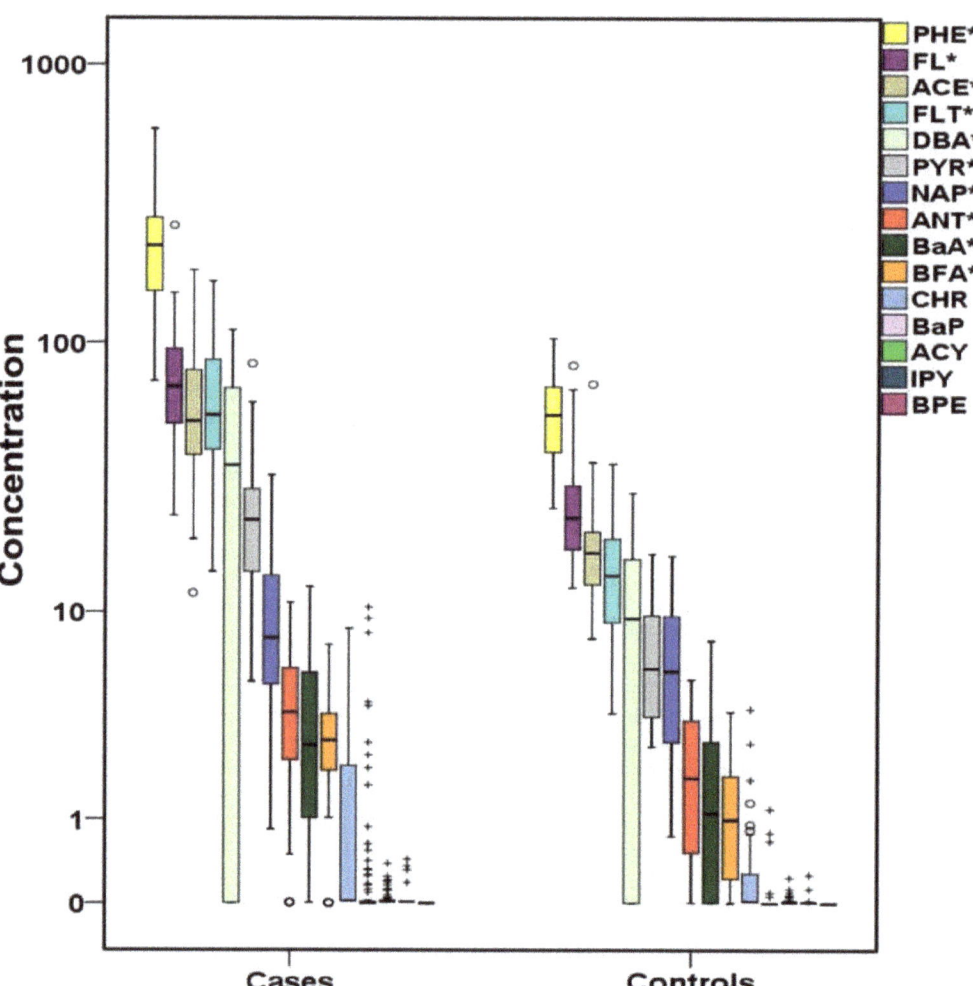

Figure 1. Variation of PAHs (µg L^{-1}) both for cases and controls ($^+$ outliers, $^\circ$ values above 3rd quartile and * p-value < 0.05) in logarithmic scale (PHE:phenanthrene, FL:fluorene, ACE:acenaphthene, FLT:fluoranthene, DBA:dibenzo[a,h]anthracene, PYR:pyrene, NAP:naphthalene, ANT:anthracene, BaA:benzo[a]anthracene, BFA:benzo[b,k]fluoranthenes, CHR:chrysene, BaP:benzo[a]pyrene, ACY:acenaphthylene, IPY:indeno[1,2,3 cd]pyrene, BPE:benzo[ghi]perylene).

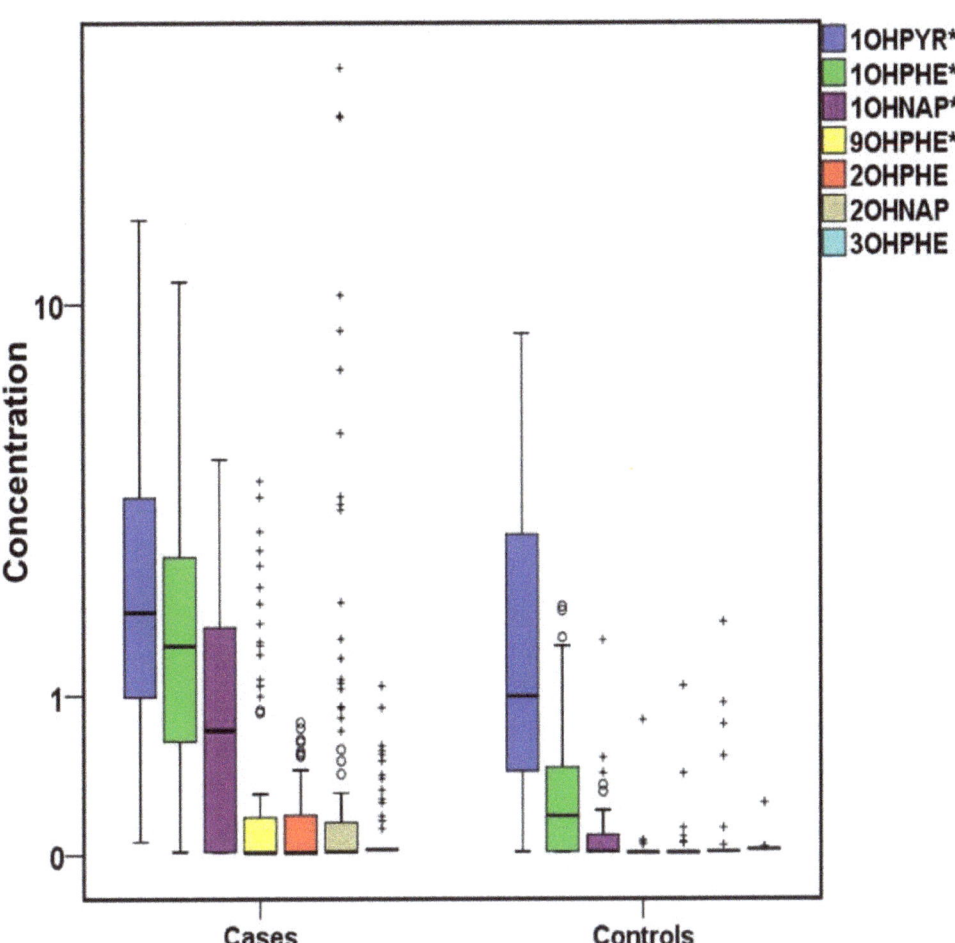

Figure 2. Variation of OHPAHs (µg L^{-1}) both for cases and controls (+ outliers, ○ values above 3rd quartile and * p-value < 0.05) in logarithmic scale (1OHPYR:1-hydroxypyrene, 1OHPHE:1-hydroxyphenanthrene, 1OHNAP:1-naphthol, 9OHPHE:9-hydroxyphenanthrene, 2OHPHE:2-hydroxyphenanthrene, 2OHNAP:2-naphthol, 3OHPHE:3-hydroxyl-phenanthrene).

High PAHs detection frequencies combined with higher percentages of LMW compared to HMW PAHs in serum have been reported in many other studies [54,69,70]. However, relatively low rates (<20%) have been also reported in a pilot study in a rural area of Egypt [71]. PAHs have been found in significantly higher concentrations in the serum of patients than those of corresponding controls' serum in the cases of bladder cancer, leukemia, and polycystic ovary syndrome [54,55,72].

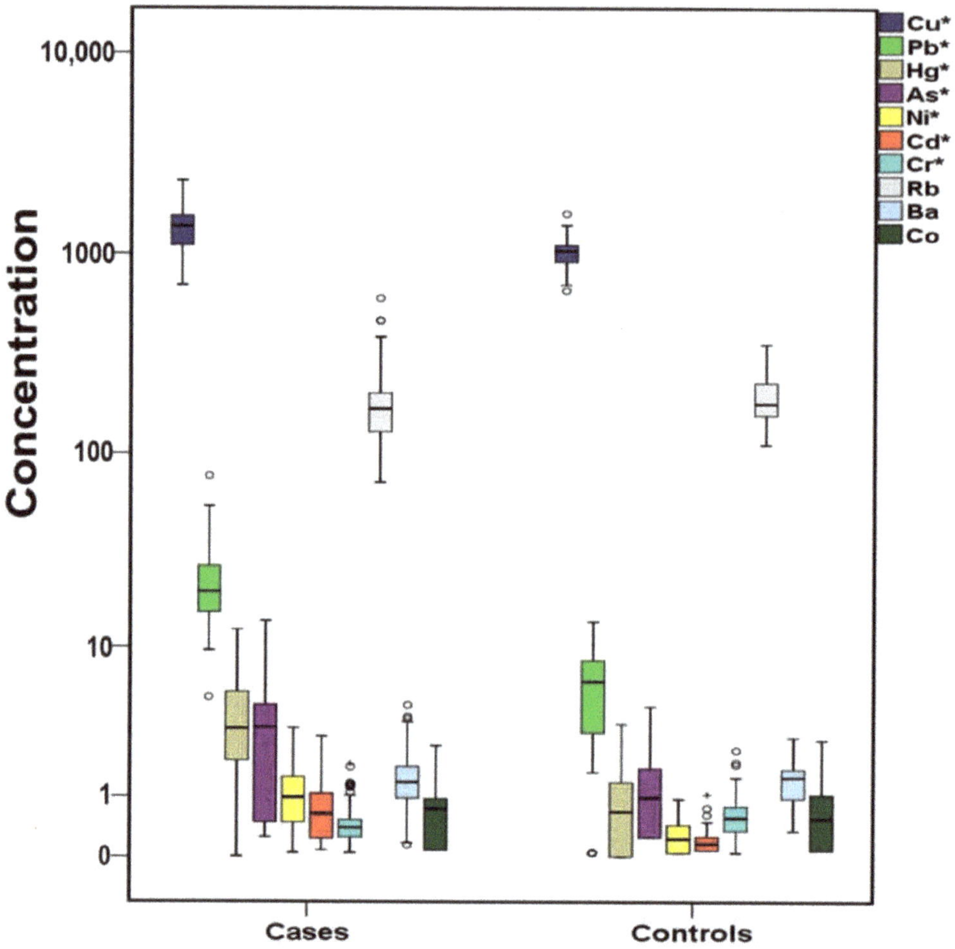

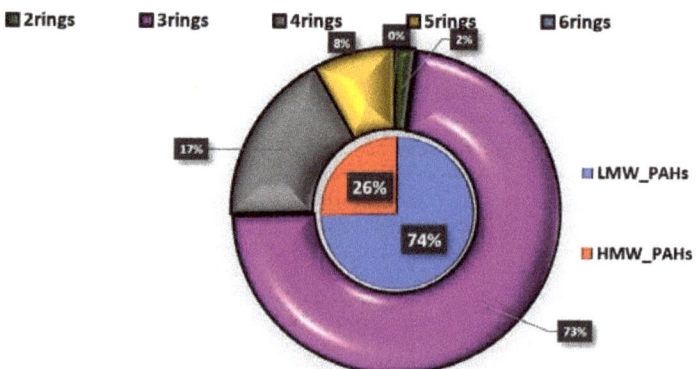

Figure 3. Variation of trace elements (μg L^{-1}) both for cases and controls (⁺ outliers, ° values above 3rd quartile and * p-value < 0.05) in logarithmic scale.

Figure 4. Distribution of HPAHs and LPAHs in cases' serum.

2.1.2. OHPAHs

Urinary OHPAHs are considered as PAH metabolites, and as a result, most studies that related PAHs exposure and CVDs have studied OHPAHs in urine [46,47,73–75]. However, concentrations of OHPAHs and/or the parent ones in urine do not only depend on the external exposure but also on difference in the metabolism and bioconversion procedures in each individual organism [76]. Therefore, noteworthy results can also be extracted from serum determination, too [54,72]. The detection frequencies of OHPAHs are shown in Table S2. In cases' serum, OHPAHs were detected at higher rates than those measured in controls', with 1-hydroxypyrene (1OHPYR) being found at all samples followed by 1-hydroxyphenanthrene (1OHPHE) and 1-napthol (1OHNAP) with respectively equal rates of 96.9 and 65.6%. The rest of the studied OHPAHs such as 9-hydroxyphenanthrene (9OHPHE), 2-hydroxyphenanthrene (2OHPHE), 2-naphthol (2OHNAP), and 3-hydroxyphenanthrene (3OHPHE) were detected at <37% of the samples. However, in controls' serum, the detection frequencies ranged from 8.57% (3OHPHE) to 77.1% (1OHPYR). Higher detection frequencies of serum OHPAHs have been found in other studies, from 70% (9OHPHE) to 100% (2OHNAP) in cases' samples and from 50% (9OHPHE) to 100% (2OHNAP) in that of controls [72]. Nevertheless, lower rates were found in our previous study (40–70% in cases and <25% in controls) [54]. In general, OHPAHs presented lower concentrations than the parent ones. Significantly ($p < 0.05$) higher concentrations in cases' serum were observed, in descending order, for 1OHPYR, 1OHPHE, 1OHNAP, 9OHPHE, as well as ΣOHPAHs (Table S2, Figure 2). 1OHPYR was the most abundant with a median of 1.87 µg L^{-1} followed by 1OHPHE and 1OHNAP with median values 1.48 and 0.71 µg L^{-1}, respectively. 2OHNAP, 2OHPHE, and 3OHPHE did not differ significantly, as in many samples, their values were found below the LoD.

Although data concerning OHPAHs serum levels are sparse, some of them appear to be significantly higher in the serum of acute leukemia patients and women with polycystic ovary syndrome than that of the respective control subjects [54,72].

2.1.3. Trace Elements

Trace elements were widely detected in the serum samples, with detection frequencies over 73.9% in cases' serum and from 57.1% in control's serum (Table S3). Significantly higher ($p < 0.05$) concentrations in cases' serum were observed for copper (Cu), lead (Pb), mercury (Hg), arsenic (As), nickel (Ni), and cadmium (Cd), while higher, but not significantly ($p > 0.05$), concentrations were found for cobalt (Co). On the contrary, chromium (Cr), rubidium (Rb), and barium (Ba) were found at higher levels in controls' serum; the difference was statistically significant ($p > 0.05$) only in the case of Cr though (Figure 3, Table S3). Recent reviews have highlighted the association of both known toxic and essential metals with CVDs development [77,78]; thus, the elements that were observed with significantly different concentrations in cases' or control's serum will be separately discussed.

Arsenic (As) is a toxic metalloid that enters the organisms through food, drinking water, cigarette smoke, and through inhalation of particle bound As [7,36,79,80]. According to the U.S Agency for Toxic Substances and Disease Registry, the safe level for As in human blood is suggested to be less than 70 µg L^{-1} [81]. The As levels presented in this work are greatly lower than the suggested value; however, As concentrations in cases' serum were approximately 3.5 times higher than those measured in controls (median of 3.39 versus 0.98 µg L^{-1}). In other studies, As was measured in the serum of 17–20-year-old students and was found at lower levels with median concentrations from 2.43 to 2.81 µg L^{-1}, with the highest As values being attributed to seafood intake [79]. Relatively higher concentrations of As were measured in whole blood samples of kids with learning disorders (mean of 12.1 ± 5.19 µg L^{-1}) which were significantly higher than the corresponding control values (9.73 ± 4.39 µg L^{-1}) [82]. Moreover, significantly higher concentrations of As were found in the serum of chronic kidney disease patients receiving continuous ambulatory peritoneal dialysis (mean of 3.79 µg L^{-1}) than those found in the respective control samples (0.52 µg L^{-1}), which was related with the increased CVD risk of the specific target group [83].

Regarding the toxic Cd, WHO has evaluated the values of 0.1 µg to 4 µg L^{-1} of the Cd blood concentrations characterizing a healthy and unexposed adult [84]. Although our findings are under the limits, the observed statistical difference between cases (median of 0.64 µg L^{-1}) and controls (median of 0.17 µg L^{-1}) can not be ruled out. In other studies, the Cd serum levels of CAD patients were found to be higher than those of the negative subjects with a mean of 2.44 µg L^{-1} versus 1.15, although the difference was not significant [62]. However, the Cd serum levels of patients on maintenance hemodialysis (HD) were significantly higher than those of the control group, which was independently associated with carotid atherosclerosis: a disease common in the specific patient group [69]. Additionally, Cd exposure was associated with atherosclerosis cardiovascular disease (ASCVD) through the strong relationship of Cd blood levels from approximately 2500 individuals with the 10-year ASCVD risk scores, using risk prediction models [85].

Hg is an easily accessible toxic metal with various intake pathways, such as air, water, food, vaccines, pharmaceuticals, and cosmetics, with its cardiotoxic effects being strongly associated with hypertension, coronary heart disease, myocardial infarction, cardiac arrhythmias, carotid artery obstruction, cerebrovascular accident, and generalized atherosclerosis [86]. The recommended, by the U.S. National Academy of Sciences (NAS), average level of Hg in human blood is below 5 µg L^{-1} [87]. In the current work, 29 out of 96 cases' samples, i.e., 30.2%, presented concentrations above the proposed level, with the median concentration, however, being below the limit (3.33 µg L^{-1}). Significantly lower concentrations ($p < 0.05$) were measured in control samples varying from below the detection limits to 3.57 µg L^{-1}, with a median value of 0.46 µg L^{-1}. Elevated concentrations were also found in the serum of CAD patients (8.19 µg L^{-1}), which were significantly higher than the respective concentrations of controls' serum (4.11 µg L^{-1}) [62]. Similar to Cd, Hg blood levels were also linked with the ASCVD risk in the Korean population [85]. Relatively high concentrations of blood Hg (mean of 102 ± 55.8 µg L^{-1}) have been reported for workers in Artisanal and Small-Scale Gold Mining (ASGM) operations in Ghana, where Hg exposure was notably elevated [88]. Although the cardiovascular implications of Hg have been evaluated [89] its biokinetics and actions in the CV system are still quite elusive [90].

Regarding Pb, there is sufficient evidence for adverse health effects in children and adults at blood levels <50 µg L^{-1} [91]. Pb in the serum samples varied from 5.18 to 77.0 µg L^{-1}, with a median of 19.8 µg L^{-1} which is significantly higher ($p < 0.05$) than those found for healthy donors' samples (median of 6.44 µg L^{-1}). In other studies, Pb has been also found in the serum of CAD patients with a mean of 8.19 µg L^{-1}, which is over two times higher than the respective control samples (mean of 3.69 µg L^{-1}) [62]. Pb competes with the essential metals, such as calcium (Ca), iron (Fe), and zinc (Zn), as it is able to bind and/or interact, with parallel ways as the latter, with the same enzymes resulting in the inhibiting of the enzyme's ability to catalyze its normal reactions [92]. The general mechanisms to CVD development, among others, include the induction of oxidative stress, impairment of the nitric oxide system, the increased generation of ROS, changes in the Ca^{+2} transport and intracellular distribution, etc. [93,94]. Pb blood levels with a geometric mean of 20.7 µg L^{-1} have been associated with the prevalence of peripheral arterial disease in the US population [95] and with the ASCVD risk in the Korean population [85]. Pb along with Cd have been also found in significantly higher concentration in the serum of acute hemorrhagic stroke patients than the corresponding controls' serum, indicating a possible association with this stroke subtype [96].

Apart from the known toxic metals, imbalanced levels of essential metals such as Cr, Co, Cu, magnesium (Mg), Ni, selenium (Se), tungsten (W), and Zn are significantly associated with CVDs [78]. As a result, the significantly higher concentrations ($p < 0.05$) observed, in this study, for Cu and Ni might have a role in heart failure development. Significantly higher concentrations of Co and Cu were also noticed in the serum of HD patients than those of controls' samples having a plausible association of increased CVD risk [69]. Elevated Cu levels in serum (median of 1175 µg L^{-1}) were found to be associated with CVDs among US adults [97]. In another study of 4035 middle-aged men, increased serum Cu levels were linked with a 30% increase in CV mortality [98], while other papers underline the relation of circulating blood levels of the essential Ni and Cr with atherosclerotic plagues in elderly [99]. However, as far as Cr is concerned, lower concentrations have been measured in the serum of CVD patients than those of the healthy controls [100], which is in agreement with the findings of this study. Particularly, Cr median values were 0.40 µg L^{-1} in cases' serum versus 0.57 µg mL^{-1} in controls ($p < 0.05$). In other works, blood and scalp hair Cr was found at lower levels of myocardial infraction patients than the controls, which was attributed to the increased urinary loss of this element [101]. The possible protective role of Cr against CVD was indicated as in a case-control study conducted involving CVD patients, where the toenail Cr levels were higher for the control participants [102].

2.2. Cases with Different Smoking Habits and Residence Areas

Data obtained from cases' serum was further classified depending on sex, smoking habit (current smokers, ex-smokers, and non-smokers) and the area of residence (urban, industrial, and rural).

2.2.1. Sex

Gender-based variation is a common classification of the data in case-control and cohort studies [103]. Most of the analytes presented higher concentrations in men (Figure 5), although only a few differences were significant ($p < 0.05$). In particular, regarding PAHs, PHE, BaA, and CHR presented significantly ($p < 0.05$) higher concentrations. OHPAHs, on the contrary, did not present any significant gender-based variation, although 1OHPHE was clearly higher in men's serum ($p = 0.054$). Among trace elements, only Pb presented statistically higher concentrations in men ($p < 0.05$), which is in agreement with a previous study involving Greek citizens from Athens Metropolitan Area [104]. Various population studies have presented higher ($p < 0.05$) concentrations in males' blood or serum. Particularly, in Korean men, blood Hg and Pb were found at significantly higher levels [105]. In Brazil, Cu and Pb were significantly correlated with gender and age [106]. In France, mean levels of blood Pb and Zn were found significantly higher in men, whereas Co and Cr were found significantly higher in women [107]. In the serum of Chinese students (17–20 years old), Pb was significantly higher in men [79]. A possible reason for the higher observed Pb in men's samples could be the higher male hematocrit levels, as lead tends to bind to erythrocytes [108].

2.2.2. Smoking Habit

Cigarette smoking is the major preventable cause of human death in the Western world being responsible for approximately 5 million annual premature deaths globally [109]. It is also a critical factor for CVD development and the second main cause of CVD mortality after high blood pressure [110]. Current and recent smokers are more vulnerable to smoking-related CVD risks than those who have quit smoking for a long time and non-smokers [111], with current or past smoking increasing the heart failure risk [112]. In Figures 6–8, cases' samples are classified in terms of the smoking habit. Among PAHs, PHE, FL, ACE, FLT, PYR, IPY (in a decreasing concentration sequence), and ΣPAHs (not shown in the figure) presented statistically higher concentrations in smokers' serum followed by ex-smokers, while BFA and BaP were measured at higher levels in ex-smokers' samples followed by smokers (Figure 6). HMW PAHs such as CHR, BaA, BaP and BPE were significantly higher in the serum of smoker leukemia patients [54]. In other studies, ACE and BaP were significantly higher in smokers' samples, although the number of smokers in that study was limited [70]. Additionally, PAHs derivatives such as 1OHPYR, 1OHPHE, 1OHNAP, and ΣOHPAHs (not shown in the figure) were significantly higher in smokers (Figure 7).

Although data about serum OHPAHs are limited, in smokers' urine, higher levels of NAP, FL, PHE, and PYR metabolites have been frequently found compared to non-smokers' urine [113,114].

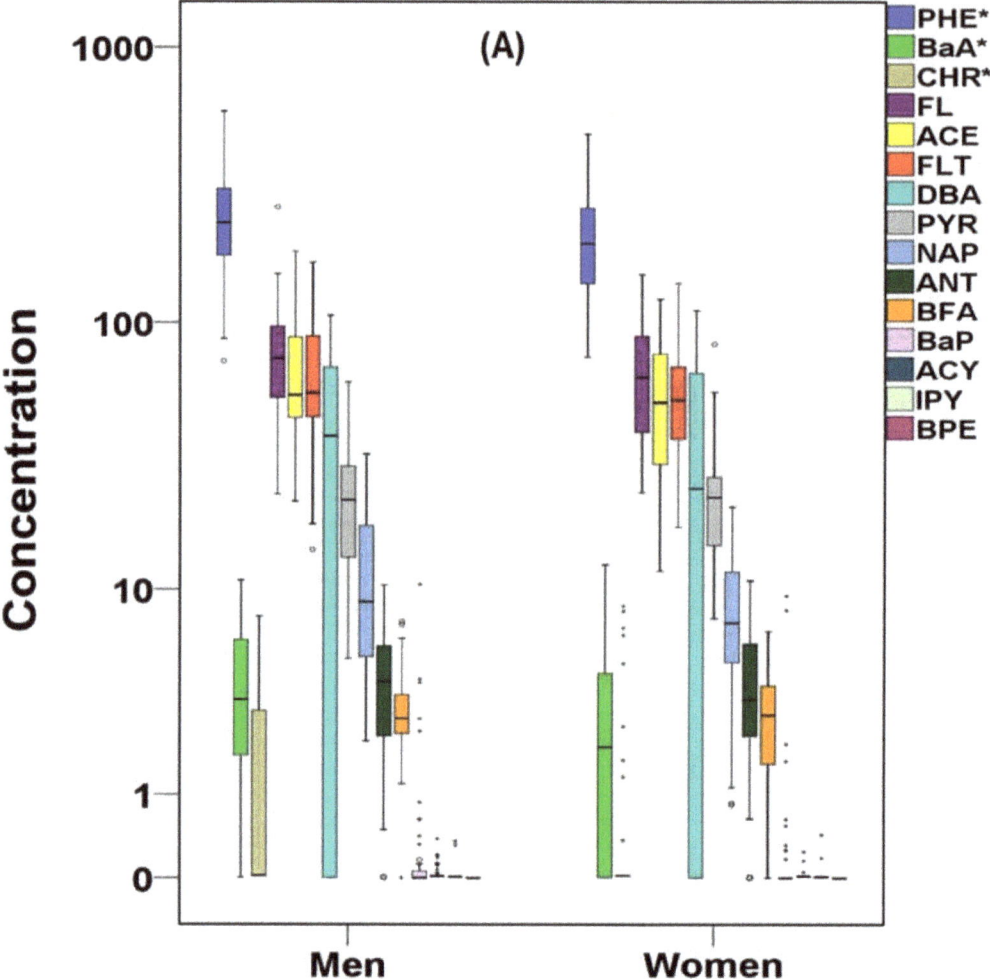

Figure 5. Cont.

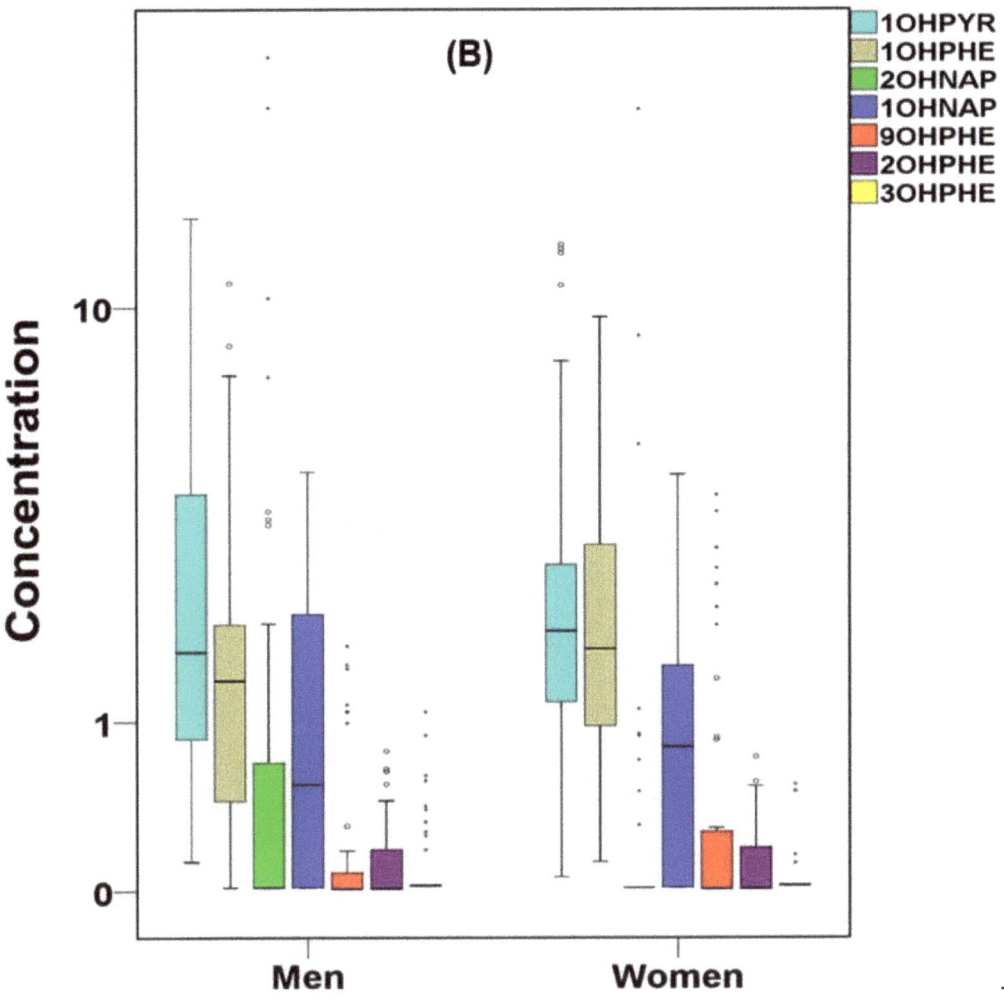

Figure 5. *Cont.*

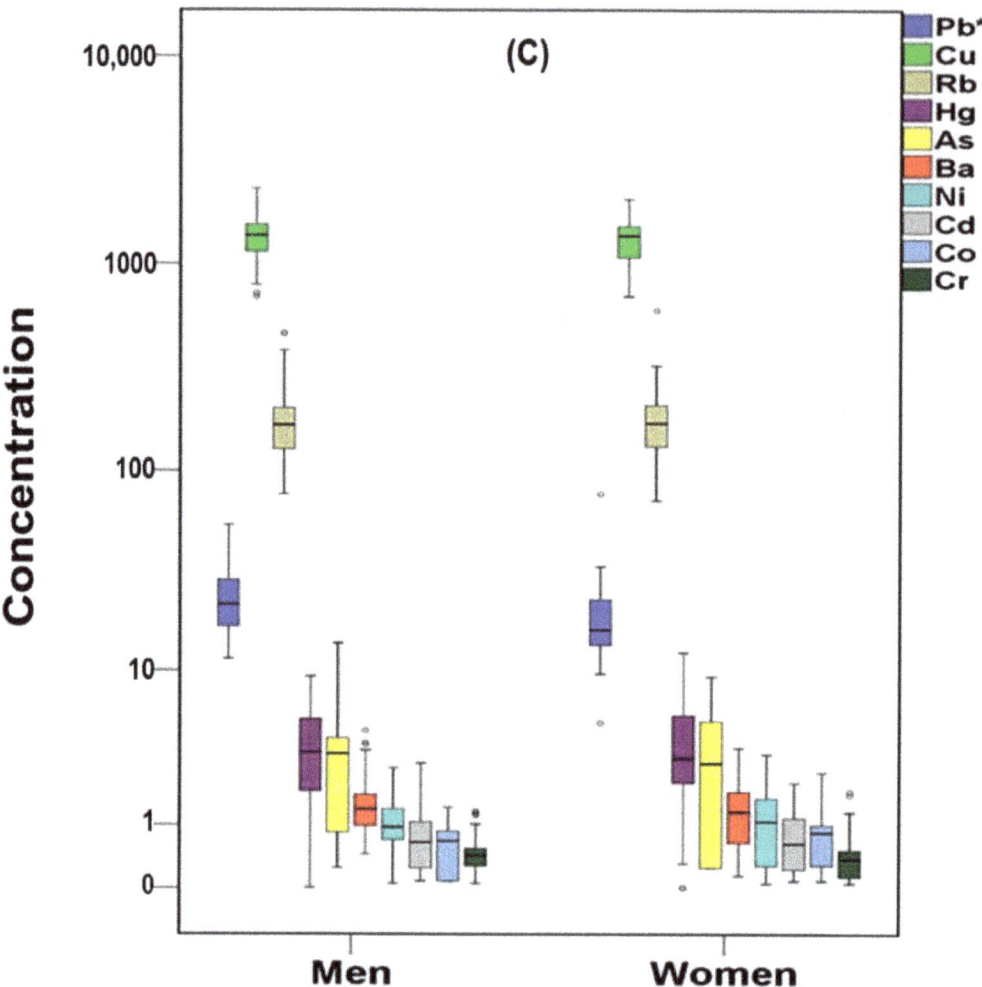

Figure 5. PAHs (**A**), OHPAHs (**B**), and trace elements' (**C**) variation according to the gender of cases in µg L^{-1} ($^+$ outliers and $^\circ$ values above 3rd quartile, * $p < 0.05$) in logarithmic scale (PHE:phenanthrene, BaA:benzo[a]anthracene, CHR:chrysene, FL:fluorene, ACE:acenaphthene, FLT:fluoranthene, DBA:dibenzo[a,h]anthracene, PYR:pyrene, NAP:naphthalene, ANT:anthracene, BFA:benzo[b,k]fluoranthenes, BaP:benzo[a]pyrene, ACY:acenaphthylene, IPY:indeno[1,2,3 cd]pyrene, BPE:benzo[ghi]perylene, 1OHPYR:1-hydroxypyrene, 1OHPHE:1-hydroxyphenanthrene, 2OHNAP:2-naphthol, 1OHNAP:1-naphthol, 9OHPHE:9-hydroxyphenanthrene, 2OHPHE:2-hydroxyphenanthrene, 3OHPHE:3-hydroxyphenanthrene).

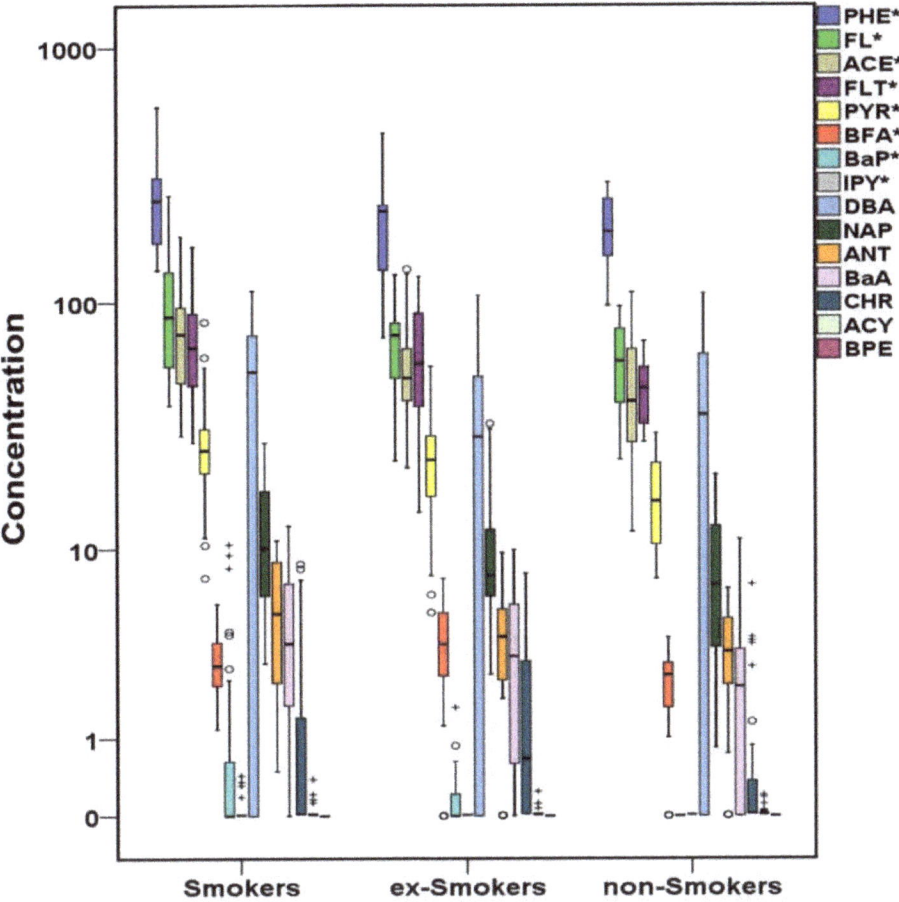

Figure 6. Variation of PAHs (µg L^{-1}) in cases' serum regarding the smoking habit ($^+$ outliers, $^\circ$ values above 3rd quartile, and * p-value <0.05) in logarithmic scale (PHE:phenanthrene, FL:fluorene, ACE:acenaphthene, FLT:fluoranthene, PYR:pyrene, BFA:benzo[b,k]fluoranthenes, BaP:benzo[a]pyrene, IPY:indeno[1,2,3 cd]pyrene, DBA:dibenzo[a,h]anthracene, NAP:naphthalene, ANT:anthracene, BaA:benzo[a]anthracene, CHR:chrysene, ACY:acenaphthylene, BPE:benzo[ghi]perylene).

Of trace elements, Hg, Ni, Cd, Co, and Cr were significantly higher in smokers' serum (Figure 8) followed by ex-smokers, while Cu, As, and Pb were found statistically higher in ex-smokers' serum followed by smokers. Ba and Rb did not present any noticeable trends. It is well known that tobacco and cigarette smoke contain plenty of trace elements including aluminum (Al), As, Ba, Beryllium (Be), Cd, Cr, Co, Cu, Fe, Hg, manganese (Mn), Ni, Pb, etc. [115]. Blood Pb and Hg levels were elevated compared to those of non-smokers [105]. A noteworthy increase of blood Cd levels have been related with smoking [114,116]. Particularly, in Greece, the blood Cd concentration of smokers has been found to be almost three times higher compared to non-smokers [104]. Nevertheless, significantly higher levels of blood Cd and Pb but lower blood Co and Hg have been also reported for smokers compared to non-smokers [107].

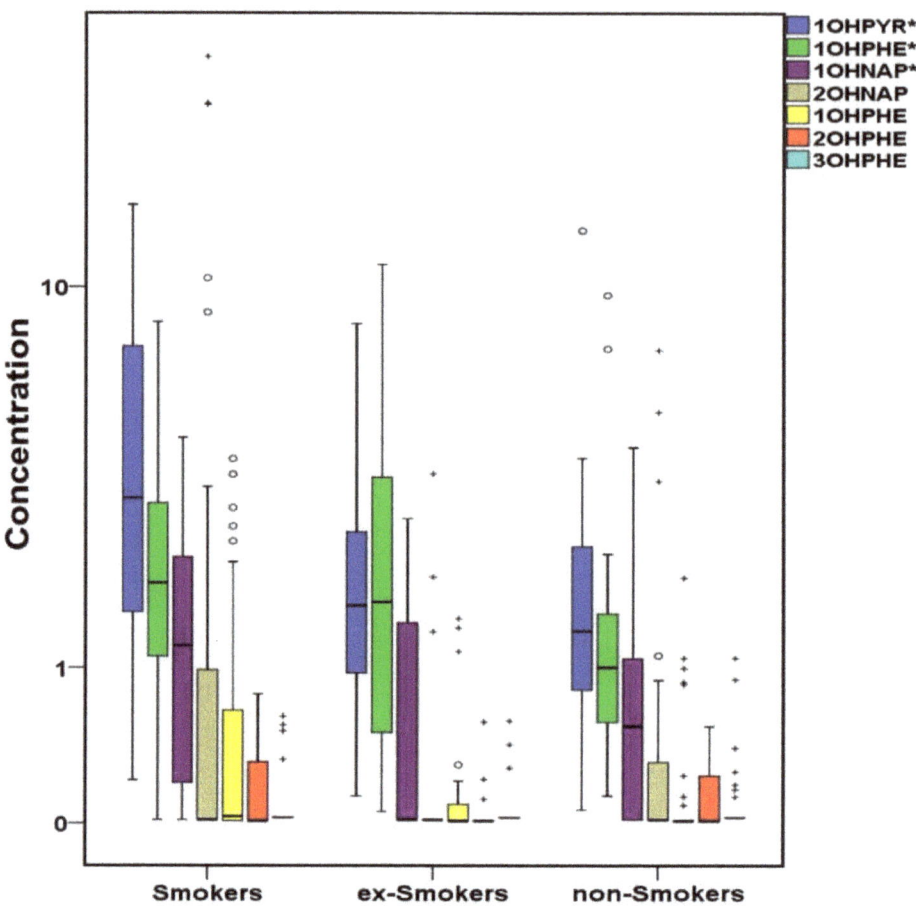

Figure 7. Variation of OHPAHs (µg L^{-1}) in cases' serum regarding the smoking habit (⁺ outliers, ° values above 3rd quartile, and * p-value < 0.05) in logarithmic scale (1OHPYR:1-hydroxypyrene, 1OHPHE:1-hydroxyphenanthrene, 1OHNAP:1-naphthol, 2OHNAP:2-naphthol, 9OHPHE:9-hydroxyphenanthrene, 2OHPHE:2-hydroxyphenanthrene, 3OHPHE:3-hydroxyphenanthrene).

2.2.3. Area of Residence

In a recent study in Greece, air pollution was associated with mortality in both urban and rural areas [117]. After the economic crisis, the air quality of Athens and Thessaloniki, the two major Greek cities where almost the 50% of the population lives, was significantly degraded, especially during the winter months, due to biomass burning for domestic heating [118–121]. During crisis, there is evidence of increased number of CVD prevalence, although the atmospheric impact on CV system was not taken into consideration [122]. Significant amounts of PM, particle bound trace elements, and gas/particle phase PAHs have been measured in the Greater Athens Area [7,123], with a recent study highlighting combustion processes emissions as the crucial contributor to the PM$_{2.5}$ and PM$_1$ mass [124]. In Figures 9–11, cases' data are classified among the area of residence. As shown in Figures 9 and 10, significantly higher ($p < 0.05$) concentrations of PHE, FL, ACE, FLT, PYR, NAP, and ANT as well as 3OHPHE from the OHPAHs group have been measured in the serum of urban site residents, followed by those of industrial sites, while in the serum of rural sites residents, the lowest PAHs concentrations were observed, except for ACY, revealing an important contribution of atmospheric PAHs yields to the serum concentrations. PAHs do not enter the human body only via inhalation, with food intake being also among the major sources [125]. As a result, the relative contribution of airborne PAHs, compared to other sources, regarding the CVD development, relies on location, dietary habits etc., although most of the PAHs that are taken up by the gastro-intestinal tract are subjected to first-path metabolism and eliminated in the liver [28]. On the other hand, inhaled BaP is absorbed in the alveolar region, enters the circulation, and reaches the heart and vasculature unmetabolized [28,126]. In our previous study, area of residence appeared to be an important contributor to the enhanced levels of ACY, BaP, IPY, and BPE in leukemia patients' serum [54].

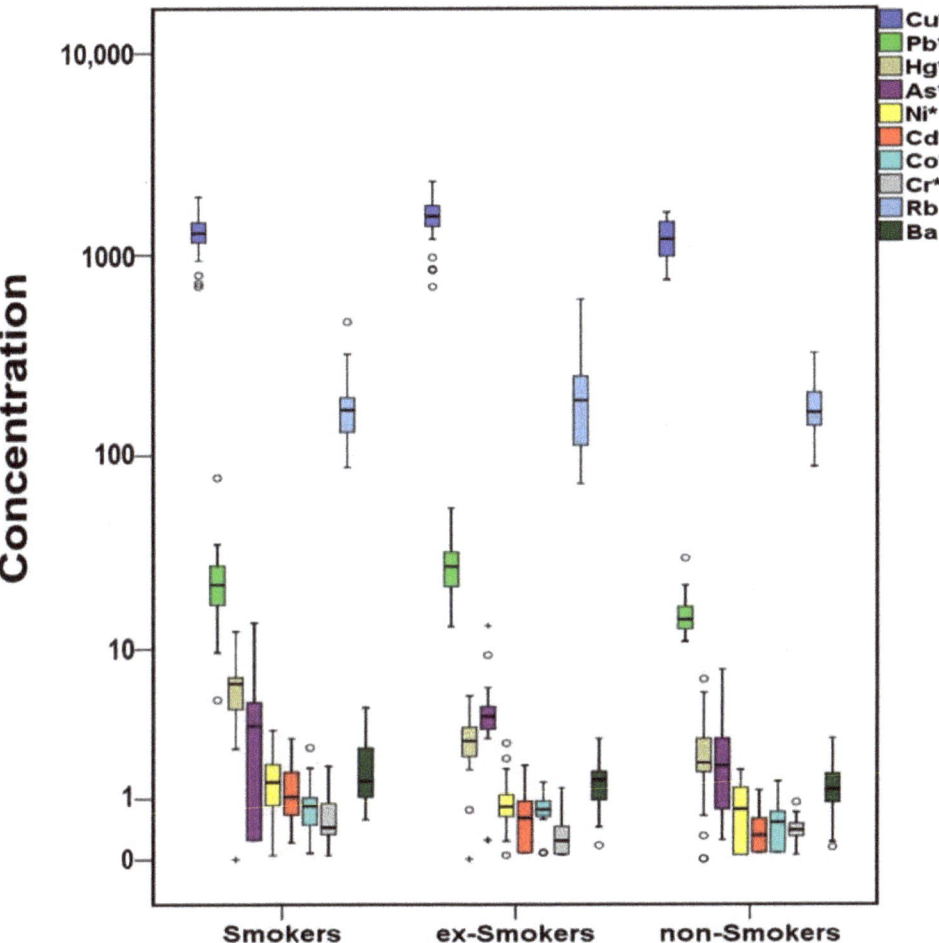

Figure 8. Variation of trace elements (μg L^{-1}) in cases' serum regarding the smoking habit (+ outliers, ° values above 3rd quartile and * p-value <0.05) in logarithmic scale.

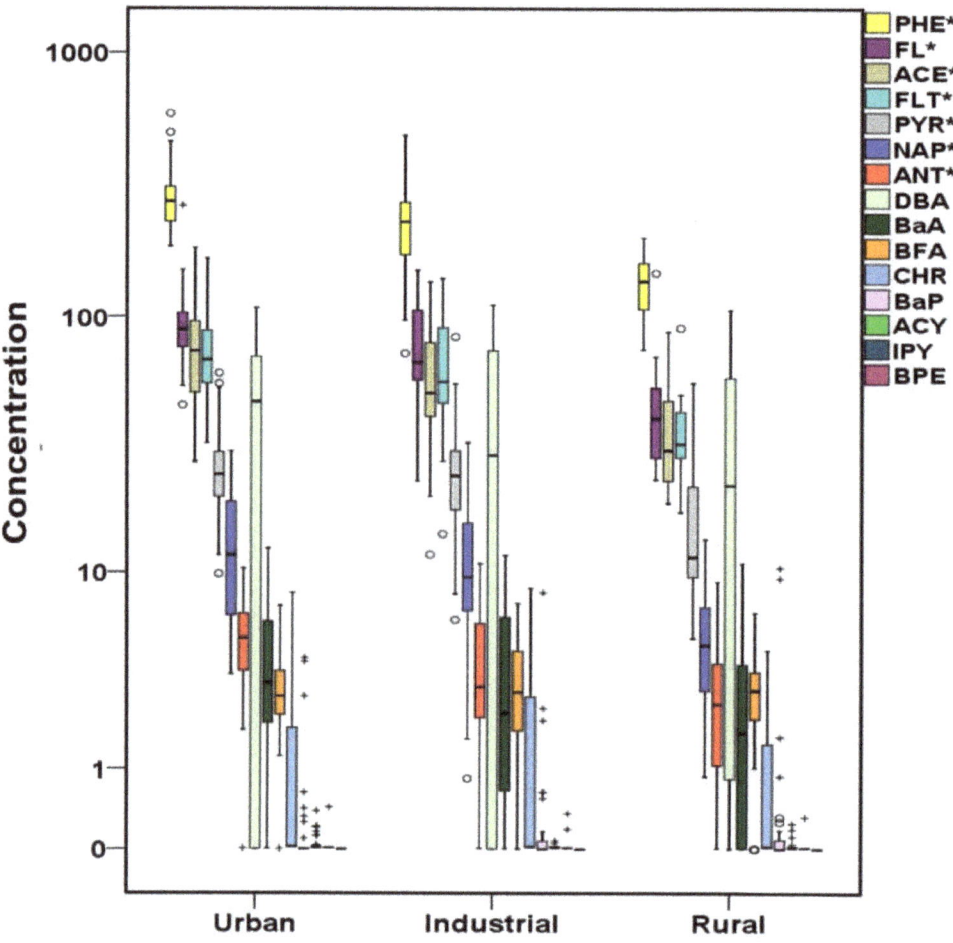

Figure 9. PAHs variation according to the area of residence of cases in µg L^{-1} (+ outliers and ° values above 3rd quartile and * p-value <0.05) in logarithmic scale (PHE:phenanthrene, FL:fluorene, ACE:acenaphthene, FLT:fluoranthene, PYR:pyrene, NAP:naphthalene, ANT:anthracene, DBA:dibenzo[a,h]anthracene, BaA:benzo[a]anthracene, BFA:benzo[b,k]fluoranthenes, CHR:chrysene, BaP:benzo[a]pyrene, ACY:acenaphthylene, IPY:indeno[1,2,3 cd]pyrene, BPE:benzo[ghi]perylene).

By contrast, the toxic Pb and Hg were found at statistically higher levels in the serum of industrial site residents (Figure 11), with the lowest concentrations of most metals observed in the samples of rural site residents, further implying the possible contribution of the local air quality on serum's concentrations. In other studies, Pb blood levels were elevated in urban areas citizens compared to those of industrial and rural, while Hg was higher in the industrial citizens' samples [127]. Significant associations between airborne trace elements and the corresponding serum levels were also found in a study conducted in China, with those who lived in urban sites, under elevated Pb and Cd emissions, presenting increased levels in their serum [79].

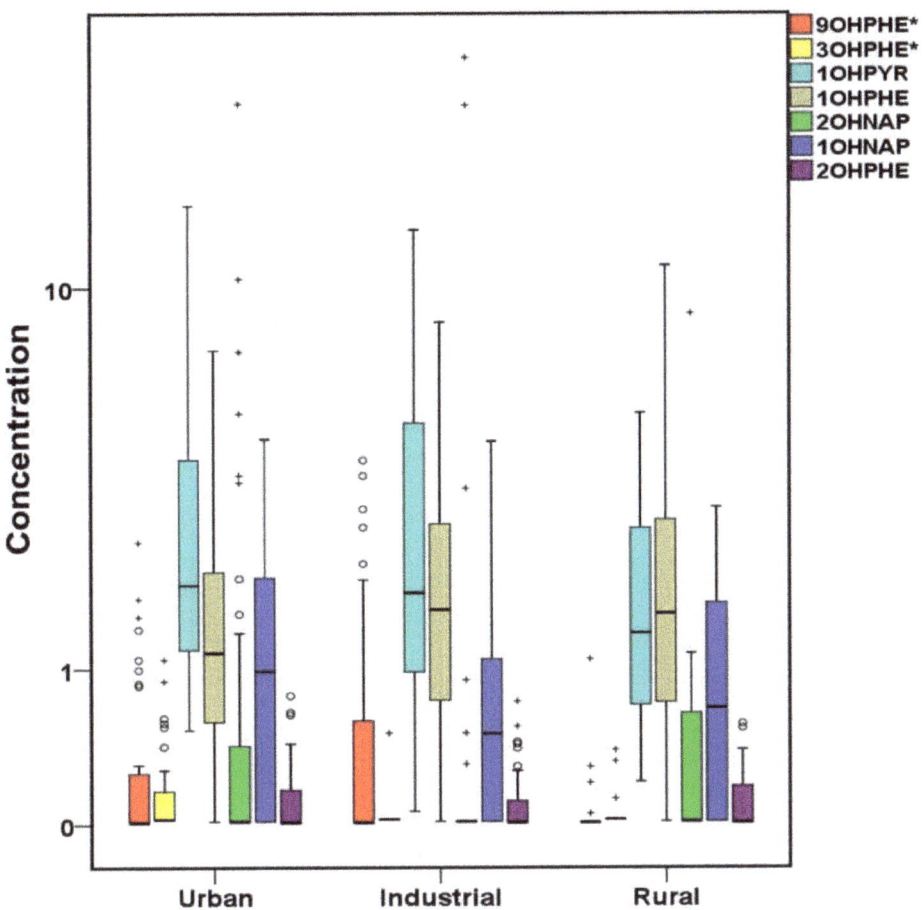

Figure 10. OHPAHs variation according to the area of residence of cases in µg L^{-1} (+ outliers and ° values above 3rd quartile and * p-value <0.05) in logarithmic scale (9OHPHE:9-hydroxyphenanthrene, 3OHPHE:3-hydroxyphenanthrene, 1OHPYR:1-hydroxypyrene, 1OHPHE:1-hydroxyphenanthrene, 2OHNAP:2-naphthol, 1OHNAP:1-naphthol, 2OHPHE:2-hydroxyphenanthrene).

2.3. Principal Component Analysis (PCA)

PCA was carried out in order to assess any potential data patterns and to evaluate the possible relationships among the analytes that may occur. PCA is widely used in environmental studies [7,128,129]; nevertheless, applying PCA in biological samples may provide useful information for discussion and further research [54,130]. In this work, PCA was applied in cases and controls datasets separately (Table 1). Three components were chosen for PCA due to the fact that components after the fourth component explained variance less than 6.5%, and they could not have an adequate meaning of the results. Particularly, three components explained 37.0 and 35.8% of the variance in the cases and controls datasets, respectively. Of cases' data, Factor 1 (18.2%) was tightly loaded with most of the LMW PAHs (except for ACY) FLT, PYR, 2OHPHE, 3OHPHRE, and OH1PYR, while the HMW PAHs were mostly clustered in Factor 3 (8.3%). The distribution of PAHs in the two factors could be explained by the different mechanisms of the compounds. For example, a recent study suggested that HMW PAHs including BFA, IPY, BaP, DBA, and others are crucial activators of AhR-mediated enzyme expressions [131]. On the contrary, LMW PAHs such as PHE are weak AhR agonists [132], having different unique cardiotoxicity mechanisms [26]. Factor 2 (10.3%) is loaded mostly by trace elements, with Cd, Co, Hg, and Pb presenting strong correlation. Regarding the metal toxicity, there are general mechanisms applied to all toxic metals, although some individual metals present additional unique mechanisms [133]. For instance, As, Cd, Hg, and Pb generate multiple reactive oxygen species, promoting oxidative stress [92,133]. In addition, Cd and Pb compete with the essential Zn, as they have similar physicochemical properties, for the binding sites of enzyme proteins [134]. As a result, the strong relation among some specific metals can be indicative of a similar mechanism, although the metal-induced cardiotoxicity mechanisms vary widely, and there is still a great matter of research [77]. Interestingly, the PCA that refers to the controls' samples is completely different, with mixed factors and less significant correlations, supporting the hypothesis that PAHs and trace elements constitute a notable factor of heart failure development.

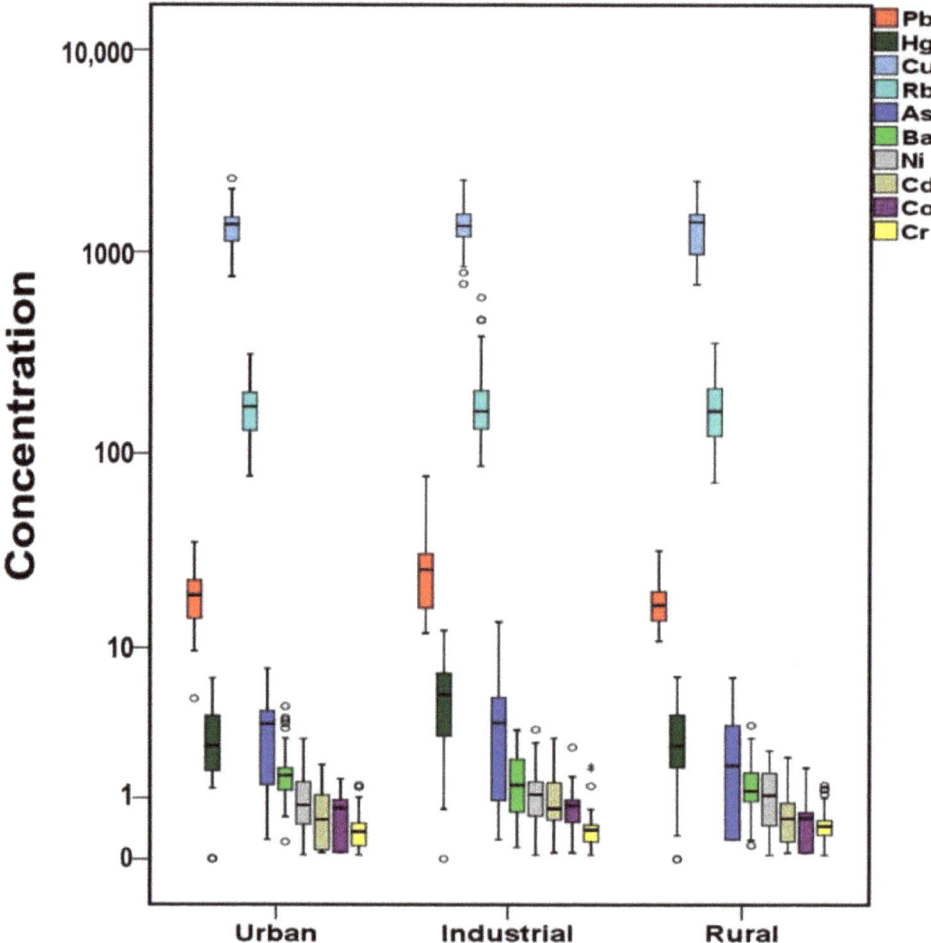

Figure 11. Trace elements variation according to the area of residence of cases in µg L^{-1} (⁺ outliers and ° values above 3rd quartile and * p-value <0.05) in logarithmic scale.

Table 1. Varimax rotated PCA for PAHs, OHPAHs, and trace elements for cases' and controls' samples. (Loadings > 0.600 appeared in bold).

	Cases			Controls		
Variance %	18.2	10.5	8.3	15.0	12.0	8.8
	1	2	3	1	2	3
PHE	**0.873**	−0.100	0.137	0.074	**0.631**	−0.481
FLT	**0.866**	0.016	0.159	0.215	0.187	−0.238
ACE	**0.804**	0.211	0.059	0.089	0.295	−0.394
NAP	**0.764**	−0.063	0.171	0.505	−0.302	−0.112
1OHPYR	**0.741**	0.362	−0.334	−0.126	**0.704**	0.247
FL	**0.735**	0.090	0.237	0.022	0.118	−0.020
PYR	**0.694**	0.177	0.289	−0.240	**0.639**	0.164
2OHPHE	**0.692**	−0.111	−0.129	0.553	0.365	−0.213
ANT	**0.676**	0.101	0.209	0.490	0.219	0.022
3OHPHE	**0.634**	−0.067	−0.252	0.016	0.268	0.589
Cd	0.139	**0.704**	0.122	−0.113	0.196	−0.365
Pb	0.138	**0.631**	0.118	**0.613**	0.089	0.203
Co	0.011	**0.624**	0.091	−0.325	−0.440	−0.016
Hg	0.015	**0.601**	0.118	0.110	−0.041	−0.002
CHR	0.061	−0.051	**0.728**	**0.604**	0.012	−0.042
BaA	0.313	−0.009	**0.614**	0.452	0.007	0.173
DBA	−0.118	−0.172	**0.610**	−0.012	**0.746**	0.146
BFA	0.189	0.073	**0.606**	0.482	0.514	0.083
ACY	0.290	0.027	0.033	0.536	−0.192	0.214
BaP	−0.034	0.360	−0.006	**0.901**	−0.024	0.057
IPY	−0.038	0.155	0.324	**0.803**	0.231	−0.176
1OHNAP	0.023	0.507	−0.038	0.005	0.261	**0.644**
2OHNAP	0.200	0.420	−0.142	−0.006	−0.094	0.477
1OHPHE	−0.098	0.473	−0.255	0.314	0.291	0.225
9OHPHE	0.368	0.264	−0.221	−0.003	−0.258	0.057
As	0.014	0.364	0.303	0.229	0.239	0.099
Ba	0.419	0.215	−0.174	0.386	0.028	−0.244
Cr	0.063	0.432	−0.188	0.166	0.252	**0.776**
Cu	0.031	0.303	−0.150	0.464	−0.487	0.256
Ni	−0.026	0.522	0.244	0.338	−0.040	−0.233
Rb	−0.197	0.048	−0.217	−0.005	0.457	−0.030

The latter hypothesis was further evaluated by applying logistic regression model, which is a tool generally used for the analysis of the relationship between individual risk/protective factors and outcomes [135]. Briefly, a new PCA was performed in the whole dataset including both cases' and controls' samples (Table S4). The regression-based factor scores of each sample for each of the three principal components (not shown), derived from the PCA, were saved as variables. Then, they were used as the input independent values and the clinical outcome (control or case occurrence) as the dependent variable for the regression analysis [136]. The Hosmer and Lemeshow Test (goodness-of-fit) indicated that the model is a satisfactory fit to the data (Table S5). Our results (Table 2) showed that every factor was statistically significant ($p < 0.05$), implying the plausible role of the studied compounds to heart failure development. It is noteworthy to mention that the strongest correlation was observed for Factor 2 (highest exp(B) value), which was mostly related with trace elements, supporting the outcome of the meta-analysis of Chowdhury et al. 2018, who highlighted the positive and approximately linear association of As, Pb, and Cd exposure with the risk of CVDs [137].

Table 2. The result of the logistic regression analysis.

	B	S.E.	Wald	Sig.	Exp(B)	95% C.I. for EXP(B)	
						Lower	Upper
Regression Scores Factor 1	3.375	1.518	4.945	0.026	29.214	1.492	571.879
Regression Scores Factor 2	10.534	3.868	7.418	0.006	37,561.068	19.167	73,6079,32.642
Regression Scores Factor 3	4.592	1650	7.747	0.005	98.668	3.890	2502.895

PCA was also performed according to the classification regarding the residence area as discussed in 2.2.3. As shown in Table S6, when the analysis refers to industrial sites, Factor 1 was dominated by the majority of PAHs. In addition, regarding the urban sites, the first two factors were loaded with LMW PAHs and their hydroxylated derivatives. Such trends were not observed in the analysis referring to rural areas, suggesting that air quality degradation could possibly be an important contributor to serum PAHs levels [54].

2.4. Strengths and Weaknesses

The present paper reports preliminary results on the relation of specific organic compounds, their metabolites, and trace elements with heart failure, using the approach of their direct measurement in human serum of cases and controls. To our knowledge, these kinds of results are reported for the first time. The findings of this study could provide useful insights about the abundance of PAHs and trace elements levels in patients serum, which are quite limited. In addition, through statistical analysis, atmospheric degradation and smoking appear to be significant contributors to the elevated serum levels of some pollutants and thus possibly enhance the development of heart failure.

However, this study does not provide evidence that PAHs, OHPAHs, and trace elements in serum are biomarkers for heart failure and it is clearly not an epidemiological study. The findings from this study should be interpreted with caution, as the population size, unknown factors related with CVDs (blood pressure, total cholesterol, or high-density lipoprotein cholesterol) and the lack of information for the dietary habits of the participants could set limits to the outcome. The findings of this study are based on a small dataset; thus, the statistics could be limited by confounding factors. Further studies are warranted regarding patients with specific disease such as CAD, or specific target groups, to estimate the possible CV risk from the elevated levels of PAHs and/or trace elements. The use of other biological matrices including whole blood, urine, and/or hair together with serum should be included in a future study.

3. Materials and Methods

3.1. Study Population

Ninety-six heart failure patients (cases) were recruited in General Hospital of Athens "Laiko" and participated on a voluntary basis. We defined incident heart failure diagnosis as the first record of heart failure in hospital admission records from any diagnostic position. Thirty-five healthy subjects (controls) were recruited as a reference. Inclusion criteria were the following: (1) the number of male and female participants should not differ significantly for both cases and controls ($p > 0.05$); (2) all participants should live in different residence areas, all over Greece, with different air quality levels; (3) the number of participants who were current smokers, ex-smokers, and non-smokers, should not differ significantly for both cases and controls ($p > 0.05$); (4) age of the participants more than 18 years; (5) long-term residence in the same area criterion was established (>5 consecutive years in the same area). The exclusion criteria were as follows: (1) all participants should not take any mineral supplement; (2) healthy subjects should not suffer for any other known disease (e.g., infectious disease). Each participant was informed in detail about the aims of the study and signed a written protocol. The study was approved by the Scientific Committee of "Laiko" Hospital, 1499/16/11/2017, according to the Helsinki Declaration. A detailed questionnaire was filled out by all participants and personal information was collected, such as gender, age, area of residence (urban, industrial, rural), and current smoking status (current smoker, ex-smoker, nonsmoker). Main socio-demographic characteristics of both controls and cases are shown in Table 3. Binomial test, for two variables, and Chi square test, for >2 variables, were carried out, respectively, with the hypothesis of equal probabilities. A value of $p < 0.05$ (95% confidence level) was considered to indicate a significant difference and thus retain the null hypothesis. The study focused on the different environmental factors, such as smoking and air quality of the residence area, which could possibly affect the analytes' serum concentration.

Table 3. Demographic characteristics and smoking status for controls and cases.

	Cases (n = 96)			Controls (n = 35)		
	N	(%)	p Value	N	(%)	p Value
		Sex				
Men	52	54.2	0.475	18	51.4	0.999
Women	44	45.8		17	48.6	
		Residence Area				
Industrial	31	32.3		15	42.9	
Urban	38	39.6	0.380	20	57.1	0.499
Rural	27	28.1		-	-	
		Age (Years)				
40–49	12	12.5		15	42.8	
50–59	26	27.1		11	31.4	
60–69	27	28.1	0.004	9	25.7	0.449
70–79	23	24.0		-	-	
80–89	8	8.3		-	-	
		Smoking Habit				
Ever	36	37.5		11	31.4	
Ex	27	28.1	0.519	11	31.4	0.892
Never	33	34.3		13	37.1	

3.2. Blood Sampling and Pretreatment

Blood samples were collected by qualified personnel at the General Hospital of Athens "Laiko", during 2018. A total of 131 samples were obtained, from which 96 refer to cases (heart failure patients) and 35 refer to controls (healthy donors). From each participant, approximately 5 mL of blood were collected. The samples were centrifuged within 30 min for 10 min at 4500 rpm in order to separate serum from the cellular components. They were stored at −67 °C, until their transfer to our laboratory for further analysis (Laboratory of Analytical Chemistry, NKUA). Storage tubes were tested for any contamination by recovery tests using same solvents as in the procedure.

3.3. PAHs and OHPAHs Analysis and Quality Control

The extraction, clean up, and derivatization procedure has been described in detail in previous work [60]. Briefly, after the addition of internal standards, extraction was performed in an ultrasonic bath, followed by a pre-concentration step with a rotary evaporator. The concentrated sample was cleaned up using glass column chromatography. The eluted fraction containing PAHs was adjusted at 0.5 mL using a gentle steam of nitrogen, while the fraction containing OHPAHs was gently evaporated until dryness followed by the addition of the derivatization reagents, i.e., 250 µL of N,O-bis(trimethylsilyl)trifluoroacetamide (BSTFA) with 1% trimethylchlorosilane (TMCS) and 50 µL of anhydrous pyridine. The reaction took place in an oven at 70 °C for 3 h. A gas chromatography/mass spectrometry system (GC/MS) (6890N/5975B, Agilent Technologies, USA) was employed for the determination of both fractions. The GC instrument was equipped with a split/splitless injector and an HP-5ms (5%-(phenyl)-methylpolysiloxane) (Agilent J&W GC Columns, Agilent Technologies, Santa Clara, CA, USA) capillary column. High-purity helium was used as carrier gas with a velocity of 1.5 mL min^{-1}. Pulsed split-less mode was used for the injection and the injector's temperature was set at 280 °C. For PAHs and OHPAHs analysis, the GC oven temperature program was 65 °C (hold for 1 min) to 320 °C at 15 °C/min with final isothermal hold for 3 min. In both cases, inlet and MS source temperatures were 280 and 230 °C, respectively. The selected ion monitoring (SIM) mode was used for the quantification of the analytes.

The PAH-determination procedure was validated using the Polynuclear Aromatic Hydrocarbons Mix (Supelco, Darmstadt, GER), which is a standard solution of the compounds studied including NAP, ACY, ACE, FL, PHE, ANTH, FTL, PYR, CHR, BaA, BFA, BaP, IPY, DBA, and BPE. In the same way, a OHPAHs mix was prepared including the following compounds: 1OHNAP, 2OHNAP, 1OHPHE, 2OHPHE, 3OHPHE, 9OHPHE, and 1OHPYR. Recovery rates and selectivity were evaluated using spiked blood serum. Blank samples, i.e., mixtures of controls' serum obtained by non-smoking residents of low polluted areas (mainly from Greek islands), were pretreated in the same way and analyzed in order to examine the potential background effect. Recovery rates, LoD, and limits of quantification (LoQ) are shown in Table S7. In general, recoveries ranged from 72.5% (1OHNAP) to 136% (ACY) and LoD varied from 0.001 (BaP) to 0.11 (NAP) µg L^{-1}.

3.4. Trace Elements' Analysis and Quality Control

All plastic materials that came into contact with the serum samples were previously washed thoroughly, soaked in dilute nitric acid (HNO$_3$) (Merck, Darmstadt, Germany), and rinsed with ultrapure water of 18.2 MΩ cm (Millipore, Bedford, MA, USA). The followed pretreatment procedures have been described elsewhere [138,139]. Briefly, 0.5 mL of serum were wet digested using a mixture of HNO$_3$ (suprapur 65%) and hydrogen peroxide (H$_2$O$_2$) (suprapur 30%) (Merck). The samples were analyzed with inductively coupled plasma mass spectrometry (ICP-MS) by a Thermo Scientific ICAP Qc (Waltham, MA, USA). Measurements were carried out in a single collision cell mode, with kinetic energy discrimination (KED) using pure He. Matrix induced signal suppressions and instrumental drift were corrected by internal standardization (^{45}Sc, ^{103}Rh).

In each batch of 10 samples, at least one laboratory blank was analyzed. In case trace element concentrations in the reagent blank were detectable, the procedure for the whole batch was repeated. In order to verify the accuracy and precision of the method, the certified reference materials (CRM) "Plasma Control lyophilized, Levels I and II" (RECIPE Chemicals + Instruments GmbH, Munich, Germany) were used. The recoveries for As, Cd, Co, Cr, Cu, Ni, and Pb ranged from 95.1 to 105% (certified values for Ba, Hg and Rb are not included in the specific certified reference materials). The USEPA method [140] was applied for the calculation of the LoD and LoQ. LOD ranged from 0.10 µg L^{-1} (Cr) to 0.7 µg L^{-1} (Ba).

3.5. Statistical Analysis

The SPSS software package (IBM SPSS statistics version 24) was employed for statistical analysis purposes. SPSS software is a common tool for statistical analysis of data from environmental including air [7,141,142], water [143], and soil [144] samples or biological [60,145] samples.

Hypothesis tests for population proportion were carried out using binomial test (2 variables) and chi-square test (3 or more variables). p values > 0.05 indicate non-significant difference. The normal distribution of the data was assessed using the Shapiro–Wilk and Kolmogorov–Smirnov tests, with a value of $p > 0.05$ indicating normal distribution. As no variable of the dataset was normally distributed, the possible statistical differences between two or more independent variables were investigated using the Mann–Whitney and Kruskal–Wallis tests, respectively, with the value of $p < 0.05$ indicating statistically significant difference. Principal Component Analysis (PCA) was used for the investigation of any possible associations of the examined parameters. PCA is a widespread multivariate statistical technique used in environmental sciences [54,141,142,145]. The application of PCA transforms the original set of variables into a smaller one of linear combinations accounting for the most of the variance of the former set. It makes the complex system more accessible, while at the same time it withholds the primary information. Varimax rotation is generally used for factor grouping in most PCA applications. The exported principal components include variables with common characteristics, which are attributed as a common source or chemical interaction. [146,147]. It should be noted that the variables used in this study were standardized before applying PCA.

Further statistical analysis includes the application of logistic regression after PCA. In logistic regression, the relationship between a binary dependent variable, for example, the occurrence of a phenomenon or not, with independent variables, which affect that phenomenon, is assessed, as generally used in medical and epidemiological studies. Although logistic regression has many similarities with linear regression, the estimation of variables' coefficients is performed by the maximum likelihood technique [148]. Under this prism, combination of logistic regression with PCA could reveal the probability of each factor to be associated with the occurrence of heart failure.

4. Conclusions

Major environmental pollutants have been measured in the serum of heart failure patients and have been compared with control samples. The statistical higher concentrations of the majority of PAHs, especially the low molecular weight, and trace elements indicate a potential link with heart failure. Smoking habit and atmospheric degradation of urban and industrial sites appeared to further elevate the analytes' serum concentrations. Possible common mechanisms related to heart failure are revealed from principal component analysis followed by logistic regression model, suggested some of the analytes as possibly significant contributors to heart failure incidence. As a future aspect, a fully designed study with a more specific patient group and a wider dataset of biochemical parameters will provide additional information for the further evaluation of the role environmental compounds to CVDs.

Supplementary Materials: The following are available online, Table S1: Detection frequencies, median, mean and ranges of PAHs concentrations ($\mu g\ L^{-1}$) in cases' and controls' samples, Table S2: Detection frequencies, median, mean and ranges of OHPAHs concentrations ($\mu g\ L^{-1}$) in cases' and controls' samples, Table S3: Detection frequencies, median, mean and ranges of trace elements' concentrations ($\mu g\ L^{-1}$) in cases' and controls' samples, Table S4: Varimax rotated PCA for PAHs, OHPAHs and trace elements for overall dataset used for logistic regression model, Table S5: Hosmer and Lemeshow Test, Table S6: Varimax rotated PCA for PAHs, OHPAHs and trace elements for cases' samples classified in terms of the residence area. (Loadings > 0.600 appeared in bold), Table S7: Analytical method recovery rates, LoD and LoQ for the determination of PAHs and OH-PAHs in human serum.

Author Contributions: E.C.: Validation, Formal analysis, Investigation, Writing—original draft, Writing—review and editing, Visualization. P.G.K.: Investigation, Validation, Writing—review and editing. K.G.K.: Validation, Investigation. A.S.: Validation. S.K.: Validation. M.M.: Validation. E.B.: Conceptualization, Methodology, Formal analysis, Writing—review and editing, Visualization, Resources, Supervision, Project administration. All authors have read and agreed to the published version of the manuscript.

Funding: This research received no external funding.

Institutional Review Board Statement: The study was conducted according to the guidelines of the Declaration of Helsinki, and approved by the Scientific Committee of "Laiko" Hospital, (1499/16/11/2017)."

Informed Consent Statement: Informed consent was obtained from all subjects involved in the study.

Data Availability Statement: The data may be available from the corresponding author on.

Conflicts of Interest: The authors declare no conflict of interest.

Sample Availability: Samples of the compounds are not available from the authors.

References

1. Kelly, F.J.; Fussell, J.C. Air pollution and public health: Emerging hazards and improved understanding of risk. *Environ. Geochem. Health* **2015**, *37*, 631–649. [CrossRef]
2. World Health Organization. Ambient (outdoor) Air Pollution. 2018. Available online: https://www.who.int/news-room/fact-sheets/detail/ambient-(outdoor)-air-quality-and-health (accessed on 1 December 2020).
3. Cohen, A.J.; Brauer, M.; Burnett, R.; Anderson, H.R.; Frostad, J.; Estep, K.; Balakrishnan, K.; Brunekreef, B.; Dandona, L.; Dandona, R.; et al. Estimates and 25-year trends of the global burden of disease attributable to ambient air pollution: An analysis of data from the Global Burden of Diseases Study 2015. *Lancet* **2017**, *389*, 1907–1918. [CrossRef]
4. Sakizadeh, M. Short-and long-term variations, spatial analysis along with cancer health risk assessment associated with 1, 3-butadiene. *Atmos. Pollut. Res.* **2020**, *11*, 755–765. [CrossRef]
5. Shuai, J.; Kim, S.; Ryu, H.; Park, J.; Lee, C.K.; Kim, G. Health risk assessment of volatile organic compounds exposure near Daegu dyeing industrial complex in South Korea. *BMC Public Health* **2018**, *18*, 528. [CrossRef]
6. Chen, T.; Gokhale, J.; Shofer, S.; Kuschner, W.G. Outdoor Air Pollution: Nitrogen Dioxide, Sulfur Dioxide, and Carbon Monoxide Health Effects. *Am. J. Med. Sci.* **2007**, *333*, 249–256. [CrossRef]
7. Koukoulakis, K.G.; Chrysohou, E.; Kanellopoulos, P.G.; Karavoltsos, S.; Katsouras, G.; Dassenakis, M.; Nikolelis, D.; Bakeas, E. Trace elements bound to airborne PM10 in a heavily industrialized site nearby Athens: Seasonal patterns, emission sources, health implications. *Atmos. Pollut. Res.* **2019**, *10*, 1347–1356. [CrossRef]
8. Mandalakis, M.; Tsapakis, M.; Tsoga, A.; Stephanou, E.G. Gas–particle concentrations and distribution of aliphatic hydrocarbons, PAHs, PCBs and PCDD/Fs in the atmosphere of Athens (Greece). *Atmos. Environ.* **2002**, *36*, 4023–4035. [CrossRef]
9. International Agency for Research on Cancer. *Air Pollution and Cancer*; Straif, K., Cohen, A., Samet, J., Eds.; Publication No. 161; IARC Scientific: Lyon, France, 2013; Available online: https://publications.iarc.fr/Book-And-Report-Series/Iarc-Scientific-Publications/Air-Pollution-And-Cancer-2013 (accessed on 10 December 2020).
10. Dockery, D.W.; Pope, C.A., III; Xu, X.; Spengler, J.D.; Ware, J.H.; Fay, M.E.; Ferris, B.G., Jr.; Speizer, F.E. An association between air pollutionand mortality in six U.S. cities. *N. Engl. J. Med.* **1993**, *329*, 1753–1759. [CrossRef]
11. Šrám, R.J.; Beneš, I.; Binková, B.; Dejmek, J.; Horstman, D.; Kotěsovec, F.; Otto, D.; Perreault, S.D.; Rubes, J.; Selevan, S.G.; et al. Teplice program–the impact of air pollution on human health. *Environ. Health Perspect.* **1996**, *104*, 699–714. [CrossRef]
12. Lippmann, M.; Cohen, M.D.; Chen, L.-C. Health effects of World Trade Center (WTC) Dust: An unprecedented disaster with inadequate risk management. *Crit. Rev. Toxicol.* **2015**, *45*, 492–530. [CrossRef]
13. Yin, P.; Brauer, M.; Cohen, A.J.; Wang, H.; Li, J.; Burnett, R.T.; Stanaway, J.D.; Causey, K.; Larson, S.; Godwin, W.; et al. The effect of air pollution on deaths, disease burden, and life expectancy across China and its provinces, 1990–2017: An analysis for the Global Burden of Disease Study 2017. *Lancet Planet. Health* **2020**, *4*, 386–398. [CrossRef]
14. Achilleos, S.; Al-Ozairi, E.; Alahmad, B.; Garshick, E.; Neophytou, A.M.; Bouhamra, W.; Yassin, M.F.; Koutrakis, P. Acute effects of air pollution on mortality: A 17-year analysis in Kuwait. *Environ. Int.* **2019**, *126*, 476–483. [CrossRef] [PubMed]
15. Lefler, J.S.; Higbee, J.D.; Burnett, R.T.; Ezzati, M.; Coleman, N.C.; Mann, D.D.; Marshall, J.D.; Bechle, M.; Wang, Y.; Robinson, A.L.; et al. Air pollution and mortality in a large, representative U.S. cohort: Multiple-pollutant analyses, and spatial and temporal decompositions. *Environ. Health* **2019**, *18*, 1–11. [CrossRef]
16. Brønnum-Hansen, H.; Bender, A.M.; Andersen, Z.J.; Sørensen, J.; Bønløkke, J.H.; Boshuizen, H.; Becker, T.; Diderichsen, F.; Loft, S. Assessment of impact of traffic-related air pollution on morbidity and mortality in Copenhagen Municipality and the health gain of reduced exposure. *Environ. Int.* **2018**, *121*, 973–980. [CrossRef]
17. Miller, M.R.; Newby, D.E. Air pollution and cardiovascular disease: Car sick. *Cardiovasc. Res.* **2020**, *116*, 279–294. [CrossRef] [PubMed]
18. Shah, A.S.V.; Langrish, J.P.; Nair, H.; Mcallister, D.A.; Hunter, A.L.; Donaldson, K.; Newby, D.E.; Mills, N.L. Global association of air pollution and heart failure: A systematic review and meta-analysis. *Lancet* **2013**, *382*, 1039–1048. [CrossRef]
19. Feng, B.; Song, X.; Dan, M.; Yu, J.; Wang, Q.; Shu, M.; Xu, H.; Wang, T.; Chen, T.; Zhang, Y.; et al. High level of source-specific particulate matter air pollution associated with cardiac arrhythmias. *Sci. Total Environ.* **2019**, *657*, 1285–1293. [CrossRef]
20. Provost, E.B.; Madhloum, N.; Panis, L.I.; Boever, P.D. Carotid Intima-Media Thickness, a Marker of Subclinical Atherosclerosis, and Particulate Air Pollution Exposure: The Meta-Analytical Evidence. *PLoS ONE* **2015**, *73*, 1–12. [CrossRef]
21. Wold, L.E.; Ying, Z.; Hutchinson, K.R.; Velten, M.; Gorr, M.W.; Velten, C.; Youtz, D.Z.; Wang, A.; Lucchesi, P.A.; Sun, Q.; et al. Cardiovascular Remodeling in Response to Long-Term Exposure to Fine Particulate Matter Air Pollution. *Circ. Heart Fail.* **2012**, *5*, 452–461. [CrossRef]
22. Tang, C.-S.; Chuang, K.-J.; Chang, T.-Y.; Chuang, H.-C.; Chen, L.-H.; Candice Lung, S.-C.; Chang, L.-T. Effects of Personal Exposures to Micro- and Nano-Particulate Matter, Black Carbon, Particle-Bound Polycyclic Aromatic Hydrocarbons, and Carbon Monoxide on Heart Rate Variability in a Panel of Healthy Older Subjects. *Int. J. Environ. Res. Public Health* **2019**, *16*, 4672. [CrossRef]
23. Huang, C.; Tang, M.; Li Huichu, L.; Wen, J.; Wang, C.; Gao, Y.; Hu, J.; Lin, J.; Chen, R. Particulate matter air pollution and reduced heart rate variability: How the associations vary by particle size in Shanghai, China. *Ecotoxicol. Environ. Saf.* **2021**, *208*, 111726. [CrossRef]
24. Cosselman, K.E.; Navas-Acien, A.; Kaufman, J.D. Environmental factors in cardiovascular disease. *Nat. Rev. Cardiol.* **2015**, *12*, 627–642. [CrossRef]

25. European Heart Network. European Cardiovascular Disease Statistics. 2017. Available online: http://www.ehnheart.org/cvd-statistics/cvd-statistics-2017.html (accessed on 2 December 2020).
26. Marris, C.R.; Kompella, S.N.; Miller, M.R.; Incardona, J.P.; Brette, F.; Hancox, J.C.; Sørhus, E.; Shiels, H.A. Polyaromatic hydrocarbons in pollution: A heart-breaking matter. *J. Physiol.* **2020**, *598*, 227–247. [CrossRef]
27. Lee, M.-S.; Magari, S.; Christiani, D.C. Cardiac autonomic dysfunction from occupational exposure to polycyclic aromatic hydrocarbons. *Occup. Environ. Med.* **2011**, *68*, 474–478. [CrossRef] [PubMed]
28. Holme, J.A.; Brinchmann, B.C.; Refsnes, M.; Låg, M.; Øvrevik, J. Potential role of polycyclic aromatic hydrocarbons as mediators of cardiovascular effects from combustion particles. *Environ. Health* **2019**, *18*, 1–18. [CrossRef] [PubMed]
29. Huang, L.; Xi, Z.; Wang, C.; Zhang, Y.; Yang, Z. Phenanthrene exposure induces cardiac hypertrophy via reducing miR-133a expression by DNA methylation. *Sci. Rep.* **2016**, *6*, 20105. [CrossRef]
30. Seinfeld, J.H.; Pandis, S.N. *Atmospheric Chemistry and Physics: From Air Pollution to Climate Change*, 2nd ed.; John Wiley & Sons: New York, NY, USA, 2006; pp. 670–671.
31. Haritash, A.K.; Kaushik, C.P. Biodegradation aspects of polycyclic aromatic hydrocarbons (PAHs): A review. *J. Hazard. Mater.* **2009**, *169*, 1–15. [CrossRef] [PubMed]
32. Rengarajan, T.; Rajendran, P.; Nandakumar, N.; Lokeshkumar, B.; Rajendran, P.; Nishigaki, I. Exposure to polycyclic aromatic hydrocarbons with special focus on cancer. *Asian. Pac. J. Trop. Biomed.* **2015**, *5*, 182–189. [CrossRef]
33. Tellez-Plaza, M.; Guallar, E.; Navas-Acien, A. Environmental metals and cardiovascular disease. *Br. Med. J.* **2018**, *362*, 1–4. [CrossRef]
34. Houston, M.C. The Role of Mercury in Cardiovascular Disease. *J. Cardiovasc. Dis. Res.* **2014**, *2*, 1–8. [CrossRef]
35. Tellez-plaza, M.; Jones, M.R.; Dominguez-Lucas, A.; Guallar, E.; Navas-Acien, A. Cadmium Exposure and Clinical Cardiovascular Disease: A Systematic Review. *Curr. Atheroscler. Rep.* **2013**, *15*, 356. [CrossRef]
36. Moon, K.; Guallar, E.; Navas-acien, A. Arsenic Exposure and Cardiovascular Disease: An Updated Systematic Review. *Curr. Atheroscler. Rep.* **2012**, *14*, 542–555. [CrossRef]
37. Navas-acien, A.; Guallar, E.; Silbergeld, E.K.; Rothenberg, S.J. Lead Exposure and Cardiovascular Disease—A Systematic Review. *Environ. Health Perspect.* **2007**, *115*, 472–482. [CrossRef] [PubMed]
38. Rayman, M.P. The importance of selenium to human health. *Lancet* **2000**, *356*, 233–241. [CrossRef]
39. Scheiber, I.; Dringen, R.; Mercer, J.F.B. Copper: Effects of deficiency and overload. In *Interrelations between Essential Metal. Ions and Human Diseases, Metal. Ions in Life Sciences*; Sigel, A., Sigel, H., Sigel, R.K.O., Eds.; Springer Science Business Media: Dordrecht, The Netherlands, 2013; Volume 13, pp. 359–387. [CrossRef]
40. Yamada, K. Cobalt: Its role in health and disease. In *Interrelations between Essential Metal Ions and Human Diseases, Metal Ions in Life Sciences*; Sigel, A., Sigel, H., Sigel, R., Eds.; Springer Science Business Media: Dordrecht, The Netherlands, 2013; Volume 13, pp. 295–320. [CrossRef]
41. Chasapis, C.T.; Spiliopoulou, C.A.; Loutsidou, A.C.; Stefanidou, M.E. Zinc and human health: An update. *Arch. Toxicol.* **2012**, *86*, 521–534. [CrossRef]
42. Venter, A.D.; van Zyl, P.G.; Beukes, J.P.; Josipovic, M.; Hendriks, J.; Vakkari, V.; Laakso, L. Atmospheric trace metals measured at a regional background site (Welgegund) in South Africa. *Atmos. Chem. Phys.* **2017**, *17*, 4251–4263. [CrossRef]
43. Thurston, G.D.; Ito, K.; Lall, R. A source apportionment of U.S. fine particulate matter air pollution. *Atmos. Environ.* **2011**, *45*, 3924–3936. [CrossRef]
44. Horemans, B.; Cardell, C.; Bencs, L.; Kontozova-Deutsch, V.; DeWael, K.; Van Grieken, R. Evaluation of airborne particles at the Alhambra monument in Granada, Spain. *Microchem. J.* **2011**, *99*, 429–438. [CrossRef]
45. Gladtke, D.; Volkhausen, W.; Bach, B. Estimating the contribution of industrial facilities to annual PM10 concentrations at industrially influenced sites. *Atmos. Environ.* **2009**, *43*, 4655–4665. [CrossRef]
46. Hu, C.; Hou, J.; Zhou, Y.; Sun, H.; Yin, W.; Zhang, Y.; Wang, X.; Wang, G.; Chen, W.; Yuan, J. Association of polycyclic aromatic hydrocarbons exposure with atherosclerotic cardiovascular disease risk: A role of mean platelet volume or club cell secretory protein. *Environ. Pollut.* **2018**, *233*, 45–53. [CrossRef]
47. Alshaarawy, O.; Elbaz, H.A.; Andrew, M.E. The association of urinary polycyclic aromatic hydrocarbon biomarkers and cardiovascular disease in the US population. *Environ. Int.* **2016**, *89–90*, 174–178. [CrossRef]
48. Clark, J.D.; Serdar, B.; Lee, D.J.; Arheart, K.; Wilkinson, J.D.; Fleming, L.E. Exposure to polycyclic aromatic hydrocarbons and serum inflammatory markers of cardiovascular disease. *Environ. Res.* **2012**, *117*, 132–137. [CrossRef] [PubMed]
49. Sun, L.; Ye, Z.; Ling, Y.; Cai, S.; Xu, J.; Fan, C.; Zhong, Y.; Shen, Q.; Li, Y. Relationship between polycyclic aromatic hydrocarbons and rheumatoid arthritis in US general population, NHANES 2003–2012. *Sci. Total Environ.* **2020**, *704*, 135294. [CrossRef]
50. Khosravipour, M.; Khosravipour, H. The association between urinary metabolites of polycyclic aromatic hydrocarbons and diabetes: A systematic review and meta-analysis study. *Chemosphere* **2020**, *247*, 125680. [CrossRef] [PubMed]
51. Santos, P.M.; del Nogal Sánchez, M.; Pavón, J.L.P.; Cordero, B.M. Determination of polycyclic aromatic hydrocarbons in human biological samples: A critical review. *Trends Analyt. Chem.* **2019**, *113*, 194–209. [CrossRef]
52. Pleil, J.D.; Stiegel, M.A.; Sobus, J.R.; Tabucchi, S.; Ghio, A.J.; Madden, M.C. Cumulative exposure assessment for trace-level polycyclic aromatic hydrocarbons (PAHs) using human blood and plasma analysis. *J. Chromatogr. B* **2010**, *878*, 1753–1760. [CrossRef]
53. Song, X.F.; Chen, Z.Y.; Zang, Z.J.; Zhang, Y.N.; Zeng, F.; Peng, Y.P.; Yang, C. Investigation of polycyclic aromatic hydrocarbon level in blood and semen quality for residents in Pearl River Delta Region in China. *Environ. Int.* **2013**, *60*, 97–105. [CrossRef] [PubMed]

54. Koukoulakis, K.G.; Kanellopoulos, P.G.; Chrysochou, E.; Koukoulas, V.; Minaidis, M.; Maropoulos, G.; Nikoleli, G.P.; Bakeas, E. Leukemia and PAHs levels in human blood serum: Preliminary results from an adult cohort in Greece. *Atmos. Pollut. Res.* **2020**, *11*, 1552–1565. [CrossRef]
55. Boada, L.D.; Henríquez-Hernández, L.A.; Navarro, P.; Zumbado, M.; Almeida-González, M.; Camacho, M.; Álvarez-León, E.E.; Valencia-Santana, J.A.; Luzardo, O.P. Exposure to polycyclic aromatic hydrocarbons (PAHs) and bladder cancer: Evaluation from a gene-environment perspective in a hospital-based case-control study in the Canary Islands (Spain). *Int. J. Occup. Med. Environ. Health* **2015**, *21*, 23–30. [CrossRef] [PubMed]
56. Singh, V.K.; Patel, D.K.; Ram, S.; Mathur, N.; Siddiqui, M.K.J.; Behari, J.R. Blood levels of polycyclic aromatic hydrocarbons in children of Lucknow, India. *Arch. Environ. Contam. Toxicol.* **2008**, *54*, 348–354. [CrossRef]
57. Tsang, H.L.; Wu, S.; Leung, C.K.M.; Tao, S.; Wong, M.H. Body burden of POPs of Hong Kong residents, based on human milk, maternal and cord serum. *Environ. Int.* **2011**, *37*, 142–151. [CrossRef]
58. Zhang, X.; Li, X.; Jing, Y.; Fang, X.; Zhang, X.; Lei, B.; Yu, Y. Transplacental transfer of polycyclic aromatic hydrocarbons in paired samples of maternal serum, umbilical cord serum, and placenta in Shangai, China. *Environ. Pollut.* **2017**, *222*, 267–275. [CrossRef]
59. Chen, Q.; Zheng, T.; Bassig, B.A.; Cheng, Y.; Leaderer, B.; Lin, S.; Holford, T.; Qiu, J.; Zhang, Y.; Shi, K.; et al. Prenatal exposure to polycyclic aromatic hydrocarbons and birth weight in China. *Open J. Air Pollut.* **2014**, *3*, 100–110. [CrossRef]
60. Heitland, P.; Koster, H.D. Human biomonitoring of 73 elements in blood, serum, erythrocytes and urine. *J. Trace. Elem. Med. Biol.* **2021**, *64*, 126706. [CrossRef]
61. Ivanenko, N.B.; Ganeev, A.A.; Solovyev, N.D.; Moskvin, L.N. Determination of Trace Elements in Biological fluids. *J. Anal. Chem.* **2011**, *66*, 784–799. [CrossRef]
62. Asgary, S.; Movahedian, A.; Keshvari, M.; Taleghani, M.; Sahebkar, A.; Sarrafzadegan, N. Serum levels of lead, mercury and cadmium in relation to coronary artery disease in the elderly: A cross-sectional study. *Chemosphere* **2017**, *180*, 540–544. [CrossRef]
63. Ari, E.; Kaya, Y.; Demir, H.; Asicioglu, E.; Keskin, S. The correlation of serum trace elements and heavy metals with carotid artery atherosclerosis in maintenance hemodialysis patients. *Biol. Trace. Elem. Res.* **2011**, *144*, 351–359. [CrossRef]
64. Koh, H.Y.; Kim, T.H.; Sheen, Y.H.; Lee, S.W.; An, J.; Kim, M.A.; Han, M.Y.; Yon, D.K. Serum heavy metal levels are associated with asthma, allergic rhinitis, atopic dermatitis, allergic multimorbidity, and airflow obstruction. *J. Allergy Clin. Immunol. Pract.* **2019**, *7*, 2912–2915.e2. [CrossRef]
65. Ohanian, M.; Telouk, P.; Kornblau, S.; Albarede, F.; Ruvolo, P.; Tidwell, R.S.S.; Plesa, A.; Kanagal-Shamanna, R.; Matera, E.L.; Cortes, J.; et al. A heavy metal baseline score predicts outcome in acute myeloid leukemia. *Am. J. Hematol.* **2020**, *95*, 422–434. [CrossRef]
66. Khuder, A.; Issa, H.; Bakir, M.A.; Habil, K.; Mohammad, A.; Solaiman, A. Major, minor, and trace elements in whole blood of patients with different leukemia patterns. *Nukleonika* **2012**, *57*, 389–399.
67. Inamdar, A.A.; Inamdar, A.C. Heart Failure: Diagnosis, Management and Utilization. *J. Clin. Med.* **2016**, *5*, 62. [CrossRef]
68. Palazzuoli, A.; Nuti, R. Heart failure: Pathophysiology and clinical picture. In *Fluid Overload: Diagnosis and Management, Contributions to Nephrology*; Ronco, C., Costanzo, M.R., Bellomo, R., Maisel, A.S., Eds.; Karger: Basel, Switzerland, 2010; Volume 164, pp. 1–10. [CrossRef]
69. Wang, B.; Jin, L.; Ren, A.; Yuan, Y.; Liu, J.; Li, Z.; Zhang, L.; Yi, D.; Wang, L.L.; Zhang, Y.; et al. Levels of polycyclic aromatic hydrocarbons in maternal serum and risk of neural tube defects in offspring. *Environ. Sci. Technol.* **2015**, *49*, 588–596. [CrossRef] [PubMed]
70. Qin, Y.Y.; Leung, C.K.M.; Lin, C.K.; Leung, A.O.W.; Wang, H.S.; Giesy, J.P.; Wong, M.H. Halogenated POPs and PAHs in blood plasma of Hong Kong residents. *Environ. Sci. Technol.* **2011**, *45*, 1630–1637. [CrossRef]
71. Loutfy, N.M.; Malhat, F.; Ahmed, M.T. Polycyclic aromatic hydrocarbons residues in blood serum and human milk, an Egyptian pilot study. *Hum. Ecol. Risk. Assess.* **2017**, *23*, 1573–1584. [CrossRef]
72. Yang, Q.; Zhao, Y.; Qiu, X.; Zhang, C.; Li, R.; Qiao, J. Association of serum levels of typical organic pollutants with polycystic ovary syndrome (PCOS): A case-control study. *Hum. Reprod.* **2015**, *30*, 1964–1973. [CrossRef] [PubMed]
73. Cao, L.; Wang, D.; Zhu, C.; Wang, B.; Cen, X.; Chen, A.; Zhou, H.; Ye, Z.; Tan, Q.; Nie, X.; et al. Polycyclic aromatic hydrocarbon exposure and atherosclerotic cardiovascular disease risk in urban adults: The mediating role of oxidatively damaged DNA. *Environ. Pollut.* **2020**, *265*, 114860. [CrossRef] [PubMed]
74. Hajir, S.; Al Aaraj, L.; Zgheib, N.; Badr, K.; Ismaeel, H.; Abchee, A.; Tamim, H.; Saliba, N.A. The association of urinary metabolites of polycyclic aromatic hydrocarbons with obstructive coronary artery disease: A red alert for action. *Environ. Pollut.* **2020**, 115967. [CrossRef]
75. Yin, W.; Hou, J.; Xu, T.; Cheng, J.; Li, P.; Wang, L.; Zhang, Y.; Wang, X.; Hu, C.; Huang, C.; et al. Obesity mediated the association of exposure to polycyclic aromatic hydrocarbon with risk of cardiovascular events. *Sci. Total Environ.* **2018**, *616–617*, 841–854. [CrossRef] [PubMed]
76. Gao, P.; da Silva, E.; Hou, L.; Denslow, N.D.; Xiang, P.; Ma, L.Q. Human exposure to polycyclic aromatic hydrocarbons: Metabolomics perspective. *Environ. Int.* **2018**, *119*, 466–477. [CrossRef]
77. Sevim, Ç.; Doğan, E.; Comakli, S. Cardiovascular disease and toxic metals. *Curr. Opin. Toxicol.* **2020**, *19*, 88–92. [CrossRef]
78. Yang, A.M.; Lo, K.; Zheng, T.Z.; Yang, J.L.; Bai, Y.N.; Feng, Y.Q.; Cheng, N.; Liu, S.M. Environmental heavy metals and cardiovascular diseases: Status and future direction. *Chronic. Dis. Transl. Med.* **2020**, *6*, 251–259. [CrossRef] [PubMed]

79. Mu, X.; Wang, Z.; Liu, L.; Guo, X.; Gu, C.; Xu, H.; Zhao, L.; Jiang, W.; Cao, H.; Mao, X.; et al. Multiple exposure pathways of first-year university students to heavy metals in China: Serum sampling and atmospheric modeling. *Sci. Total Environ.* **2020**, *746*, 141405. [CrossRef] [PubMed]
80. Pappas, R.S.; Fresquez, M.R.; Martone, N.; Watson, C.H. Toxic metal concentrations in mainstream smoke from cigarettes available in the USA. *J. Anal. Toxicol.* **2014**, *38*, 204–211. [CrossRef]
81. Agency for Toxic Substances and Disease Registry. Medical Management Guidelines for Arsenic (As) and Inorganic Arsenic Compounds. 2014. Available online: https://www.atsdr.cdc.gov/MMG/MMG.asp?id=1424&tid=3 (accessed on 1 December 2020).
82. Yousef, S.; Eapen, V.; Zoubeidi, T.; Kosanovic, M.; Mabrouk, A.A.; Adem, A. Learning disorder and blood concentration of heavy metals in the United Arab Emirates. *Asian. J. Psychiatr.* **2013**, *6*, 394–400. [CrossRef] [PubMed]
83. Xiang, S.; Jin, Q.; Xu, F.; Yao, Y.; Liang, W.; Zuo, X.; Ye, T.; Ying, C. High serum arsenic and cardiovascular risk factors in patients undergoing continuous ambulatory peritoneal dialysis. *J. Trace. Elem. Med. Biol.* **2019**, *52*, 1–5. [CrossRef]
84. World Health Organization. Environmental Health Criteria 135: Cadmium: Environmental Aspects. 1992. Available online: http://www.inchem.org/documents/ehc/ehc/ehc135.htm (accessed on 11 December 2020).
85. Choi, S.; Kwon, J.; Kwon, P.; Lee, C.; Jang, S.I. Association between blood heavy metal levels and predicted 10-year risk for a first atherosclerosis cardiovascular disease in the general korean population. *Int. J. Environ. Res. Public Health* **2020**, *17*, 2134. [CrossRef]
86. Genchi, G.; Sinicropi, M.S.; Carocci, A.; Lauria, G.; Catalano, A. Mercury exposure and heart diseases. *Int. J. Environ. Res. Public Health* **2017**, *14*, 74. [CrossRef]
87. Hightower, J.M.; Moore, D. Mercury levels in high-end consumers of fish. *Environ. Health Perspect.* **2003**, *111*, 604–608. [CrossRef]
88. Adimado, A.A.; Baah, D.A. Mercury in Human Blood, Urine, Hair, Nail, and Fish from the Ankobra and Tano River Basins in Southwestern Ghana. *Bull. Environ. Contam. Toxicol.* **2002**, *68*, 339–346. [CrossRef]
89. Houston, M.C. Role of mercury toxicity in hypertension, cardiovascular disease, and stroke. *J. Clin. Hypertens.* **2011**, *13*, 621–627. [CrossRef]
90. Liu, S.; Tsui, M.T.; Lee, E.; Fowler, J.; Jia, Z. Uptake, efflux, and toxicity of inorganic and methyl mercury in the endothelial cells (EA.hy926). *Sci. Rep.* **2020**, *10*, 9023. [CrossRef] [PubMed]
91. Agency for Toxic Substances and Disease Registry. Case Studies in Environmental Medicine (CSEM), Lead Toxicity. 2017. Available online: https://www.atsdr.cdc.gov/csem/lead/docs/csem-lead_toxicity_508.pdf (accessed on 2 December 2020).
92. Sharma, B.; Singh, S.; Siddiqi, N.J. Biomedical Implications of Heavy Metals Induced Imbalances in Redox Systems. *Biomed. Res. Int.* **2014**, *2014*, 640754. [CrossRef] [PubMed]
93. Vaziri, N.D. Mechanisms of lead-induced hypertension and cardiovascular disease. *Am. J. Physiol. Heart Circ. Physiol.* **2008**, *295*, 454–465. [CrossRef]
94. Patrick, L. Lead toxicity part II: The role of free radical damage and the use of antioxidants in the pathology and treatment of lead toxicity. *Altern. Med. Rev.* **2006**, *11*, 114–127. [PubMed]
95. Navas-acien, A.; Selvin, E.; Sharrett, A.R.; Calderon-Aranda, E.; Silbergeld, E.; Guallar, E. Lead, Cadmium, Smoking, and Increased Risk of Peripheral Arterial Disease. *Circulation* **2004**, *109*, 3196–3201. [CrossRef]
96. Karadas, S.; Sayin, R.; Aslan, M.; Gonullu, H.; Kati, C.; Dursun, R.; Duran, L.; Gonullu, E.; Demir, H. Serum levels of trace elements and heavy metals in patients with acute hemorrhagic stroke. *J. Membr. Biol.* **2014**, *247*, 175–180. [CrossRef]
97. Ford, E.S. Serum copper concentration and coronary heart disease among US adults. *Am. J. Epidemiol.* **2000**, *151*, 1182–1188. [CrossRef]
98. Leone, N.; Courbon, D.; Ducimetiere, P.; Zureik, M. Zinc, copper, and magnesium and risks for all-cause, cancer, and cardiovascular mortality. *Epidemiology* **2006**, *17*, 308–314. [CrossRef] [PubMed]
99. Lind, P.M.; Olsén, L.; Lind, L. Circulating levels of metals are related to carotid atherosclerosis in elderly. *Sci. Total Environ.* **2012**, *416*, 80–88. [CrossRef]
100. Alissa, E.M.; Bahjri, S.M.; Ahmed, W.H.; Al-Ama, N.; Ferns, G.A.A. Chromium status and glucose tolerance in Saudi men with and without coronary artery disease. *Biol. Trace. Elem. Res.* **2009**, *131*, 215–228. [CrossRef]
101. Afridi, H.; Kazi, T.G.; Kazi, N.; Sirajuddin, U.; Kandhro, G.A.; Baig, J.A.; Shah, A.Q.; Jamali, M.K.; Arain, M.B.; Wadhwa, S.K.; et al. Chromium and manganese levels in biological samples of Pakistani myocardial infarction patients at different stages as related to controls. *Biol. Trace. Elem. Res.* **2011**, *142*, 259–273. [CrossRef]
102. Gutierrez-Bedmar, M.; Martinez-Gonzalez, M.A.; Munoz-Bravo, C.; Ruiz-Canela, M.; Mariscal, A.; Salas-Salvado, J.; Estruch, R.; Corella, D.; Aros, F.; Fito Colomer, M.; et al. Chromium exposure and risk of cardiovascular disease in high cardiovascular risk subjects: Nested case-control study in the prevention with mediterranean diet (PREDIMED) study. *Circ. J.* **2017**, *81*, 1183–1190. [CrossRef] [PubMed]
103. Keir, J.L.A.; Cakmak, S.; Blais, J.M.; White, P.A. The influence of demographic and lifestyle factors on urinary levels of PAH metabolites—empirical analyses of Cycle 2 (2009–2011) CHMS data. *J. Expo. Sci. Environ. Epidemiol.* **2020**, *31*, 386–397. [CrossRef]
104. Sakellari, A.; Karavoltsos, S.; Kalogeropoulos, N.; Theodorou, D.; Dedoussis, G.; Chrysohoou, C.; Dassenakis, M.; Scoullos, M. Predictors of cadmium and lead concentrations in the blood of residents from the metropolitan area of Athens (Greece). *Sci. Total Environ.* **2016**, *568*, 263–270. [CrossRef]

105. Lee, J.W.; Lee, C.K.; Moon, C.S.; Choi, I.J.; Lee, K.J.; Yi, S.M.; Jang, B.K.; Yoon, B.J.; Kim, D.S.; Peak, D.; et al. Korea National Survey for Environmental Pollutants in the Human Body 2008: Heavy metals in the blood or urine of the Korean population. *Int. J. Hyg. Environ. Health* **2012**, *215*, 449–457. [CrossRef]
106. Nunes, J.A.; Batista, B.L.; Rodrigues, J.L.; Caldas, N.M.; Neto, J.A.G.; Barbosa, F. A simple method based on ICP-MS for estimation of background levels of arsenic, cadmium, copper, manganese, nickel, lead, and selenium in blood of the Brazilian population. *J. Toxicol. Environ. Health Part. A* **2010**, *73*, 878–887. [CrossRef] [PubMed]
107. Nisse, C.; Tagne-Fotso, R.; Howsam, M.; Richeval, C.; Labat, L.; Leroyer, A. Blood and urinary levels of metals and metalloids in the general adult population of Northern France: The IMEPOGE study, 2008–2010. *Int. J. Hyg. Environ. Health* **2017**, *220*, 341–363. [CrossRef] [PubMed]
108. Vahter, M.; Berglund, M.; Akesson, A.; Liden, C. Metals and women's health. *Environ. Res.* **2002**, *88*, 145–155. [CrossRef] [PubMed]
109. Mathers, C.D.; Loncar, D. Projections of global mortality and burden of disease from 2002 to 2030. *PLoS Med.* **2006**, *3*, 2011–2030. [CrossRef] [PubMed]
110. Wong, N.D. Epidemiological studies of CHD and the evolution of preventive cardiology. *Nat. Rev. Cardiol.* **2014**, *11*, 276–289. [CrossRef]
111. Pirie, K.; Peto, R.; Reeves, G.K.; Green, J.; Beral, V. The 21st century hazards of smoking and benefits of stopping: A prospective study of one million women in the UK. *Lancet* **2013**, *381*, 133–141. [CrossRef]
112. Gopal, D.M.; Kalogeropoulos, A.P.; Georgiopoulou, V.V.; Smith, A.L.; Bauer, D.C.; Newman, A.B.; Kim, L.; Bibbins-Domingo, K.; Tindle, H.; Harris, T.B.; et al. Cigarette smoking exposure and heart failure risk in older adults: The Health, Aging, and Body Composition Study. *Am. Heart J.* **2012**, *164*, 236–242. [CrossRef]
113. Wang, Y.; Wong, L.; Meng, L.; Pittman, E.N.; Trinidad, D.A.; Hubbard, K.L.; Etheredge, A.; Del Valle-Pinero, A.Y.; Zamoiski, R.; van Bemmel, D.M.; et al. Urinary concentrations of monohydroxylated polycyclic aromatic hydrocarbons in adults from the U.S. Population Assessment of Tobacco and Health (PATH) Study Wave 1 (2013–2014). *Environ. Int.* **2019**, *123*, 201–208. [CrossRef] [PubMed]
114. Jain, R.B. Trends and concentrations of selected polycyclic aromatic hydrocarbons in general US population: Data from NHANES 2003–2008. Cogent. *Environ. Sci.* **2015**, *110*, 1–16. [CrossRef]
115. Pappas, R.S. Toxic elements in tobacco and in cigarette smoke: Inflammation and sensitization. *Metallomics* **2011**, *3*, 1181–1198. [CrossRef] [PubMed]
116. Bonberg, N.; Pesch, B.; Ulrich, N.; Moebus, S.; Eisele, L.; Marr, A.; Arendt, M.; Jöckel, K.H.; Brüning, T.; Weiss, T. The distribution of blood concentrations of lead (Pb), cadmium (Cd), chromium (Cr) and manganese (Mn) in residents of the German Ruhr area and its potential association with occupational exposure in metal industry and/or other risk factors. *Int. J. Hyg. Environ. Health* **2017**, *220*, 998–1005. [CrossRef] [PubMed]
117. Kasdagli, M.I.; Katsouyanni, K.; de Hoogh, K.; Lagiou, P.; Samoli, E. Associations of air pollution and greenness with mortality in Greece: An ecological study. *Environ. Res.* **2020**, 110348. [CrossRef]
118. Fourtziou, L.; Liakakou, E.; Stavroulas, I.; Theodosi, C.; Zarmpas, P.; Psiloglou, B.; Sciare, J.; Maggos, T.; Bairachtari, K.; Bougiatioti, A.; et al. Multi-tracer approach to characterize domestic wood burning in Athens (Greece) during wintertime. *Atmos. Environ.* **2017**, *148*, 89–101. [CrossRef]
119. Gratsea, M.; Liakakou, E.; Mihalopoulos, N.; Adamopoulos, A.; Tsilibari, E.; Gerasopoulos, E. The combined effect of reduced fossil fuel consumption and increasing biomass combustion on Athens' air quality, as inferred from long term CO measurements. *Sci. Total Environ.* **2017**, *592*, 115–123. [CrossRef] [PubMed]
120. Paraskevopoulou, D.; Liakakou, E.; Gerasopoulos, E.; Mihalopoulos, N. Sources of atmospheric aerosol from long-term measurements (5years) of chemical composition in Athens, Greece. *Sci. Total Environ.* **2015**, *527–528*, 165–178. [CrossRef]
121. Saffari, A.; Daher, N.; Samara, C.; Voutsa, D.; Kouras, A.; Manoli, E.; Karagiozidou, O.; Vlachokostas, C.; Moussiopoulos, N.; Shafer, M.M.; et al. Increased biomass burning due to the economic crisis in Greece and its adverse impact on wintertime air quality in Thessaloniki. *Environ. Sci. Technol.* **2013**, *47*, 13313–13320. [CrossRef]
122. Michas, G.; Karvelas, G.; Trikas, A. Cardiovascular disease in Greece; the latest evidence on risk factors. *Hellenic J. Cardiol.* **2019**, *60*, 271–275. [CrossRef] [PubMed]
123. Vasilakos, C.; Levi, N.; Maggos, T.; Hatzianestis, J.; Michopoulos, J.; Helmis, C. Gas-particle concentration and characterization of sources of PAHs in the atmosphere of a suburban area in Athens, Greece. *J. Hazard. Mater.* **2007**, *140*, 45–51. [CrossRef]
124. Pateraki, S.; Asimakopoulos, D.N.; Maggos, T.; Assimakopoulos, V.D.; Bougiatioti, A.; Bairachtari, K.; Vasilakos, C.; Mihalopoulos, N. Chemical characterization, sources and potential health risk of PM2.5 and PM1 pollution across the Greater Athens Area. *Chemosphere* **2020**, *241*, 125026. [CrossRef] [PubMed]
125. Srogi, K. Monitoring of environmental exposure to polycyclic aromatic hydrocarbons: A review. *Environ. Chem. Lett.* **2007**, *5*, 169–195. [CrossRef]
126. Gerde, P.; Muggenburg, B.A.; Lundborg, M.; Dahl, A.R. The rapid alveolar absorption of diesel soot-adsorbed benzo[a]pyrene: Bioavailability, metabolism and dosimetry of an inhaled particle-borne carcinogen. *Carcinogenesis* **2001**, *22*, 741–749. [CrossRef]
127. Laamech, J.; Bernard, A.; Dumont, X.; Benazzouz, B.; Lyoussi, B. Blood lead, cadmium and mercury among children from urban, industrial and rural areas of Fez Boulemane Region (Morocco): Relevant factors and early renal effects. *Int. J. Occup. Med. Environ. Health* **2014**, *27*, 641–659. [CrossRef]

128. Kim, L.; Jeon, H.J.; Kim, Y.C.; Yang, S.H.; Choi, H.; Kim, T.O.; Lee, S.E. Monitoring polycyclic aromatic hydrocarbon concentrations and distributions in rice paddy soils from Gyeonggi-do, Ulsan, and Pohang. *Appl. Biol. Chem.* **2019**, *62*, 18. [CrossRef]
129. Wang, Z.; Liu, Z.; Xu, K.; Mayer, L.M.; Zhang, Z.; Kolker, A.S.; Wu, W. Concentrations and sources of polycyclic aromatic hydrocarbons in surface coastal sediments of the northern Gulf of Mexico. *Geochem. Trans.* **2014**, *15*, 1–12. [CrossRef]
130. Neta, G.; Goldman, L.R.; Barr, D.; Sjodin, A.; Apelberg, B.J.; Witter, F.R.; Halden, R.U. Distribution and determinants of pesticide mixtures in cord serum using principal component analysis. *Environ. Sci. Technol.* **2010**, *44*, 5641–5648. [CrossRef] [PubMed]
131. Andrysík, Z.; Vondráček, J.; Marvanová, S.; Ciganek, M.; Neča, J.; Pěnčíková, K.; Mahadevan, B.; Topinka, J.; Baird, W.M.; Kozubík, A.; et al. Activation of the aryl hydrocarbon receptor is the major toxic mode of action of an organic extract of a reference urban dust particulate matter mixture: The role of polycyclic aromatic hydrocarbons. *Mutat. Res.* **2011**, *714*, 53–62. [CrossRef]
132. Barron, M.G.; Heintz, R.; Rice, S.D. Relative potency of PAHs and heterocycles as aryl hydrocarbon receptor agonists in fish. *Mar. Environ. Res.* **2004**, *58*, 95–100. [CrossRef]
133. Solenkova, N.V.; Newman, J.D.; Berger, J.S.; Thurston, G.; Hochman, J.S.; Lamas, G.A. Metal pollutants and cardiovascular disease: Mechanisms and consequences of exposure. *Am. Heart J.* **2014**, *168*, 812–822. [CrossRef] [PubMed]
134. Duruibe, J.O.; Ogwuegbu, M.O.; Egwurugwu, J.N. Heavy metal pollution and human biotoxic effects. *Int. J. Phys. Sci.* **2007**, *2*, 112–118. [CrossRef]
135. Huang, T.; Li, J.; Zhang, W. Application of principal component analysis and logistic regression model in lupus nephritis patients with clinical hypothyroidism. *BMC Med. Res. Methodol.* **2020**, *20*, 99. [CrossRef] [PubMed]
136. DiStefano, C.; Zhu, M.; Mîndrilǎ, D. Understanding and Using Factor Scores: Considerations for the Applied Researcher. *Pract. Assess. Res. Eval.* **2009**, *14*. [CrossRef]
137. Chowdhury, R.; Ramond, A.; O'Keeffe, L.M.; Shahzad, S.; Kunutsor, S.K.; Muka, T.; Gregson, J.; Willeit, P.; Warnakula, S.; Khan, H.; et al. Environmental toxic metal contaminants and risk of cardiovascular disease: Systematic review and meta-analysis. *BMJ* **2018**, *362*, k3310. [CrossRef] [PubMed]
138. Batáriová, A.; Spěváčková, V.; Beneš, B.; Čejchanová, M.; Šmíd, J.; Černá, M. Blood and urine levels of Pb, Cd and Hg in the general population of the Czech Republic and proposed reference values. *Int. J. Hyg. Environ. Health* **2006**, *209*, 359–366. [CrossRef] [PubMed]
139. Jin, L.; Liu, J.; Ye, B.; Ren, A. Concentrations of selected heavy metals in maternal blood and associated factors in rural areas in Shanxi Province, China. *Environ. Int.* **2014**, *66*, 157–164. [CrossRef]
140. United States Environmental Protection Agency. Definition and Procedures for the Determination of the Method Detection Limit. App. B, Part 136:343–345. 40 CFR Ch (7–1–12 edition). 2016. Available online: https://www.epa.gov/sites/production/files/2016-12/documents/mdl-procedure_rev2_12-13-2016.pdf (accessed on 10 December 2020).
141. Kanellopoulos, P.G.; Verouti, E.; Chrysochou, E.; Koukoulakis, K.; Bakeas, E. Primary and secondary organic aerosol in an urban/industrial site: Sources, health implications and the role of plastic enriched waste burning. *J. Environ. Sci.* **2021**, *99*, 222–238. [CrossRef]
142. Manoli, E.; Kouras, A.; Karagkiozidou, O.; Argyropoulos, G.; Voutsa, D.; Samara, C. Polycyclic aromatic hydrocarbons (PAHs) at traffic and urban background sites of northern Greece: Source apportionment of ambient PAH levels and PAH-induced lung cancer risk. *Environ. Sci. Pollut. Res.* **2016**, *23*, 3556–3568. [CrossRef] [PubMed]
143. Santos, E.; Souza, M.R.R.; Vilela Junior, A.R.; Soares, L.S.; Frena, M.; Alexandre, M.R. Polycyclic aromatic hydrocarbons (PAH) in superficial water from a tropical estuarine system: Distribution, seasonal variations, sources and ecological risk assessment. *Mar. Pollut. Bull.* **2018**, *127*, 352–358. [CrossRef]
144. Hussain, K.; Hoque, R.R. Seasonal attributes of urban soil PAHs of the Brahmaputra Valley. *Chemosphere* **2015**, *119*, 794–802. [CrossRef] [PubMed]
145. Alhamdow, A.; Lindh, C.; Albin, M.; Gustavsson, P.; Tinnerberg, H.; Broberg, K. Early markers of cardiovascular disease are associated with occupational exposure to polycyclic aromatic hydrocarbons. *Sci. Rep.* **2017**, *7*, 1–11. [CrossRef]
146. Ravindra, K.; Sokhi, R.; Van Grieken, R. Atmospheric polycyclic aromatic hydrocarbons: Source attribution, emission factors and regulation. *Atmos. Environ.* **2008**, *42*, 2895–2921. [CrossRef]
147. Jollife, I.T.; Cadima, J. Principal component analysis: A review and recent developments. *Phil. Trans. R. Soc. A* **2016**, *374*, 20150202. [CrossRef] [PubMed]
148. Pathak, A.K.; Sharma, M.; Katiyar, S.K.; Katiyar, S.; Nagar, P.K. Logistic regression analysis of environmental and other variables and incidences of tuberculosis in respiratory patients. *Sci. Rep.* **2020**, *10*, 21843. [CrossRef]

Article

Study on the Joint Toxicity of BPZ, BPS, BPC and BPF to Zebrafish

Ying Han [1,2,*], Yumeng Fei [1,2], Mingxin Wang [1,*], Yingang Xue [1], Hui Chen [1] and Yuxuan Liu [1]

1. School of Environmental & Safety Engineering, Changzhou University, Changzhou 213164, China; feiyumeng1997@163.com (Y.F.); yzxyg@126.com (Y.X.); huichen@cczu.edu.cn (H.C.); dinoice0401@163.com (Y.L.)
2. Jiangsu Engineering Research Center of Petrochemical Safety and Environmental Protection, Changzhou 213164, China
* Correspondence: hanying@cczu.edu.cn (Y.H.); wmx@cczu.edu.cn (M.W.)

Citation: Han, Y.; Fei, Y.; Wang, M.; Xue, Y.; Chen, H.; Liu, Y. Study on the Joint Toxicity of BPZ, BPS, BPC and BPF to Zebrafish. *Molecules* 2021, 26, 4180. https://doi.org/10.3390/molecules26144180

Academic Editor: Tomasz Tuzimski

Received: 16 June 2021
Accepted: 7 July 2021
Published: 9 July 2021

Publisher's Note: MDPI stays neutral with regard to jurisdictional claims in published maps and institutional affiliations.

Copyright: © 2021 by the authors. Licensee MDPI, Basel, Switzerland. This article is an open access article distributed under the terms and conditions of the Creative Commons Attribution (CC BY) license (https://creativecommons.org/licenses/by/4.0/).

Abstract: Bisphenol Z (BPZ), bisphenol S (BPS), bisphenol C (BPC), and bisphenol F (BPF) had been widely used as alternatives to bisphenol A (BPA), but the toxicity data of these bisphenol analogues were very limited. In this study, the joint toxicity of BPZ, BPS, BPC, and BPF to zebrafish (*Danio rerio*) was investigated. The median half lethal concentrations (LC50) of BPZ, BPS, BPC, and BPF to zebrafish for 96 h were 6.9×10^5 µM, 3.9×10^7 µM, 7.1×10^5 µM, and 1.6×10^6 µM, respectively. The joint toxicity effect of BPF–BPC (7.7×10^5–3.4×10^5 µM) and BPZ–BPC (3.4×10^5–3.5×10^5 µM) with the same toxic ratio showed a synergistic effect, which may be attributed to enzyme inhibition or induction theory. While the toxicity effect of the other two bisphenol analogue combined groups and multi-joint pairs showed an antagonistic effect due to the competition site, other causes need to be further explored. Meanwhile, the expression levels of the estrogen receptor genes (ERα, ERβ1) and antioxidant enzyme genes (SOD, CAT, GPX) were analyzed using a quantitative real-time polymerase chain reaction in zebrafish exposure to LC_{50} of BPZ, BPS, BPC, and BPF collected at 24, 48, 72, and 96 h. Relative expression of CAT, GPX, and ERβ1 mRNA declined significantly compared to the blank control, which might be a major cause of oxidant injury of antioxidant systems and the disruption of the endocrine systems in zebrafish.

Keywords: bisphenol analogues; zebrafish; joint toxicity; gene expression

1. Introduction

Bisphenol analogues are typical environmental endocrine disruptors and are prone to accumulate in water bodies. With the limitation of bisphenol A (BPA), the production and application of BPA substitutes such as such as bisphenol S (BPS), bisphenol F (BPF), bisphenol C (BPC) and bisphenol Z (BPZ) has gradually increased, and these substitutes release large amounts of bisphenol analogues into the water, increasing the safety risk of the growth and development of fish-based aquatic organisms [1,2]. BPF can be used instead of BPA to make epoxy resin. BPS is mainly used as a color fixing agent, as a resin flame retardant, and as materials in color photography. BPC is often used in the preparation of flame retardant. BPZ can be applied in the manufacturing of chemical compounds [3]. The estrogen activities and octyl alcohol water distribution coefficients of BPZ, BPS, BPC, and BPF are similar to those of BPA or may be even higher, which may pose risks to aquatic ecosystems and human health [4].

Bisphenol analogues have been found in different media [5]. Liao et al. collected indoor dust samples from the United States, China, Japan, and South Korea. The total content of eight bisphenols in the dust ranged from 0.026 to 111 µg·g^{-1} (mean 2.3 µg·g^{-1}). BPA, BPS, and BPF were the three main bisphenols, accounting for 98% of the total concentrations [6]. The content of seven bisphenol analogues in surface water and sediments from Lake Taihu (range: 81–3.0×10^3 ng·L^{-1}) was higher than that of Lake Luoma (range:

1.5×10^2–1.9×10^3 ng·L^{-1}) [7]. Chen et al. determined the concentrations of 7 kinds of bisphenol analogues in the urine of 283 samples of children aged 3–11 years in South China and found that the total concentrations of 7 bisphenol analogues in urine ranged from 0.43 µg·L^{-1} to 32 µg·L^{-1} [8]. In addition, Kha et al. also investigated the genotoxicity of BPA, BPS, BPF, BPAF, and their combinations [9]. Mu et al. conducted an acute toxicity test of four bisphenol analogues on zebrafish embryos, in which the half lethal concentration (LC$_{50}$) value of BPF for 96 h was 3.9×10^6 µM [10].

Zebrafish (*Danio rerio*), a type of small tropical fish, has organs and systems that were initially used for neurodevelopmental and genetic related research. As a new type of vertebrate model organism, zebrafish have gradually entered into more fields [11,12]. Currently, studies on the toxic effects of bisphenol analogues on zebrafish mainly focus on single toxicity and mechanisms. Zhao et al. exposed male zebrafish to 2.5×10^2 µM and 2.5×10^3 µM BPS solutions for 28 days and found that the plasma insulin level of male zebrafish was significantly reduced, thus impeding its physiological effect on glucose metabolism, leading to increased liver glucose output and decreased glucose metabolism and storage [13]. Sangwoo et al. studied the interfering effects of BPF, BPS, and BPZ on the thyroid hormone in zebrafish embryo larva, finding that they can damage the thyroid function in juvenile fish, and their damaging effects may be greater than that of BPA [14]. The exposure of zebrafish embryos to different concentrations of BPS (0, 0.1, 1, 10, and 100 µg·L^{-1}) for 75 days was performed and adverse effects on the endocrine system such as decreasing gonadosomatic index, decreasing plasma 17β-estradiol, and increasing hepatosomatic index were observed [15]. Ji et al. discovered significantly decreasing concentrations of testosterone with the *cyp19a* upregulated gene and the *cyp17* and *17βhsd* downregulated genes [16] Niederberger et al. showed that exposure to 10 µM of BPA, BPS, and BPF reduced the expression of *Insl3* in tests with cultured mouse fetuses [17]. In addition to BPS and BPF, other BPs were also detected at very low concentrations. However, the enrichment effect and acute toxicity of these BPs were higher than those of BPS and BPF [18].

In this study, we investigated 96-h single and joint acute toxicity experiments of BPZ, BPS, BPC, and BPF on zebrafish and evaluated the expression of key genes that vary with the exposure of BPZ, BPS, BPC, and BPF. The aim of the study was to identify the what impacts emitted bisphenol analogues other than BPA may have on zebrafish and to assess the impacts of these bisphenol analogues on aquatic ecosystems.

2. Results and Discussion

2.1. Single Acute Toxicity of BPZ, BPC, BPF and BPS to Zebrafish

The structures and physicochemical properties of BPZ, BPS, BPC, and BPF are shown in Table 1. The increasing molecular weight ranking order was BPF, BPS, BPC, and BPZ. The octanol-water partition coefficient (Log Kow) of BPS, BPF, BPC, and BPZ had similar trends to their bioaccumulation factors (Log BAF). The Log Kow value of BPZ was the highest, followed by that of BPC, BPF, and BPS. Stronger Log Kow values of the bisphenol analogues implied a stronger biological enrichment ability.

No death or abnormality in the zebrafish in the blank and solvent control groups were observed. The linear regression analysis results of toxicity test data for zebrafish were shown in Figure 1. The LC$_{50}$ of BPZ, BPC, BPF, and BPS was 6.9×10^5 µM, 7.1×10^5 µM, 1.6×10^6 µM, and 3.9×10^7 µM, respectively. The toxicity of these four bisphenol analogues was BPZ > BPC > BPF > BPS. According to relevant studies, the acute toxicity of the compounds to the fish was divided into 5 grades: LC$_{50}$ < 1 mg·L^{-1} was extremely hazardous, $1 \leq$ LC$_{50}$ < 100 mg·L^{-1} was highly hazardous, $100 \leq$ LC$_{50}$ < 1000 mg·L^{-1} was moderately hazardous, $1000 \leq$ LC$_{50}$ < 10,000 mg·L^{-1} was slightly hazardous, and LC$_{50} \geq$ 10,000 mg·L^{-1} was mildly toxic [19]. Therefore, the toxicity of BPF, BPZ, and BPC belong to the highly hazardous group, while BPS was considered to be moderately hazardous.

Table 1. Physicochemical properties of BPC, BPZ, BPF, and BPS.

Analytes	Molecular Weight	CAS	Log Kow	Log BAF	Structures
BPC	256.34	79-97-0	4.74	2.79	
BPZ	268.35	843-55-0	5.48	3.28	
BPF	200.23	620-92-8	3.06	1.59	
BPS	250.27	80-09-1	1.65	0.75	

Ren et al. chose adult zebrafish and embryos as test organisms and did not feed or change the water during their experiment. The LC_{50} values of BPZ in adult zebrafish and embryos after 96 h were 7.1×10^5 μM and 7.9×10^5 μM, respectively. For BPF and BPS, the LC_{50} values in adult zebrafish were 1.9×10^6 μM and 8.6×10^7 μM, and in embryo, they were 1.5×10^6 μM and 8.1×10^7 μM [20]. Relevant research on BPC was limited. The toxicity of BPZ, BPF, and BPS was BPZ > BPF > BPS, which was consistent with this study. The toxicity of BPS was two orders of magnitude smaller than the other three bisphenol analogues. The numerical differences between the results may be related to the growth environment, the biological state of test organism, and the experimental conditions.

2.2. Dual Joint Acute Toxicity of BPZ, BPC, BPF and BPS to Zebrafish

The pairwise LC_{50} values of BPZ, BPC, BPF, and BPS with equal toxicity ratios to zebrafish after 96 h and the dual combined effect of these four compounds were calculated and evaluated. As shown in Table 2, the dual combined effect of BPF–BPC (3.4×10^5–7.7×10^5 μM) with an additive index (AI) value of 0.05 and BPZ–BPC (3.4×10^5–3.5×10^5 μM) with an AI value of 0.01 showed synergistic effect, while the rest of the groups with negative AI values showed an antagonistic effect, which can also be seen from Figure 2.

As shown in Figure 2, a clear trend was not observed for the decrease in the percentage for the lethality rate in BPF–BPC and BPZ–BPC. However, there was sudden drop in the performance of another four joint groups, meaning that the antagonistic effect may exist. A small number of studies were carried out to explore the combined toxic effects of BPZ, BPC, BPF and BPS to zebrafish, which may promote the inhibition or induction of relative genes expression. Enzyme inhibition or induction theory states that one compound converts to the enzyme inducer of another one in a hybrid system, which may promote the detoxification or activation the toxicity of another compound, presenting a synergy [21]. The mechanisms of the combined toxic effects need to be further identified.

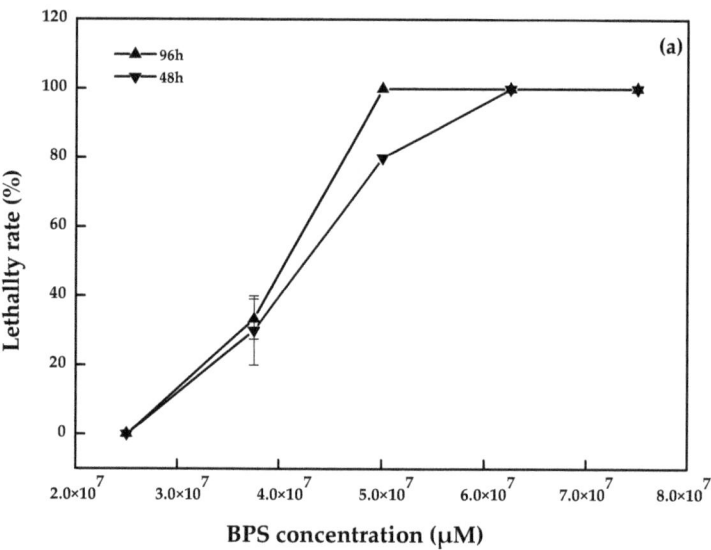

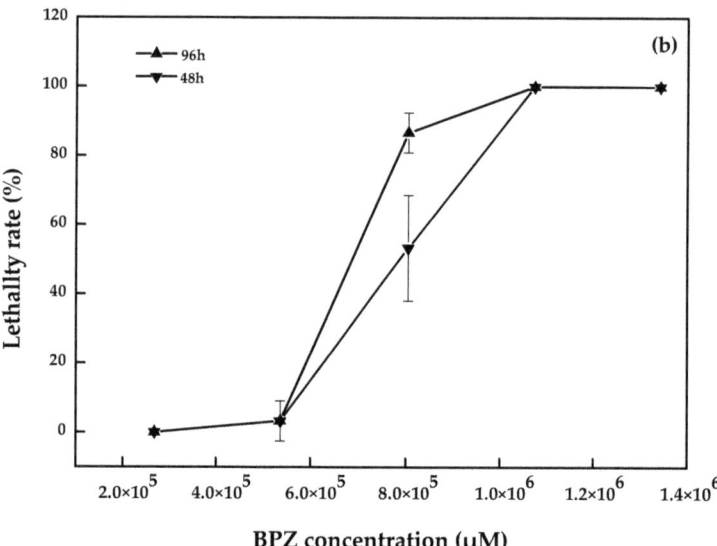

Figure 1. Cont.

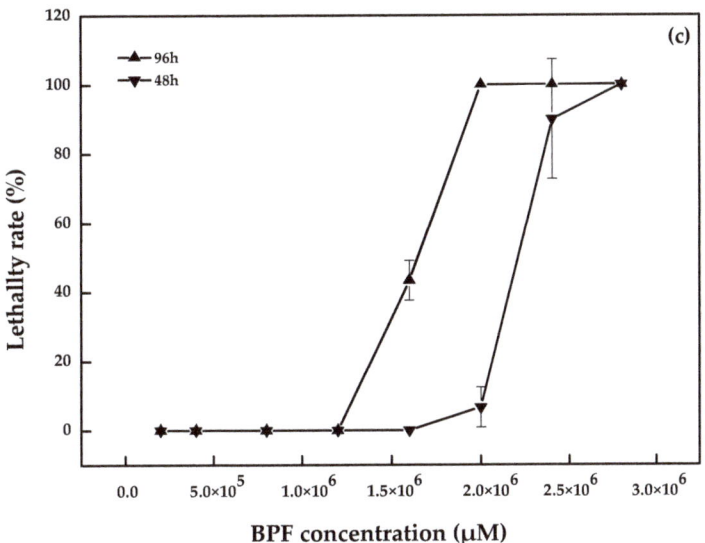

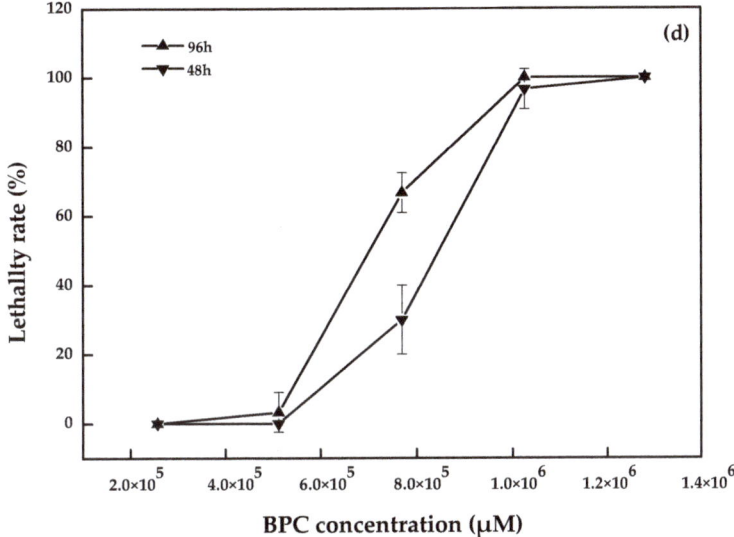

Figure 1. BPS (**a**), BPZ (**b**), BPF (**c**), and BPC (**d**) exposure on zebrafish after 48 h and 96 h (values represent mean ± S.D., sample size: 3).

Lei et al. reported that the joint toxicity of BPA and nitrophenol on loach as an antagonism effect was due to the nitrophenol reactive ion occupied in the limited binding sites in the cells, reducing the combination efficiency of the BPA reactive ion [22]. Fang et al. found three cresol isomers with similar chemical properties appearing to interfere during a short period of time and reduce the toxicity to the zebrafish [23]. BPZ, BPC, BPF, and BPS had two hydroxyl phenyl groups with different substituents on the carbon bridge. The antagonistic effect may be caused by the competing effects of bisphenol analogues on the active sites of the cell surface and the metabolic system. One substance occupied the binding site on the cell surface and reduced the binding of other substances. The toxicity was inhibited and resulted in an overall antagonistic effect.

Table 2. Combined toxicity of BPZ, BPC, BPF, and BPS to zebrafish after 96 h.

Target	$LC_{50}/(\mu M)$	S [1]	AI	Action
BPF–BPS	$8.7 \times 10^5 (7.6 \times 10^5 \sim 9.7 \times 10^5)$–$2.1 \times 10^7 (1.8 \times 10^7 \sim 2.3 \times 10^7)$	1.07	−0.07	antagonism
BPF–BPZ	$8.7 \times 10^5 (7.7 \times 10^5 \sim 9.6 \times 10^5)$–$3.7 \times 10^5 (3.3 \times 10^5 \sim 4.1 \times 10^5)$	1.08	−0.08	antagonism
BPF–BPC	$7.7 \times 10^5 (6.4 \times 10^5 \sim 8.5 \times 10^5)$–$3.4 \times 10^5 (2.8 \times 10^5 \sim 3.7 \times 10^5)$	0.95	0.05	synergism
BPS–BPZ	$2.2 \times 10^7 (2.0 \times 10^7 \sim 2.5 \times 10^7)$–$4.0 \times 10^5 (3.6 \times 10^5 \sim 4.5 \times 10^5)$	1.15	−0.15	antagonism
BPS–BPC	$2.4 \times 10^7 (2.2 \times 10^7 \sim 2.8 \times 10^7)$–$4.4 \times 10^5 (4.0 \times 10^5 \sim 5.2 \times 10^5)$	1.25	−0.25	antagonism
BPZ–BPC	3.4×10^5–3.5×10^5	0.99	0.01	synergism

[1] The sum of the additive effects of biological toxicity.

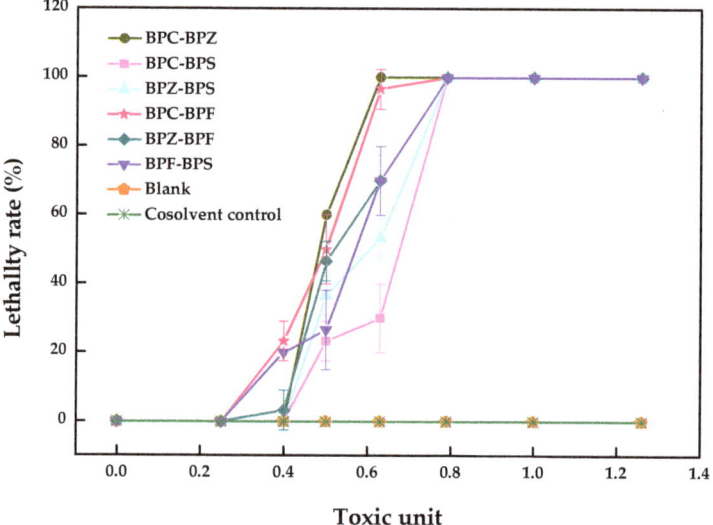

Figure 2. Combined toxicity of BPZ, BPC, BPF, and BPS on zebrafish after 96 h (values represent mean ± S.D., sample size: 3).

2.3. Multi-Joint Acute Toxicity of BPZ, BPC, BPF and BPS to Zebrafish

It can be seen from Table 3 and Figure 3 that the AI values of the multi-joint acute toxicity of BPZ, BPC, BPF, and BPS with equal toxic ratios to zebrafish were all negative, which showed an antagonistic effect and is consistent with most conclusions obtained from dual combined acute toxicity test.

Table 3. Multi-joint acute toxicity of BPZ, BPC, BPF, and BPS on zebrafish after 96 h.

Target	LC$_{50}$/(μM)	S	AI	Action
BPF–BPS–BPZ	6.0×10^5–1.4×10^7–2.6×10^5	1.11	−0.11	antagonism
BPF–BPZ–BPC	$6.6 \times 10^5 (3.3 \times 10^5$~$7.6 \times 10^5)$–$2.8 \times 10^5 (1.4 \times 10^5$~$3.2 \times 10^5)$–$2.9 \times 10^5 (1.5 \times 10^5$~$3.3 \times 10^5)$	1.22	−0.22	antagonism
BPS–BPZ–BPC	1.4×10^7–2.4×10^5–2.5×10^5	1.05	−0.05	antagonism
BPS–BPC–BPF	1.7×10^7–3.1×10^5–7.0×10^5	1.30	−0.30	antagonism
BPF–BPS–BPZ–BPC	$4.8 \times 10^5 (2.5 \times 10^5$~$5.6 \times 10^5)$–$1.1 \times 10^7 (5.9 \times 10^6$~$1.3 \times 10^7)$–$2.0 \times 10^5 (1.1 \times 10^5$~$2.4 \times 10^5)$–$2.1 \times 10^5 (1.1 \times 10^5$~$2.4 \times 10^5)$	1.18	−0.18	antagonism

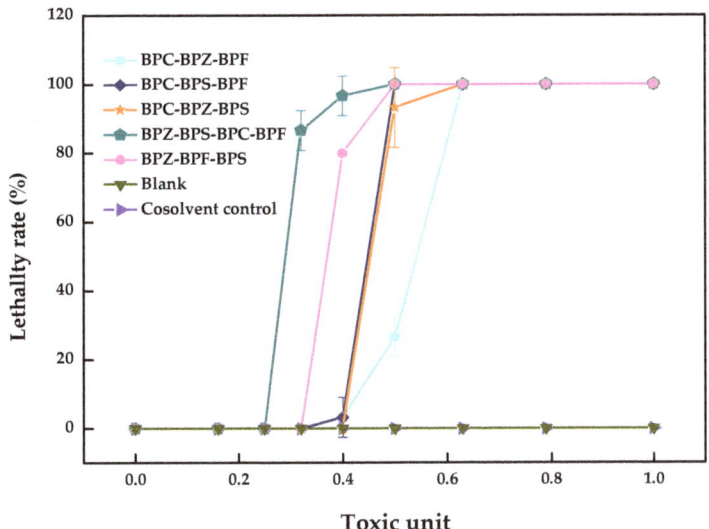

Figure 3. Multi-joint acute toxicity of BPZ, BPC, BPF, and BPS on zebrafish after 96 h (values represent mean ± S.D., sample size: 3).

The mechanism and function of each component in the activation site differences leads to the different joint toxicity effects. Each component's contribution to the ultimate toxic effects was different, especially for mixtures containing more than three components. The appearance of an antagonistic effect meant that the decisive composition in the zebrafish had an inconsistent toxicity effect or competition site. In addition to the above reasons, different exposure times and matching methods would also change the determination of joint toxicity. At present, data on the toxicity of the combined action of bisphenol analogues are limited even though the mechanism of the biological toxicity of bisphenol analogues is multi-faceted and very complex. The specific mechanism needs to be further studied and discussed.

2.4. Effects of BPZ, BPC, BPF and BPS Exposure on Relevant Gene Expression

Bisphenol analogues have interfering effects on both endogenous hormones determined by the expression of catalase (CAT), peroxide dismutase (SOD) and glutamine peroxidase (GPX) as well as the anti-oxidation system that is influenced by the expression of estrogen receptor (ERα and ERβ1) [24–26]. Consequently, the determination of CAT, SOD, GPX, ERα and ERβ1 expression during exposure to LC$_{50}$ of BPC, BPZ, BPF, and BPS on zebrafish was investigated on the basis of the reference gene ribosomal protein 17 (*rp17*). The blank sample was set as control. The threshold cycle (Ct) was used to evaluate the variations. Previous studies focused on relationships between gene expression and different concentrations of chemical exposure groups. For instance, Xu et al. investigated the immunotoxicity of dibutyl phthalate to zebrafish, which upregulated the expression of *rag1/2* [27]. In the present study, concentration-dependent groups were not created, but there were different exposure time-dependent groups instead, which were limited. Based on our results, the differences in the relative expression of both SOD and ERα was nonsignificant, and the relative gene expression of CAT, GPX, and ERβ1 was significant.

As shown in Figure 4, a description of CAT, GPX, and ERβ1 expression was illustrated. Bars represent the relative change in mRNA expression in the treated group compared to the control group. With an increasing exposure time (24 h, 48 h, 72 h, and 96 h) of zebrafish to BPC, BPZ, BPF, and BPS, there was a fluctuating relative expression of CAT, GPX, and ERβ1. However, compared to the control group, relative expression of CAT, GPX, and ERβ1 was significantly decreased in all of the treated groups. Salahinejad et al. evaluated the expression of CAT in zebrafish exposure to concentration-dependent groups of BPS, discovering the obviously down-regulated CAT expression [28], which was consistent with this study. It can be seen from Figure 4a that the relative expression of CAT declined markedly with exposure time in the BPS treated group. Meanwhile, the relative expression of GPX and ERβ1 also apparently decreased in the BPZ and BPS treated groups, respectively, which can be seen in Figure 4b,c. In the BPC, BPZ, and BPF treated groups, the CAT and ERβ1 expression showed similar trends. Expression of CAT and ERβ1 declined in the period of 24 h to 48 h and from 72 h to 96 h but slightly increased from 48 h to 72 h (Figure 4a,c). As shown in Figure 4b, the expression of GPX in the BPC, BPF, and BPS treated groups marginally increased from after 48 h to 72 h of exposure and decreased during other exposure time periods. Pikulkaew et al. detected low expression levels of ERβ1 after 8 h post-fertilization, which then increased from 24 h to 48 h post-fertilization in zebrafish [29] A significant decrease of CAT and GPX in the brain and liver of zebrafish was reported by Gyimah et al., which may cause damage to the organs of zebrafish [30].

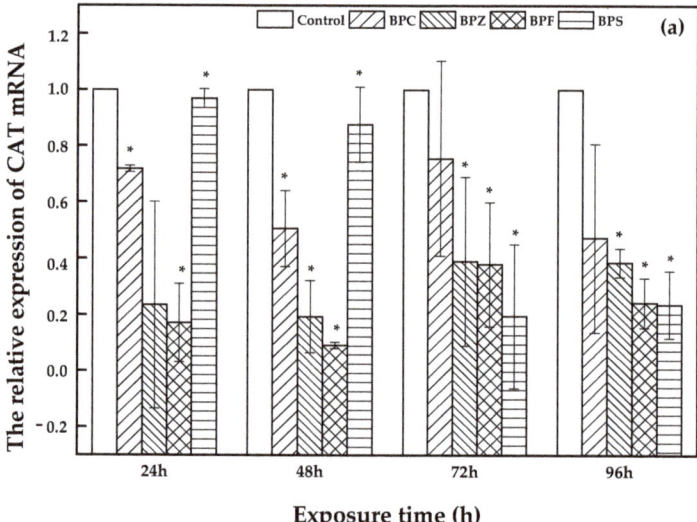

Figure 4. *Cont.*

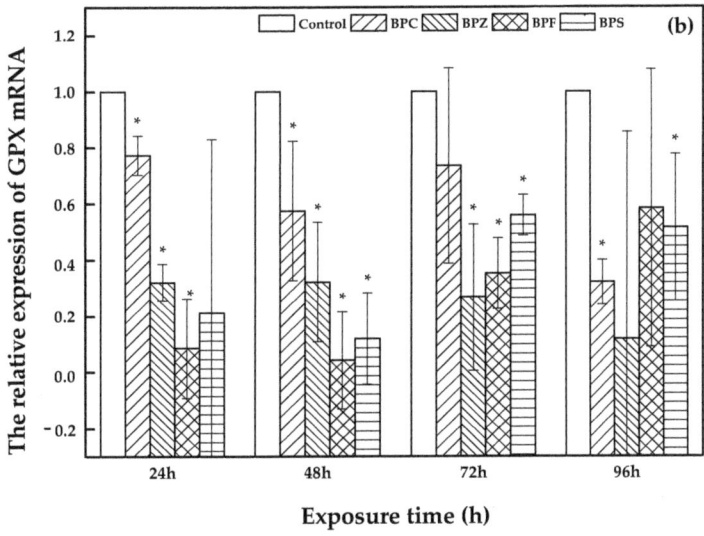

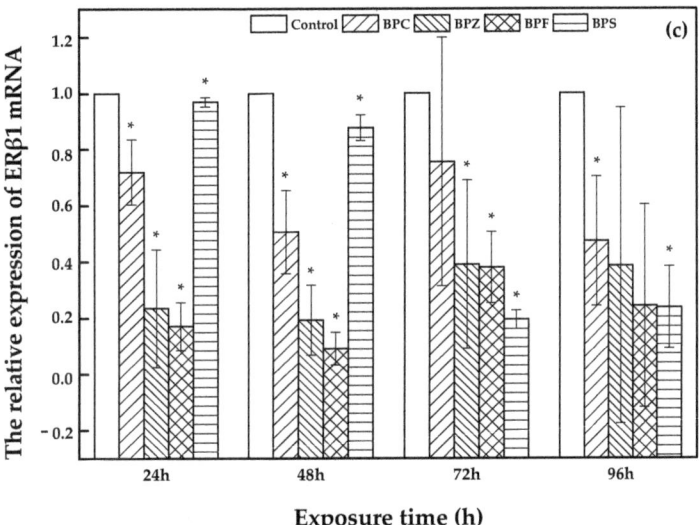

Figure 4. Relative expression of CAT (**a**), GPX (**b**), and ERβ1 (**c**) in treated and control groups (values represent mean ± S.D.; sample size: 3; * significant difference compared to the blank control ($p < 0.05$)).

3. Materials and Methods

3.1. Chemicals and Materials

Zebrafish (type AB) used in this experiment were all purchased from Shanghai Jiayu Aquarium, with an average body length of 2.5 ± 0.5 cm and an average weight of 0.17 g. After body surface disinfection with 5% sodium chloride solution, the zebrafish were domesticated in tap water which had been dechlorinated after 72 h of aeration. The pH of the test water was 7.74–7.83. The hardness of water is 91–108 mg·L^{-1} (based on CaCO$_3$); the concentration of dissolved oxygen was 7.45–7.60 mg·L^{-1}; the temperature was controlled at 25 ± 0.5 °C; and the time distribution was a 14:10 h day night cycle. Zebrafish were domesticated in the laboratory for more than 7 days, during which they were fed with commercial feed once a day, and they could not be used in the following experiments until the mortality rate within 7 days was less than 5%.

BPF (98%) was purchased from Shanghai Maclin Biochemical Technology Co., Ltd., Shanghai, China. BPS (99%), BPC (>98.0%), and BPZ (≥98.0%) were purchased from Shanghai Aldin Reagent Co., Ltd., Shanghai, China. Dimethyl sulfoxide (DMSO) was purchased from Shanghai Lingfeng Chemical Reagent Co., Ltd., Shanghai, China. Trizol Reagent was purchased from Thermo Fisher Technology Co., Ltd., Waltham, MA, USA. TaKaRa AMV Kit was purchased from Takara Biomedical Technology (Beijing) Co., Ltd., Beijing, China. dATP, dTTP, dCTP, and dGTP were purchased from Thermo Fisher Technology Co., Ltd., USA. SYBR® Premix Ex Taq™ II (Perfect Real Time) was purchased from Takara Biomedical Technology (Beijing) Co., Ltd., Beijing, China. DNase I was purchased from New England Biolabs Co., Ltd., Ipswich, MA, USA.

Preparation of mother liquor: mother liquor of 10^3–10^5 mg·L^{-1} was prepared with DMSO (stored in a refrigerator at 4 °C and kept away from light). DMSO was used as solvent (0.01–0.5%) for all gradient dilution operations.

3.2. Experimental Instruments

The durometer was purchased from the Hach Co., Ltd., Loveland, CO, USA. The dissolution oxygen tester was purchased from the Hach Co., Ltd., USA. The electronic analytical balance was purchased from the Shimadsu Co., Ltd., Kyoto, Japan. The ultrasonic curing machine was purchased from the Longjie Ultrasonic Electric Appliance Co., Ltd., Shenzhen, China. The ultra-pure water machine was purchased from the Ultra-Pure Technology Co., Ltd., Sichuan, China. The centrifuge was purchased from Eppendorf Co., Ltd., Hamburg, Germany. The Real-Time PCR System was purchased from Thermo Fisher Technology Co., Ltd., USA. The thermal cycle was purchased from Beijing Dongsheng Innovation Biotechnology Co., Ltd., Beijing, China. The gel electrophoresis and imaging system were purchased from Bio-Rad Laboratories, Inc., Hercules, CA, USA. The Trace UV Nucleic Acid Quantitative System was purchased from Thermo Fisher Technology Co., Ltd., USA.

3.3. Experiment Design

3.3.1. Toxicity Test

Zebrafish with sensitive response, normal appearance, and uniform body shape, aged between two and three months old, were selected for the toxicity test. A total length of 2.0–3.0 cm of the zebrafish was selected to be exposed. Glass beakers were selected as the poisoning test container. The volume of each beaker was 3 L, in which 10 zebrafish were placed. There was a blank control, a solvent control (0.01–0.50% dimethyl sulfoxide, DMSO), and a series of bisphenol analogue treatments. For each concentration, three parallel beakers were set up, and all beakers were placed in a water tank with temperature controlled at (25 ± 0.5) °C on a 14:10 light and dark cycle, and the pH value was 7.7–7.8. The experiment period was 96 h, and the test method was static. During the test, there was no feeding or water changes. The test water was dechlorinated tap water that had been aerated for more than 72 h. The number of dead zebrafish was recorded at 6 h, 24 h, 48 h, 72 h, and 96 h, respectively, and the dead fish in the container were removed to avoid affecting the following experiment.

3.3.2. Single Toxicity Test

Before the formal test, preliminary experiments of BPZ, BPC, BPF, and BPS concentrations in a wide range series (0, 1, 1, 10, 100, and 1000 mg·L^{-1}) were done. First, 3 L aerated tap water was imported to glass beaker together with reservoir fluid, with no parallel groups. 10 zebrafish were put in each experimental container. During the test, water changes or feeding was permitted, the number of dead fish was then observed and recorded. According to the results of the pre-experiment, 5–7 concentration gradients were set with 3 parallel gradients for each concentration, and blank control experiments were carried out at the same time. The details are as follows: BPF: 1, 2, 4, 6, 8, 10, 12, 14 mg·L^{-1}; BPS: 100, 150, 200, 250, 300 mg·L^{-1}; BPC: 1, 2, 3, 4, 5 mg·L^{-1}; BPZ: 1, 2, 3, 4, 5 mg·L^{-1}, and the control group was set as tap water without any drug. The state of the fish was observed at 6 h, 24 h, 48 h, 72 h, and 96 h. After lightly touching the tail of the fish with a glass rod, fish with no reaction were judged to be dead, and the corresponding number of dead fish was recorded. At the end of the experiment, the death rate of the blank control group was less than 10%.

3.3.3. Joint Acute Toxicity Test

Taking the two-element combination as an example, the LC$_{50}$ value of a single compound for 96 h was taken as a toxicity unit, and 6 groups of different experimental concentrations were set using to the 1:1 toxicity ratio with equal logarithmic spacing. The design of the ternary and quaternary combinations was the same. The procedure for the joint acute toxicity test of the combinations was the same as the one conducted for the single compounds. DPS 9.01 software was used to calculate the corresponding LC$_{50}$ value and 95% confidence interval.

3.3.4. Joint Toxicity Assessment Method

At present, there is no unified international standard method for the joint toxicity test of aquatic organisms. In this paper, the additive index method for joint effect of aquatic toxicology by Xiu et al. was used, which was improved from the additive index method of Marking and had been widely used in comprehensive toxicity testing of aquatic toxicology in China [31,32].

Taking the dual combination as an example, after the LC$_{50}$ value of single and combined toxicity was calculated by DPS, S was obtained by Equation (1):

$$S = \frac{A_m}{A_1} + \frac{B_m}{B_1} \tag{1}$$

In Equation (1), A_1 and B_1 are LC$_{50}$ values of the single toxicity of poisons A and B, respectively. A_m and B_m are LC$_{50}$ values of each poison in the mixture, respectively. Converting S to an additive exponent AI, that is, when $S \leq 1$, AI = $(1/S) - 1$; When $S > 1$, AI = $-S + 1$. Finally, AI was used to evaluate the combined effect of the toxicants, when AI > 0, the joint toxicity effect is more obvious than the additive effect, that is, the synergistic effect; when AI < 0, the joint toxicity effect is less than the additive effect, that is, antagonism; when AI = 0, the effect of joint toxicity is the addition.

3.3.5. Gene Expression Analysis

Half lethal concentrations of BPZ, BPC, BPS, and BPF were selected in previous experiments, and samples were collected 24 h, 48 h, 72 h, and 96 h after exposure and placed on ice. After homogenization, samples were snap frozen and stored at −80 °C until they were assayed. To indicate the possible mechanisms of bisphenol analogue exposure to zebrafish, we also evaluated the gene expression of β-actin, rp17, CAT, SOD, GPX, ERα, and ERβ1 in zebrafish using quantitative real-time polymerase chain reaction (PCR). β-actin and rp17 were selected as internal references, and the other five genes were targets. First, blocks of frozen tissue were quickly ground in liquid nitrogen. 1 mL of Trizol reagent was added after grinding. Total RNA was extracted and isolated using the kit, following the manufacturer's instructions, meanwhile, cDNA templates were synthesized by the reverse transcription of RNA. For the quantification of PCR results, the Ct value was determined. The Ct value of rp17 was lower than β-actin and much more easily reproducible. rp17 was stably expressed across all tested samples. Therefore, rp17 was chosen for all subsequent sample correction. Primer sequences of the genes are shown in Table 4. The relative gene expression levels were detected by PCR. PCR reaction mixtures (25 μL) contained 12.5 μL of SYBR® Premix Ex Taq™ II, 8.5 μL of nuclease free water, 1 μL of forward primer, 1 μL of reverse primer, and 2 μL of cDNA template. The thermal cycle profile was 95 °C for 2 min, followed by 40 cycles of denaturation for 10 s at 96 °C and 30 s at 60 °C, The PCR was repeated three times for each RNA to generate biological replicates.

Table 4. Primer sequences of candidate reference genes.

Gene Symbol	Accession NO.	Primer Sequence(5'-3')	Product Length (bp)
β-actin	AF025305.1	F: CGAGCTGTCTTCCCATCCA R: TCACCAACGTAGCTGTCTTTCTG	86
rp17	NM_213213644.2	F: CAGAGGTATCAATGGTGTCAGCCC R: TTCGGAGCATGTTGATGGAGGC	119
CAT	AF170069.1	F: CTCCTGATGTGGCCCGATAC R: TCAGATGCCCGGCCATATTC	126
SOD	BX055516	F: GTCCGCACTTCAACCCTCA R: TCCTCATTGCCACCCTTCC	217
GPX	AW232474	F: AGATGTCATTCCTGCACACG R: AAGGAGAAGCTTCCTCAGCC	94
ERα	AF268283	F: CCC ACA GGA CAA GAG GAA GA R: CCT GGT CAT GCA GAG ACA GA	250
ERβ1	AJ414566	F: GGG GAG AGT TCA ACC ACG GAG R: GCT TTC GGA CAC AGG AGG ACG	89

3.4. Statistical Analysis

DPS 9.01 software was used to analyze the obtained data and the experimental results were expressed as mean ± S.D. The *t*-test method was used to test the statistical difference between the two groups of data. * $p < 0.05$ indicates significant difference. Ct values for each gene of interest were normalized to *rp17* by using the $2^{-\Delta\Delta Ct}$ method [33].

4. Conclusions

Bisphenol analogues enter the aquatic environment through different pathways and coexist in water. Multiple substances that act on organisms together may cause different toxic reactions than those of a single substance. In the present study, zebrafish exposure to BPZ, BPS, BPC, and BPF presented different toxic effects. The toxicity of BPZ, BPC, BPF, and BPS decreased sequentially. During the joint toxicity exposure groups, BPF–BPC and BPZ–BPC showed synergistic effects. However, the other joint and multi-joint toxicity exposure groups all exhibited antagonistic effects. Additionally, the estrogen receptor gene (ERβ1) and antioxidant enzyme gene (CAT and GPX) apparently downregulated, indicating that potential risks may be posed by bisphenol analogues. Thus, more studies should be conducted to evaluate the toxicity mechanisms of bisphenol analogues, which would be useful for assessing the health risks associated with bisphenol analogues in the aquatic environment.

Author Contributions: Y.H. was responsible for project administration, funding acquisition, supervision, visualization, review, and editing; Y.F. was responsible for data curation, formal analysis, visualization, and writing the original draft; M.W. provided help for resources, supervision, visualization, review, and editing; Y.X. provided help with methodology and software; H.C. provided help with the methodology; Y.L. was responsible for the investigation. All authors have read and agreed to the published version of the manuscript.

Funding: This research was funded by National Natural Science Foundation of China (No. 21906009), Jiangsu High-level Innovative and Entrepreneurial Talent Introduction Program Special Fund (No. SCZ2010300004), and The Research Start-up Funding of Changzhou University (No. ZMF19020389).

Institutional Review Board Statement: Not applicable.

Informed Consent Statement: Not applicable.

Data Availability Statement: The authors declare that all data generated or analyzed during this study are included in the published article.

Acknowledgments: The authors acknowledge the zebrafish used for experiments.

Conflicts of Interest: The authors declare no conflict of interest.

Sample Availability: Samples of the compounds are available from the authors.

References

1. Shigeyuki, K.; Tomoharu, S.; Seigo, S.; Ryuki, K.; Norimasa, J.; Kazumi, S.; Shin'Ichi, Y.; Nariaki, F.; Hiromitsu, W.; Shigeru, O. Comparative study of the endocrine-disrupting activity of bisphenol A and 19 related compounds. *Toxicol. Sci. Off. J. Soc. Toxicol.* **2005**, *84*, 249–259.
2. Ruth, R.J.; Louise, B.A. Bisphenol S and F: A Systematic review and comparison of the hormonal activity of bisphenol A substitutes. *Environ. Health Perspect.* **2015**, *123*, 643–650.
3. Becerra, V.; Odermatt, J. Interferences in the direct quantification of bisphenol S in paper by means of thermochemolysis. *J. Chromatogr. A* **2013**, *1275*, 70–77. [CrossRef] [PubMed]
4. Cao, L.Y.; Ren, X.M.; Li, C.H.; Zhang, J.; Qin, W.P.; Yang, Y.; Wan, B.; Guo, L.H. Bisphenol AF and bisphenol B exert higher estrogenic effects than bisphenol A via g protein-coupled estrogen receptor pathway. *Environ. Sci. Technol.* **2017**, *51*, 11423–11430. [CrossRef]
5. Chen, D.; Kannan, K.; Tan, H.; Zheng, Z.G.; Widelka, M. Bisphenol analogues other than BPA: Environmental occurrence, human exposure, and toxicity—A review. *Environ. Sci. Technol.* **2016**, *50*, 5438–5453. [CrossRef]
6. Liao, C.; Liu, F.; Ying, G.; Moon, H.B.; Nakata, H.; Wu, Q.; Kannan, K. Occurrence of eight bisphenol analogues in indoor dust from the United States and several Asian countries: Implications for human exposure. *Environ. Sci. Technol.* **2012**, *46*, 9138–9145. [CrossRef]
7. Yan, Z.; Liu, Y.; Yan, K.; Wu, S.; Han, Z.; Guo, R.; Chen, M.; Yang, Q.; Zhang, S.; Chen, J. Bisphenol analogues in surface water and sediment from the shallow Chinese freshwater lakes: Occurrence, distribution, source apportionment, and ecological and human health risk. *Chemosphere* **2017**, *184*, 318–328. [CrossRef]
8. Chen, Y.; Fang, J.; Ren, L.; Fan, R.; Zhang, J.; Liu, G.; Zhou, L.; Chen, D.; Yu, Y.; Lu, S. Urinary bisphenol analogues and triclosan in children from south China and implications for human exposure. *Environ. Pollut.* **2018**, *238*, 299. [CrossRef]
9. Kha, B.; Smc, D.; Mf, A.; Sd, E.; Lka, E.; Be, A. Genotoxic activity of bisphenol A and its analogues bisphenol S, bisphenol F and bisphenol AF and their mixtures in human hepatocellular carcinoma (HepG2) cells. *Sci. Total Environ.* **2019**, *687*, 267–276.
10. Mu, X.; Huang, Y.; Li, X.; Lei, Y.; Teng, M.; Li, X.; Wang, C.; Li, Y. Developmental effects and estrogenicity of bisphenol A alternatives in a zebrafish embryo model. *Environ. Sci. Technol.* **2018**, *52*, 3222–3231. [CrossRef] [PubMed]
11. Zheng, X.M.; Liu, H.L.; Wei, S.; Wei, S.; Giesy, J.P.; Yu, H.X. Effects of perfluorinated compounds on development of zebrafish embryos. *Environ. Sci. Pollut. Res.* **2011**, *19*, 2498–2505. [CrossRef]
12. Khan, A.; Zaman, T.; Fahad, T.M.; Akther, T.; Bari, M. Perspectives of zebrafish as an animal model in the biomedical science. In Proceedings of the International Conference on Science and Technology for Celebrating the Birth Centenary of Bangabandhu (ICSTB-2021), Dhaka, Bangladesh, 11–13 March 2021.
13. Zhao, F.; Jiang, G.; Wei, P.; Wang, H.; Ru, S. Bisphenol S exposure impairs glucose homeostasis in male zebrafish (*Danio rerio*). *Ecotoxicol. Environ. Saf.* **2017**, *147*, 794–802. [CrossRef] [PubMed]
14. Lee, S.; Kim, C.; Shin, H.; Kho, Y.; Choi, K. Comparison of thyroid hormone disruption potentials by bisphenols A, S, F, and Z in embryo-larval zebrafish. *Chemosphere* **2019**, *221*, 115–123. [CrossRef] [PubMed]
15. Naderi, M.; Wong, M.; Gholami, F. Developmental exposure of zebrafish (*Danio rerio*) to bisphenol-S impairs subsequent reproduction potential and hormonal balance in adults. *Aquat. Toxicol.* **2014**, *148*, 195–203. [CrossRef]
16. Ji, K.; Hong, S.; Kho, Y.; Choi, K. Effects of bisphenol s exposure on endocrine functions and reproduction of zebrafish. *Environ. Sci. Technol.* **2013**, *47*, 8793–8800. [CrossRef] [PubMed]
17. Niederberger, C. Re: A new chapter in the bisphenol a story: Bisphenol S and bisphenol F are not safe alternatives to this compound editorial comment. *J. Urol.* **2015**, *194*, 1368–1370. [CrossRef]
18. Chen, M.Y. Acute toxicity, mutagenicity, and estrogenicity of bisphenol-A and other bisphenols. *Environ. Toxicol.* **2010**, *17*, 80–86. [CrossRef] [PubMed]
19. Busquet, F.; Halder, B.T.; Braunbeck, T.; Gourmelon, L.A.; Kleensang, A.; Belanger, S.; Carr, G.; Walter-rohde, S. *OECD Guidlines for the Testing of Chemicals 236—Fish Embryo Acute Toxicity (FET) Test*; The OECD Observer; Organisation for Economic Co-operation and Development: Paris, France, 2013.
20. Ren, W.J.; Wang, Z.; Yang, X.H.; Liu, J.N.; Yang, Q.; Chen, Y.W.; Shen, S.B. Acute toxicity effect of bisphenol A and its analogues on adult and embryo of zebrafish. *J. Ecol. Rural Environ.* **2017**, *33*, 372–378.
21. Hernández, A.F.; Gil, F.; Lacasaña, M. Toxicological interactions of pesticide mixtures: An update. *Arch. Toxicol.* **2017**, *91*, 3211–3223. [CrossRef] [PubMed]

22. Lei, X.; Sun, Z.H.; Yan, Z.L.; Yuan, C.X. Joint toxicity of bisphenol A and p-Nitrophenol on Misgurnus anguillicadatus. *J. Northwest A F Univ.* **2012**, *40*, 93–1600.
23. Fang, N.; Zhang, S.L.; Guo-Peng, J.U. Joint toxicity effect of cresol isomers on zebrafish (*Brachydanio rerio*). *J. Baoji Univ. Arts Sci. (Nat. Sci. Ed.)* **2017**, *37*, 43–46.
24. Fol, V.L.; Ait-Aissa, S.; Sonavane, M.; Porcher, J.M.; Balaguer, P.; Cravedi, J.P.; Zalko, D.; Brion, F. In vitro and in vivo estrogenic activity of BPA, BPF and BPS in zebrafish-specific assays. *Ecotoxicoogyl. Environ. Saf.* **2017**, *142*, 150–156. [CrossRef] [PubMed]
25. Qiu, W.; Shao, H.; Lei, P.; Zheng, C.; Qiu, C.; Ming, Y.; Yi, Z. Immunotoxicity of bisphenol S and F are similar to that of bisphenol A during zebrafish early development. *Chemosphere* **2017**, *194*, 1–8. [CrossRef] [PubMed]
26. Gu, J.; Wu, J.; Xu, S.; Zhang, L.; Fan, D.; Shi, L.; Wang, J.; Ji, G. Bisphenol F exposure impairs neurodevelopment in zebrafish larvae (*Danio rerio*). *Ecotoxicol. Environ. Saf.* **2020**, *188*, 109870.109871–109870.109877. [CrossRef] [PubMed]
27. Xu, H.; Dong, X.; Zhang, Z.; Yang, M.; Wu, X.; Liu, H.; Lao, Q.; Li, C. Assessment of immunotoxicity of dibutyl phthalate using live zebrafish embryos. *Fish Shellfish Immunol.* **2015**, *45*, 286–292. [CrossRef] [PubMed]
28. Salahinejad, A.; Attaran, A.; Naderi, M.; Meuthen, D.; Niyogi, S.; Chivers, D.P. Chronic exposure to bisphenol S induces oxidative stress, abnormal anxiety, and fear responses in adult zebrafish (*Danio rerio*). *Sci. Total Environ.* **2020**, *750*, 141633. [CrossRef]
29. Pikulkaew, S.; Nadai, A.D.; Belvedere, P.; Colombo, L.; Valle, L.D. Expression analysis of steroid hormone receptor mRNAs during zebrafish embryogenesis. *Gen. Comp. Endocrinol.* **2010**, *165*, 215–220. [CrossRef]
30. Gyimah, E.; Dong, X.; Qiu, W.; Zhang, Z.; Xu, H. Sublethal concentrations of triclosan elicited oxidative stress, DNA damage, and histological alterations in the liver and brain of adult zebrafish. *Environ. Sci. Pollut. Res.* **2020**, *27*, 17329–17338. [CrossRef] [PubMed]
31. Xiu, R.Q.; Xu, Y.X.; Gao, S.R. Joint toxicity test of arsenic with cadmium and zinc ions to zebrafish, brachynanio rerio. *China Environ. Sci.* **1998**, *18*, 349–352.
32. Marking, L.L. Method for assessing additive toxicity of chemical mixtures. In *Aquatic Toxicology and Hazard Evaluation*; ASTM International: West Conshohocken, PA, USA, 1977.
33. Schmittgen, T.D. Analyzing real-time PCR data by the comparative CT method. *Nat. Protoc.* **2008**, *3*, 1101–1108. [CrossRef] [PubMed]

Article

Stable and Effective Online Monitoring and Feedback Control of PCDD/F during Municipal Waste Incineration

Shijian Xiong [1], Fanjie Shang [2], Ken Chen [1], Shengyong Lu [1,*], Shaofu Tang [2], Xiaodong Li [1] and Kefa Cen [1]

[1] State Key Laboratory of Clean Energy Utilization, Zhejiang University, Hangzhou 310027, China; 11827072@zju.edu.cn (S.X.); 21927009@zju.edu.cn (K.C.); lixd@zju.edu.cn (X.L.); kfcen@zju.edu.cn (K.C.)
[2] Zhejiang Fuchunjiang Environmental Technology Research Co., Ltd., Hangzhou 311401, China; shangfanjie@163.com (F.S.); shaofutang263@163.com (S.T.)
* Correspondence: lushy@zju.edu.cn

Abstract: For the long-term operation of municipal solid waste incineration (MSWI), online monitoring and feedback control of polychlorinated dibenzo-*p*-dioxin and dibenzofuran (PCDD/F) can be used to control the emissions to national or regional standards. In this study, 500 PCDD/F samples were determined by thermal desorption gas chromatography coupled to tunable-laser ionization time-of-flight mass spectrometry (TD-GC-TLI-TOFMS) for 168 h. PCDD/F emissions range from 0.01 ng I-TEQ/Nm^3 to 2.37 ng I-TEQ/Nm^3, with 44% of values below 0.1 ng I-TEQ/Nm^3 (the national standard). In addition, the temperature of the furnace outlet, bed pressure, and oxygen content are considered as key operating parameters among the 13 operating parameters comprising four temperature parameters, four pressure parameters, four flow parameters, and oxygen content. More specifically, maintaining the furnace outlet temperature to be higher than 800 °C, or bed pressure higher than 13 kPa, or the oxygen content stably and above 10% are effective methods for reducing PCDD/F emissions. According to the analysis of the Pearson coefficients and maximal information coefficients, there is no significant correlation between operating parameters and PCDD/F I-TEQ. Only when there is a significant change in one of these factors will the PCDD/F emissions also change accordingly. The feedback control of PCDD/F emissions is realized by adjusting the furnace outlet temperature, bed temperature, and bed pressure to control the PCDD/F to be less than 0.1 ng I-TEQ/Nm^3.

Keywords: online monitoring; diagnosis; PCDD/F; incineration; feedback control

1. Introduction

Nowadays, incineration is considered to be the preferred technology to dispose of municipal solid waste (MSW) in China [1]. In the future, all MSW in in Zhejiang province, China, will be disposed of by incineration. However, the toxic pollutants emitted from MSW incineration, especially polychlorinated dibenzo-p-dioxins and dibenzofurans (PCDD/F), are risky to the environment and human health. Furthermore, the measurement of PCDD/F involves sampling, extraction, purification, and analysis with high-resolution gas chromatography/high-resolution mass spectrometry (HRGC/HRMS) [2,3]. Hence, the traditional analysis procedure for PCDD/F from industrial incineration takes at least a week. Additionally, the cost of the traditional measurement method is high, resulting in a measurement cycle of once a year. Nonetheless, the long-term sampling of PCDD/F has been verified as an effective monitoring method for the total PCDD/F emission over a month [4]. Though the standard method and long-term sampling ensure the accuracy of measurements, the time lag hardly meets the public concern and prevents the development of rapid feedback control [5].

Due to the diversity of the 210 PCDD/F congeners and the parts per quadrillion concentrations, online monitoring of PCDD/F concentrations can be realized by measuring the indicators with good correlation and higher concentrations. Based on a previous study [6], the indicators included chlorophenols, chlorobenzene [7], and polycyclic aromatic hydrocarbons (PAH) [8]. In addition, the above indicators have been successfully determined by resonance-enhanced multi-photo ionization coupled to time-of-flight mass spectrometry (REMPI-TOFMS) [9,10], with a good correlation (R-square > 0.7) with PCDD/F emissions. Considering the difficulty of ionization and the correlation with PCDD/F emissions, 1,2,4-trichlorobenzene (1,2,4-TrCBz) is selected as the best indicator with a high correlation coefficient (R-square > 0.9) [11,12] with the international toxic equivalent quantity (I-TEQ). Hence, the online monitoring of PCDD/F emissions was realized by measuring the 1,2,4-TrCBz concentration in flue gas during MSW incineration, through thermal desorption gas chromatography (TD-GC)-REMPI-TOFMS [5,13]. Moreover, an accurate correlation model between 1,2,4-TrCBz and I-TEQ was constructed regardless of the change of waste composition and operating parameters [14]. Though the specific PCDD/F congeners were detected by vacuum ultraviolet (VUV) single-photon ionization (SPI) ion trap (IT) (VUV-SPI-IT)-TOFMS [15] or resonance ionization with multi-mirror photo accumulation (RIMMPA)-TOFMS [16], the measurement cycle was 2–6 h, resulting in limited feedback responsiveness. Therefore, the method using 1,2,4-TrCBz as an indicator is considered to be the most promising way to realize online monitoring of PCDD/F emissions from industrial incineration.

The stable and continuous online monitoring of PCDD/F is crucial for the MSW incinerator rather than the experiments for 2–3 days in previous studies [5,13]. Due to the fluctuation of operation parameters during MSW incineration (MSWI), PCDD/F emissions can increase or decrease at short notice. As for the long-term operation of the MSW incinerator, there is no study reporting a change of PCDD/F emissions in a short time (5–10 min) rather than the average value of 2–4 h. An accurate correlation between operating parameters and PCDD/F emissions is key to the diagnosis and feedback control of PCDD/F but difficult to be constructed, due to being limited by the defects of the standard method. As for the industrial managers and government, the optimal operating parameters for PCDD/F emissions are crucial for controlling the emissions to national or regional standards (0.1 ng I-TEQ/Nm^3 or 0.05 ng I-TEQ/Nm^3). Furthermore, numerous studies constructed the control strategy to reduce PCDD/F emissions by adding inhibitors [17]/active carbon [18], cleaning the accumulated fly ash [19], and reducing the chlorine content of waste [20]. However, the effect of temperature and oxygen content on the formation of PCDD/F was only investigated on a reactor in the laboratory [21,22] or pilot scale sintering plant [23] rather than a full-scale incinerator. For the full-scale MSW incinerator, the cost for sampling PCDD/F is high, and adjusting the operating parameters is difficult; it is both difficult and meaningful to investigate the correlation between operating parameters and PCDD/F emissions.

In the present study, continuous online PCDD/F measurement was carried out at the stack of a full-scale MSW incinerator, lasting 168 h. Firstly, calibration and validation of the equipment were constructed to ensure accuracy and effectiveness. Then, a total of 500 stack gas samples were collected and determined by TD-GC-REMPI-TOFMS. The effect of operating parameters on the PCDD/F formation was investigated, which include temperature, pressure, oxygen content, and mass transfer rate. Moreover, statistical methods were applied to analyze the correlation in depth. Finally, the diagnosis and feedback control of PCDD/F were realized in the MSW incinerator with the same air pollution control devices (APCD). The study is conducive to the development and application of the online monitoring of PCDD/F. In addition, PCDD/F emissions can be controlled to the national standard in time by rapid feedback control.

2. Materials and Methods

2.1. Experimental Procedure

First, the continuous and online measurement of PCDD/F was carried out at the stack of a municipal solid waste incinerator in Zhejiang Province. The capacity of the circulating fluidized bed (CFB) incinerator was 400 tons/day, and the mixing ratio of coal to raw refuse was 2–8. The air pollution control devices (APCD) included selective non-catalytic reduction (SNCR), semi-dry desulfurization (SDD), active carbon injection (ACI), and bag filter (BF). The schematic of the CFB and the sampling details are described in Figure 1. In order to ensure the continuous and stable operation of TD-GC-TLI-TOFMS, a special room was built to maintain constant temperature and humidity (temperature of 23 °C, relative humidity of 40%) and a certain degree of cleanliness (100,000 level) so as to avoid the impact of the external environment. Then, the diagnosis and feedback control experiments were carried out in another CFB, with 1200 tons/day capacity, which was fed by the refuse-derived fuel (RDF). The APCD of the newly built CFB was the same as the previous incinerator.

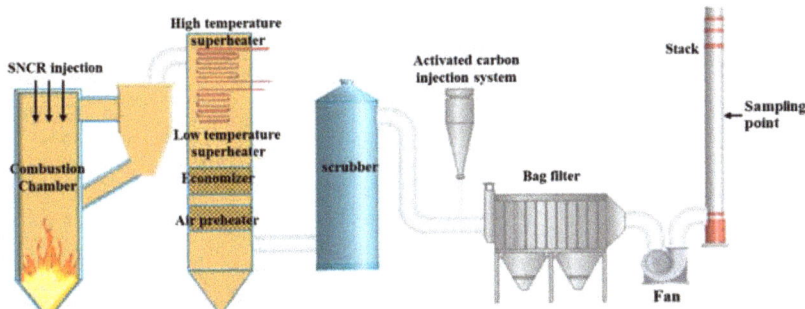

Figure 1. Schematic diagram of the MSW incinerator: sampling point at the stack.

As for the stable and continuous monitoring of PCDD/F emissions, the international toxicity equivalent (I-TEQ) was selected to describe PCDD/F emissions at the stack. The experiments started at 9:00 a.m. on 19 March and ended at 9:00 a.m. on 26 March, for a total of 168 h. During this period, the TD-GC-TLI-TOFMS system performed a backflush cleaning procedure for approximately 1 h every 8 h, and ran continuously the rest of the time. The sampling time for each flue gas sample was 15 min, and the sampling flow rate was 10 mL/min. As for the diagnosis and feedback control of PCDD/F, the sampling time was 15 min. The operating parameters of the incinerator were directly collected from the Distributed Control System (DCS) and Continuous Emission Monitoring System (CEMS). The operating parameters included the temperature, pressure, flow, oxygen content, and feeding conditions.

2.2. Analysis Method

The flue gas was collected into the purification module at a flow rate of 3–10 L/min by the sampling pump. After the heating filter and Nafion drying tube, most of the dust particles and moisture were removed. Then, it flowed to the multipored valve group, discharging to the atmosphere. In this period, the Nafion tube was heated to about 100 °C, and all other sampling pipelines and heating filters were heated to about 200 °C to avoid condensation of 1,2,4-TrCBz or adsorption on the tube wall.

An accurate 1,2,4-TrCBz concentration was detected by the homemade instrument named tunable-laser ionization (TLI)-TOFMS, based REMPI. The instrument consists of a pulsed Nd: YAG optical parametric oscillator (OPO) laser with frequency doubling module (RADIANT, OPOTEK) and time-of-flight mass spectrometer (CTF10, STEFAN KAESDORF). The primary component of TD-GC-TLI-TOFMS is illustrated in Figure S1 (Supplementary Materials). First, the stack gas was sampled via a heated 1/4" stainless steel tube (T = 180 °C) by a sampling pump. Second, after particles were removed by the sampling probe, a fraction (50 mL/min) was extracted through a heated 1/4" stainless steel tube (T = 180 °C) from the mainstream. Third, the stack gas was absorbed by the TD device, with the operating condition that the absorbing and desorbing temperature of the trap was 30–300 °C. Fourth, the desorbed gas went to GC (Trace1300, Thermo Fisher, Waltham,

MA, USA) through a stainless capillary tube at 200 °C. Fifth, the gas separated after GC was extracted into the TOFMS through a pulsed microfluidic valve, with a supersonic molecular beam. Sixth, the ionization and detection of 1,2,4-TrCBz were carried out by a (1 + 1′) TLI process (284 and 213 nm). Then, the ions were separated for the purpose of the mass-to-charge ratio during the process to the microchannel plate (MCP) detector; the digitized MCP signals and their corresponding times were acquired by the data acquisition. Finally, the detection signal was transferred to a computer, and then the concentration values of the 1,2,4-TrCBz and the corresponding PCDD/F I-TEQ were calculated through processing and analysis with LabVIEW 2017. PCDD/F I-TEQ is calculated by the equation "y = 0.2752x + 0.0398", where y represents the PCDD/F I-TEQ and x represents the concentration of 1,2,4-TrCBz.

2.3. Quality Control and Quality Assurance

The TD-GC-TLI-TOFMS system was calibrated daily by the 1,2,4-TrCBz calibration standard, with the concentration of 1,2,4-TrCBz ranging from 2.5 parts per billion by volume (ppbv) to 10 ppbv. The Relative Standard Deviation (RSD) of thermal desorption introduction was less than 2%, and the RSD of repeat measurements of the system was less than 5%. Moreover, the detection limit of 1,2,4-TrCBz in the stack gas was 4 parts per trillion by volume, on the condition that the sample volume was 1000 mL at a signal-to-noise ratio (S/N) of 3 [13,24]. To ensure the accuracy of the measurements, TD-GC-TLI-TOFMS was calibrated before, during, and after the experiment.

Furthermore, the validation between the prediction and the standard measurement was carried out before the continuous measurement. The standard measurement for PCDD/F of the flue gas was constructed by the EPA 23a, considered to be "offline measurement" due to a long lag time (more than 1 week). For the measurement of PCDD/F by the traditional method (EPA 23), the flue gas was collected utilizing an isokinetic sampler (Model KNJ23, KNJ, Korea). The sampling, pretreatment, and analysis followed the EPA method 1613. The details of the analysis method have been described in previous studies [3,25]. The relative error between online measurement and offline measurement was kept within 30% (Figure 2). Therefore, the accuracy of online detection and prediction of PCDD/F was reliable.

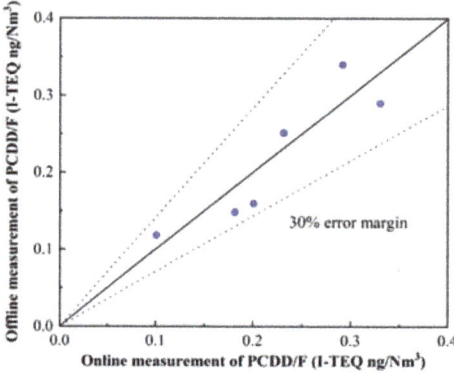

Figure 2. Comparison between online measurement and offline measurement of PCDD/F.

2.4. Statistical Analysis

The Pearson coefficient is universally used to measure linear correlation using the following equation:

$$r = \frac{\sum_{i=1}^{n}(X_i - \overline{X})(Y_i - \overline{Y})}{\sqrt{\sum_{i=1}^{n}(X_i - \overline{X})^2}\sqrt{\sum_{i=1}^{n}(Y_i - \overline{Y})^2}} \quad (1)$$

The maximal information coefficient [26] is used to measure the dependence between two discrete variables whether linear or not, with fairness and symmetry.

$$I(x,y) = \int p(x,y) \log_2 \frac{p(x,y)}{p(x)p(y)} dxdy \approx \sum_{X,Y} p(X,Y) \log_2 \frac{p(X,Y)}{p(X)p(Y)} s \quad (2)$$

$$MIC(X,Y) = \max_{|X||Y|<B} \frac{I(X,Y)}{\log_2(\min(|X|,|Y|))} \quad (3)$$

where $I(X,Y)$ is the mutual information for two random variables X and Y. B is a parameter to limit the interval grid number, approximately 0.6 times the amount of data. The Pearson coefficient and maximal information coefficient between PCDD/F emissions and operating parameters were calculated in MATLAB 2020b. Furthermore, to investigate the relationship between operating parameters and PCDD/F I-TEQ, principal component analysis (PCA) was adopted.

3. Results and Discussion

3.1. PCDD/F Concentration

Figure 3 shows all the results for PCDD/F I-TEQ during the continuous 168 h experiment. A total of 500 samples of stack gas were collected and measured throughout the procedure of the experiment. The average value of the calculated PCDD/F I-TEQ was 0.272 ng I-TEQ/Nm^3. Among them, the calculated PCDD/F I-TEQ of 226 samples was less than 0.1 ng I-TEQ/Nm^3 (the legislation limit of the national standard), accounting for approximately 44%. A drastic fluctuation of PCDD/F emissions could be observed. The results indicate that the concern regarding PCDD/F emissions needs to be transferred to long-term online monitoring rather than the measurement for a certain period.

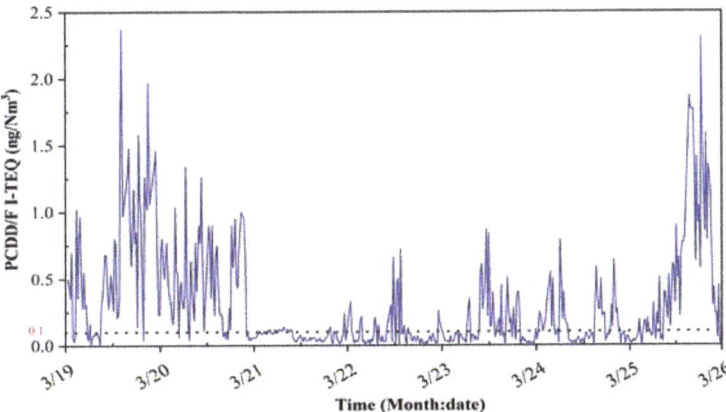

Figure 3. PCDD/F I-TEQ during the 168 h experiment.

The feedstock suspension happened frequently, and the intermission varied from minutes to hours. Remarkably, the longest suspension occurred from 07:30 a.m. on 21 March to 0:05 a.m. on 22 March, resulting in drastically changing operating parameters. Operating conditions of the incinerator such as the temperature of the furnace outlet, the oxygen content of the furnace outlet, and bed temperature were recorded every 10 min. The statistics of operating parameters is shown in Table 1. The average temperature of the furnace outlet was 835.6 °C, ranging from 686.9 °C to 967.1 °C. The average bed temperature was 780.7 °C, ranging from 261.1 °C to 908.4 °C. The trend of the furnace outlet temperature is similar to bed temperature. The average oxygen content of the furnace outlet was 8.82%, ranging from 0 to 20.02%. The results indicate the fluctuating operating conditions and the unstable combustion during the experiment. High levels of PCDD/F can be formed during experiments due to the failure of continuous and stable waste feeding.

Table 1. Statistics of operating parameters and PCDD/F concentrations.

Statistics	Min	Max	Mean	Standard Deviation
Bed temperature (°C)	724.56	921.42	835.59	29.02
Bed pressure (kPa)	−0.04	14.65	10.01	1.44
Flow of primary air (m^3/h)	4186.58	82,527.50	58,456.06	14,329.15
Pressure of primary air (kPa)	0.00	15.40	11.31	1.37
Flow of secondary air (m^3/h)	5316.99	43,428.07	35,527.49	5364.39
Pressure of secondary air (kPa)	−0.20	2.42	1.74	0.46
Pressure of return air (kPa)	0.00	19.21	17.86	1.39
Oxygen content (%)	0.32	19.75	8.82	2.97
Temperature of furnace outlet (°C)	274.21	894.21	780.59	66.89
Temperature of flue gas (°C)	158.44	270.78	193.05	15.20
Flow of feed water (m^3/h)	0.00	84.53	56.64	9.86
Flow of steam (m^3/h)	15.90	70.72	55.21	8.13
Temperature of steam (°C)	375.73	490.58	460.77	9.91
Pressure of steam (kPa)	−0.99	5.36	5.01	0.40
PCDD/F (ng I-TEQ/Nm^3)	0.01	2.37	0.30	0.39

Figure 4 shows the comparison between the hypothetical results to simulate offline measurement and the results by online measurement. The hypothetical values were obtained by calculating the average within 6 consecutive hours, which were measured online by TD-GC-TLI-TOFMS. The reason for this is that, in general, three hypothetical parallel samples were collected continuously within 6 h in the offline measurement. For example, the hypothetical value in Figure 4 is the average of the 15 values measured by the online method. Deduced from Figure 4, offline measurement cannot describe the change of PCDD/F emissions with the change of operating parameters. The hypothetical value for the high level of PCDD/F is 1.02 ng I-TEQ/Nm^3. On the other hand, the hypothetical value for the low level of PCDD/F is only 0.07 ng I-TEQ/Nm^3. It is obvious that the emissions measured by the offline method can only represent the average of PCDD/F emissions during the sampling period. They cannot reflect the dramatic changes of PCDD/F emissions from MSWI. Furthermore, the online measurement of PCDD/F provides the PCDD/F emission value in 15 min, which truly reflects the change of dioxin emissions over time. Therefore, it is the basis for the realization of feedback control for PCDD/F emissions.

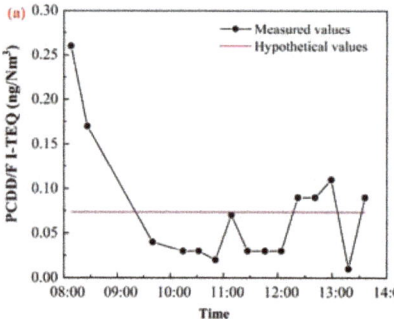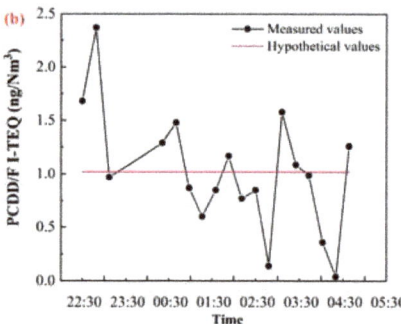

Figure 4. Comparison of hypothetical values of offline measurement and online measurement; hypothetical values represent the average values in continuous 6 h by TD-GC-REMPI-TOFMS. (**a**) low emissions under stable operation (**b**) high emissions under fluctuant operation.

3.2. Effect of Operating Parameters

Oxygen content, temperature, and pressure have been reported as the key parameters that influence PCDD/F formation [27]. A previous study reported the effect of oxygen on the formation of PCDD/F, including three aspects [28]: (1) promoting the formation of C-Cl bonds; (2) promoting the carbon-oxygen complexes by reducing the activation energy; (3) providing highly active chlorine in the Deacon reaction. In addition, the higher oxygen level was conducive to PCDD/F formation in laboratory-scale and full-scale incinerators [29,30]. As for temperature and pressure, the reaction rate and activation energy are decided by the two parameters if the reaction is fixed based on the Arrhenius equation. The effects of the above parameters on the formation of PCDD/F for the full-scale incinerator are discussed below.

3.2.1. Effect of Temperature

As shown in Figure 5, the fluctuating ranges of the furnace outlet temperature (647 °C) and bed temperature (280 °C) are larger than those of the flue gas temperature (58 °C) and the steam temperature (130 °C). The comparison of the statistics for the temperature of the furnace outlet, flue gas, bed, and steam is illustrated in Figure 6. The distributions for the temperature of the flue gas, bed, and steam are Gaussian, excluding the temperature of the furnace outlet. All outliers of the furnace outlet temperatures are less than 75% of the mean. The result indicates that part of the minimum values of the furnace outlet temperatures below 75% of the mean are abnormal conditions, resulting in high levels of pollutants. Corresponding to PCDD/F emissions, the trends of furnace outlet temperature and bed temperature are opposite to the trend of PCDD/F emissions on 20 March. As for flue gas temperature and steam temperature, there is almost no correspondence with PCDD/F emissions. These results indicate that furnace outlet temperature and bed temperature can be considered to be the key operating parameters controlling PCDD/F emissions.

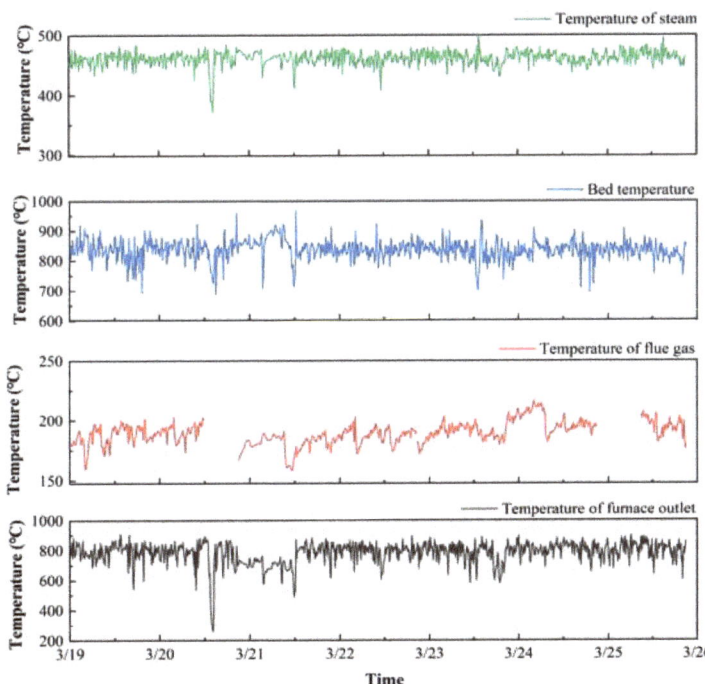

Figure 5. The trend for the temperatures of furnace outlet, bed, steam, and flue gas.

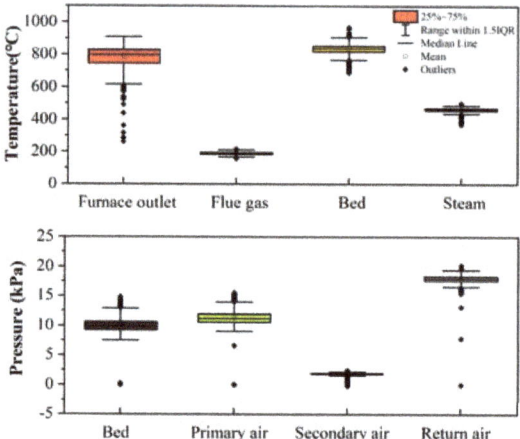

Figure 6. Comparison of the statistics for the temperature of furnace outlet, flue gas, bed, and steam, the pressure of bed, primary air, secondary air, and return air.

For specific daily analysis (Table S1, Supplementary Materials), it can be deduced that the PCDD/F emissions decrease under the condition that the temperature of furnace outlet is maintained to be below 800 °C during the following times: 19 March (10:50–11:40, 12:10–12:20, 20:00–20:06); 20 March (22:30–22:50); 22 March (9:50–12:00, 13:30–15:30, 16:30–16:50); 23 March (22:00–22:30); 25 March (9:00–11:00, 13:10–13:50). On the other hand, PCDD/F emissions increase under the condition that the temperature of the furnace outlet is less than 700 °C. Therefore, maintaining the temperature of the furnace outlet to be higher than 800 °C is an effective method to reduce PCDD/F emissions.

3.2.2. Effect of Pressure

As shown in Figure 7, the trend of bed pressure is consistent with that of primary air pressure. The pressure of the secondary air was kept stably with a mean of 2 kPa. The pressure of the return air was generally kept stably with a mean of 18 kPa, excluding two outliers (0.5 kPa). These results indicate that bed pressure is determined by the primary air pressure. The comparison of the statistics for the pressure of the bed, primary air, secondary air, and return air is illustrated in Figure 6. The distribution of bed pressure is the same as primary air pressure, with vast outliers larger than 125% of the mean. The distributions of the secondary air pressure and the return air pressure are both the left distribution. Theoretically, bed pressure can reflect the flow rate of air through the waste in the furnace. Hence, it is better to consider bed pressure to be the key operating parameter rather than primary air pressure.

Apart from the analysis of key operating parameters, the effect of bed pressure on the PCDD/F emissions is discussed below. Corresponding to the PCDD/F emissions on 21 March, the PCDD/F emissions were less than 0.1 ng I-TEQ/Nm^3 when the bed pressure was larger than 13 kPa. This result indicates that a high bed pressure with a high mass transfer rate between air and waste can reduce PCDD/F emissions. However, the PCDD/F emissions were larger than 1 ng I-TEQ/Nm^3 under the condition that the bed pressure was occasionally above the mean. This result indicates that the bed pressure is a more important parameter than the decisive parameter.

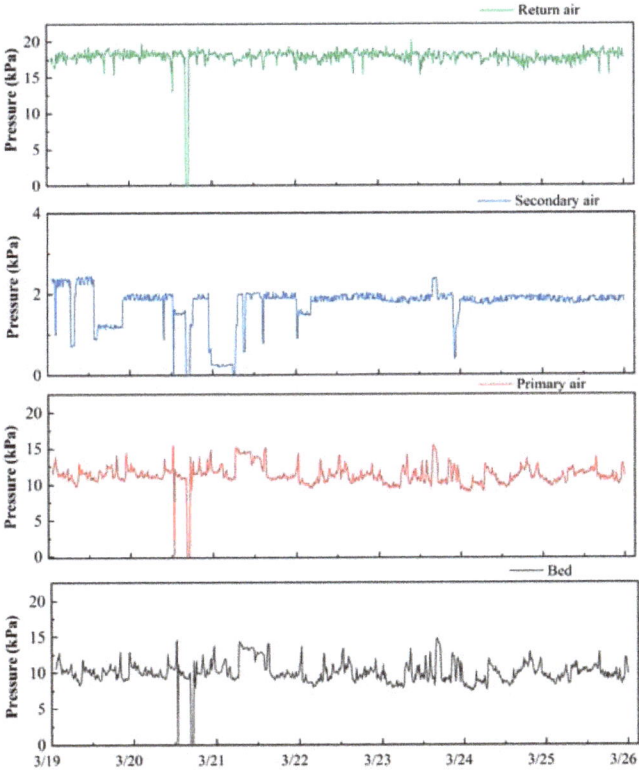

Figure 7. The trend for the pressure of the bed, primary air, secondary air, and return air.

3.2.3. Effect of Oxygen Content and Mass Transfer Rate

Figure 8 shows the trend for oxygen content, the flow of primary air and secondary air, feed water, and steam. Due to the long-term failure of the screw feeder, it is reasonable for the flow rates of the primary air and the feed water to decrease to zero. As shown in Figures 3 and 8, a significant delay can be observed for the decrease of PCDD/F when the flow of the secondary air decreases. Additionally, the change of the flow of primary air lags behind the change of the flow of secondary air. The comparison of the statistics for the oxygen content, flow of primary air and secondary air, feed water, and steam are illustrated in Figure 9. The distribution of the flow of primary air is the same as that of secondary air. Additionally, the primary air flow is directly correlated with the mass transfer rate between waste and air. These results indicate that the flow of primary air is a better operating parameter for controlling PCDD/F emissions than the flow of secondary air. Regarding the PCDD/F emissions on 21 March, the oxygen content was maintained stably and above 10%, resulting in low PCDD/F emissions ($\leq$0.1 ng I-TEQ/Nm^3). However, at other times, the oxygen content fluctuated greatly from 0% to 20%, which corresponds to the great fluctuation of PCDD/F emissions and the flow of feed water and steam.

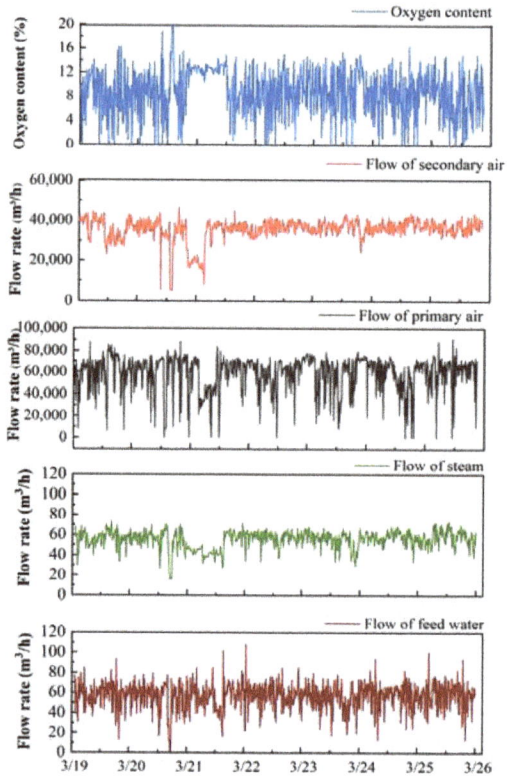

Figure 8. The trend for oxygen content, flow of primary air and secondary air, feed water, and steam.

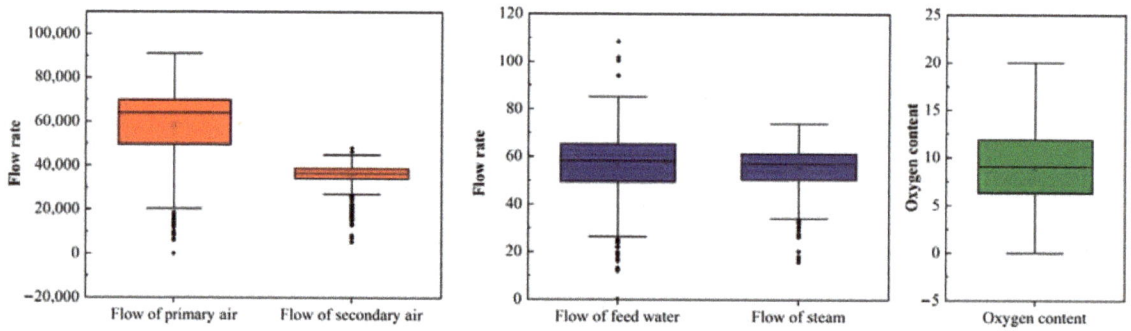

Figure 9. Comparison of the statistics for oxygen content, the flow of primary air and secondary air, feed water, and steam.

For specific daily analysis (Table S1, Supplementary Materials), it can be found that the PCDD/F emissions increase in the condition that the oxygen content fluctuate sharply or below 5% during the following times: 20 March (22:10–22:30); 21 March (3:30–6:30); 22 March (12:00–13:00, 15:30–16:20, 19:50–20:30); 23 March (21:30–22:00); 24 March (12:30–15:00); and 25 March (15:20–15:30, 16:10–17:00, 18:00–21:00). This is because low oxygen results in a high level of incomplete combustion soot, thereby promoting PCDD/F formation in de novo synthesis. On the other hand, PCDD/F emissions decreased to 0.1 I-TEQ/Nm^3 under the condition that the oxygen content was stable and above 8%. Since high oxygen content promotes the formation of PCDD/F in the laboratory-scale incinerator, the influence of incomplete combustion products is larger than that of the oxygen content.

3.3. Correlation Analysis between PCDD/F Emissions and Operational Parameters

The correlation between PCDD/F emissions and operational parameters is analyzed by the Pearson coefficient and the maximal information coefficient in Table 2. There are no significant correlated variables with PCDD/F I-TEQ. All Pearson coefficients are less than 0.2. This result indicates that operational parameters have a weak linear correlation with PCDD/F emissions. As for the correlations analyzed by the maximal information coefficient, all variables have higher coefficients with PCDD/F emissions than the correlation analyzed by the Pearson coefficient. Moreover, the maximal information coefficients increase from 0.011 to 0.203, from −0.003 to 0.231 for bed pressure and pressure of secondary air, respectively. This result indicates that the relationship between operational parameters and PCDD/F emissions is more possible with a nonlinear relationship rather than a linear one. Correlation analysis indicates that there is no significant strong correlation (linear or nonlinear) between operating parameters and PCDD/F emissions. This is mainly because the formation of PCDD/F is directly related to many factors, such as temperature, oxygen content, pressure, and mass transfer rate. Only when there is a significant change in one of these factors will the PCDD/F emissions also change accordingly.

Table 2. Correlation between PCDD/F emissions and operational parameters.

Variables	Bed Temperature	Bed Pressure	Flow of Primary Air	Pressure of Primary Air	Flow of Secondary Air	Pressure of Secondary Air	Pressure of Return Air
Pearson Coefficient	−0.143	0.011	0.060	0.037	0.014	−0.003	0.100
Maximal Information Coefficient	0.166	0.203	0.160	0.203	0.184	0.231	0.127
Variables	Oxygen Content	Temperature of Furnace Outlet	Temperature of Flue Gas	Flow of Feed Water	Flow of Steam	Temperature of Steam	Pressure of Steam
Pearson Coefficient	−0.176	0.102	0.078	0.037	0.135	0.054	−0.054
Maximal Information Coefficient	0.183	0.182	0.231	0.158	0.181	0.171	0.134

The relationship between 14 operating parameters and PCDD/F I-TEQ was investigated by PCA (Figure 10). The total contribution of the first three factors is only 60.36%, and the contribution of Factor 1 is 29.78%. This result is consistent with the 80% loading value of the first five factors in a previous study [31]. The results show that the first three factors do not represent most of the information of the sample, and the self-correlation between variables is small. In addition, the flow of steam has a high correlation with the temperature of the furnace outlet. A similar correlation between the pressure of secondary air and the flow of secondary air can be observed. Furthermore, the bed temperature, the furnace outlet temperature, steam pressure, and oxygen content were located with PCDD/F I-TEQ, indicating the close relationship. Moreover, the flow of secondary air, the flow of primary air, bed pressure, and pressure of secondary air were located far from PCDD/F I-TEQ. This outcome indicates the weak relationship between the above parameters and PCDD/F emissions.

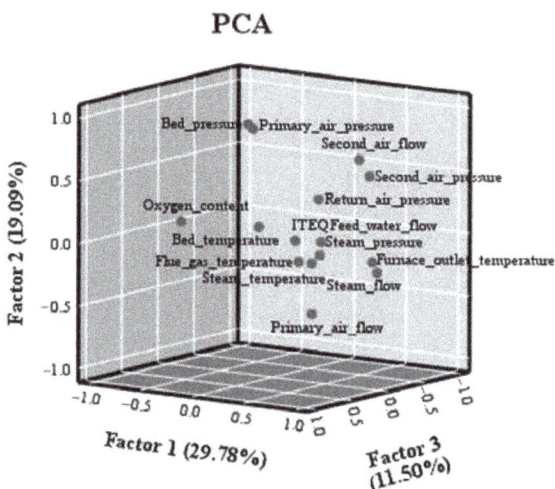

Figure 10. PCA analysis for 14 operating parameters and PCDD/F I-TEQ.

To investigate the correlation between the temperature and PCDD/F emissions, the plummet of the operating parameters was introduced to describe the fluctuation. The plummet is defined as the temperature is 10% being below the mean in a short time. Generally, the plummet of the temperature of a furnace outlet is caused by the insufficiency of air or waste, resulting in incomplete combustion and a large number of organic pollutants. There are 43 minimum points for furnace outlet temperature illustrated in Figure 3. The plummet in the temperature of the furnace outlet signals the poor operation of the incinerator. Among the 43 minimums of the furnace outlet temperature, 29 values (67.4%) directly correspond to the maximum PCDD/F I-TEQ in Figure 3. This result indicates that the plummet is positively correlated to the increase in PCDD/F emissions. However, the other 14 plummets of the furnace outlet temperature do not match the maximum PCDD/F I-TEQ. This abnormal outcome may be the effect of APCD, which reduces PCDD/F emissions. In addition, the total count of the plummet of bed temperature is 13. Among them, 6 plummets (46%) correspond to the maximum of PCDD/F emissions. The other plummets do not correspond to the change in PCDD/F emissions. Therefore, compared with bed temperature, the furnace outlet temperature can better reflect the change of the incinerator operation.

3.4. Diagnosis and Feedback Control of PCDD/F

As shown in Figure 11, PCDD/F I-TEQ increases from 0.008 ng I-TEQ/Nm^3 to 0.14 ng I-TEQ/Nm^3 under the condition that the temperature of the furnace outlet decreases from 108 °C to 870 °C within 10 min. This result indicates that the plummet of the furnace outlet temperature can promote the formation of PCDD/F. The plummet of the furnace outlet temperature can be caused by the suspension of MSW feedstock, resulting in the incomplete combustion of the carbon matrix. A previous study reported that abundant PCDD/F was formed from the soot or carbon matrix [20], especially on the effect of metal catalysts ($CuCl_2$, Fe_2O_3, etc.). Therefore, it is reasonable for the PCDD/F emissions to increase. Remarkably, the delay of the increase in PCDD/F emissions can be observed by comparing the increase of PCDD/F emissions and the decrease of the temperature. More specifically, when the temperature of the furnace outlet dropped to 870 °C, the PCDD/F emission simultaneously rose to 0.08 ng I-TEQ/Nm^3 rather than the highest emission (0.14 ng I-TEQ/Nm^3), which occurred later. This result indicates that the formation of PCDD/F is affected by the reaction time. Furthermore, the design of feedback control should consider the delay effect.

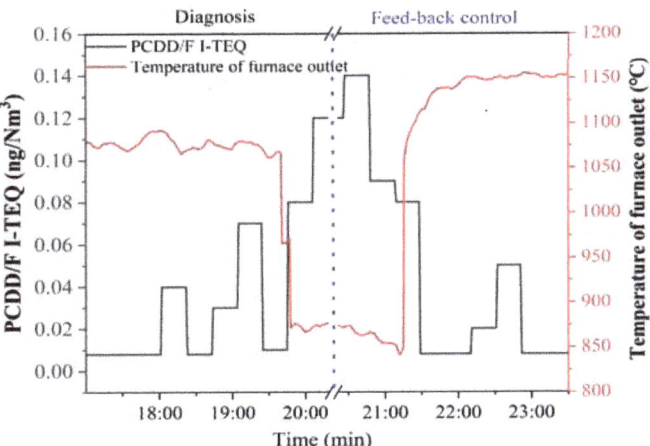

Figure 11. Diagnosis and feedback control of PCDD/F based on the furnace outlet temperature.

As shown in Figure 12, PCDD/F emissions increase from 0.06 ng I-TEQ/Nm^3 to 0.14 ng I-TEQ/Nm^3 under the condition that the bed temperature decreases from 800 °C to 774 °C within 4 h. This result indicates that an increase in bed temperature over 800 °C can reduce the formation of PCDD/F. As shown in Figure 13, PCDD/F I-TEQ increases from 0.02 ng I-TEQ/Nm^3 to 0.12 ng I-TEQ/Nm^3 under the condition that the bed pressure decreases from 70 kPa to 60 kPa within 10 min. This result indicates that the plummet of bed pressure can promote the formation of PCDD/F. As for CFB, the low bed pressure results in a low air concentration gradient. Hence, the diffusion coefficient for the air to the waste is low, according to the Fick Law. The increase in PCDD/F emissions is reasonable for the decrease in the bed pressure.

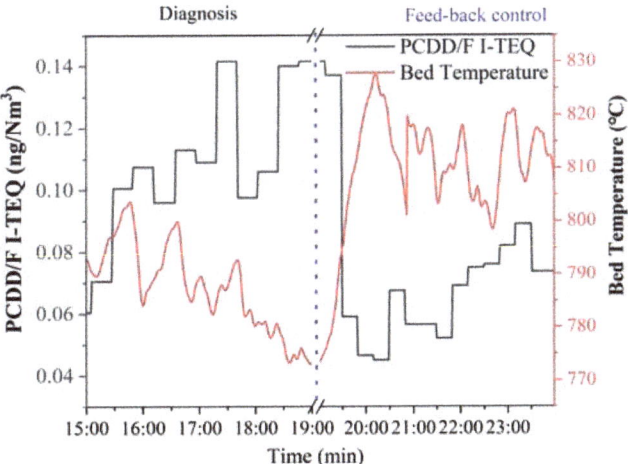

Figure 12. Diagnosis and feedback control of PCDD/F based on bed temperature.

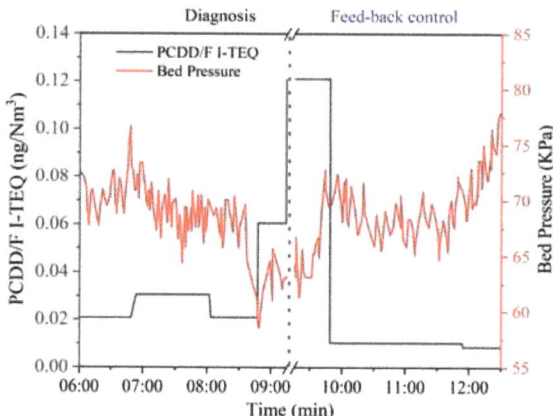

Figure 13. Diagnosis and feedback control of PCDD/F based on bed pressure.

Apart from the diagnosis of operational parameters, the feedback control of PCDD/F was constructed based on correlation analysis and equipment for the online monitoring of PCDD/F. As shown in Figure 11, the PCDD/F emissions decreased from 0.14 ng I-TEQ/Nm^3 to 0.008 ng I-TEQ/Nm^3 when the temperature of the furnace outlet returned to 1152 °C. Therefore, the equipment for the online monitoring of PCDD/F emissions can realize the feedback control of PCDD/F emissions. As a result, the reduction of PCDD/F emissions can be achieved. Thus, excessive emissions over the national standard can be prevented by adjusting the operating parameters. Moreover, due to the bed temperature outlet increasing from 774 °C to 826 °C within 1 h, PCDD/F emissions descended steeply from 0.14 ng I-TEQ/Nm^3 to 0.05 ng I-TEQ/Nm^3. When the bed pressure increased from 60 kPa to 77.5 kPa, PCDD/F emissions decreased from 0.12 ng I-TEQ/Nm^3 to 0.008 ng I-TEQ/Nm^3. These results indicate that adjusting the temperature and pressure parameters to the levels corresponding to PCDD/F emissions lower than 0.1 ng I-TEQ/Nm^3 can be an effective method for controlling PCDD/F emissions under the national standard.

The above results show that the diagnosis and feedback control of PCDD/F emissions can be realized by the TD-GC-REMPI-TOFMS. As for industrial managers, the real-time PCDD/F emissions can be obtained through the online monitoring dioxin system. Thus, it is feasible to adjust operating parameters to reduce PCDD/F emissions in time. The optimization and guidance of incineration operating conditions have enabled the newly built CFB to meet the national standard of PCDD/F emissions.

4. Conclusions

To realize stable and effective online monitoring and feedback control of PCDD/F, the continuous 168 h measurement was carried out at the stack of a full-scale MSW incinerator. Then, the diagnosis and feedback control of PCDD/F was constructed on a newly built MSW incinerator. A drastic fluctuation of PCDD/F emissions could be observed during the long-term operation of MSWI, with 44% of PCDD/F I-TEQ below 0.1 ng/Nm^3. Among the operating parameters, the furnace outlet temperature, bed pressure, and oxygen content are key operating parameters for controlling PCDD/F emissions. Maintaining the furnace outlet temperature to be higher than 800 °C, or the bed pressure higher than 13 kPa, or the oxygen content stably and above 10% are effective methods for reducing PCDD/F emissions. According to the analysis of Pearson coefficients and maximal information coefficients, there is no significant correlation between operating parameters and PCDD/F I-TEQ. Only when there is a significant change in one of these factors will the PCDD/F emissions also change accordingly. The feedback control of PCDD/F emissions was realized by adjusting the furnace outlet temperature, bed temperature, and bed pressure to control the PCDD/F to be less than 0.1 ng I-TEQ/Nm^3. This study can provide effective and quantized guidance for controlling PCDD/F emissions under the national standard during the operation of MSWI. In the future, the control models between operating parameters and PCDD/F emissions will be established to achieve real-time automatic control of PCDD/F emissions to the national standard.

Supplementary Materials: The following are available online, Figure S1: The analysis equipment for 1,2,4-TrCBz (TD-GC-TLI-TOFMS), Table S1: Original data of Oxygen content and Temperature of furnace outlet, and I-TEQ from 19 March to 25 March.

Author Contributions: S.X. was responsible for data curation, formal analysis, visualization, and writing the original draft; F.S. was responsible for resources and software; K.C. (Ken Chen) provided help with methodology and software; S.L. provided help for resources, supervision, visualization, review, and editing; S.T. provided help with the methodology; X.L. was responsible for the investigation and supervision. K.C. (Kefa Cen) was responsible for the investigation. All authors have read and agreed to the published version of the manuscript.

Funding: This research is supported by the National Key Research and Development Program of China (2020YFE0107600).

Data Availability Statement: The authors declare that all data generated or analyzed during this study are included in the published article.

Acknowledgments: The authors appreciate the technical advice given by Xuan Cao. In addition, we thank the plant personnel for their contributions.

Conflicts of Interest: The authors declare no conflict of interest.

Sample Availability: Samples of the compounds are available from the authors.

References

1. Lu, J.-W.; Zhang, S.; Hai, J.; Lei, M. Status and perspectives of municipal solid waste incineration in China: A comparison with developed regions. *Waste Manag.* **2017**, *69*, 170–186. [CrossRef] [PubMed]
2. Ministry of Environmental Protection of the People's Republic of China. *Ambient Air and Waste Gas Determination of Polychlorinated Dibenzo-p-dioxins(PCDDs) and Polychlorinated Dibenzofurans (PCDFs) Isotope Dilution HRGC/HRMS*; HJ 77.2-2008; Environmental Science Press: Beijing, China, 2008.
3. EPA. U. S. Method 23: Determination of Polychlorinated Dibenzop-Dioxins and Polychlorinated Dibenzofurans from Stationary Sources 2017. Available online: https://www.epa.gov/emc/method-23-dioxins-and-furans (accessed on 16 May 2021).
4. Mayer, J.; Linnemann, H.; Becker, E.; Rentschler, W.; Jockel, W.; Wilbring, P.; Gerchel, B. Certification of a long-term sampling system for PCDFs and PCDDs in the flue gas from industrial facilities. *Chemosphere* **2000**, *40*, 1025–1027. [CrossRef]
5. Gullett, B.K.; Oudejans, L.; Tabor, D.; Touati, A.; Ryan, S. Near-real-time combustion monitoring for PCDD/PCDF indicators by GC-REMPI-TOFMS. *Environ. Sci. Technol.* **2012**, *46*, 923–928. [CrossRef]
6. Lavric, E.D.; Konnov, A.A.; Ruyck, J.D. Surrogate compounds for dioxins in incineration. A review. *Waste Manag.* **2005**, *25*, 755–765. [CrossRef] [PubMed]
7. Kaune, A.; Lenoir, D.; Nikolai, U.; Kettrup, A. Estimating concentrations of polychlorinated dibenzo-p-dioxins and dibenzofurans in the stack gas of a hazardous waste incinerator from concentrations of chlorinated benzenes and biphenyls. *Chemosphere* **1994**, *29*, 2083–2096. [CrossRef]
8. Yan, M.; Li, X.; Zhang, X.; Liu, K.; Yan, J.; Cen, K. Correlation between PAHs and PCDD/Fs in municipal solid waste incinerators. *J. Zhejiang Univ. Eng. Sci.* **2010**, *44*, 1118.
9. Zimmermann, R.; Heger, H.J.; Blumenstock, M.; Dorfner, R.; Schramm, K.W.; Boesl, U.; Kettrup, A. On-line measurement of chlorobenzene in waste incineration flue gas as a surrogate for the emission of polychlorinated dibenzo-p-dioxins/furans (I-TEQ) using mobile resonance laser ionization time-of-flight mass spectrometry. *Rapid. Commun. Mass Spectrom.* **1999**, *13*, 307–314. [CrossRef]
10. Heger, H.J.; Zimmermann, R.; Dorfner, R.; Beckmann, M.; Griebel, H.; Kettrup, A.; Boesl, U. On-line emission analysis of polycyclic aromatic hydrocarbons down to pptv concentration levels in the flue gas of an incineration pilot plant with a mobile resonance enhanced multiphoton ionization time-of-flight mass spectrometer. *Anal. Chem.* **1999**, *71*, 46–57. [CrossRef]
11. Wang, T.; Chen, T.; Lin, B.; Lin, X.; Zhan, M.; Li, X. Emission characteristics and relationships among PCDD/Fs, chlorobenzenes, chlorophenols and PAHs in the stack gas from two municipal solid waste incinerators in China. *RSC Adv.* **2017**, *7*, 44309–44318. [CrossRef]
12. Tanaka, M.; Kobayashi, Y.; Fujiyoshi, H.; Halazawa, S.; Nagano, H.; Iwasaki, T. Evaluation of substituted indexes for chlorobenzenes using dioxin precursor analyzer. *Organohalogen Compd.* **2001**, *54*, 222–225.

13. Cao, X.; Stevens, W.R.; Tang, S.; Lu, S.; Li, X.; Lin, X.; Tang, M.; Yan, J. Atline measurement of 1,2,4-trichlorobenzene for polychlorinated dibenzo-p-dioxin and dibenzofuran International Toxic Equivalent Quantity prediction in the stack gas. *Environ. Pollut.* **2019**, *244*, 202–208. [CrossRef]
14. Xiong, S.; Lu, S.; Shang, F.; Li, X.; Yan, J.; Cen, K. Online predicting PCDD/F emission by formation pathway identification clustering and Box-Cox Transformation. *Chemosphere* **2021**, *274*, 129780. [CrossRef] [PubMed]
15. Tsuruga, S.; Suzuki, T.; Takatsudo, Y.; Seki, K.; Takatsudo, Y.; Seki, K.; Yamuchi, S.; Kuribayashi, S.; Morii, S. On-line monitoring system of P5CDF homologues in waste incineration plants using VUV-SPI-IT-TOFMS. *Environ. Sci. Technol.* **2007**, *41*, 3684–3688. [CrossRef]
16. Kirihara, N.; Haruaki, Y.; Mizuho, T.; Kenji, T.; Norifumi, K.; Yasuo, S.; Toru, S. Development of a RIMMPA-TOFMS. Isomer selective soft ionization of PCDDs/DFs. In Proceedings of the Dioxin 2004: 24. International Symposium on Halogenated Environmental Organic Pollutants and POPs, Berlin, Germany, 6–10 September 2004.
17. Lu, S.; Xiang, Y.; Chen, Z.; Chen, T.; Lin, X.; Zhang, W.; Li, X.; Yan, J. Development of phosphorus-based inhibitors for PCDD/Fs suppression. *Waste Manag.* **2021**, *119*, 82–90. [CrossRef] [PubMed]
18. Chang, Y.M.; Hung, C.Y.; Chen, J.H.; Chang, C.T.; Chen, C.H. Minimum feeding rate of activated carbon to control dioxin emissions from a large-scale municipal solid waste incinerator. *J. Hazard. Mater.* **2009**, *161*, 1436–1443. [CrossRef] [PubMed]
19. Cheruiyot, N.K.; Yang, H.H.; Wang, L.C.; Lin, C.C. Feasible and effective control strategies on extreme emissions of chlorinated persistent organic pollutants during the start-up processes of municipal solid waste incinerators. *Environ. Pollut.* **2020**, *267*, 115469. [CrossRef]
20. Hell, K.; Stieglitz, L.; Dinjus, E. Mechanistic Aspects of the De-Novo Synthesis of PCDD/PCDF on Model Mixtures and MSWI Fly Ashes Using Amorphous 12C- and 13C-Labeled Carbon. *Environ. Sci. Technol.* **2001**, *35*, 3892–3898. [CrossRef]
21. Aurell, J.; Fick, J.; Marklund, S. Effects of Transient Combustion Conditions on the Formation of Polychlorinated Dibenzo-p-Dioxins, Dibenzofurans, and Benzenes, and Polycyclic Aromatic Hydrocarbons During Municipal Solid Waste Incineration. *Environ. Eng. Sci.* **2009**, *26*, 509–520. [CrossRef]
22. Grandesso, E.; Gullett, B.; Touati, A.; Tabor, D. Effect of moisture, charge size, and chlorine concentration on PCDD/F emissions from simulated open burning of forest biomass. *Environ. Sci. Technol.* **2011**, *45*, 3887–3894. [CrossRef]
23. Chen, Y.-C.; Tsai, P.-J.; Mou, J.-L. Determining Optimal Operation Parameters for Reducing PCDD/F Emissions (I-TEQ values) from the Iron Ore Sintering Process by Using the Taguchi Experimental Design. *Environ. Sci. Technol.* **2008**, *42*, 5298–5303. [CrossRef]
24. Williams, B.A.; Tanada, T.N.; Cool, T.A. Resonance ionization detection limits for hazardous emissions. *Symp. Int. Combust.* **1992**, *24*, 1587–1596. [CrossRef]
25. Yan, J.H.; Chen, T.; Li, X.D.; Zhang, J.; Lu, S.Y.; Ni, M.J.; Cen, K.F. Evaluation of PCDD/Fs emission from fluidized bed incinerators co-firing MSW with coal in China. *J. Hazard. Mater.* **2006**, *135*, 47–51. [CrossRef]
26. Reshef, D.N.; Reshef, Y.A.; Finucane, H.K.; Grossman, S.R.; McVean, G.; Turnbaugh, P.J.; Lander, E.S.; Mitzenmacher, M.; Sabeti, P.C. Detecting Novel Associations in Large Data Sets. *Science* **2011**, *334*, 1518–1524. [CrossRef] [PubMed]
27. Yang, J.; Li, X.D.; Yan, M.; Chen, T.; Lu, S.Y.; Yan, J.H.; Olie, K.; Buekens, A. De novo tests on medical waste incineration fly ash: Effect of oxygen and temperature. *Fresenius Environ. Bull.* **2015**, *24*, 587–595.
28. Stieglitz, L. Selected topics on the de novo synthesis of PCDD/PCDF on fly ash. *Environ. Eng. Sci.* **1998**, *15*, 5–18. [CrossRef]
29. Pekarek, V.; Grabic, R.; Marklund, S.; Puncochar, M.; Ullrich, J. Effects of oxygen on formation of PCB and PCDD/F on extracted fly ash in the presence of carbon and cupric salt. *Chemosphere* **2001**, *43*, 777–782. [CrossRef]
30. Zhang, H.J.; Ni, Y.W.; Chen, J.P.; Zhang, Q. Influence of variation in the operating conditions on PCDD/F distribution in a full-scale MSW incinerator. *Chemosphere* **2008**, *70*, 721–730. [CrossRef]
31. Bunsan, S.; Chen, W.Y.; Chen, H.W.; Chuang, Y.H.; Grisdanurak, N. Modeling the dioxin emission of a municipal solid waste incinerator using neural networks. *Chemosphere* **2013**, *92*, 258–264. [CrossRef]

Article

Development of a Water Matrix Certified Reference Material for Volatile Organic Compounds Analysis in Water

Liping Fang, Linyan Huang, Gang Yang, Yang Jiang, Haiping Liu, Bingwen Lu, Yaxian Zhao and Wen Tian *

State Environmental Protection Key Laboratory of Environmental Pollutant Metrology and Reference Materials, Institute for Environmental Reference Materials, Environmental Development Centre of the Ministry of Ecology and Environment, Beijing 100029, China; fang.liping@ierm.com.cn (L.F.); huang.linyan@ierm.com.cn (L.H.); yang.gang@ierm.com.cn (G.Y.); jiang.yang@ierm.com.cn (Y.J.); liu.haiping@ierm.com.cn (H.L.); lu.bingwen@ierm.com.cn (B.L.); zhao.yaxian@ierm.com.cn (Y.Z.)
* Correspondence: tian.wen@ierm.com.cn; Tel.: +86-10-8466-5743; Fax: +86-10-8464-3412

Abstract: Water matrix certified reference material (MCRM) of volatile organic compounds (VOCs) is used to provide quality assurance and quality control (QA/QC) during the analysis of VOCs in water. In this research, a water MCRM of 28 VOCs was developed using a "reconstitution" approach by adding VOCs spiking, methanol solution into pure water immediately prior to analysis. The VOCs spiking solution was prepared gravimetrically by dividing 28 VOCs into seven groups, then based on ISO Guide 35, using gas chromatography-mass spectrometry (GC-MS) to investigate the homogeneity and long-term stability. The studies of homogeneity and long-term stability indicated that the batch of VOCs spiking solution was homogeneous and stable at room temperature for at least 15 months. Moreover, the water MCRM of 28 VOCs was certified by a network of nine competent laboratories, and the certified values and expanded uncertainties of 28 VOCs ranged from 6.2 to 17 µg/L and 0.5 to 5.3 µg/L, respectively.

Keywords: volatile organic compounds (VOCs); reference materials (RMs); quality control (QC); water analysis

Citation: Fang, L.; Huang, L.; Yang, G.; Jiang, Y.; Liu, H.; Lu, B.; Zhao, Y.; Tian, W. Development of a Water Matrix Certified Reference Material for Volatile Organic Compounds Analysis in Water. *Molecules* **2021**, *26*, 4370. https://doi.org/10.3390/molecules26144370

Academic Editor: Wenbin Liu

Received: 2 June 2021
Accepted: 12 July 2021
Published: 20 July 2021

Publisher's Note: MDPI stays neutral with regard to jurisdictional claims in published maps and institutional affiliations.

Copyright: © 2021 by the authors. Licensee MDPI, Basel, Switzerland. This article is an open access article distributed under the terms and conditions of the Creative Commons Attribution (CC BY) license (https://creativecommons.org/licenses/by/4.0/).

1. Introduction

Volatile organic compounds (VOCs), due to their toxicity and persistence in the environment, are one group of particularly important pollutants [1,2]. Many of these substances are toxic, and some are considered to be carcinogenic, mutagenic or teratogenic [3,4]. As reported, VOCs have been widely detected at trace levels in surface water and groundwater [2,5]. Therefore, during VOCs detection in water, references with VOCs are needed to calibrate the instrument, verify analysis procedure or control the analysis quality according to QA/QC (quality assurance and quality control) guidelines. Validation of the entire analytical procedure requires the use of an matrix certified reference material (MCRM), which is a homogeneous and well-defined matrix that contains known amounts of the target compounds [6–8]. However, there is currently no MCRM available for the measurement of VOCs in water. Therefore, in this research, we prepared and characterized a water MCRM of 28 VOCs.

There are several reasons for the current absence of MCRMs for organic materials in water. Specifically, the relatively low concentrations of most organic contaminants in water necessitates that large volumes of water be manipulated and transferred for analysis, which may lead to problems such as cross-contamination or improper and inaccurate dilution. Additionally, there is a tendency for target compounds to be adsorbed onto suspended or colloidal particles because of their low water solubility or their high lipophilic characteristics as expressed by the high octanol-water partition coefficient (K_{ow}). Furthermore, the instability of some organic materials causes problems, as they may have a tendency to undergo hydrolysis, metabolization or photochemical reactions. Finally, the residual biological activity of samples can cause severe stability problems [9].

At present, the only commercially available water matrix reference material (MRM) is IRMM-428, released by the Institute for Reference Materials and Measurements (IRMM). IRMM-428 is prepared by spiking PFASs in methanol into a known volume of drinking water. Other attempts have been made to prepare proficiency testing (PT) samples for organics in water employing the widely used "reconstitution" approach. In this approach, a solution of analytes of interest in an organic solvent (miscible with water) is spiked into a water sample in the laboratory immediately prior to analysis [8,10,11] or at the producer's premises just before shipping [12,13]. The rest of the approach is the "immobilization" of analytes (in this case, pesticides) on a solid-phase extraction (SPE) cartridge [8]. Water MRMs that consisted of pesticides stored on SPE cartridges were used in a collaborative study including 15 laboratories, and the observed reproducibility was 26.7% [10]. However, this approach is not sound from a metrological point of view because it was not possible to confirm that the true value equaled the initial concentration in the percolated water sample.

To the best of our knowledge, no water MRMs for VOCs are currently available for quality control. Therefore, in this research, we prepared and certified a water MCRM containing 28 VOCs. The water MCRM was prepared using a "reconstitution" approach by spiking the solution of VOCs in methanol into pure water in the laboratory immediately prior to analysis. The preparation and characterization of 28 VOCs spiking solutions were comprehensively studied, and the water MCRM containing all 28 VOCs was certified using a network of nine competent laboratories.

2. Results and Discussion

2.1. Methodology Study

The 28 VOCs in the spiking solution were determined by GC-MS, and the total ion chromatography is shown in Figure 1. With the exception of p-xylene and m-xylene, 26 other VOCs, IS1 and IS2 were separated well on the chromatographic column.

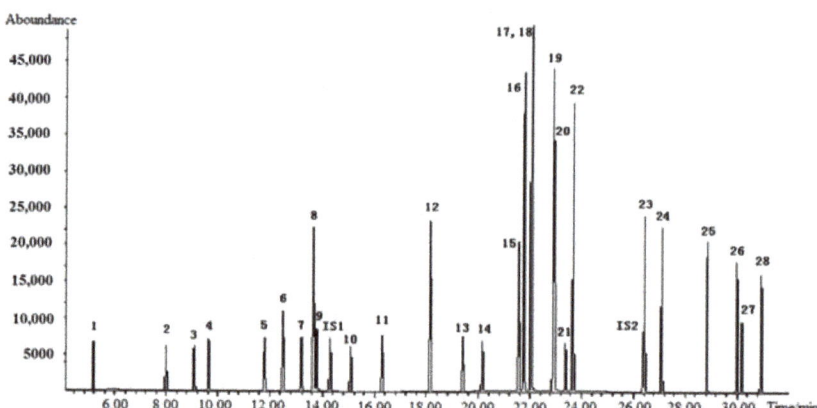

Figure 1. Total ion chromatography of 28 VOCs determined by GC-MS (1. vinyl chloride; 2. 1,1-dichloroethylene; 3. dichloromethane; 4. trans-1,2-dichloroethene; 5. cis-1,2-dichloroethene; 6. trichloromethane; 7. carbon tetrachloride; 8. benzene; 9. 1,2-dichloroethane; 10. trichloroethylene; 11. bromodichloromethane; 12. toluene; 13. tetrachloroethylene; 14. chlorodibromomethane; 15. chlorobenzene; 16. ethylbenzene; 17,18. p-xylene/m-xylene; 19. o-xylene; 20. styrene; 21. bromoform; 22. cumene; 23. 1,4-dichlorobenzene; 24. 1,2-dichlorobenzene; 25. 1,3,5-trichlorobenzene; 26. 1,2,4-trichlorobenzene; 27. hexachloro-1,3-butadiene; 28. 1,2,3-trichlorobenzene; IS1. fluorobenzene; IS2. 1,4-dichlorobenzene D4).

The precision of the method was investigated by analyzing each reference material of 10 µg/mL 28 VOCs in methanol 10 times, and the results were presented in Table 1.

The within laboratory RSDs of the method were between 0.14% and 2.55%, indicating good instrumental repeatability that could meet the requirements of homogeneity and stability studies. The detection limits of 28 VOCs in methanol were calculated based on three standard deviations, and the results were between 0.014 and 0.225 µg/mL. These levels were obviously lower than the concentrations of the 28 VOCs in the spiking solution, which ranged from 6.0 to 18 µg/mL. With the exception of hexachloro-1,3-butadiene, which had an r value of 0.9989, the linear correlation r values of the other 27 VOCs were higher than 0.9990, indicating good linear relationships.

Table 1. Precision and detection limits of 28 VOCs in methanol using GC-MS.

No.	Composition	RSD(%)	Detection Limit/ (µg/mL)	Linear Correlation r
1	vinyl chloride	2.55	0.1	1.000
2	1,1-dichloroethylene	1.53	0.1	0.9999
3	dichloromethane	0.78	0.07	1.000
4	trans-1,2-dichloroethene	1.03	0.07	0.9999
5	cis-1,2-dichloroethene	0.34	0.03	1.000
6	trichloromethane	0.26	0.03	1.000
7	carbon tetrachloride	0.74	0.04	0.9998
8	benzene	0.44	0.02	1.000
9	1,2-dichloroethane	0.50	0.04	1.000
10	trichloroethylene	0.72	0.2	1.000
11	bromodichloromethane	0.63	0.02	0.9998
12	toluene	0.53	0.03	0.9998
13	tetrachloroethylene	0.33	0.1	1.000
14	chlorodibromomethane	0.94	0.05	0.9997
15	chlorobenzene	0.45	0.02	0.9990
16	ethylbenzene	0.34	0.03	0.9999
17	p-xylene	0.43	0.04	0.9997
18	m-xylene	0.43	0.04	0.9997
19	o-xylene	0.43	0.03	0.9999
20	styrene	0.50	0.03	1.000
21	bromoform	0.39	0.03	0.9998
22	cumene	0.27	0.04	1.000
23	1,4-dichlorobenzene	0.18	0.01	0.9991
24	1,2-dichlorobenzene	0.14	0.02	0.9992
25	1,3,5-trichlorobenzene	0.27	0.02	0.9992
26	1,2,4-trichlorobenzene	0.28	0.01	0.9996
27	hexachloro-1,3-butadiene	0.50	0.06	0.9989
28	1,2,3-trichlorobenzene	0.31	0.02	0.9995

2.2. Purity Test

The purity was used to correct the gravimetric preparation of the standard solutions, and therefore ensure the metrological traceability of the water MCRM. Detailed information regarding the manufacturer, labeled purities and the uncertainties of the 28 commercial standards are given in Table 1. The labeled purities of the commercial standards were verified by GC-FID using the peak area normalization method, and the results are presented in Table 1. The measured purities of all commercial standards were within the range of their labeled uncertainties. Therefore, the labeled purities and uncertainties were used during gravimetric preparation and calculation of the uncertainty in the characterization study of water MCRM, respectively, in consideration of the limitations of the peak area normalization method.

The purity test results of the batch of methanol showed that the methanol purity was acceptable. Additionally, no VOCs were detected, indicating that methanol was suitable for use as the solvent for the 28 VOCs spiking solution.

2.3. Homogeneity Assessment

Homogeneity is an important property of a reference material. Nevertheless, it is a relative concept closely related to the distribution of components in the material, sample size and the number of samples that have been selected to measure homogeneity [14].

In the present study, homogeneity was assessed by selecting 15 ampoules of the VOCs spiking solution using a stratified random samplings scheme covering the entire batch, after which three sub-samples of each ampoule were analyzed. The homogeneity study was evaluated using ANOVA [15], and the results are presented in Figure 2. All calculated F values were below or equal to the critical value $F_{0.05}(14,30) = 2.04$, indicating no significant difference within bottles. The homogeneity analysis confirmed that the batch had a good agreement among its units (ampoules) for each of the 28 analytes in methanol, and the VOCs spiking solution was regarded as homogeneous.

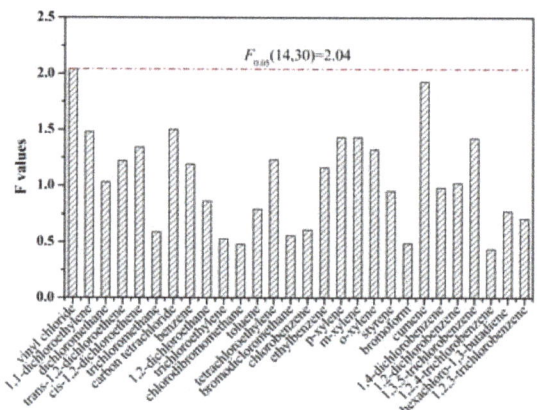

Figure 2. Results of homogeneity study of 28 VOCs in spiking solution.

It should be noted that the calculated F value of vinyl chloride was equal to the critical value $F_{0.05}(14, 30) = 2.04$. The relatively high F value of vinyl chloride might relate to its high volatility and the relatively poor method repeatability using GC-MS, leading to the calculated MS_{among} being relatively higher than the MS_{within}.

2.4. Stability Assessment

Stability testing is crucial to the certification of reference materials. The long-term stability study was based on linear regression [16], and the results measured after 0, 1, 3, 6, 9, 12 and 15 months of storage in different temperatures are shown in Figure 3 and Tables 2–4. At a confidence level of 95%, the observed slope for all results was $|b_1| < t_{0.95,n-2} \times s(b_1)$, indicating that the 28 analytes in the VOCs spiking solution stored at room temperature, 4 and −18 °C were stable after 15 months of storage. Therefore, the spiking solution, including the 28 VOCs, could be stored at room temperature for convenience, and the shelf-life of the VOCs spiking solution was at least 15 months.

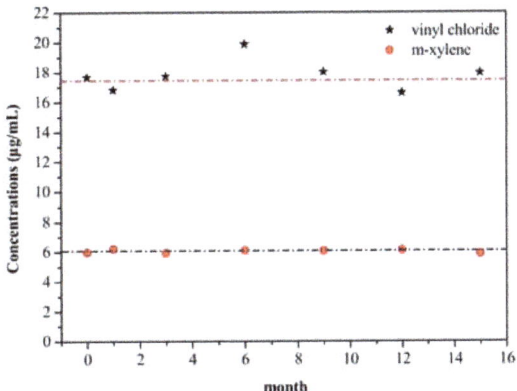

Figure 3. Results of vinyl chloride and m-xylene in spiking solution stored at room temperature at intervals of 15 months.

Table 2. Stability study of VOCs spiking solution stored at room temperature.

No.	Composition	b_0 (µg/mL)	b_1 (µg/mL)	$s(b_1)$ (µg/mL)	$t_{0.95,n-2}$	$t_{0.95,n-2} \times s(b_1)$
1	vinyl chloride	18.04	−0.165	0.220	2.57	0.565
2	1,1-dichloroethylene	12.74	−0.048	0.080	2.57	0.207
3	dichloromethane	12.56	0.011	0.056	2.57	0.145
4	trans-1,2-dichloroethene	10.56	0.021	0.069	2.57	0.178
5	cis-1,2-dichloroethene	10.27	0.000	0.041	2.57	0.106
6	trichloromethane	13.72	0.112	0.160	2.57	0.410
7	carbon tetrachloride	12.00	0.069	0.074	2.57	0.190
8	benzene	8.654	−0.019	0.047	2.57	0.120
9	1,2-dichloroethane	10.83	0.059	0.067	2.57	0.172
10	trichloroethylene	14.00	0.049	0.109	2.57	0.281
11	chlorodibromomethane	14.49	0.117	0.100	2.57	0.257
12	toluene	8.580	−0.003	0.023	2.57	0.059
13	tetrachloroethylene	12.21	0.046	0.061	2.57	0.158
14	bromodicloromethane	12.09	0.051	0.109	2.57	0.280
15	chlorobenzene	9.873	−0.041	0.037	2.57	0.095
16	ethylbenzene	7.701	−0.028	0.020	2.57	0.050
17	p-xylene	6.092	−0.007	0.009	2.57	0.023
18	m-xylene	6.092	−0.007	0.009	2.57	0.023
19	o-xylene	7.191	−0.020	0.013	2.57	0.034
20	styrene	8.533	−0.021	0.017	2.57	0.043
21	bromoform	15.43	0.015	0.055	2.57	0.141
22	cumene	7.752	−0.010	0.005	2.57	0.012
23	1,4-dichlorobenzene	10.33	−0.053	0.029	2.57	0.074
24	1,2-dichlorobenzene	9.287	−0.032	0.026	2.57	0.066
25	1,3,5-trichlorobenzene	9.433	−0.019	0.029	2.57	0.073
26	1,2,4-trichlorobenzene	9.275	−0.040	0.040	2.57	0.102
27	hexachloro-1,3-butadiene	10.80	−0.040	0.057	2.57	0.147
28	1,2,3-trichlorobenzene	8.463	−0.006	0.025	2.57	0.063

Table 3. Stability study for VOCs spiking solution stored at 4 °C.

No.	Composition	b_0 (μg/mL)	b_1 (μg/mL)	$s(b_1)$ (μg/mL)	$t_{0.95,n-2}$	$t_{0.95,n-2} \times s(b_1)$
1	vinyl chloride	17.06	0.108	0.158	2.57	0.405
2	1,1-dichloroethylene	12.11	0.059	0.050	2.57	0.128
3	dichloromethane	12.56	0.041	0.028	2.57	0.071
4	trans-1,2-dichloroethene	10.83	0.025	0.037	2.57	0.095
5	cis-1,2-dichloroethene	10.13	0.000	0.048	2.57	0.124
6	trichloromethane	14.77	0.048	0.044	2.57	0.113
7	carbon tetrachloride	12.11	0.079	0.041	2.57	0.105
8	benzene	8.330	0.010	0.017	2.57	0.045
9	1,2-dichloroethane	11.10	0.057	0.024	2.57	0.063
10	trichloroethylene	14.64	0.021	0.068	2.57	0.174
11	chlorodibromomethane	15.03	0.075	0.063	2.57	0.163
12	toluene	8.543	0.002	0.025	2.57	0.064
13	tetrachloroethylene	12.36	0.037	0.033	2.57	0.084
14	bromodicloromethane	11.83	0.073	0.092	2.57	0.238
15	chlorobenzene	10.03	−0.049	0.035	2.57	0.090
16	ethylbenzene	7.805	−0.032	0.020	2.57	0.050
17	p-xylene	6.156	−0.007	0.010	2.57	0.027
18	m-xylene	6.156	−0.007	0.010	2.57	0.027
19	o-xylene	7.259	−0.023	0.011	2.57	0.029
20	styrene	8.410	−0.009	0.011	2.57	0.028
21	bromoform	15.75	−0.023	0.041	2.57	0.107
22	cumene	7.753	−0.021	0.008	2.57	0.022
23	1,4-dichlorobenzene	10.21	−0.041	0.027	2.57	0.068
24	1,2-dichlorobenzene	9.324	−0.034	0.019	2.57	0.049
25	1,3,5-trichlorobenzene	9.432	−0.019	0.025	2.57	0.065
26	1,2,4-trichlorobenzene	9.275	−0.040	0.040	2.57	0.102
27	hexachloro-1,3-butadiene	10.28	−0.008	0.041	2.57	0.105
28	1,2,3-trichlorobenzene	8.578	−0.016	0.019	2.57	0.048

Table 4. Stability study for VOCs spiking solution stored at −18 °C.

No.	Composition	b_0 (μg/mL)	b_1 (μg/mL)	$s(b_1)$ (μg/mL)	$t_{0.95,n-2}$	$t_{0.95,n-2} \times s(b_1)$
1	vinyl chloride	18.50	−0.018	0.139	2.57	0.358
2	1,1-dichloroethylene	12.48	0.021	0.052	2.57	0.134
3	dichloromethane	12.79	0.021	0.035	2.57	0.090
4	trans-1,2-dichloroethene	10.99	0.022	0.035	2.57	0.091
5	cis-1,2-dichloroethene	10.18	0.023	0.031	2.57	0.079
6	trichloromethane	14.82	0.051	0.043	2.57	0.110
7	carbon tetrachloride	12.21	0.073	0.049	2.57	0.126
8	benzene	8.434	0.011	0.021	2.57	0.054
9	1,2-dichloroethane	11.12	0.058	0.029	2.57	0.074
10	trichloroethylene	14.79	0.023	0.061	2.57	0.156
11	chlorodibromomethane	15.03	0.090	0.064	2.57	0.165
12	toluene	8.516	0.012	0.028	2.57	0.071
13	tetrachloroethylene	12.33	0.048	0.031	2.57	0.080
14	bromodicloromethane	11.61	0.091	0.060	2.57	0.154
15	chlorobenzene	10.093	−0.059	0.035	2.57	0.089
16	ethylbenzene	7.836	−0.034	0.020	2.57	0.051
17	p-xylene	6.168	0.001	0.011	2.57	0.028
18	m-xylene	6.168	0.001	0.011	2.57	0.028
19	o-xylene	7.243	−0.020	0.008	2.57	0.020
20	styrene	8.378	0.000	0.009	2.57	0.022
21	bromoform	15.64	−0.027	0.057	2.57	0.147
22	cumene	7.710	−0.012	0.007	2.57	0.017
23	1,4-dichlorobenzene	7.739	−0.021	0.006	2.57	0.015
24	1,2-dichlorobenzene	9.331	−0.032	0.018	2.57	0.047
25	1,3,5-trichlorobenzene	9.508	−0.032	0.022	2.57	0.057
26	1,2,4-trichlorobenzene	8.942	−0.018	0.023	2.57	0.060
27	hexachloro-1,3-butadiene	10.25	−0.013	0.042	2.57	0.107
28	1,2,3-trichlorobenzene	8.628	−0.028	0.016	2.57	0.042

Interestingly, the dispersions of the stability of the 28 VOCs under the same storage conditions were quite different. The stability of vinyl chloride and m-xylene after storage at room temperature for different lengths are shown as examples in Figure 3. The stability of vinyl chloride fluctuated with storage time around the preparation value in a relatively large range, although vinyl chloride in methanol was confirmed to be stable. However, the stability of m-xylene fluctuated closely around the preparation value with storage time. The obvious differences in stability results between vinyl chloride and m-xylene might be related to their characterizations. In the former methodology study, the within-laboratory RSD of vinyl chloride was 2.55%, while that of m-xylene was 0.43%. The lower boiling point of vinyl chloride than m-xylene might lead to significant differences in method precision [17], which could then lead to the different dispersions of stable results. Similar results were also observed for the other 26 VOCs, with the stable values of VOCs having higher boiling points generally being closer to their certified values than those having lower boiling points.

2.5. Characterization and Uncertainty Study

The certification of the water MCRM was conducted by nine competent laboratories. Laboratory 4 used the headspace GC-MS method for certification, while the remaining eight laboratories used the purge-and-trap GC-MS method. The reported values of the 28 VOC analytes were expressed after diluting 1000 times in water. Statistical analysis was conducted for the received data using Grubb's test, the Cochran test and Dixon's test, and the mean values of the retained data were calculated as certified values. Figure 4 shows the reported mean values and standard deviations of the 28 VOCs in water from the nine laboratories. Most of the values were close to the certified values and within their expended uncertainties. However, some values from one or two laboratories showed obvious deviations from the certified values, such as the results reported from laboratory 3. Because both headspace GC-MS and purge-and-trap GC-MS are commonly used for the determination of VOCs in water, there were no obvious discrepancies in the results from the nine laboratories using the two different detection methods.

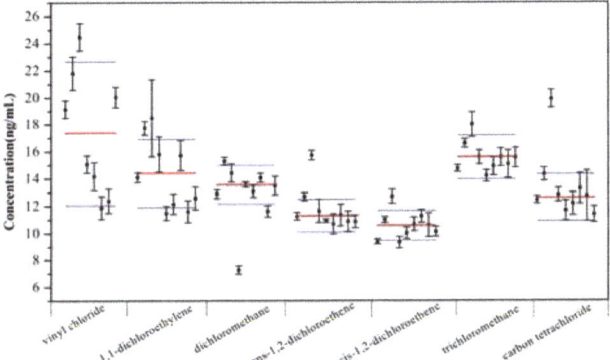

Figure 4. Cont.

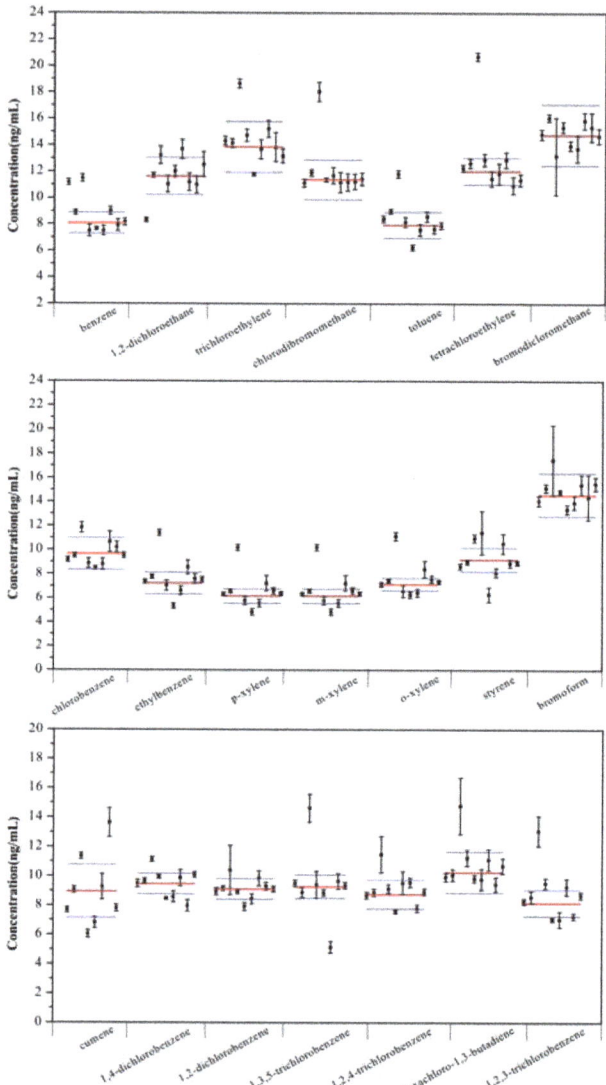

Figure 4. Measurement of 28 VOCs in water from nine laboratories. Thick red lines represent certified values, thin blue lines represent expended uncertainties.

The results of the certified values and the expanded uncertainties of the water MCRM expressed at 95% confidence (with the coverage factor k = 2) are presented in Table 5. The certified values of the 28 VOCs in the water MCRM were in the range of 6.162 to 17.37 µg/L. The uncertainty of the water MCRM was calculated by combining the uncertainty of inhomogeneity (u_{bb}), instability (u_{lst}) and characterization (u_{char}) [18], and the u_{bb} and u_{lst} were those of VOCs spiking solution. Some of the u_{bb} were calculated through $u_{bb} = s_{bb} = \sqrt{\frac{MS_{among} - MS_{within}}{n}}$ when $MS_{among} > MS_{within}$ for some VOCs, while others were calculated through $u_{bb}' = \sqrt{\frac{MS_{within}}{n}} \sqrt[4]{\frac{2}{v_{MS_{within}}}}$ when $MS_{among} < MS_{within}$. As shown in Table 5, the u_{bb} of the 28 VOC analytes ranged from 0.18% to 3.73%, which was lower than the reported

CRM of BTEX in methanol [19]. The u_{char} and u_{lst} of the 28 VOC analytes ranged from 0.87% to 9.44% and 0.91% to 11.4%, respectively. The expanded uncertainty of the 28 VOC analytes in the water MCRM ranged from 0.5 to 5.3 µg/L with a coverage factor $k = 2$ under an approximately 95% confidence level. The expanded uncertainty of o-xylene as 0.5 µg/L was lowest, and the expanded uncertainty of vinyl chloride as 5.3 µg/L was highest. The relatively high expanded uncertainty of vinyl chloride might result from two aspects. On the one hand, the highly volatile nature and early elution time on chromatography could result in a relatively high operation and instrument effect on quantification. On the other hand, the tested value of commercial vinyl chloride standard solution was not consistent with the labeled value, and large discrepancies existed among commercial standard solutions from different producers, making precise quantification difficult.

Table 5. Certified values and expended uncertainties of 28 VOCs in the water MCRM.

No.	Composition	u_{char} (%)	u_{bb} (%)	u_{lst} (%)	Certified Value (µg/L)	u_{CRM} ($k = 2$) (µg/L)
1	vinyl chloride	9.44	3.73	11.4	1.7×10^1	5.3
2	1,1-dichloroethylene	5.75	1.22	6.20	1.4×10^1	2.5
3	dichloromethane	2.87	1.89	4.07	1.4×10^1	1.4
4	trans-1,2-dichloroethene	2.04	0.84	4.75	1.1×10^1	1.2
5	cis-1,2-dichloroethene	2.24	0.81	4.48	1.1×10^1	1.1
6	trichloromethane	2.42	1.52	4.25	1.6×10^1	1.6
7	carbon tetrachloride	2.61	1.71	5.80	1.3×10^1	1.7
8	benzene	2.99	0.86	3.67	8.1×10^0	0.8
9	1,2-dichloroethane	4.49	1.02	3.77	1.2×10^1	1.4
10	trichloroethylene	2.68	1.20	6.09	1.4×10^1	1.9
11	chlorodibromomethane	0.87	1.40	6.17	1.1×10^1	1.5
12	toluene	3.74	0.46	4.81	7.9×10^0	1.0
13	tetrachloroethylene	2.18	0.82	3.69	1.2×10^1	1.0
14	bromodicloromethane	2.23	1.23	7.38	1.5×10^1	2.3
15	chlorobenzene	3.65	0.50	5.38	9.7×10^0	1.3
16	ethylbenzene	4.65	0.30	3.94	7.2×10^0	0.9
17	p-xylene	4.25	0.31	2.62	6.2×10^0	0.6
18	m-xylene	4.25	0.31	2.62	6.2×10^0	0.6
19	o-xylene	3.46	0.31	1.64	7.1×10^0	0.5
20	styrene	5.44	0.34	1.56	9.2×10^0	1.0
21	bromoform	1.87	1.41	5.54	1.5×10^1	1.8
22	cumene	9.87	0.42	1.32	9.0×10^0	1.8
23	1,4-dichlorobenzene	3.15	0.18	0.91	9.5×10^0	0.7
24	1,2-dichlorobenzene	1.74	0.22	3.00	9.1×10^0	0.7
25	1,3,5-trichlorobenzene	1.56	0.25	3.60	9.3×10^0	0.8
26	1,2,4-trichlorobenzene	3.76	0.39	3.98	8.8×10^0	1.0
27	hexachloro-1,3-butadiene	1.83	0.77	6.14	1.0×10^1	1.4
28	1,2,3-trichlorobenzene	4.72	0.31	2.91	8.2×10^0	0.9

3. Materials and Methods

The Institute for Environmental Reference Materials, Ministry of Environmental Protection (IERM) has a quality management system based on ISO Guide 35 and ISO/IEC 17025, which is accredited by the China National Accreditation Service for Conformity Assessment (CNAS). The preparation and certification of the CRMs for environmental monitoring have been carried out according to the technical requirements of ISO Guide 35 [20].

3.1. Chemicals and Instruments

The commercial standards of 28 VOCs were purchased from several manufacturers. Detailed information regarding the manufacturers, purities and uncertainties of the purities are given in Table 6. Pesticide residue grade methanol was purchased from J.T. Baker, USA. The stock standard solutions were 27 mixed VOCs standard solution (IRMM, 100 µg/mL) and vinyl chloride standard solution (2000 µg/mL, Supelo, USA). The internal stock standard solutions were fluorobenzene and 1,4-dichlorobenzene D_4 (IRMM, 1000 and 1000 µg/mL, respectively).

Table 6. Purities of 28 commercial VOC standards.

No.	Composition	Manufacturer	Labeled Purity (%)	Labeled Uncertainty (%)	Measured Purity (%)
1	vinyl chloride	gmgas, China	99.999	0.5	99.99
2	1,1-dichloroethylene	ChemService,USA	99.5	0.5	99.83
3	dichloromethane	ChemService,USA	99.5	0.5	99.74
4	trans-1,2-dichloroethene	ChemService,USA	99.3	0.5	99.37
5	cis-1,2-dichloroethene	ChemService,USA	99.5	0.5	99.78
6	trichloromethane	ChemService,USA	99.5	0.5	99.34
7	carbon tetrachloride	ChemService,USA	99.5	0.5	99.85
8	benzene	ChemService,USA	99.5	0.5	99.95
9	1,2-dichloroethane	ChemService,USA	99.5	0.5	99.93
10	trichloroethylene	ChemService,USA	99.5	0.5	99.83
11	chlorodibromomethane	Fluka,USA	98.8	0.5	98.88
12	toluene	ChemService,USA	99.5	0.5	99.99
13	tetrachloroethylene	ChemService,USA	99.5	0.5	99.94
14	bromodichloromethane	Fluka,USA	99.5	0.5	99.86
15	chlorobenzene	ChemService,USA	99.5	0.5	99.91
16	ethylbenzene	ChemService,USA	99.5	0.5	99.48
17	p-xylene	ChemService,USA,USA	99.5	0.5	99.77
18	m-xylene	ChemService,USA	99.4	0.5	99.83
19	o-xylene	ChemService,USA	99.0	0.5	99.27
20	styrene	ChemService,USA	99.4	0.5	99.86
21	bromoform	ChemService,USA	99.5	0.5	99.53
22	cumene	ChemService,USA	99.5	0.5	99.93
23	hexachloro-1,3-butadiene	AccuStandard,USA	98.3	1.0	97.82
24	1,4-dichlorobenzene	ChemService,USA	99.5	0.5	99.98
25	1,2-dichlorobenzene	ChemService,USA	99.5	0.5	99.73
26	1,2,4-trichlorobenzene	ChemService,USA	99.5	0.5	99.58
27	1,2,3-trichlorobenzene	ChemService,USA	99.5	0.5	99.93
28	1,3,5-trichlorobenzene	ChemService,USA	99.5	0.5	99.11

The VOCs spiking solution was prepared gravimetrically using a calibrated Mettler Toledo analytical balance (AE-240, 205 g capacity, resolution of 0.01 mg, Switzerland). The purities of the 28 VOC commercial standards were verified by a calibrated gas chromatography with flame ionization detection (Agilent 7890A GC-FID, USA). Homogeneity and stability studies of VOCs spiking solution were performed on a calibrated Agilent 7890A gas chromatograph coupled with an Agilent 5975C mass spectrometer (Agilent 7890A GC-5975C MS, USA). Analysis of the water MCRM was performed on a calibrated purge-and-trap Agilent 7890A GC-5975C MS.

3.2. Purity Test

The purities of the 28 VOC commercial standards were determined in-house using a GC-FID with a DB-1 column (30 m × 320 μm ID × 0.25 μm film). The oven temperature program started at 70 °C, then increased to 150 °C at 5 °C/min. The injection volume was 1 μL, and the injection was performed in split mode (30:1). The carrier gas was high purity nitrogen (1.0 mL/min), and the temperature of the injector and detector were 220 and 230 °C, respectively. The final purities of the 28 commercial standards were calculated by the peak area normalization method.

The purity of methanol (pesticides residue grade) selected as the solvent was checked by gas chromatography-mass spectrometry (GC-MS) prior to the preparation of CRM.

3.3. Determination of VOCs

3.3.1. Determination of VOCs in the Spiking Solution

Measurement of the 28 VOCs in the spiking solution was performed on a GC-MS [21] equipped with a DB-624 (60 m × 250 μm ID × 1.4 μm film, Agilent, Santa Clara, CA, USA) capillary column. The oven temperature was programmed as follows: 35 °C for 2 min, followed by 5 °C/min to 120 °C, then 10 °C/min to 220 °C, where it was held for 3 min. The injection volume was 1 μL, and the injection was performed in split mode (30:1). The carrier gas was helium (1.0 mL/min), and the temperature of the injector, transfer line and ion source was 220, 260 and 230 °C, respectively. Data acquisition was performed under selected ion monitoring (SIM) mode.

3.3.2. Determination of VOCs in Water MCRM

Analysis of the 28 VOCs in water MCRM was performed on a purge-and-trap GC-MS [21]. The conditions of the purge-and-trap were as follows: purge time: 11 min; purge rate: 40 mL/min; dry purge time: 1 min; desorption time: 2 min; desorption temperature: 190 °C; baking time: 6 min; baking temperature: 200 °C. The conditions of GC-MS were the same as for the determination of 28 VOCs in the spiking solution.

3.4. Preparation of the VOCs Spiking Solution

The 28 VOCs spiking solution was prepared gravimetrically. Briefly, the 28 VOCs were divided into seven groups during weighing and dissolution, then mixed into a certain volume. Among the 28 VOC commercial standards, only vinyl chloride standard is gaseous at room temperature. Therefore, vinyl chloride was placed in its own group, while the remaining 27 VOCs were divided into six groups according to their characteristics (Table 7). According to the labeled purities of the 28 VOC commercial standards, the stock solution of vinyl chloride in methanol was prepared by drawing a moderate volume of vinyl chloride using an airtight syringe and then adding it to methanol. For the other six groups, the stock solution of each group was prepared gravimetrically in the sequence of their polarity from weak to strong. After all stock solutions were prepared, a moderate volume of stock solution from each group was transferred into the same 1 L flask and then diluted to 1 L with methanol. The mass fractions of target compositions of the spiked solution were between 5.0 and 20 mg/L. Approximately 1 L of the VOC mixture was subdivided into 2 mL amber glass ampoules with 1.2 mL per ampoule using an ampoule filling machine. During the process of subdivision, 15 ampoules were sampled for a homogeneity study using a stratified random sampling strategy. After confirmation that the batch was homogeneous, the sealed ampoules were packed and divided into three parts, then stored at room temperature, 4 and −18 °C.

Table 7. Grouping of 28 VOCs during preparation of VOCs spiking solution.

Group	Composition
1	vinyl chloride
2	dichloromethane, trichloroethylene, bromoform, chlorodibromomethane
3	carbon tetrachloride, bromodichloromethane, tetrachloroethylene
4	trichloromethane, 1,2-dichloroethane, 1,4-dichlorobenzene, hexachloro-1,3-butadiene
5	toluene, ethylbenzene, o-xylene, m-xylene, p-xylene, styrene, cumene
6	chlorobenzene, 1,2-dichlorobenzene, 1,2,4-trichlorobenzene, 1,2,3-trichlorobenzene, 1,3,5-trichlorobenzene
7	1,1-dichloroethylene, benzene, trans-1,2-dichloroethene, cis-1,2-dichloroethene

3.5. Homogeneity Testing

The VOCs spiking solution was subjected to a homogeneity study in which both the homogeneity between and within ampoules was evaluated. Fifteen ampoules of VOC spiking solution were selected using a stratified random sampling scheme covering the whole batch, and three sub-samples were analyzed in each ampoule. The internal standard was spiked into the solution for the QA/QC process, and each sample was analyzed in triplicate by GC-MS. Measurement sequences were randomized to be able to minimize possible trends in both the filling sequence and the analytical sequence.

The homogeneity was evaluated by one-way analysis of variance (ANOVA) as described by ISO Guide 35 [20]. The between-bottle standard deviation (s_{bb}) and within-bottle standard deviation (s_{wb}) were also calculated. Possible inhomogeneity was expressed as the uncertainty due to the between-bottle inhomogeneity of the material (u_{bb}) and quantified as:

$$u_{bb} = s_{bb} = \sqrt{\frac{MS_{among} - MS_{within}}{n}} \quad (1)$$

In cases in which $MS_{among} < MS_{within}$ (indicating that the study set-up and/or method repeatability were not sufficient), the maximum heterogeneity that could be hidden by the method variability, the influence of analytical variation on the standard deviation between units u_{bb}' was calculated and used for estimates of in-homogeneity. The u_{bb}' was calculated as:

$$u_{bb}' = \sqrt{\frac{MS_{within}}{n}} \sqrt[4]{\frac{2}{v_{MS_{within}}}} \quad (2)$$

where MS_{within} is the mean square within groups determined from ANOVA, n is the number of replicates per bottle and $v_{MS_{within}}$ represents the degrees of freedom of MS_{within}.

The instrumental repeatability of the measurement of 28 VOCs was determined by conducting 10 replicate GC-MS analyses of 10 µg/mL reference material containing 28 VOCs.

3.6. Stability Testing

A long-term stability study was conducted to ensure the shelf-life of the VOCs spiking solution. A long-term stability study was conducted based on a classical stability study. Stability monitoring was performed for each analyte of the spiking solution after 0, 1, 3, 6, 9, 12 and 15 months of storage at room temperature, 4 and −18 °C. For each round of analysis, three ampoules were sampled randomly from those samples stored under each storage condition.

A linear regression model was utilized for processing data, namely assuming component values (Y) of time (X) varying linear equation as $Y = b_0 + b_1 X$, where b_0 and b_1 are the regression coefficients. The estimated standard deviation of b_1 is then given by:

$$S(b_1) = \frac{s}{\sqrt{\sum_{i=1}^{n}(X_i - \overline{X})^2}} \quad (3)$$

where

$$s^2 = \frac{\sum_{i=1}^{n}(Y_i - b_0 - b_1 X_i)^2}{n - 2} \quad (4)$$

and t_{cal} is given by

$$t_{cal} = \frac{|b_1|}{S(b_1)} \quad (5)$$

According to ISO Guide 35, the long-term instability is estimated as $u_{lts} = S(b_1) \times t$, with u_{lts} being the uncertainty of long-term instability, $S(b_1)$ is the standard error of the slope and t is the selected duration.

3.7. Characterization and Uncertainty Study

The certified values of 28 VOCs in the water MCRM were determined by the average of the results obtained from nine competent laboratories that each received six ampoules selected at random. Before analysis, the samples were diluted 1000 times with pure water, and the average concentrations of analytes after dilution were used.

Data sets were checked to ensure they followed approximately normal distributions and that variances for each compound were homogeneous. Grubb's test was applied to evaluate within laboratory parallel data. Between-laboratory outliers of variance were detected using the Cochran test, while outliers of average were detected using Dixon's test. Generally, the outliers detected from Grubb's test were retained. When the RSD of within-laboratory parallel data was smaller than 15%, the stragglers and outliers from Cochran's test were retained. The outliers (95% confidence interval) detected from Dixon's test were removed.

The standard uncertainty of the certified values included collaborating characterization uncertainty (u_{char}), between-bottle inhomogeneity uncertainty (u_{bb}) and long-term instability uncertainty (u_{lts}). The expanded uncertainty (U) of the certified property value was calculated as: $U = k \times \sqrt{u_{bb}^2 + u_{lts}^2 + u_{char}^2}$, where k is the coverage factor (usually set $k = 2$, approximately the 95% confidence level), u_{bb} is the uncertainty due to inhomogeneity of the material, u_{its} is the uncertainty due to instability of the material and u_{char} is uncertainty in the characterization of the property value.

4. Conclusions

In conclusion, a water MCRM of 28 VOCs was developed using a "reconstitution" approach by adding the prepared VOCs spiking methanol solution into pure water directly prior to analysis. The 28 VOCs in methanol as a spiking solution was prepared gravimetrically by dividing the VOCs into seven groups. The batch of spiking solution was homogeneous and stable at room temperature, 4 and $-18\ °C$ for at least 15 months. The certification of the water MCRM was established in a study involving nine competent laboratories applying purge-and-trap GC-MS or headspace GC-MS. The certified values of the 28 VOC analytes in the MCRM ranged from 6.162 to 17.37 µg/L with expanded uncertainties in the range of 0.5 to 5.3 µg/L. The prepared water MCRM could be used for quality control during VOC analysis in water and for developing or verifying measurement methods for VOCs monitoring in water.

Author Contributions: Conceptualization, L.F. and W.T.; methodology, L.F. and G.Y.; software, B.L.; validation, Y.J., H.L., and Y.Z.; investigation, H.L. and L.F.; data curation, L.H.; writing—original draft preparation, L.F. and L.H.; writing—review and editing, L.F., L.H. and W.T.; visualization, L.H.; supervision, W.T.; project administration, W.T. All authors have read and agreed to the published version of the manuscript.

Funding: This study was funded by the National Key Scientific Instrument and Equipment Development Project, China (2012YQ-060027).

Conflicts of Interest: The authors declare no conflict of interest.

References

1. Chary, N.S.; Fernandez-Alba, A.R. Determination of volatile organic compounds in drinking and environmental waters. *Trac-Trend. Anal. Chem.* **2012**, *32*, 60–75. [CrossRef]
2. Marczak, M.; Wolska, L.; Chrzanowski, W.; Namieśnik, J. Microanalysis of volatile organic compounds (VOCs) in water samples–methods and instruments. *Microchimica Acta* **2006**, *155*, 331–348. [CrossRef]
3. Feron, V.J.; Til, H.P.; de Vrijer, F.; van Bladeren, P.J. Toxicology of volatile organic compounds in indoor air and strategy for further research. *Indoor Environ.* **1992**, *1*, 69–81.
4. Chang, C.T.; Chen, B.Y.J. Toxicity assessment of volatile organic compounds and polycyclic aromatic hydrocarbons in motorcycle exhaust. *Hazard. Mater.* **2008**, *153*, 1262–1269. [CrossRef]
5. Altalyan, H.N.; Jones, B.; Bradd, J.; Long, D.N.; Alyazichi, Y.M.J. Removal of volatile organic compounds (VOCs) from groundwater by reverse osmosis and nanofiltration. *Water Process Eng.* **2016**, *9*, 9–21. [CrossRef]
6. Emons, H.; Linsinger, T.P.J.; Gawlik, B.M. Trac-Trend. Reference materials: Terminology and use. Can't one see the forest for the trees? *Anal. Chem.* **2004**, *23*, 442–449.
7. Quevauviller, P. Challenges for achieving traceability of environmental measurements. *Trac-Trend. Anal. Chem.* **2004**, *23*, 171–177. [CrossRef]
8. Deplagne, J.; Vial, J.; Pichon, V.; Lalere, B.; Hervouet, G.; Hennion, M.C. Feasibility study of a reference material for water chemistry: Long term stability of triazine and phenylurea residues stored in vials or on polymeric sorbents. *J. Chromatogr. A.* **2006**, *1123*, 31–37. [CrossRef]
9. Bercaru, O.; Manfred, G.; Awlik, B.; Ulberth, F.; Vandecasteele, C.J. Reference materials for the monitoring of the aquatic environment-a review with special emphasis on organic priority pollutants. *Environ. Monitor.* **2003**, *5*, 697–705. [CrossRef]
10. Mrabet, K.E.; Poitevin, M.; Vial, J.; Pichon, V.; Amarouche, S.; Hervouet, G.; Lalere, B.J. An interlaboratory study to evaluate potential matrix reference materials for herbicides in water. *Chromatogr. A.* **2006**, *1134*, 151–161. [CrossRef]

11. Kreeke, J.V.D.; Calle, B.D.L.; Held, A.; Bercaru, O.; Ricci, M.; Shegunova, P.; Taylor, P. IMEP-23: The eight EU-WFD priority PAHs in water in the presence of humic acid. *Trend. Anal. Chem.* **2010**, *29*, 928–937. [CrossRef]
12. Elordui-Zapatarietxe, S.; Fettig, I.; Philipp, R.; Gantois, F.; Lalère, B.; Swart, C.; Petrov, P.; Goenaga-Infante, H.; Vanermen, G.; Boom, G.; et al. Novel concepts for preparation of reference materials as whole water samples for priority substances at nanogram-per-liter level using model suspended particulate matter and humic acids. *Anal. Bioanal. Chem.* **2015**, *407*, 3055–3067. [CrossRef]
13. Elordui-Zapatarietxe, S.; Fettig, I.; Richter, J.; Philipp, R.; Gantois, F.; Lalère, B.; Swart, C.; Emteborg, H. Interaction of 15 priority substances for water monitoring at ng L−1 levels with glass, aluminium and fluorinated polyethylene bottles for the containment of water reference materials. *Accredit. Qual. Assur.* **2015**, *20*, 447–455. [CrossRef]
14. Guimarães, E.D.F.; Rego, E.C.P.D.; Cunha, H.C.M.; Rodrigues, J.M.; Figueroa-Villar, J.D. Certified reference material for traceability in environmental analysis: PAHs in toluene. *J. Braz. Chem. Soc.* **2013**, *25*, 351–360. [CrossRef]
15. van der Veen, A.M.H.; Linsinger, T.P. Pauwels, Uncertainty calculations in the certification of reference materials. 2. Homogeneity study. *J. Accredit. Qual. Assur.* **2001**, *6*, 26–30. [CrossRef]
16. van der Veen, A.M.H.; Linsinger, T.P.; Lamberty, A. Pauwels, Uncertainty calculations in the certification of reference materials 3. Stability study. *J. Accredit. Qual. Assur.* **2001**, *6*, 257–263. [CrossRef]
17. Sun, S.P.; Duan, J.P.; Zhou, X.M.; Xing, D.R. Determination of VOCs in air with capillary gas chromatography. *J. Environ. Health* **2007**, *24*, 446–448.
18. van der Veen, A.M.H.; Linsinger, T.P.J.; Schimmel, H.; Lamberty, A. Pauwels, Uncertainty calculations in the certification of reference materials 4. Characterisation and certification. *J. Accredit. Qual. Assur.* **2001**, *6*, 290–294. [CrossRef]
19. Neves, L.A.; Almeida, R.R.; Rego, E.P.; Rodrigues, J.M.; de Carvalho, L.J.; Goulart, A.L.D.M. Certified reference material of volatile organic compounds for environmental analysis: BTEX in methanol. *Anal. Bioanal. Chem.* **2015**, *407*, 3225–3229. [CrossRef]
20. Ellison, S.L.R. Homogeneity studies and ISO Guide 35:2006. *Accredit. Qual. Assur.* **2015**, *20*, 519–528. [CrossRef]
21. Fang, L.P.; Yang, G.; Jiang, Y.; Wang, W.; Liu, H.P.; Lu, B.W.; Tian, W. Development of the certified reference material of volatile organic compounds. *Environ. Chem.* **2016**, *35*, 688–696.

Article

Effect of Pore Size Distribution and Amination on Adsorption Capacities of Polymeric Adsorbents

Wei Liu *, Yuxi Zhang, Shui Wang, Lisen Bai, Yanhui Deng and Jingzhong Tao

Jiangsu Key Laboratory of Environmental Engineering, Jiangsu Academy of Environmental Sciences, 176# Jiangdong Beilu Road, Nanjing 210036, China; zhangyuxiNJU@163.com (Y.Z.); ws@vip.sina.com (S.W.); bailisen23@163.com (L.B.); jsshkylw@163.com (Y.D.); jshbtao@126.com (J.T.)
* Correspondence: liuwei@smail.nju.edu.cn or jshblw@126.com; Tel.: +86-025-58527708

Abstract: Polymeric adsorbents with different properties were synthesized via suspension polymerization. Equilibrium and kinetics experiments were then performed to verify the adsorption capacities of the resins for molecules of various sizes. The adsorption of small molecules reached equilibrium more quickly than the adsorption of large molecules. Furthermore, the resins with small pores are easy to lower their adsorption capacities for large molecules because of the pore blockage effect. After amination, the specific surface areas of the resins decreased. The average pore diameter decreased when the resin was modified with either primary or tertiary amines, but the pore diameter increased when the resin was modified with secondary amines. The phenol adsorption capacities of the amine-modified resins were reduced because of the decreased specific area. The amine-modified resins could more efficiently adsorb reactive brilliant blue 4 owing to the presence of polar functional groups.

Keywords: resin; adsorption; pore size distribution; chemical modification; water treatment

1. Introduction

Nowadays, large amounts of wastewater containing pollutants such as pesticides, dyes, heavy metals, and pharmaceuticals have been produced because of increased agricultural applications and the growth of the pharmaceutical and chemical industries [1]. Effluent containing even low concentrations of organic contaminants can be highly discernible and recalcitrant [2]. The large-scale production and widespread use of chemical substances can cause serious environmental problems, making it an important public concern [3]. Thus, the effective removal of organic pollutants from water sources is necessary for environmental security and public health.

Adsorption is a low-energy solid phase extraction technique that has been widely applied in industry in recent years. Various materials have been used as adsorbents, including mesoporous silica, natural zeolites and activated carbons [4,5]. However, these materials have some disadvantages that limit their application, such as unfavorable selective adsorption or poor regeneration [6].

Since the 1970s, functional polymers that have adsorption and separation capability have rapidly developed, and adsorbent resins have been widely used in various fields. Resins can adsorb organic matter through non-covalent bonding between molecules. The adsorbed molecules can then be eluted, thereby regenerating the resin to achieve the enrichment, separation and recovery of organic matter in wastewater [7–9]. For example, Zhu Shiyun et al. [10] used the strong anionic resin Amberlite IRA 402-OH to remove acetaldehyde from petrochemical production wastewater. The acetaldehyde removal efficiency from wastewater was 86%.

Adsorbent resins have been widely used to remove in dyes chemicals (such as benzene and naphthalene), pharmaceutical intermediates, and other organic compounds from wastewater [11]. Although existing adsorption resins can efficiently remove micromolecules and low water-soluble organic matter from wastewater, they are not effective for the adsorption of large molecules or organic matter with complex molecular structures. Therefore, a series of resins with different pore size distributions and high-adsorption capacity chemical groups were synthesized to address this problem. The adsorption behavior and effects on adsorption properties were studied.

Adsorbent resins can be synthesized via addition polymerization and condensation polymerization. The design of the pore structure in the resin synthesis is important and includes the pore volume, pore size, pore distribution, and specific surface area [12]. Many factors can affect the pore structure. For instance, when synthesizing macroporous adsorbents, the polarity, pore size, pore distribution, and specific surface area of the resin can be adjusted by controlling the polymerization conditions, such as the crosslinking agent, amount and type of porogen, and monomer composition [13,14]. The pore structure plays an important role in the adsorption ability of resins.

In addition to the pore diameter, the chemical structure of the resin also has an important effect on its adsorption properties. Li et al. [15] carbonized XAD-4 resin and found that the modified resin had good adsorption of phenolic compounds. Huang et al. [16] synthesized a new diethylenetriamine-modified hyper-cross-linked polystyrene resin with improved adsorptive removal of phenol. Li et al. [17] synthesized anion-exchange resins with different trialkylammonium groups, which significantly improved the nitrate adsorption capacity and improved the humic acid anti-fouling performance. Thus, we modified synthesized resins with amines and increase the resins adsorption for stronger polar and bigger molecules.

2. Materials and Methods

2.1. Chemicals

Divinylbenzene (DVB, 83.5%) and divinylbenzene (DVB, 50.0%) was purchased from Shandong Dongda Chemicals Company (Shandong, China) which were used as monomers. Toluene (>99.7%, Shanghai Chemical Reagent Company), liquid wax (>99.5%, Shanghai Chemical Reagent Company), n-heptane (>99.7%, Shandong Dongda Chemicals Company) and hexadecanol (>99.7%, Shanghai Chemical Reagent Company) were used as porogen. Gelatin (Photographic grade) was purchased from Yancheng Dafeng Gelatin Company (Jiangsu, China). Benzoyl peroxide (>99.7%, Shanghai Chemical Reagent Company) was used as an initiator. Zinc chloride (>99.7%) as the catalyst for chloromethylation were purchased from Shanghai Chemical Reagent Company. Chloromethyl ether (>99.7%, Shanghai Chemical Reagent Company), chloroform (>99.7%, Nanjing Chemical Reagent Company) and methylal (>99.7%, Shanghai Chemical Reagent Company) were used as the swelling agent. Hexamethylenetetramine (>99.7%), methylamine aqueous solution (25%) and dimethylamine aqueous solution (33%) used for amino-modified were all purchased from Nanjing Chemical Reagent Company. HCl and NaOH were obtained from Shanghai Chemical Reagent Company. Phenol (>99.7%), 1-amino-4- bromoanthraquinone-2sulfonic acid (ABS acid, >99.7%), and reactive brilliant blue 4 (RBB4, >99.5%) used as adsorbates were also purchased from Nanjing Chemical Reagent Company. Their structure and molecular sizes of them were estimated by WinMopac7.21, and the results are shown in Figure S1.

2.2. Preparation

An aqueous phase (200.0 g) solution consisting of 1% gelatin and an oil phase consisting of divinylbenzene (DVB, 83.5% or 50.0%), benzoyl peroxide (initiator) and different porogen were added to a 500-mL flask. The stirring speed was 300 rpm. The temperature was gradually increased to 80 °C for 3 h, then to 85 °C for 4 h, and finally to 95 °C for 4 h. After cooling for 2 h, the polymer beads were filtered with glass sand core vacuum filtration apparatus (50 mesh) and washed with hot water (60 °C). The obtained resins were NDA-1, NDA-2, NDA-3, NDA-4 and NDA-1800 for different monomer concentrations, porogen proportion and monomer: porogen ratio. All the polymer beads were then extracted with acetone in a Soxhlet extractor for 8 h, following by vacuum drying for 8 h at 60 °C under a

10 mmHg vacuum, respectively. The reaction process, final product and corresponding reagents are shown in Figure 1 and Table S1.

Then, the primary aminated resin (NDA-1801), the secondary aminated resin (NDA-1802) and the tertiary aminated resin (NDA-1803) were prepared, respectively. NDA-1800 resin (300.0 g) was chloromethylated with chloromethyl ether (1800.0 g) by using zinc chloride (180.0 g) as the catalyst. For NDA-1801 preparation, the chloromethylated NDA-1800 resin (80.0 g) was then emerged in chloroform (120 mL) for 2 h, and then hexamethylenetetramine aqueous solution (50.0%, 331.7 g) was added. The pH value was adjusted to 12.0 with NaOH (10 mol/L). For the preparation of NDA-1802, NDA-1800 resin (80.0 g) was swelled in methylal (120 g) 2 h, and then the methylamine aqueous solution (25%) was added. The pH was adjusted to 12 with NaOH (10 mol/L). The solution was then incubated at 45 °C for 10 h. For the preparation of NDA-1803, NDA-1800 resin (80.0 g) was also swelled in methylal (120 g) 2 h. Then, the dimethylamine aqueous solution (33%) was added, and the pH was adjusted to 12 with NaOH (40%). The solution was incubated at 45 °C for 10 h. After the amination, all the obtained beads were washed with deionized water until the pH was neutral.

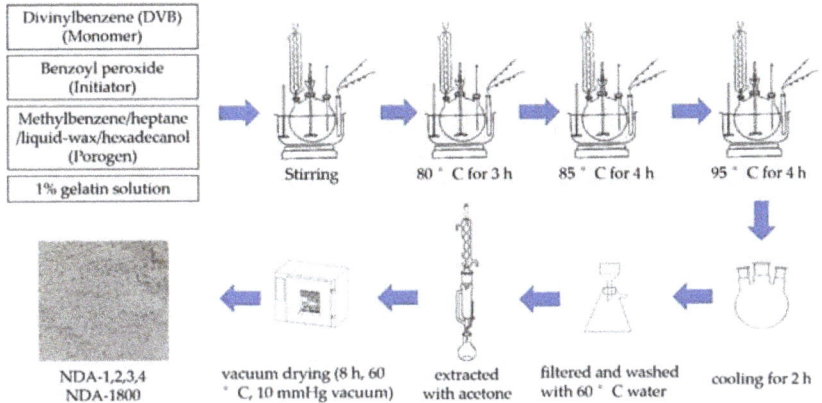

Figure 1. The flow diagram of resin synthesis.

2.3. Characterization

The BET surface area (S_{BET}) of the resin was calculated using the isothermal adsorption and desorption experiments of N_2 at 77 K. The micropore volume (V_{micro}) and the micropore area (S_{micro}) were determined by the t-plot method [18]. When the relative pressure is close to 1 (0.98 in this experiment), the N_2 adsorption amount (V_t) corresponds to the total adsorption amount of micropores and mesopores in the resin. The mesopore volume (V_{meso}) of the resin can be obtained by subtracting V_{micro} from the adsorption amount at this time. The average pore diameter in the mesoporous region was obtained through a BJH desorption experiment. The whole process was completed automatically by an accelerated surface area and porosimeter system (ASAP 2010, Micromeritics, Georgia USA). The exchange capacity was determined by acid-base titration. The pretreated resin was soaked in the HCl aqueous solution, and the capacity was obtained by NaOH titration of the residual HCl in the solution with phenolphthalein [19]. Fourier transform infrared (FTIR) spectroscopy was performed with an FTIR spectrophotometer (Nicolet 170 SX, Georgia USA) via KBr pressed-disk technique.

2.4. Adsorption Kinetics

A 0.1000 g aliquot of resin was accurately weighed into a 250-mL conical flask, and 100 mL of preheated (303 K) 500 mg·L^{-1} aqueous solution of adsorbate was added. The flask was sealed and placed on a constant temperature shaker set at 303 K and shaken at 130 rpm. Samples were removed at regular intervals to measure the concentration of the adsorbates via a high-performance liquid chromatography (Waters 600) and a UV-vis spectrophotometer (GBC UV/VIS 916).

2.5. Static Equilibrium Adsorption

A 0.1000 g aliquot of resin was weighed into a 250-mL conical flask. Different concentrations of adsorbate solution (100 mL) were added, and the mixtures were shaken for 48 h on a constant temperature shaker at 303 K and 130 rpm. The solute concentrations in the equilibrated solutions were then analyzed.

2.6. Analysis

Phenol was determined by a high-performance liquid chromatography (Waters 600). ABS acid and RBB4 were determined by a UV-vis spectrophotometer (GBC UV/VIS 916) at 486 nm and 605 nm, respectively.

The equilibrium adsorption capacity (Q_e (mmol·g^{-1})) was calculated as follows:

$$Q_e = \frac{V_1(C_0 - C_e)}{MW} \quad (1)$$

where, V_1 (L) is the volume of the solution, W (g) is the mass of the dry resin, and M is the molar mass of the adsorbate molecule.

3. Results and Discussion

3.1. Characterization

The specific surface area and pore structure of the synthesized resins (NDA-1, NDA-2, NDA-3, and NDA-4) were measured. Their specific surface areas ranged from 448.0 to 565.1 m^2·g^{-1}.

The characterization results of the resins are shown in Table 1. The mesopore diameter distribution was calculated using the Barret-Joyner-Halenda (BJH) method, as shown in Figure 2. NDA-2 had the largest specific surface area, and the other three resins had similar specific surface areas. The pore diameter distributions of NDA-1 and NDA-2 were relatively narrow, but those of NDA-3 and NDA-4 were relatively wide.

Table 1. Characterization of resins.

Number	Specific Surface Area (m^2·g^{-1})			S_{ext}%	Pore Volume (cm^3·g^{-1})			V_{meso}%	Average Pore Diameter (nm)
	S_{BET}	S_{micro}	S_{ext}		V_t	V_{micro}	V_{meso}		
NDA-1	448.0	37.2	410.8	91.7	0.310	0.0098	0.3002	96.8	3.80
NDA-2	565.1	51.8	513.3	90.8	0.474	0.0148	0.4592	96.9	4.74
NDA-3	452.4	43.1	409.3	90.5	0.736	0.0213	0.7147	97.1	6.87
NDA-4	486.6	32.4	454.2	93.3	0.767	0.0091	0.7579	98.8	6.72

Note: S_{BET}, S_{micro} and S_{ext} are the BET specific surface area, micropore area and non-micropore area, respectively. V_t, V_{micro} and V_{meso} are the pore volume, micropore volume and meso-pore volume, respectively, when the relative pressure was 0.98. S_{ext}% and V_{meso}% are the percentage of the specific surface area of non-micropores and the percentage of the mesopore volume, respectively.

The N$_2$ adsorption isotherms of the synthetic resins were mostly II and IV hybrid adsorption isotherms (see Figure 3). Micropore filling occurred when relative pressure (p/p$_0$) was between 0.0 to 0.2 and the adsorption was low. When the relative pressure was slightly higher, the hysteresis curve appeared between the adsorption and desorption curves owing to the occurrence of multilayer adsorption and capillary condensation, indicating that most pores in the resins were mesopores [20]. The low range isotherms (at the logarithmic scale (p/p$_0$) scale) were also presented in supporting information (Figure S2).

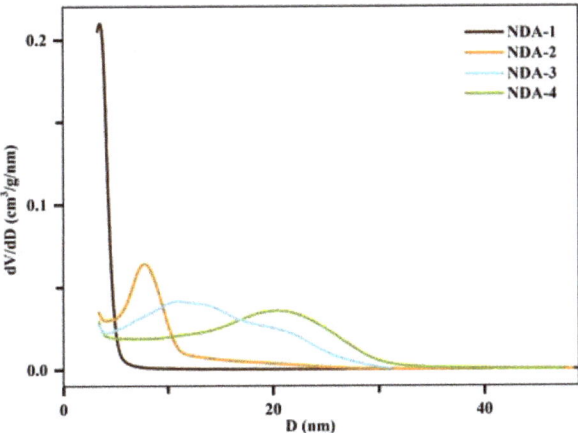

Figure 2. Mesopore diameter distribution of NDA-1, NDA-2, NDA-3, and NDA-4.

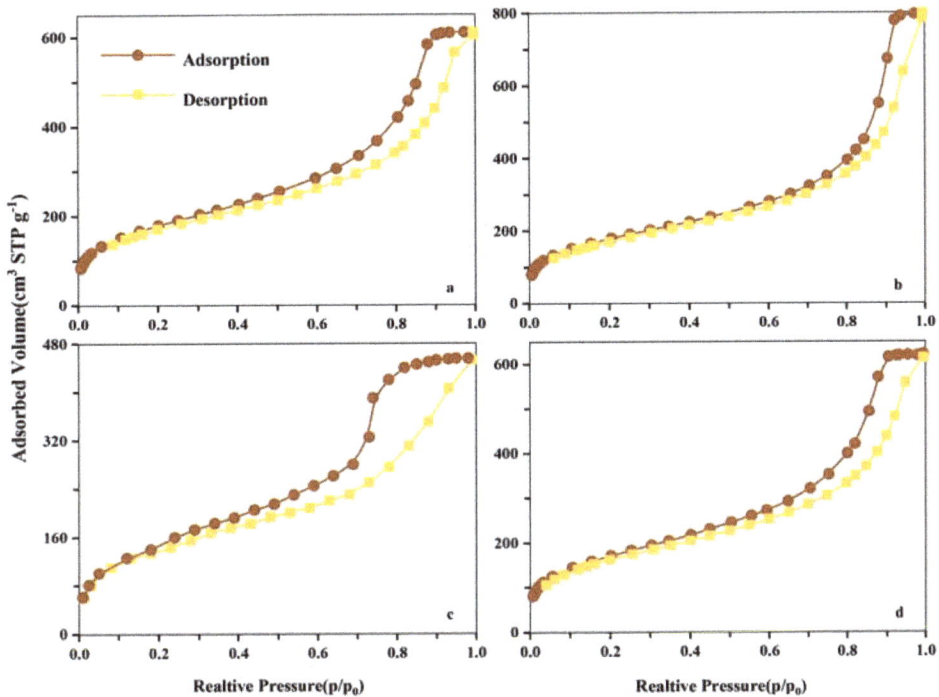

Figure 3. N_2 adsorption isotherms of NDA-1 (**a**), NDA-2 (**b**), NDA-3 (**c**) and NDA-4 (**d**) resins at 77 K.

After modification of NDA-1800 with amines, their characterizations were determined. The infrared spectra of NDA-1800, NDA-1801, NDA-1802, and NDA-1803 resins are shown in Figure S3. Compared with the NDA-1800 resin, the resins modified by primary amines, secondary amines and tertiary amines had a very small peak around 672 cm^{-1}; this was attributed to the chloromethyl peak. The peaks appearing near 1040 cm^{-1} and 1118 cm^{-1} were designated as C-N stretching vibration peaks, and they were not present for NDA-1800. In the modified resins, the peaks around 3445 cm^{-1} were N-H stretching vibration peaks, which indicated that amine groups had been successfully introduced to the three modified resins. Moreover, the chloride content of the resin after amination was decreased, indicating that the chlorine had been partly replaced by the amino groups (Table 2). The exchange capacities of the three modified resins were 1.3, 1.49, and 5.4 mmol·g^{-1} for NDA-1801, NDA-1802, and NDA-1803, effectively, showing that the number of exchangeable amino groups increased on the resin skeleton after reaction. Therefore, when contacted by adsorbate, the resins could provide exchange sites and electrostatic adsorption sites in addition to the π-π interactions of the resin skeleton.

Table 2. Characterization of the modified resins.

Number	Specific Surface Area (m^2·g^{-1})		Pore Volume (cm^3·g^{-1})		Average Pore Diameter (nm)	Modification Methods	Polarity	Mean Particle Size (mm)	Residual Chlorine Content (%)	Total Exchange Capacity (mmol·g^{-1})
	S_{BET}	S_{micro}	V_t	V_{micro}						
NDA-1800	852.3	93.3	1.470	0.022	7.47	/	nonpolar	0.4–0.6	/	/
NDA-1801	630.6	84.1	0.944	0.021	5.99	Primary amine	polar	0.4–0.6	6.8	1.13
NDA-1802	628.4	54.7	1.230	0.009	7.86	Secondary amine	polar	0.4–0.6	7.1	1.49
NDA-1803	602.6	59.0	0.961	0.010	6.38	Tertiary amine	polar	0.4–0.6	5.4	5.40

The data in Table 2 show that compared with NDA-1800, the specific surface areas of the resins modified with primary amines, secondary amines and tertiary amines were 26.0%, 26.3%, and 29.3% smaller, respectively. This was because during chloromethylation some part of the resin pores collapsed owing to the swelling action of chloromethyl ether, resulting in a decreased specific surface area. Figure 4 shows that the pore diameter of the resins also changed after modification with amine groups. The concentration pore diameters of resins modified with primary amines and tertiary amines slightly decreased to approximately 14 nm, while that of the resin modified with secondary amines slightly increased.

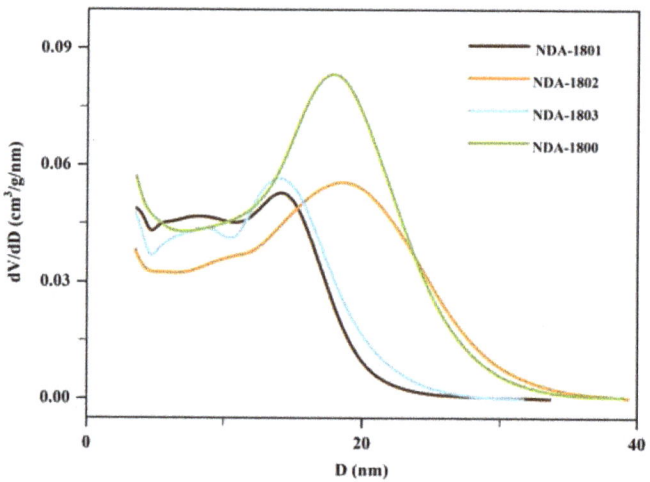

Figure 4. Mesopore diameter distribution of NDA-1800, NDA-1801, NDA-1802, and NDA-1803.

3.2. Adsorption Kinetics

The adsorption kinetic data were analyzed using pseudo-first-order and pseudo-second-order rate equations via nonlinear fitting [21,22]; Table 3 lists the corresponding parameters. The correlation coefficient (R^2) suggested that the pseudo-second-order rate equation was a better fitting model.

Figure 5 shows the adsorption kinetic curves of the three adsorbates on NDA-2 and NDA-3. Phenol rapidly reached equilibrium in approximately 0.5 h, ABS acid reached equilibrium in approximately 8 h, and RBB4 reached equilibrium in approximately 20 h. The kinetics of the different adsorbates varied because the adsorption process on the inner surface of the resin is often very fast, and is mainly controlled by internal diffusion into the resin pores [23]. Owing to the relative size of the resin pores and the molecular size of phenol, the diffusion of phenol at the resin pore interface is similar to molecular diffusion in which there is free diffusion and a relatively high diffusion rate. However, RBB4 has a larger molecular volume, which leads to more frequent collisions with the pore wall, greater diffusion resistance, and a slower diffusion rate.

The rate constant, k_2, of phenol adsorption was in the following order: NDA-2 > NDA-3, while this was the opposite for RBB4. The equilibrium adsorption capacities of NDA-2 for phenol and ABS acid were greater than those of NDA-3, while this was the opposite for RBB4. The pore diameter of NDA-2 was smaller than that of NDA-3, leading to a higher equilibrium adsorption capacity for micromolecules such as phenol and ABS acid. In contrast, NDA-3 had higher equilibrium adsorption capacity for the macromolecule RBB4. Therefore, the pore diameter distribution of the resin not only affected the adsorption rate, but also influenced the adsorption capacity of the resin for an adsorbate.

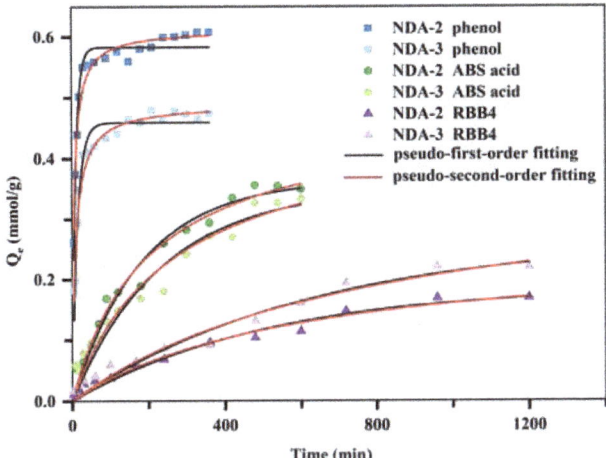

Figure 5. Adsorption kinetic curves of the three adsorbates.

Table 3. The corresponding parameters of the adsorption kinetic curves.

	Pseudo-First-Order Rate Equation			Pseudo-Second-Order Rate Equation		
	Q_e (L·mmol^{-1})	K_1 (L·mmol^{-1})	R^2	Q_e (L·mmol^{-1})	K_2 (L·mmol^{-1})	R^2
NDA-2 phenol	0.583	0.102	0.963	0.611	0.284	0.969
NDA-3 phenol	0.459	0.069	0.949	0.489	0.213	0.963
NDA-2 ABS acid	0.364	0.005	0.974	0.477	0.011	0.980
NDA-3 ABS acid	0.350	0.004	0.939	0.467	0.008	0.951
NDA-2 RBB4	0.267	0.002	0.957	0.384	0.003	0.958
NDA-3 RBB4	0.189	0.002	0.972	0.260	0.006	0.976

3.3. Static Adsorption Equilibrium

We used a three-parameter multilayer adsorption model to describe the adsorption data [24]. The equation was as follows:

$$Q_e = \frac{Q_m K_1 C_e}{(1 - K_2 C_e)[1 + (K_1 - K_2)C_e]} \qquad (2)$$

where C_e (mmol·L^{-1}) is the equilibrium concentration of adsorbate, Q_e (mmol·g^{-1}) is the adsorption capacity of resin, Q_m (mmol·g^{-1}) is the maximum first-layer adsorption capacity of resin, K_1 (L·mmol^{-1}) is the first-layer adsorption equilibrium constant, and K_2 (L·mmol^{-1}) is the multilayer adsorption equilibrium constant. The three-parameter multilayer adsorption model can be used to describe almost all S-type adsorption isotherms, including the BET isotherm and the Freundlich isotherm, which are typical for gas adsorption and organic matter adsorption. The equilibrium adsorption isotherms are shown in Figure 6, and Table 4 summarizes the corresponding parameters. The adsorption isotherms of phenol were approximately straight lines in the concentration range of our study and could not be simulated by this model.

Figure 6a shows the adsorption isotherms of phenol on the four resins. The order of adsorption capacity was NDA-2 > NDA-1 > NDA-3 > NDA-4. The NDA-2 resin had the highest adsorption capacity because it had the largest specific surface area. Although NDA-1, NDA-3 and NDA-4 had similar specific surface areas, the pore diameter distribution of NDA-1 was mainly concentrated in the small mesopore region. This resin is suitable for the adsorption of micromolecules such as phenol. However, the opposite was true for NDA-4, which had lowest adsorption capacity for phenol.

Figure 6b shows the adsorption isotherm of ABS acid on the four resins. The order of adsorption capacity was NDA-2 > NDA-3 > NDA-4 > NDA-1. Different from phenol, the adsorption capacity of NDA-1 for ABS acid was much lower than that of the other three resins. This was because the adsorption of ABS acid to NDA-1 was affected by pore hindrance, making part of the pore volume unusable. One reason for this phenomenon is that the pore size of NDA-1 was smaller than that of the adsorbate molecule; thus, the molecule cannot be adsorbed within the resin pores. Second, in the process of diffusion from the surface to the inner channels of the resin, the adsorbate molecules are partly adsorbed onto the pore walls. This can cause the channels become blocked, preventing the adsorbate from passing through the channels. As can be seen from the pore diameter distribution diagram, the pore diameter of NDA-1 was mainly distributed around approximately 3 nm, this is 2–3 times larger than the diameter of ABS acid. Therefore, the adsorption of this molecule caused pore obstruction.

Figure 6c shows the adsorption isotherm of RBB4 on the four resins. The order of adsorption capacity was NDA-3 > NDA-4 > NDA-2 > NDA-1. Although NDA-2 had the largest specific surface area, its adsorption capacity was far lower than that of NDA-3 and NDA-4. This was because the NDA-2 resin was also affected by pore hindrance. This effect was enhanced by the molecular aggregation of RBB4 [25]. Walker et al. [26] showed that the degree to which some acid dyes aggregate on the surface of activated carbon and bone charcoal could be greater than 10 molecules. Accordingly, it is understandable that the macromolecule RBB4 would experience pore hindrance on, NDA-1 and NDA-2 owing to small pore diameters of these resins.

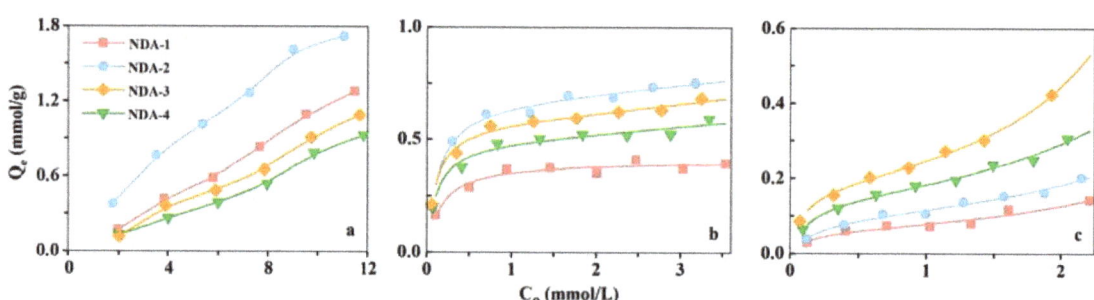

Figure 6. (a) Adsorption isotherms of phenol; (b) Adsorption isotherms of ABS acid; (c) Adsorption isotherms of RBB4.

As shown in Figure 6, although NDA-4 and NDA-3 had similar specific surface areas, pore volumes and average pore diameters, the adsorption capacity of NDA-4 was lower than that of NDA-3 for the three adsorbates. This was because the pore diameter of NDA-4 was larger than that of the adsorbates. Thus, pore size that is either too small or too large is not conducive to improving the adsorption capacity of a resin, and there was an optimal adsorption pore size for specific adsorbates.

Table 4. The corresponding parameters of three-parameter multilayer adsorption model.

Resins	ABS Acid				RBB4			
	Q_m (mmol·g^{-1})	K_1 (L·mmol^{-1})	K_2 (L·mmol^{-1})	R^2	Q_m (mmol·g^{-1})	K_1 (L·mmol^{-1})	K_2 (L·mmol^{-1})	R^2
NDA-1	0.407	6.239	0.040	0.971	0.066	6.727	0.252	0.971
NDA-2	0.683	7.862	0.038	0.995	0.113	4.054	0.200	0.995
NDA-3	0.576	10.830	0.050	0.994	0.239	6.015	0.154	0.999
NDA-4	0.493	10.080	0.047	0.985	0.166	5.884	0.212	0.995

The size distribution of the pores greatly influences the adsorption capacity of a resin. Only when the resin has a proper pore diameter can the adsorbate achieve good diffusion into the channels and be absorbed effectively. If the pore diameter is too large and the specific surface area is small, the resin will have a low adsorption capacity. If the pore diameter is too small, it will limit the diffusion of adsorbates and solvents, which enhances the shielding effect for molecules with larger diameters [27]. When the pore diameter of a micropore is less than several times the diameter of the adsorbate molecule, a potential shielding effect can occur. This can increase the interaction between the solid surface and gas molecules, thus enhancing adsorption [28]. Macropores, mesopores and micropores are typically categorized by pore diameters greater than 50 nm, 2–50 nm, and less than 2 nm, respectively. The surface of macropores contributes little to adsorption and only provides a channel for the diffusion of adsorbates and solvent. Mesopores can adsorb macromolecules and help micromolecules pass through the micropores.

Adsorption is related to the chemical structure of resin, regardless of the adsorption method. Regarding the adsorption mechanism, non-polar adsorbent resins adsorb molecules through van der Waals forces. By contrast, adsorption typically occurs via dipole–dipole and electrostatic interactions (including hydrogen bond and donor–acceptor interactions) for medium and strongly polar resins.

3.4. Adsorption Properties of the Modified Resins

The adsorption properties of the three amino-modified resins (NDA-1801, NDA-1802, and NDA-1803) for phenol, ABS acid and RBB4 were investigated at 288 K, 303 K, and 318 K and were compared with NDA-1800. The results are presented in Figure 7 and Table S2 summarizes the corresponding parameters.

The phenol adsorption capacities of the four resins were similar at the three temperatures. The order of adsorption capacity was NDA-1800 > NDA-1801 ≈ NDA-1803 ≈ NDA-1802. The adsorption of micromolecules such as phenol is mainly achieved through π-π and dipole–dipole interactions. The resin adsorption capacity was greatly affected by the surface of the resin, and no ion exchange occurred. Compared with NDA-1800, the specific surface areas of the three amine-modified resins were decreased; thus, the adsorption capacities were lower. However, the specific surface area of the three amino-modified resins were similar, and their adsorption capacities were not significantly different.

The ABS acid adsorption capacity of the four resins was in the order of NDA-1801 ≈ NDA-1803 > NDA-1800 > NDA-1802 at all three temperatures. Compared with NDA-1800, the adsorption capacities of NDA-1801 and NDA-1803 were increased because of the increased polarity of the resins and the enhanced dipole-dipole and ion exchange interactions that occurred between the resin and the adsorbate, even though the specific surface areas were reduced after amination. The pore diameter distribution was slightly reduced, which was suitable for the adsorption of ABS acid molecules and increased the adsorption capacity. The adsorption capacity of the resin modified with secondary amines decreased slightly owing to the increased pore size distribution and the decreased specific surface area.

The RBB4 adsorption capacity of the four resins at all three temperatures was in the order of NDA-1802 > NDA-1803 ≈ ND-1801 > NDA-1800. Compared with NDA-1800, the adsorption capacities of the amine-modified resins markedly improved. The adsorption capacity of the secondary amine resin (NDA-1802) increased the most. This was because the presence of the polar functional group could be used for ion exchange and because the resin pore size remained large. The adsorption capacities of NDA-1801 and NDA-1803 increased owing to the presence of polar functional groups.

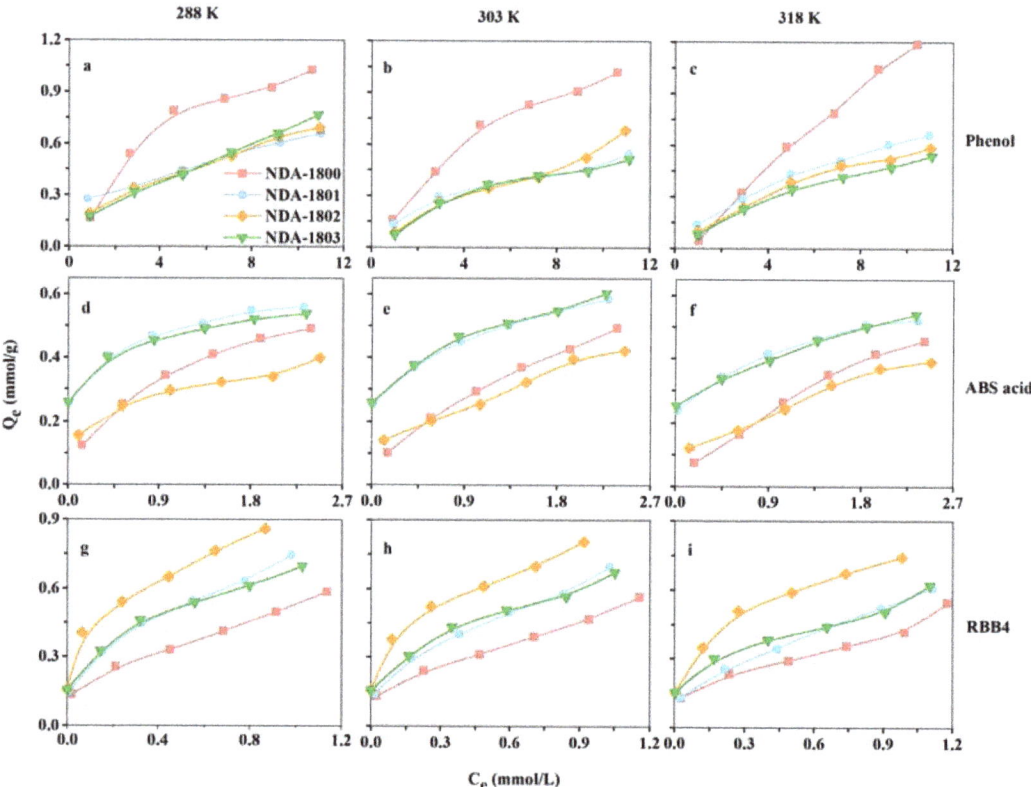

Figure 7. (a) Adsorption isotherms of phenol at 288 K; (b) Adsorption isotherms of phenol at 303 K; (c) Adsorption isotherms of phenol at 318 K; (d) Adsorption isotherms of ABS acid at 288 K; (e) Adsorption isotherms of ABS acid at 303 K; (f) Adsorption isotherms of ABS acid at 318 K; (g) Adsorption isotherms of RBB4 at 288 K; (h) Adsorption isotherms of RBB4 at 303 K; (i) Adsorption isotherms of RBB4 at 318 K.

4. Conclusions

We have synthesized a series of mesoporous resins with different pore size distributions via suspension polymerization. These resins, which were suitable for the adsorption of macromolecules. The adsorption kinetics data were well-fitted by a pseudo-second-order rate equation, and the adsorption equilibrium data were described by a three-parameter multilayer adsorption model. The adsorption of small molecules reached equilibrium quicker than large molecules, and resins with smaller pores had a lower adsorption capacity for large molecules because of the pore blockage effect. NDA-1 and NDA-2 could efficiently adsorb phenol and ABS acid, while NDA-3 could efficiently adsorb RBB4. The resins were modified by amine groups. The presence of polar functional groups led to more efficient RBB4 adsorption, while the decreased specific surface areas led to decreased phenol adsorption capacities. Thus, a composite functional adsorbent resin with an improved ability to adsorb the macromolecule RBB4 was obtained. These mesoporous resins are promising candidates for application in removing organic contaminants from wastewater.

Supplementary Materials: The following are available online, Figure S1: Characterization of the adsorbates, Table S1: The reagents used of synthetic resins, Figure S2: N2 adsorption isotherms of NDA-1 (a), NDA-2 (b), NDA-3 (c) and NDA-4 (d) resins at 77K (logarithmic scale (p/p0)), Figure S3: The infrared spectra of the four resins, Table S2: The corresponding parameters of three-parameter multilayer adsorption model.

Author Contributions: Conceptualization, W.L. and Y.D.; methodology, W.L. and J.T.; validation, S.W.; formal analysis, Y.Z.; investigation, W.L.; writing—original draft preparation, Y.Z.; writing—review and editing, Y.Z. and W.L.; visualization, Y.Z. and W.L.; supervision, L.B.; project administration, L.B. and S.W. All authors have read and agreed to the published version of the manuscript.

Funding: This work was supported by The National Key Research and Development Program of China (2018YFC1801606), National Natural Science Foundation of China (51808267) and Independent Research Project of Key Laboratory of Environmental Engineering of Jiangsu Province (ZZ2020002).

Institutional Review Board Statement: Not applicable.

Informed Consent Statement: Not applicable.

Data Availability Statement: Not applicable.

Conflicts of Interest: The authors declare no conflict of interest.

References

1. Vijayaraghavan, K.; Ashokkumar, T. Characterization and evaluation of reactive dye adsorption onto Biochar Derived from Turbinaria conoides Biomass. *Environ. Prog. Sustain. Energy* **2019**, *38*, 13143. [CrossRef]
2. Jothirani, R.; Kumar, P.S.; Saravanan, A.; Narayan, A.S.; Dutta, A. Ultrasonic modified corn pith for the sequestration of dye from aqueous solution. *J. Ind. Eng. Chem.* **2016**, *39*, 162–175. [CrossRef]
3. Malik, A.; Rahman, M.; Ansari, M.I.; Masood, F.; Grohmann, E. Environmental protection strategies: An overview. *Environ. Prot. Strateg. Sustain. Dev.* **2012**, 1–34. [CrossRef]
4. Fu, Y.; Jiang, J.; Chen, Z.; Ying, S.; Wang, J.; Hu, J. Rapid and selective removal of Hg(II) ions and high catalytic performance of the spent adsorbent based on functionalized mesoporous silica/poly(m-aminothiophenol) nanocomposite. *J. Mol. Liq.* **2019**, *286*, 110746. [CrossRef]
5. Yadav, V.B.; Gadi, R.; Kalra, S. Clay based nanocomposites for removal of heavy metals from water: A review. *J. Environ. Manag.* **2019**, *232*, 803–817. [CrossRef]
6. Issabayeva, G.; Aroua, M.K.; Sulaiman, N.M. Study on palm shell activated carbon adsorption capacity to remove copper ions from aqueous solutions. *Desalination* **2010**, *262*, 94–98. [CrossRef]
7. Tsyurupa, M.P.; Blinnikova, Z.K.; Davidovich, Y.A.; Lyubimov, S.E.; Naumkin, A.V.; Davankov, V.A. On the nature of "functional groups" in non-functionalized hypercrosslinked polystyrenes. *React. Funct. Polym.* **2012**, *72*, 973–982. [CrossRef]
8. Xiao, G.; Wen, R.; Wei, D.; Wu, D. Effects of the steric hindrance of micropores in the hyper-cross-linked polymeric adsorbent on the adsorption of p-nitroaniline in aqueous solution. *J. Hazard. Mater.* **2014**, *280*, 97–103. [CrossRef] [PubMed]
9. Sun, Y.; Chen, J.; Li, A.; Liu, F.; Zhang, Q. Adsorption of resorcinol and catechol from aqueous solution by aminated hypercrosslinked polymers. *React. Funct. Polym.* **2005**, *64*, 63–73. [CrossRef]
10. Salehi, E.; Shafie, M. Adsorptive removal of acetaldehyde from water using strong anionic resins pretreated with bisulfite: An efficient method for spent process water recycling in petrochemical industry. *J. Water Process. Eng.* **2020**, *33*, 101025. [CrossRef]
11. Mansha, M.; Waheed, A.; Ahmad, T.; Kazi, I.W.; Ullah, N. Synthesis of a novel polysuccinimide based resin for the ultrahigh removal of anionic azo dyes from aqueous solution. *Environ. Res.* **2020**, *184*, 109337. [CrossRef] [PubMed]
12. Tsyurupa, M.P.; Davankov, V.A. Porous structure of hypercrosslinked polystyrene: State-of-the-art mini-review. *React. Funct. Polym.* **2006**, *66*, 768–779. [CrossRef]
13. Kuang, W.; Li, H.; Huang, J.; Liu, Y.-N. Tunable porosity and polarity of the polar hyper-cross-linked resins and the Enhanced adsorption toward phenol. *Ind. Eng. Chem. Res.* **2016**, *55*, 12213–12221. [CrossRef]

14. Wu, Y.; Li, Z.; Xi, H. Influence of the microporosity and surface chemistry of polymeric resins on adsorptive properties toward phenol. *J. Hazard. Mater.* **2004**, *113*, 131–135. [CrossRef]
15. Li, A.; Zhang, Q.; Chen, J.; Fei, Z.; Long, C.; Li, W. Adsorption of phenolic compounds on Amberlite XAD-4 and its acetylated derivative MX-4. *React. Funct. Polym.* **2001**, *49*, 225–233. [CrossRef]
16. Huang, J.; Zha, H.; Jin, X.; Deng, S. Efficient adsorptive removal of phenol by a diethylenetriamine-modified hypercrosslinked styrene–divinylbenzene (PS) resin from aqueous solution. *Chem. Eng. J.* **2012**, *195–196*, 40–48. [CrossRef]
17. Li, Q.; Lu, X.; Shuang, C.; Qi, C.; Wang, G.; Li, A.; Song, H. Preferential adsorption of nitrate with different trialkylamine modified resins and their preliminary investigation for advanced treatment of municipal wastewater. *Chemosphere* **2019**, *223*, 39–47. [CrossRef] [PubMed]
18. Shields, S.L.E. *Powder Surface Area and Porosity*; Chapman and Hall: New York, NY, USA, 1991; p. 119.
19. Shuang, C.; Pan, F.; Zhou, Q.; Li, A.; Li, P.; Yang, W.J.I.; Research, E.C. Magnetic polyacrylic anion exchange resin: Preparation, characterization and adsorption behavior of humic acid. *Ind. Eng. Chem. Res.* **2012**, *51*, 4380–4387. [CrossRef]
20. Horikawa, T.; Do, D.D.; Nicholson, D. Capillary condensation of adsorbates in porous materials. *Adv. Colloid Interface Sci.* **2011**, *169*, 40–58. [CrossRef] [PubMed]
21. Lagergren, S.K. About the theory of so-called adsorption of solution substances. *Sven. Vetenskapsakad. Handingarl* **1898**, *24*, 1–39.
22. Ho, Y.S.; McKay, G.; Pollution, S. Sorption of Copper(II) from aqueous solution by peat. *Water Air Soil Pollut.* **2004**, *158*, 77–97. [CrossRef]
23. Zhang, J.; Zhu, C.; Sun, H.; Peng, Q. Separation of glycolic acid from glycolonitrile hydrolysate using adsorption technology. *Colloids Surf. A Physicochem. Eng. Asp.* **2017**, *520*, 391–398. [CrossRef]
24. Wang, J.; Huang, C.P.; Allen, H.E.; Cha, D.K.; Kim, D.-W. Adsorption characteristics of dye onto sludge particulates. *J. Colloid Interface Sci.* **1998**, *208*, 518–528. [CrossRef]
25. Yang, W.; Xue, X.; Zheng, F.; Yang, X. Importance of surface area and pore size distribution of resin for organic toxicants adsorption. *Sep. Sci. Technol.* **2011**, *46*, 1321–1328. [CrossRef]
26. Walker, G.M.; Weatherley, L. Adsorption of dyes from aqueous solution—The effect of adsorbent pore size distribution and dye aggregation. *Chem. Eng. J.* **2001**, *83*, 201–206. [CrossRef]
27. Huang, J. Molecular sieving effect of a novel hyper-cross-linked resin. *Chem. Eng. J.* **2010**, *165*, 265–272. [CrossRef]
28. Long, C.; Lu, J.; Li, A.; Hu, D.; Liu, F.; Zhang, Q. Adsorption of naphthalene onto the carbon adsorbent from waste ion exchange resin: Equilibrium and kinetic characteristics. *J. Hazard. Mater.* **2008**, *150*, 656–661. [CrossRef]

Article

Distributions of Polychlorinated Naphthalenes in Sediments of the Yangtze River, China

Zhitong Liu [1], Ke Xiao [1,*], Jingjing Wu [1], Tianqi Jia [1,2], Rongrong Lei [1,2] and Wenbin Liu [1,2,3]

[1] State Key Laboratory of Environmental Chemistry and Ecotoxicology, Research Center for Eco-Environmental Sciences, Chinese Academy of Sciences, Beijing 100085, China; ztliu@rcees.ac.cn (Z.L.); jingwoo@rcees.ac.cn (J.W.); tqjia@126.com (T.J.); leirongr@163.com (R.L.); liuwb@rcees.ac.cn (W.L.)
[2] College of Resources and Environment, University of Chinese Academy of Sciences, Beijing 100049, China
[3] Hangzhou Institute for Advanced Study, University of Chinese Academy of Sciences, Hangzhou 310024, China
* Correspondence: kexiao@rcees.ac.cn; Tel.: +86-10-6284-9870; Fax: +86-10-62849339

Abstract: The pollution status of polychlorinated naphthalenes (PCNs) in the sediment of the Yangtze River Basin, Asia's largest river basin, was estimated. The total concentrations of PCNs (mono- to octa-CNs) ranged from 0.103 to 1.631 ng/g. Mono-, di-, and tri-PCNs—consisting of CN-1, CN-5/7, and CN-24/14, respectively, as the main congeners—were the dominant homolog groups. Combustion indicators and principal component analysis showed that the emissions from halowax mixtures were the main contributor to PCNs in sediment, among most of the sampling sites. The mean total toxic equivalent (TEQ) was calculated to be 0.045 ± 0.077 pg TEQ/g, which indicates that the PCNs in sediments were of low toxicity to aquatic organisms. This work will expand the database on the distribution and characteristics of PCNs in the river sediment of China.

Keywords: PCNs; spatial distribution; congener; environmental risk; sediment

1. Introduction

Polychlorinated naphthalenes (PCNs) have been synthesized since the 1930s, with 75 congeners based on the number and position of the chlorine(s) in the naphthalene ring system, including 2 mono-chlorinated (CN-1–CN-2), 10 di-chlorinated (CN-3–CN-12), 14 tri-chlorinated (CN-13–CN-26), 22 tetra-chlorinated (CN-27–CN-48), 14 penta-chlorinated (CN-49–CN-62), 10 hexa-chlorinated (CN-64–CN-72), 2 hepta-chlorinated (CN-73–CN-74), and 1 octa-chlorinated (CN-75) [1]. Because of the properties of low water solubility, low vapor pressure, and resistance to degradation, PCNs were widely used in various industries including cable insulation, wood preservation, graphite electrode lubrication, masking compounds for electroplating, dye manufacturing, capacitors, and refracting index testing oils [1,2]. In parallel, an increasing number of toxicological studies have demonstrated that PCNs exhibited the potential risk to a variety of organisms, including dioxin-like toxic effects on mammals [3–5]. PCNs have been listed in Annexes A and C of the Stockholm Convention on Persistent Organic Pollutants since May 2015, owing to their potential toxicity, persistence, bioaccumulation, and long-range transport [6,7]. Although, the production of PCNs was banned in the 1980s [2], unfortunately, products containing PCNs have been unlawfully used so far [8]. There are mainly four pathways of PCNs discharging into the environment, including historical use, historical and present use of polychlorinated biphenyl (PCBs), thermal and other processes, and landfills [9]. Moreover, PCNs may also be formed unintentionally and emitted into the environment via thermal processes involved in nonferrous metallurgical facilities, metal refineries, waste incineration, and cooking industries [9–11]. The possible sources of PCNs released into the environment could be identified by analyzing homolog profiles and ratios of several PCN congeners used as indicators of particular emission activities [11].

PCNs have been found in various media such as soil, sediment, water, air, and biota, even in human breastmilk [12,13]. Sediment, as one of the most important pollutant sinks

Citation: Liu, Z.; Xiao, K.; Wu, J.; Jia, T.; Lei, R.; Liu, W. Distributions of Polychlorinated Naphthalenes in Sediments of the Yangtze River, China. *Molecules* **2021**, *26*, 5298. https://doi.org/10.3390/molecules26175298

Academic Editor: James Barker

Received: 12 August 2021
Accepted: 27 August 2021
Published: 31 August 2021

Publisher's Note: MDPI stays neutral with regard to jurisdictional claims in published maps and institutional affiliations.

Copyright: © 2021 by the authors. Licensee MDPI, Basel, Switzerland. This article is an open access article distributed under the terms and conditions of the Creative Commons Attribution (CC BY) license (https://creativecommons.org/licenses/by/4.0/).

among the environmental media, can hold the contemporary PCNs from aquatic and terrestrial sources and release them to environment again. Thus, it is essential to monitor PCNs in sediment as secondary pollution sources in order to examine their ecological risk. In fact, detectable PCNs have been reported in numerous sediment samples collected in rivers, lakes, and coastal waters of China [7,14–18], even though PCNs have not been intentionally produced in China. For example, in Zhang's study, the PCN concentration was up to 4610 ng/g, highest among all the sediment samples of China reported [16]. Moreover, these pollutants tend to long-range transport in the atmosphere and finally deposit to sediments. Therefore, it is necessary to conduct comprehensive research on the whole valley for accurate assessment of the extent of pollute.

The Yangtze River, Asia's longest river and the third longest river in the world, serves as an important resource for drinking water, aquaculture and industrial use. With a rapid increase in population and economic development around the river, there are numerous inputs from industrial wastewater, municipal sewage, atmospheric deposition, as well as agricultural soils containing fertilizers, pesticides, herbicides, and heavy metals [19]. Preserving the river's water quality is critical for sustainable development as well as the health and survival of residents. In this study, sediment samples of the Yangtze River Basin were collected from the cradle to the estuary, including the reservoirs, industrial zones and residential areas. The samples spanned more than 6000 km across six provinces at different levels of economic development. Our aims were to provide fundamental data for quantifying PCN concentrations in the sediments, establishing their distributions and characteristics, and evaluating their potential ecological risks.

2. Results and Discussion

2.1. Spatial Distributions of PCNs in Sediments

The PCN concentrations of the sampling sites from the Yangtze River basin are shown in Figure 1.

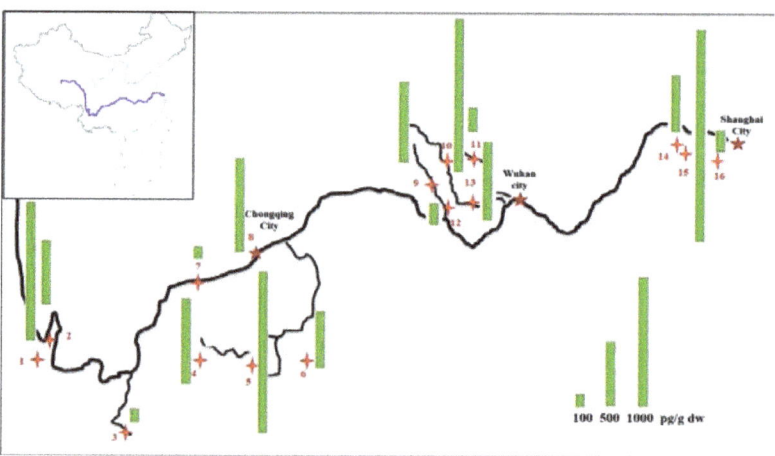

Figure 1. The PCN concentrations of the sampling sites from the Yangtze River Basin.

The PCN concentrations in the sediment samples ranged from 0.103 to 1.631 ng g^{-1}. The lowest concentration of PCNs was found at S7 located in Minjiang River which was served as a water reservoir. The highest concentration of PCNs was found at S15, collected from Xijiu River in Yixing City, a residential urban area. In the upstream zone (samples S1–S3), the relatively higher PCN concentration was found at S1 located near a famous tourist resort. Among the upper- and middle-stream zone samples, the PCN concentrations of S8 and S5 were relatively high. This was mainly due to the fact that

S8 was in a large city, and S5 was close to an industrial development zone. Within the sample from the middle and lower zones, the concentration of PCNs was relatively high at S10, which was near the outlet of a sewage treatment plant. As mentioned in the previous study [20], the input of wastewater and sewage sludge from urban treatment plants might be the most significant source of PCNs in the sediment. In the downstream zones, the highest concentration of PCNs was at S15, which was the inlet of Taihu Lake. The lake has become a pollutant sink with rapid urbanization and subsequently increasing amount of agricultural and industrial waste [16]. In our study, S3, S7, S12 and S16 could all be treated as background with lower concentrations of PCNs (0.1–0.2 ng g^{-1}). Unlike results reported from Li et al. [7], the trend in PCN concentrations at various spots above did not increase from cradle to the estuary. Therefore, PCN concentrations in sediments were more closely related to local human activities and industrial sources rather than geographically locations.

2.2. Comparison with Reported PCN Levels of Other Studies

PCN levels in this study were compared with those reported from other parts of the world (Table 1). Sample S16 in our study in Suzhou City was near the sampling point from the study of Zhang et al. [16], who reported that the PCN concentrations in the sediment was up to 4610 ng/g·dw in development zone of major industrial plants located in downtown Suzhou City. In contrast, our measured value (0.17 ng/g·dw) was five orders of magnitude lower than the above mentioned, while the sampling sites were at such a close distance about 25 miles away. It might because S16 was collected from a drinking water source in the suburb of Suzhou City. This result confirmed that local human activities and industrial sources strongly influence the PCN concentrations.

Compared with other locations, the PCN concentrations in our study were the same order of magnitude as in the Venice Lagoon in Venice (0.03–1.15 ng/g·dw), the Gulf of Bothnia in Sweden (0.088–1.9 ng/g·dw), the Liaojie River in Taiwan (0.029–0.987 ng/g·dw), the Tokyo Bay in Japan (1.81 ng/g·dw), the Qingdao Coastal Sea in China (1.1–1.2 ng/g·dw), and Laizhou Bay in China (0.12–5.10 ng/g·dw) [8,18,21–24]. Results herein were also lower than those in the Liao River basin in China (0.33–12.5 ng/g·dw), the Yellow River in China (0.18–130 ng/g·dw), and the Danube catchment in Czech Republic (0.05–29.2 ng/g·dw) [7,25,26]. They were significantly lower than that in the Bitterfeld and coastal Georgia, with chloralkali industries nearby as a source of PCNs [27,28]. The PCN concentrations in the sediments of the Yangtze River Basin were relatively lower, for the reason that some samples were collected from the water source protection area or the rural areas.

Table 1. Polychlorinated naphthalene levels reported in other parts of the world.

Countries/River Side	Sediment Concentration (ng/g·dw)	TEQ * (pg/g)	Reference
Georgia coastal	23400	3.71 × 10^6	[27]
Bitterfeld, Germany	2540	/	[28]
Swedish Lake	0.1–7.6	/	[29]
Gdansk Basin, Baltic Sea	6.7	/	[22]
Venice Lagoon, Venice	0.03–1.15	0.01–0.22	[21]
Tokyo Bay, Japan	1.81	/	[8]
Gulf of Bothnia, Sweden	0.088–1.9	/	[30]
Baltic Sea	0.32–1.9	0.19–0.23	[31]
Qingdao Coastal Sea, China	0.212–1.21	0.04–0.38	[24]
Barcelona, Spain	0.17–6.56	/	[32]
Lake Ontario	21–38	17 ± 4	[33]
Laizhou Bay, China	0.12–5.10	/	[18]
Jiaozhou Bay, China	0.0039–0.00564	<0.1	[14]
River Chenab, Pakistan	8.94–414	0.1–57	[34]
Danube Catchment, Czech Republic	0.05–29.2	0.02–17	[26]
Yangtze River Delta, East China	0.6–4600	0.0014–2160	[16]
Liao River, China	0.33–12.5	/	[7]
Yellow River, China	0.618–130	/	[24]
East China Sea	0.002–261.71	0–212	[35]
Liaojie River, Taiwan	0.029–0.987	3.4 × 10^{-3}–0.71	[23]
Yangtze River, China	0.103–1.631	0.010–0.304	This study

* TEQ: total toxic equivalent.

2.3. PCN Homolog Profiles

PCN homolog profiles for the environmental samples may be of great help to qualitatively identify the sources. Figure 2 shows the PCN homolog profiles of the sediment samples from the Yangtze River Basin. All of 75 PCN congeners were detected by high-resolution gas chromatography/high-resolution mass spectrometry. In other studies, PCNs homologs were usually reported the results of tri- to octa-CNs, with the dominant homologs of tri-and tetra-CNs, such as the Liaohe River basin of China, Laizhou Bay of China, and the Yellow River of China [7,18,25]. By contrast, in our study, mono-, di-, and tri-CNs were the dominant homolog groups in most of the samples, with the CN-1, CN-5/7 (i.e., complex of CN-5 and CN-7), and CN-24/14 being dominant congeners. Moreover, the proportion of PCN homolog groups decreased with an increasing number of chlorine atoms. Experimental results showed that S8 dominated by hepta-and octa-CN homologs, while S11 and S12 were dominated by tri-and tetra-CN homologs. S8 was collected from the stem of the Yangtze River, which was near the developed cities Chongqing City. This sampling site was surrounded by large commercial streets, busy transport hubs, and a garment industry. S11 was collected in Yunshui River near a fertilizer plant. S12 was collected from Changhu Lake surrounded by the woodland near Jingmen City. Although there was a sewage treatment plants that may discharge waste into the lake, the Jingmen City gathered a few industries including battery factories, petroleum machinery factories, etc. Therefore, we speculated that the significant difference in the distribution of PCN congeners in S11 and S12 samples may be due to the transport and deposition of surrounding industrial sources. The local hydrologic monitoring stations reported that the flow rate in S12 was much lower than in S11. Therefore, the different distribution of the PCN congeners in these three sites may result from human activities, industry discharges as well as the hydrological characteristics, deserving further investigations.

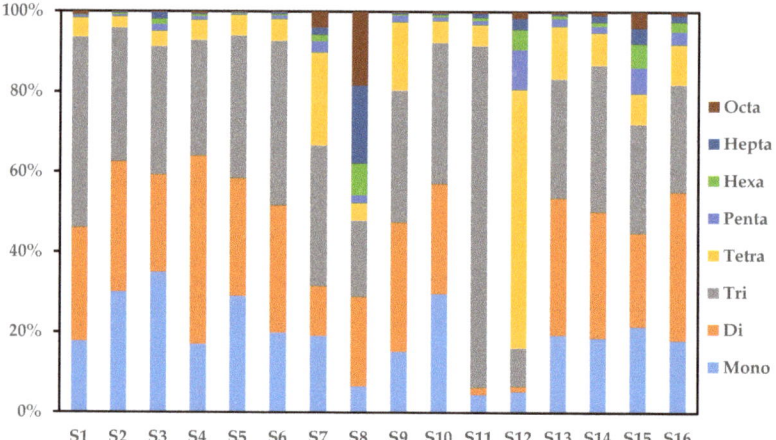

Figure 2. Homolog profiles of polychlorinated naphthalene in the sediment samples from the Yangtze River basin.

2.4. Source Identification

To further identify the potential sources of PCNs in the samples, we conducted principal component analysis (PCA) on PCN homologs in the sediment samples of the Yangtze River. Figure 3 shows the results of the PCA score plot of PCN among different samples. The extraction variance contributions of PC1 and PC2 accounted for 77.7% and 10.6% of the total variance, respectively, with a cumulative value of 88.3%. Except for two samples (S8 and S12), the homolog of PCNs in other samples was relatively consistent (marked in the red circle), indicating that they may be from the same types of emission sources.

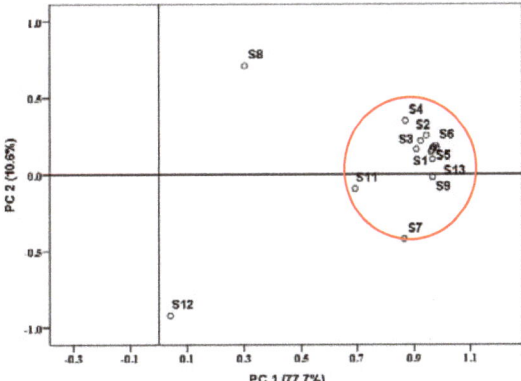

Figure 3. PCA score plot of the polychlorinated naphthalene profile of sediment samples.

Several congeners—such as CN-17/25, -36/45, -39, -35, -52/60, -50, -51, -53, and -66/67—have been identified as combustion indicators (PCN$_{com}$) [3,9,15,36]. The ratio of $\sum$PCN$_{com}$ to $\sum$PCNs was usually calculated to estimate primary sources [36]. Values of $\sum$PCN$_{com}$/$\sum$PCNs < 0.11 suggested emission from the halowax mixture, values of $\sum$PCN$_{com}$/$\sum$PCNs > 0.5 indicated combustion-related source emissions, and 0.11 < $\sum$PCN$_{com}$/$\sum$PCNs < 0.5 was assumed for the indication of emissions from combustion sources and halowax products [15]. In this study, the values of $\sum$PCN$_{com}$/$\sum$PCNs were all lower than 0.11, except for S12 (0.19). Thus, it was speculated that the dominant sources of the PCNs were mainly from the emission of the halowax mixture. Besides, the concentration of PCNs in S12 were also affected by combustion-related source emissions. Even though PCN mixtures were never historically produced and are not currently in commercial use in China, the historic usage of halowax mixture such as paintings and rubber materials could still be a big contributor to PCNs of the sediment in the Yangtze River Basin [18].

2.5. Ecological Risks of PCNs in Sediments

Some PCN congeners have toxic effects similar to those of 2,3,7,8-tetrachlorobenzo-*p*-dioxin (TCDD) in terms of their biological actions in animals [37]. Based on the relative potency factors [3], the calculated PCN-corresponding total toxic equivalent (TEQ) values in sediments ranged from 0.009 to 0.250 pg TEQ/g, as presented in Figure 4. Herein, the TEQ values were significantly lower than those in the interim sediment and quality guidelines in Canada and the USA (0.85 pg TEQ/g and 2.5 pg TEQ/g) [17], meaning that PCNs in the sediments expressed very low toxicity to aquatic life. Highest TEQ values were found at S8 and S15, which were nearly ten times higher than those of the other samples. CN-73 concentrations contributed the most to the TEQs, with 54.8 and 37.4 pg/g·dw, respectively.

Compared with previous studies, the TEQ values in this study (Table 1) were several orders of magnitude lower than those reported for the Danube catchment (17 pg TEQ/g), Lake Ontario (17 ± 4 pg TEQ/g), the River Chenab in Pakistan (57.1 pg TEQ/g), and the Yangtze River Delta in China (2160 pg TEQ/g) [16,26,33,34]. Rather, our TEQ values were similar to those of the Venice Lagoon of Venice, the Qingdao Coastal Sea of China, Jiaozhou Bay of China, and the Liaojie River of Taiwan [14,21,23,24].

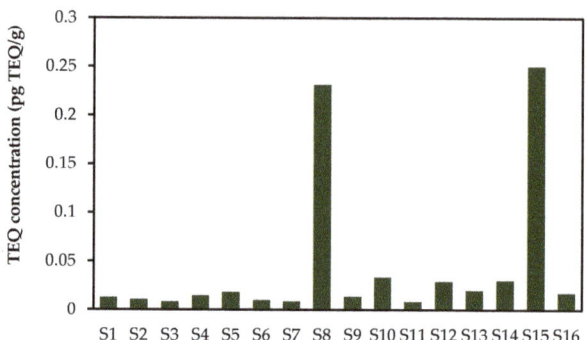

Figure 4. TEQ concentrations of the sediments in the Yangtze River Basin.

3. Materials and Methods

3.1. Sampling

To investigate the distribution of PCNs in the sediments along the Yangtze River Basin, a series of sediments was collected from 16 typical sites, including rural, urban and industrial areas based on their surrounding environment and levels of economic development nearly (Table 2).

Table 2. Geographical information and site types for the 16 sampling sites on the Yangtze River Basin.

Number	Sites Name	Latitude	Longitude	Type of Sample Site
S1	Erhai	25°37′	100°15′	city
S2	Sansu River	26°59′	100°13′	rural
S3	Dianchi	24°54′	102°43′	rural
S4	Xiangshui River	26°34′	104°56′	city
S5	Hongfenghu Reservoir	26°34′	106°26′	industrial areas
S6	Qingshui River	27°31′	109°35′	city
S7	Min River	28°45′	104°38′	city
S8	Stem Stream of Yangtze River	29°32′	106°34′	developed city
S9	Zhanghe Reservoir	30°58′	112°04′	rural
S10	Man River	31°35′	112°16′	city
S11	Runshui River	31°43′	113°20′	city
S12	Chang Lake	30°23′	112°30′	woodland
S13	Hanbei River	30°40′	113°08′	rural
S14	Tao Lake	31°61′	119°55′	rural
S15	Xi Jiu River	31°21′	119°48′	city
S16	Cheng Lake	31°18′	120°51′	rural

This picture just showed Level I and II tributaries. With the direction of the river, the samples could also be divided in four major zones: the upstream (Samples S1–S3), the upper and middle-stream (Samples S4–S8), the middle and lower (Samples S9–S13), and the downstream (Samples S14–S16) (Figure 5). According to the sampling method reported [21], the surface sediments (from the top 0–20 cm) were collected using a grab sampler, transported to the laboratory, and stored at $-20\ °C$ until analysis.

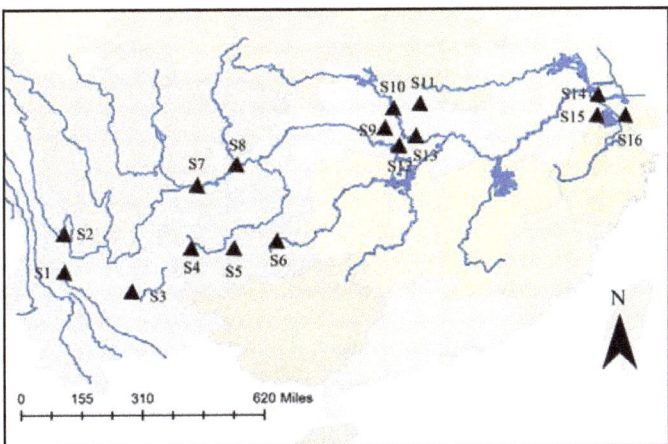

Figure 5. Yangtze River sampling sites in this study.

3.2. Sample Preparation and Analysis

The PCNs were analyzed according to our previously described method [20]. Briefly, the sediments were freeze-dried for approximately 48 h and passed through a 100-mesh (150 µm) stainless steel sieve. Approximately 10 g of sediments was weighed and spiked with a mixture of $^{13}C_{10}$-labeled PCN internal standards (ECN-5102, tetra-octa PCNs Mixture composed of $^{13}C_{10}$-CN-27, -42, -52, -67, -73, and -75; Cambridge Isotope Laboratories, Andover, MA, USA) and then mixed with 40 g (dry weight) of diatomaceous earth, and extracted by accelerated solvent extraction (ASE 350; Thermo-Fisher Scientific, Waltham, MA, USA) at 120 °C with a mixture extraction solvent of hexane and dichloromethane (1:1, v:v). The extraction solvents were first cleaned using acid silica column, followed by multilayer silica gel column and then basic alumina column. The elution fraction was then concentrated to 20 µL by rotatory evaporation and a gentle nitrogen gas stream. Finally, $^{13}C_{10}$-labeled PCN (ECN-5260: $^{13}C_{10}$-CN-64, Cambridge Isotope Laboratories, United States) was spiked for the calculation of recoveries before analysis.

PCNs were analyzed by high resolution gas chromatography coupled with a high-resolution mass spectrometer (Thermo Fisher Scientific, Waltham, MA, USA). A DB-5 fused silica capillary column (60 m × 0.25 mm × 0.25 µm; Agilent Technologies, Santa Clara, CA, USA) was used for the separation of PCN congeners. The injection volume was 1 µL, the flow rate of helium as carrier gas was 1 mL min^{-1}, and GC inlet temperature was set at 270 °C. The temperature program was initiated at 80 °C (for 2 min) and increased to 180 °C at 20 °C min^{-1} (hold for 1 min), 280 °C at 2.5 °C min^{-1} (for 2 min), and 300 °C at 10 °C min^{-1} (for 5 min). The HRMS was tuned and operated at approximately 10,000 resolutions with 45 eV EI energy.

3.3. Quality Assurance and Quality Control (QA/QC)

A procedural blank sample was evaluated to assess the possible contamination and instrumental stability. Only a small amount of monochlorinated polychlorinated naphthalenes was detected in the blank samples and was 10% lower than the concentrations in sediment samples. The PCN concentrations in the samples were, thus, not corrected using the values from the blanks. The recoveries of the $^{13}C_{10}$-labeled congeners ranged from 62% to 98%. The instrumental detection limits were assimilated when the signal-to-noise ratio was three times. PCNs were quantified using a relative response factor of the labeled congener at the same level of chlorination and a similar retention time.

4. Conclusions

We evaluated the distribution, composition, and ecological risks of PCNs by analyzing 16 sediment samples of the Yangtze River basin from the cradle to the estuary. Their concentrations and TEQs were less than 2 ng g^{-1} and 0.3 pg TEQ/g, respectively. These levels were lower than those of most of the previous reports, demonstrating that there was nearly no alarm to aquatic life with toxic aspects. In our study, the relatively higher PCN concentrations and TEQs were generally related with frequent human activities and nearby industrial sources. Further research is needed, however, to elucidate the relationship between concentrations of PCN congeners and human activities, providing new insights into understanding the environmental and health risk of exposure to PCN at low level.

Author Contributions: Conceptualization and writing—original draft, Z.L.; writing—review and editing, K.X.; investigation, J.W.; visualization, T.J.; validation, R.L.; funding acquisition, W.L. All authors have read and agreed to the published version of the manuscript.

Funding: This work was funded by the National Key Research and Development Plan (2018YFC1801602), the National Natural Science Foundation of China (22076207).

Institutional Review Board Statement: Not applicable.

Informed Consent Statement: Not applicable.

Data Availability Statement: Not applicable.

Conflicts of Interest: The authors declare no conflict of interest.

Sample Availability: Samples of the compounds are not available from the authors.

References

1. Hayward, D. Identification of bioaccumulating polychlorinated naphthalenes and their toxicological significance. *Environ. Res.* **1998**, *76*, 1–18. [CrossRef]
2. Noma, Y.; Yamamoto, T.; Sakai, S.I. Congener-specific composition of polychlorinated naphthalenes, coplanar PCBs, dibenzo-p-dioxins, and dibenzofurans in the halowax series. *Environ. Sci. Technol.* **2004**, *38*, 1675–1680. [CrossRef]
3. Villeneuve, D.L.; Khim, J.S.; Kannan, K.; Giesy, J.P. In vitro response of fish and mammalian cells to complex mixtures of polychlorinated naphthalenes, polychlorinated biphenyls, and polycyclic aromatic hydrocarbons. *Aquat. Toxicol.* **2001**, *54*, 125–141. [CrossRef]
4. Vinitskaya, H.; Lachowicz, A.; Kilanowicz, A.; Bartkowiak, J.; Zylinska, L. Exposure to polychlorinated naphthalenes affects GABA-metabolizing enzymes in rat brain. *Environ. Toxicol. Pharmacol.* **2005**, *20*, 450–455. [CrossRef] [PubMed]
5. Sisman, T.; Geyikoglu, F. The teratogenic effects of polychlorinated naphthalenes (PCNs) on early development of the zebrafish (*Danio rerio*). *Environ. Toxicol. Pharmacol.* **2008**, *25*, 83–88. [CrossRef]
6. Stockholm Convention. Available online: http://chm.pops.int/TheConvention/ThePOPs/TheNewPOPs/tabid/2511/Default.aspx (accessed on 15 May 2015).
7. Li, F.; Jin, J.; Gao, Y.; Geng, N.; Tan, D.; Zhang, H.; Ni, Y.; Chen, J. Occurrence, distribution and source apportionment of polychlorinated naphthalenes (PCNs) in sediments and soils from the Liaohe River Basin, China. *Environ. Pollut.* **2016**, *211*, 226–232. [CrossRef] [PubMed]
8. Yamashita, N.; Kannan, K.; Imagawa, T.; Villeneuve, D.L.; Hashimoto, S.; Miyazaki, A.; Giesy, J.P. Vertical profile of polychlorinated dibenzo-p-dioxins, dibenzofurans, naphthalenes, biphenyls, polycyclic aromatic hydrocarbons, and alkylphenols in a sediment core from Tokyo Bay, Japan. *Environ. Sci. Technol.* **2000**, *34*, 3560–3567. [CrossRef]
9. Falandysz, J. Polychlorinated naphthalenes: An environmental update. *Environ. Pollut.* **1998**, *101*, 77–90. [CrossRef]
10. Takasuga, T.; Inoue, T.; Ohi, E.; Kumar, K.S. Formation of polychlorinated naphthalenes, dibenzo-*p*-dioxins, dibenzofurans, biphenyls, and organochlorine pesticides in thermal processes and their occurrence in ambient air. *Arch. Environ. Contam. Toxicol.* **2004**, *46*, 419–431. [CrossRef]
11. Jansson, S.; Fick, J.; Marklund, S. Formation and chlorination of polychlorinated naphthalenes (PCNs) in the post-combustion zone during MSW combustion. *Chemosphere* **2008**, *72*, 1138–1144. [CrossRef] [PubMed]
12. Bidleman, T.F.; Helm, P.A.; Braune, B.M.; Gabrielsen, G.W. Polychlorinated naphthalenes in polar environments—A review. *Sci. Total Environ.* **2010**, *408*, 2919–2935. [CrossRef]
13. Li, C.; Zhang, L.; Li, J.; Min, Y.; Yang, L.; Zheng, M.; Wu, Y.; Yang, Y.; Qin, L.; Liu, G. Polychlorinated naphthalenes in human milk: Health risk assessment to nursing infants and source analysis. *Environ. Int.* **2020**, *136*, 105436. [CrossRef] [PubMed]

14. Pan, J.; Yang, Y.; Taniyasu, S.; Yeung, L.W.; Falandysz, J.; Yamashita, N. Comparison of historical record of PCDD/Fs, dioxin-like PCBs, and PCNs in sediment cores from Jiaozhou Bay and coastal Yellow Sea: Implication of different sources. *Bull. Environ. Contam. Toxicol.* **2012**, *89*, 1240–1246. [CrossRef] [PubMed]
15. Wang, Y.; Cheng, Z.N.; Li, J.; Luo, C.L.; Xu, Y.; Li, Q.L.; Liu, X.; Zhang, G. Polychlorinated naphthalenes (PCNs) in the surface soils of the Pearl River Delta, South China: Distribution, sources, and air-soil exchange. *Environ. Pollut.* **2012**, *170*, 1–7. [CrossRef] [PubMed]
16. Zhang, L.L.; Zhang, L.F.; Dong, L.; Huang, Y.R.; Li, X.X. Concentrations and patterns of polychlorinated naphthalenes in surface sediment samples from Wuxi, Suzhou, and Nantong, in East China. *Chemosphere* **2015**, *138*, 668–674. [CrossRef]
17. Zhao, X.F.; Zhang, H.J.; Fan, J.F.; Guan, D.M.; Zhao, H.D.; Ni, Y.W.; Li, Y.; Chen, J.P. Dioxin-like compounds in sediments from the Daliao River Estuary of Bohai Sea: Distribution and their influencing factors. *Mar. Pollut. Bull.* **2011**, *62*, 918–925. [CrossRef] [PubMed]
18. Pan, X.; Tang, J.; Chen, Y.; Li, J.; Zhang, G. Polychlorinated naphthalenes (PCNs) in riverine and marine sediments of the Laizhou Bay area, North China. *Environ. Pollut.* **2011**, *159*, 3515–3521. [CrossRef]
19. Gao, L.R.; Huang, H.T.; Liu, L.D.; Li, C.; Zhou, X.; Xia, D. Polychlorinated dibenzo-*p*-dioxins, dibenzofurans, and dioxin-like polychlorinated biphenyls in sediments from the Yellow and Yangtze Rivers, China. *Environ. Sci. Pollut. Res.* **2015**, *22*, 19804–19813. [CrossRef] [PubMed]
20. Guo, L.; Zhang, B.; Xiao, K.; Zhang, Q.H.; Zheng, M.H. Levels and distributions of polychlorinated naphthalenes in sewage sludge of urban wastewater treatment plants. *Chin. Sci. Bull.* **2008**, *53*, 508–513. [CrossRef]
21. Eljarrat, E.; Caixach, J.; Jimenez, B.; Gonzalez, M.J.; Rivera, J. Polychlorinated naphthalenes in sediments from the Venice and Orbetello lagoons, Italy. *Chemosphere* **1999**, *38*, 1901–1912. [CrossRef]
22. Falandysz, J.; Strandberg, L.; Bergqvist, P.A.; Kulp, S.E.; Strandberg, B.; Rappe, C. Polychlorinated naphthalenes in sediment and biota from the Gdansk Basin, Baltic Sea. *Environ. Sci. Technol.* **1996**, *30*, 3266–3274. [CrossRef]
23. Dat, N.D.; Chang, K.S.; Wu, C.P.; Chen, Y.J.; Tsai, C.L.; Chi, K.H.; Chang, M.B. Measurement of PCNs in sediments collected from reservoir and river in northern Taiwan. *Ecotoxicol. Environ. Saf.* **2019**, *174*, 384–389. [CrossRef]
24. Pan, J.; Yang, Y.L.; Xu, Q.; Chen, D.Z.; Xi, D.L. PCBs, PCNs and PBDEs in sediments and mussels from Qingdao coastal sea in the frame of current circulations and influence of sewage sludge. *Chemosphere* **2007**, *66*, 1971–1982. [CrossRef] [PubMed]
25. Li, Q.; Cheng, X.; Wang, Y.; Cheng, Z.; Guo, L.; Li, K.; Su, X.; Sun, J.; Li, J.; Zhang, G. Impacts of human activities on the spatial distribution and sources of polychlorinated naphthalenes in the middle and lower reaches of the Yellow River. *Chemosphere* **2017**, *176*, 369–377. [CrossRef] [PubMed]
26. Kukucka, P.; Audy, O.; Kohoutek, J.; Holt, E.; Kalabova, T.; Holoubek, I.; Klanova, J. Source identification, spatio-temporal distribution and ecological risk of persistent organic pollutants in sediments from the upper Danube catchment. *Chemosphere* **2015**, *138*, 777–783. [CrossRef]
27. Kannan, K.; Imagawa, T.; Blankenship, A.L.; Giesy, J.P. Isomer-specific analysis and toxic evaluation of polychlorinated naphthalenes in soil, sediment, and biota collected near the site of a former chlor-alkali plant. *Environ. Sci. Technol.* **1998**, *32*, 2507–2514. [CrossRef]
28. Brack, W.; Kind, T.; Schrader, S.; Moder, M.; Schuurmann, G. Polychlorinated naphthalenes in sediments from the industrial region of Bitterfeld. *Environ. Pollut.* **2003**, *121*, 81–85. [CrossRef]
29. Jaernberg, U.; Asplund, L.; de Wit, C.; Grafstroem, A.K.; Haglund, P.; Jansson, B.; Lexen, K.; Strandell, M.; Olsson, M.; Jonsson, B. Polychlorinated biphenyls and polychlorinated naphthalenes in Swedish sediment and biota: Levels, patterns, and time trends. *Environ. Sci. Technol.* **1993**, *27*, 1364–1374. [CrossRef]
30. Lundgren, K.; Tysklind, M.; Ishaq, R.; Broman, D.; Van Bavel, B. Polychlorinated naphthalene levels, distribution, and biomagnification in a benthic food chain in the Baltic Sea. *Environ. Sci. Technol.* **2002**, *36*, 5005–5013. [CrossRef]
31. Lundgren, K.; Tysklind, M.; Ishaq, R.; Broman, D.; van Bavel, B. Flux estimates and sedimentation of polychlorinated naphthalenes in the northern part of the Baltic Sea. *Environ. Pollut.* **2003**, *126*, 93–105. [CrossRef]
32. Castells, P.; Parera, J.; Santos, F.J.; Galceran, M.T. Occurrence of polychlorinated naphthalenes, polychlorinated biphenyls and short-chain chlorinated paraffins in marine sediments from Barcelona (Spain). *Chemosphere* **2008**, *70*, 1552–1562. [CrossRef]
33. Helm, P.A.; Gewurtz, S.B.; Whittle, D.M.; Marvin, C.H.; Fisk, A.T.; Tomy, G.T. Occurrence and biomagnification of polychlorinated naphthalenes and non- and mono-ortho PCBs in Lake Ontario sediment and biota. *Environ. Sci. Technol.* **2008**, *42*, 1024–1031. [CrossRef] [PubMed]
34. Mahmood, A.; Malik, R.N.; Li, J.; Zhang, G. Congener specific analysis, spatial distribution and screening-level risk assessment of polychlorinated naphthalenes in water and sediments from two tributaries of the River Chenab, Pakistan. *Sci. Total. Environ.* **2014**, *485*, 693–700. [CrossRef] [PubMed]
35. Liu, A.; Jia, J.; Lan, J.; Zhao, Z.; Yao, P. Distribution, composition, and ecological risk of surface sedimental polychlorinated naphthalenes in the East China Sea. *Mar. Pollut. Bull.* **2018**, *135*, 90–94. [CrossRef] [PubMed]
36. Lee, S.C.; Harner, T.; Pozo, K.; Shoeib, M.; Wania, F.; Muir, D.C.G.; Barrie, L.A.; Jones, K.C. Polychlorinated naphthalenes in the Global Atmospheric Passive Sampling (GAPS) study. *Environ. Sci. Technol.* **2007**, *41*, 2680–2687. [CrossRef]
37. Falandysz, J.; Fernandes, A.; Gregoraszczuk, E.; Rose, M. The Toxicological Effects of Halogenated Naphthalenes: A Review of Aryl Hydrocarbon Receptor-Mediated (Dioxin-like) Relative Potency Factors. *J. Environ. Sci. Health Part C* **2014**, *32*, 239–272. [CrossRef] [PubMed]

Article

Investigation on Distribution and Risk Assessment of Volatile Organic Compounds in Surface Water, Sediment, and Soil in a Chemical Industrial Park and Adjacent Area

Rongrong Lei [1,2], Yamei Sun [3,*], Shuai Zhu [4], Tianqi Jia [1,2], Yunchen He [1,2], Jinglin Deng [1,2] and Wenbin Liu [1,2,5,*]

1. Research Center for Eco-Environmental Sciences, Chinese Academy of Sciences, Beijing 100085, China; leirongr@163.com (R.L.); tqjia@126.com (T.J.); yunchenhe@yeah.net (Y.H.); dengjlin@yeah.net (J.D.)
2. University of Chinese Academy of Sciences, Beijing 100049, China
3. Chinese Academy of Environmental Planning, Beijing 100012, China
4. National Research Center for Geoanalysis, Beijing 100037, China; zhu15131215153@126.com
5. Hangzhou Institute for Advanced Study, UCAS, Hangzhou 310024, China
* Correspondence: sunym@caep.org.cn (Y.S.); liuwb@rcees.ac.cn (W.L.); Tel.: +86-10-62849356 (W.L.); Fax: +86-10-62923563 (Y.S.)

Abstract: The occurrences, distributions, and risks of 55 target volatile organic compounds (VOCs) in water, sediment, sludge, and soil samples taken from a chemical industrial park and the adjacent area were investigated in this study. The Σ_{55}-VOCs concentrations in the water, sediment, sludge, and soil samples were 1.22–5449.21 µg L^{-1}, ND–52.20 ng g^{-1}, 21.53 ng g^{-1}, and ND–11.58 ng g^{-1}, respectively. The main products in this park are medicines, pesticides, and novel materials. As for the species of VOCs, aromatic hydrocarbons were the dominant VOCs in the soil samples, whereas halogenated aliphatic hydrocarbons were the dominant VOCs in the water samples. The VOCs concentrations in water samples collected at different locations varied by 1–3 orders of magnitude, and the average concentration in river water inside the park was obviously higher than that in river water outside the park. However, the risk quotients for most of the VOCs indicated a low risk to the relevant, sensitive aquatic organisms in the river water. The average VOCs concentration in soil from the park was slightly higher than that from the adjacent area. This result showed that the chemical industrial park had a limited impact on the surrounding soil, while the use of pesticides, incomplete combustion of coal and biomass, and automobile exhaust emissions are all potential sources of the VOCs in the environmental soil. The results of this study could be used to evaluate the effects of VOCs emitted from chemical production and transportation in the park on the surrounding environment.

Keywords: VOCs; distribution; environmental media; risk assessment; chemical industrial park

1. Introduction

Volatile organic compounds (VOCs) are a class of organic chemicals that easily vaporize and enter the environment at room temperature. There is great concern about the potential harm they can cause to human health and the environment [1–5]. VOCs are mainly emitted from biological and anthropogenic sources [6,7] and are widespread in the environment because of their volatility. Anthropogenic VOCs are important precursors in the production of ozone and secondary organic aerosols, which have a significant impact on the atmospheric environment [8–10]. VOCs are extensively released from fuels, paints, spray, solvents, deodorants, combustion exhausts, etc. [11]. They are released into the environment during their production, storage, handling, and use and can infiltrate surface water, soil, and sediments through various physical and chemical processes from many sources [12]. Many VOCs are classified as toxic and carcinogenic pollutants, and exposure to VOCs can increase the risk of illness, congenital malformation, neurocognitive impairment, and cancer in humans [11].

The emission of VOCs from anthropogenic sources is a major issue in China. In 2012, 29.85 million tons of VOCs were emitted in China, and the emissions from industry sources in 2018 totaled 12.7 million tons [13,14]. As a key source of VOCs in China, the chemical industry accounts for more than 40% of the total emissions from industrial sources nationally [15]. The chemical industry plays an important role in economic development. However, chemical industrial parks contain many plants that manufacture multiple products and generate associated, complex pollutants and emissions, which are difficult to monitor and regulate. As an important emission source of various pollutants, chemical industrial parks will have an adverse impact on the surrounding environment. Studies have shown that the concentrations and risks of pollutants, such as pesticides, pharmaceutical pollutants, polycyclic aromatic hydrocarbons, and some VOCs, are relatively high in various environmental media surrounding chemical industrial parks [16–18].

Previous studies on VOCs mainly focus on air samples and few on environmental media such as water and soil. To evaluate the effects of VOCs emissions from chemical production and transportation in chemical industrial parks on the surrounding environment, we comprehensively investigated the occurrences, distributions, and risks of 55 target VOCs in water, soil, and sediment samples from a selected park and an adjacent area. There are more than 100 chemical plants in the park, and the main products are medicines, pesticides, and novel materials. The analysis of the concentration levels and composition characteristics of the VOCs in the samples suggests a source-to-sink relationship, with the chemical industrial park as the main emission source. The impact of production and transportation in the chemical park on the surrounding environment was studied, and the ecological environmental risk of VOCs was evaluated. The results will contribute to the knowledge of potential influences of VOCs from chemical industrial parks on the environment, and these are also helpful for risk management and the control of pollutants in chemical industrial parks.

2. Materials and Methods

2.1. Chemicals and Reagents

A total of 55 compounds (Table S1, in the Supplementary Materials) were monitored. VOC mixed standards, surrogate standards (fluorobenzene and 4-bromofluorobenzene), and internal standards (dibromofluoromethane, toluene-d_8, and 1,4-dichlorobenzene-d_4) referenced to US EPA Method 8260D [19] were purchased from o2si Smart Solutions (North Charleston, SC, USA). Methanol and dichloromethane were pesticide residue grade and were purchased from J.T. Baker Chemical Company (Phillipsburg, NJ, USA).

2.2. Sampling

Rudong Yangkou Chemical Industrial Park has an area of 11.6 km^2 and is located in Nantong, Jiangsu Province, China. In November 2018, we collected a series of water, soil, and sediment samples in the park and the surrounding area. A map of the sampling sites is presented in Figure 1.

The differences between the VOCs concentrations in samples collected inside and outside of the park could reflect the influence of the production activities in the chemical industrial park on the nearby environment. To compare the VOCs concentrations and compositions, different samples collected inside and outside the chemical industrial park were analyzed. Five sewage samples (labeled W1–W5) were collected from the sewage treatment plant (STP) at the water inlet (W3–W5), the regulating tank (W1), and the water outlet (W2). One sludge sample was collected from the STP. Eight surface water samples (labeled W6–W13) were collected from the river water inside the chemical industrial park (InCP) and the river water outside the chemical industrial park (OuCP). Because the river bank and bottom were sealed with cement, only three sediment samples (labeled Se1–Se3) were collected from a depth of 0–20 cm at different sites in the rivers using a stainless-steel grab sampler. All the water samples were collected in 40 mL pre-cleaned brown glass sampling bottles. The sediment and sludge samples were stored in 60 mL screw-top wide-mouth jars.

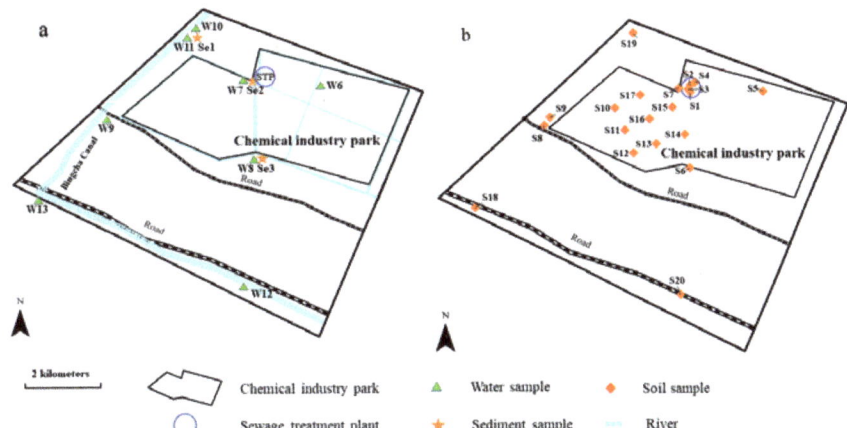

Figure 1. Map of the sampling sites. (**a**: water and sediment samples, **b**: soil samples).

Twenty surface soil samples (labeled S1–S20) were collected in the park and the adjacent area from a depth of 0–20 cm using a stainless-steel scoop (Figure 1b). Each soil sample was a composite of five samples collected from the four corners and the center of a square (5 m × 5 m) at each sampling site. A portion of the mixed sample was used to fill a 60 mL screw-top wide-mouth jar, and the jar was rapidly sealed and retained for VOC analysis.

To prevent effusion and photolysis of the analytes, all the samples were collected into brown glass sampling bottles that were filled completely without headspace. The samples were stored in the dark at 4 °C in a portable refrigerator and analyzed within 48 h.

2.3. Sample Analysis

The VOCs were analyzed using a gas chromatograph/mass spectrometer (GC/MS) (Agilent 7890B/5977A, Agilent Technologies Inc., Santa Clara, CA, USA) that was equipped with a purge and trap (P&T) concentrator (Tekmar-3100, Tekmar Dohrmann, Mason, OH, USA). The sample analysis followed EPA Method 8260D [19]. The water samples were directly analyzed, but the soil and sediment samples were pre-treated before analysis. First, approximately 5 g of the sample was accurately weighed into a 40 mL brown bottle, and then 1 g of sodium bisulfate, 5 mL of ultrapure water, and a magnetic stirring bar were added. Surrogate and internal standards (1 µL, 30 mg L^{-1}) were added to the sample by the autosampler during analysis.

The parameters of the P&T were set as follows: sample volume, 5 mL; purge gas, helium (99.999%); flow rate, 40 mL min^{-1}; purge time, 11 min; purge temperature, 20 °C; desorption temperature, 190 °C; desorption time, 2 min; baking temperature, 210 °C; and hold time, 10 min.

A capillary column (Agilent DB-624, 30 m × 0.25 mm, 1.4 µm) was used for the GC/MS. The injection temperature was 190 °C, and the split ratio was 20:1. Helium (99.99%) was used as the carrier gas at a flow rate of 1.73 mL min^{-1}. The oven temperature was held at 45 °C for 2 min, increased to 120 °C at 6 °C min^{-1}, and then increased to 220 °C at 12 °C min^{-1} and held for 5 min. Electron ionization was performed at 70 eV with an ion source temperature of 200 °C. The temperature of the quadrupole rods was 150 °C, and the interface temperature was 210 °C. The MS was operated in scan mode (45.00–280.00 amu).

2.4. Quality Assurance and Quality Control

Strict quality assurance and quality control measures were performed for field sampling and laboratory analyses. The recovery for surrogates was maintained in the range of 72–118%. The concentrations of target compounds in laboratory blanks and procedural blanks were below the method detection limits. The recovery range of spiked sample matrices was 83–114%.

2.5. Environmental Risk Assessment

The environmental risks assessment of the VOCs in the water to the aquatic organisms followed the method used in previous research [5,20]. The risk quotient (RQ) model can be used to quantify the risk of exposure of a specific species to the chemicals in the surrounding natural environment [4]. The measured environmental concentration (MEC, µg L^{-1}) of an individual compound and the predicted no-effect concentration (PNEC, µg L^{-1}) to organisms are used to calculate the RQ as follows:

$$RQ_i = \frac{MEC_i}{PNEC_i} \quad (1)$$

PNECs are derived from toxicity test data (acute and chronic toxicity, such as lethal concentration (LC$_{50}$), effective concentration (EC$_{50}$), and chronic value (ChV) and the assessment factor (AF). In this study, the minimum ChV of the individual chemicals for fish, *Daphnia*, and green algae were used to calculate the PNECs. As the toxicity data of three trophic levels were selected here, the AF was set to 10, according to the literature (Van Leeuwen and Vermeire, 2007). The PNECs were calculated using the following equation:

$$PNEC_i \equiv \frac{min(ChV_{i,fish},\ ChV_{i,\ Daphnia},\ ChV_{i,\ green\ algea})}{10} \quad (2)$$

The ChVs (µg L^{-1}) were obtained using ECOSAR V2.0 software [21], and the relevant data are given in Table S1 (in the Supplementary Materials). The sum of the RQ (RQ$_{total}$) for all the detected compounds was calculated to estimate the total toxicity using the following equation:

$$RQ_{total} = RQ_1 + RQ_2 + \cdots + RQ_i \quad (3)$$

3. Results and Discussion

3.1. VOCs in Water, Sediment, and Sludge

The Σ$_{55}$-VOCs concentrations in the water, sediment, and sludge samples collected from different sites within and around the chemical industrial park were analyzed for detection rates, total concentrations, and compositions of VOCs (Table 1). The detection rate of VOC species in the water samples (2–74%) extended to a higher value than in the sludge (19%) or sediment (0–22%) samples. The VOCs concentrations in the water samples from different locations varied by 1–3 orders of magnitude. The VOCs concentrations in the rivers from the study area were obviously higher than those in other environmental rivers in China, such as Dongjiang Lake (2.93 to 4.69 µg/L) [3], Yangtze River (0–4.03 µg/L), Huaihe River (0–22.21 µg/L), Yellow River (1.94–42.78 µg/L), Haihe River (0.48–42.78 µg/L), and Liaohe River (1.10–20.85 µg/L) [4].

The VOCs concentrations at some points in the chemical park are very high, which may be caused by unorganized emissions or leakage in the process of chemical production and transportation. Therefore, the unorganized discharge of VOCs from some enterprises in the parks should be controlled. Meanwhile, measures should be taken to minimize the leakage of VOCs during transportation.

Table 1. VOCs concentrations and compositions in different samples from sewage treatment plant (STP), inside the park (InCP), and outside the park (OuCP).

Sample No.	Location	Count of Detected VOCs	Concentration ($\mu g\ L^{-1}$ or $ng\ g^{-1}$)	Composition			
				Alkyl Halide	Alkenyl Halide	Aryl Halide	Aromatic Hydrocarbon
W1	STP	40	5449.21	97.5%	1.6%	0.6%	0.3%
W2	STP	19	147.74	61.8%	6.0%	4.3%	27.9%
W3	STP	10	88.86	16.3%	74.9%	4.2%	4.6%
W4	STP	5	36.19	92.9%	7.1%	/	/
W5	STP	4	19.08	/	49.9%	26.1%	24.0%
W6	InCP	8	1905.99	99.5%	0.1%	0.1%	0.3%
W7	InCP	4	10.23	91.7%	/	/	8.3%
W8	InCP	6	1512.09	99.7%	0.2%	/	0.1%
W9	InCP	3	30.19	100.0%	/	/	/
W10	OuCP	5	54.36	96.6%	/	/	3.4%
W11	OuCP	1	1.22	/	/	/	100.0%
W12	OuCP	3	3.45	71.0%	/	/	29.0%
W13	OuCP	20	57.07	31.7%	9.8%	22.1%	36.4%
Slu *	STP	10	21.53	32.1%	30.0%	/	37.8%
Se1	OuCP	0	ND	/	/	/	/
Se2	InCP	1	1.09	/	/	/	100.0%
Se3	InCP	12	52.20	4.9%	44.8%	6.4%	43.9%
S1	STP	1	2.54	/	/	/	100.0%
S2	STP	1	2.00	/	/	/	100.0%
S3	STP	1	1.75	/	/	/	100.0%
S4	STP	0	ND	/	/	/	/
S5	InCP	0	ND	/	/	/	/
S6	InCP	3	5.62	/	/	/	100.0%
S7	InCP	4	5.69	/	/	/	100.0%
S8	InCP	3	6.15	/	/	/	100.0%
S9	InCP	0	ND	/	/	/	/
S10	InCP	0	ND	/	/	/	/
S11	InCP	3	7.70	/	/	/	100.0%
S12	InCP	5	9.24	/	/	/	100.0%
S13	InCP	0	ND	/	/	/	/
S14	InCP	0	ND	/	/	/	/
S15	InCP	5	6.00	/	/	/	100.0%
S16	InCP	0	ND	/	/	/	/
S17	InCP	5	9.16	/	/	/	100.0%
S18	OuCP	1	2.43	/	/	/	100.0%
S19	OuCP	1	3.09	/	/	/	100.0%
S20	OuCP	4	11.58	/	/	/	100.0%

ND: not detected. * Slu is sludge sample from STP.

3.1.1. Samples from the Sewage Treatment Plant

The STP is a sink of pollutants in the chemical industrial park, and the treatment efficiency of pollutants in the STP is important in regulating the impact of pollutants on the surrounding environment. Water samples taken from the inlet, the regulating tank, and the outlet of the STP were analyzed in this study. Sewage discharged by specific enterprises in the inlet of the STP represented the VOCs concentrations and emission characteristics of these enterprises. Samples W3–W5 were collected from sewage originating from enterprises that produce polymer materials and drug intermediates; agricultural chemicals; and pesticide intermediates and formulations, respectively. The VOCs compositions in these samples were different. Samples W3 and W4 were dominated by alkenyl halides (75%) and alkyl halides (93%), respectively. The dominant components in sample W5 were alkenyl halides (50%) as well as aryl halides and aromatic hydrocarbons (50%). The differences in the VOCs compositions in these samples could be used as a reference for source analysis of the VOCs in the environment.

The sample taken from the regulating tank represents the overall level and characteristics of the VOCs in the chemical industrial park. The difference in the VOCs concentrations between the regulating tank (W1) and the outlet (W2) reflects the removal efficiency of the VOCs in the STP (Figure 2). As expected, W1 contained the most VOCs (40 species) at the highest concentrations (up to 5449.21 µg L^{-1}). The dominant compound in the sample from the regulating tank was dichloromethane (87%), and the total contribution of alkyl halides and alkenyl halides was up to 99% of all compounds in the sample. These results could be attributed to the fact that the main products manufactured in the studied chemical industrial park are medicines, pesticides, and novel materials, and alkyl halides and alkenyl halides are important compounds used in the synthesis of drugs and polymer materials. Compared with the regulating tank sample, the sample from the effluent outlet of the STP had a much lower concentration (147.74 µg L^{-1}) and a variety of VOCs. According to our results, the VOC-removal rate of the STP was up to approximately 97%. However, although most of the VOCs were reduced in concentration or removed from the sewage after treatment, the concentration of benzene significantly increased. This could be attributed to the fact that most aromatic hydrocarbons and aryl halides are degraded and converted into benzene during sewage treatment, but they are not completely decomposed and removed. Ten different VOCs were detected in the sludge from the STP, and the total VOCs concentration was 21.54 ng g^{-1}. These results imply that VOCs adsorbed in the sludge are degraded during sewage treatment.

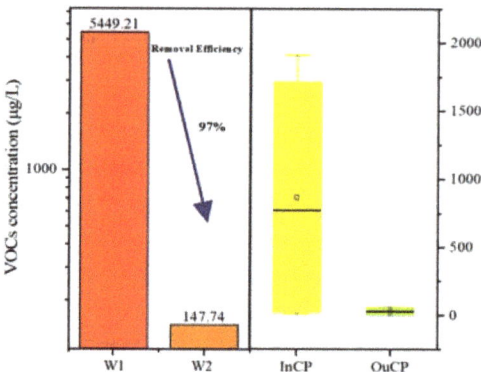

Figure 2. VOCs concentrations in the water samples collected at the sewage treatment plant (W1, W2), inside the park (InCP), and outside the park (OuCP).

3.1.2. Comparison of VOCs in River Water Inside and Outside the Chemical Industrial Park

The average VOCs concentrations in the InCP samples were obviously higher than those in the OuCP samples (Figure 2). However, for individual samples, the VOCs concentrations in some InCP samples were lower than those in some of the OuCP samples. Among the OuCP samples, the highest VOCs concentrations were found in the water from the Bingcha Canal (site W13), which is located about 3 km from the chemical industrial park. Moreover, the sample taken from this waterway also contained more VOC species than any of the other samples, except for the sample from the STP regulating tank. The high VOCs concentrations and number of species in this waterway may have been caused by the intensive use of river transportation and human activities.

The top five abundant VOC species in the different types of samples are shown in Figure 3. The most abundant VOC species was dichloromethane in the samples from the STP and the InCP and bromochloromethane in the samples from the OuCP. The results were different from the other research where chloroform was the compound with the highest concentration in the surface water from a chemical industrial park [22]. This indicated that the VOCs released by the chemical industrial processes directly affected the environment. However, chemical industrial park's industrial processes are not the only source of VOCs for samples collected outside a chemical industrial park. Benzene, toluene, ethylbenzene, and xylene (BTEX) have received much attention due to frequent detection in the environment, with the widespread use of petroleum products as their main source [22,23]. The BTEX concentration ranges in the InCP and the OuCP samples in the present study were ND–3.97 µg L^{-1} (mean: 1.50 µg L^{-1}) and 1.00–3.55 µg L^{-1} (mean: 1.91 µg L^{-1}), respectively. Unlike the Σ_{55}-VOCs, the BETX concentration in the OuCP samples was slightly higher than that in the InCP samples. This is because the production and the transportation of chemical products in the chemical industrial park are the only sources of BTEX for the InCP samples. In the case of the OuCP samples, the BTEX concentration is affected by a combination of chemical production, coal and biomass burning, automobile exhaust emissions, and other industrial activities such as iron and steel production [1].

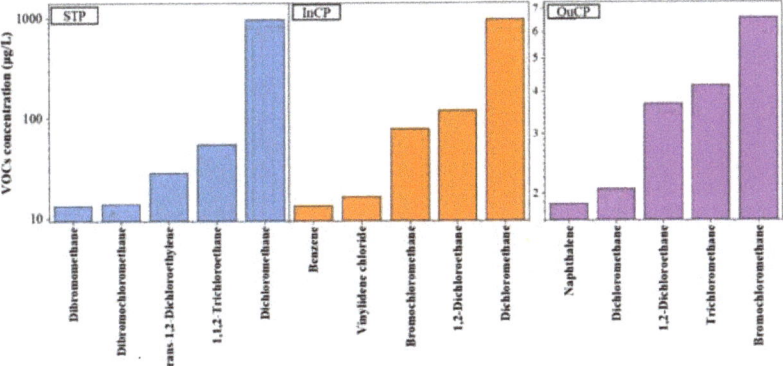

Figure 3. The top five VOC species in water samples collected at the sewage treatment plant (STP), inside the park (InCP), and outside the park (OuCP).

3.2. VOCs in Soil

Soil samples were collected from different sites, including inside the chemical industrial park, around the STP, and outside the park. The Σ_{55}-VOCs concentration range in the soil samples was ND–11.58 ng g^{-1}, which was lower than that in other chemical industrial parks [22]. The distribution of total VOCs concentrations in the soil at the study site is shown in Figure 4. The average VOCs concentration in the soil from inside the park was slightly higher than that in the soil from outside the park, but the VOCs were not detected in some samples. The Σ_{55}-VOCs concentration in the soil from inside the park had a wide range (ND–9.16 ng g^{-1}), and in the samples where VOCs were detected, the concentrations were generally higher, and more species were found than in the soil from outside the park. These results implied that the chemical production processes affected the concentrations and compositions of VOCs in the soil inside the chemical park. The highest Σ_{55}-VOCs concentration was found in a sample (S20) from outside the park (11.58 ng g^{-1}), which was collected in a small vegetable garden near the road. The largest contributor to the Σ_{55}-VOCs concentration in this sample was *tert*-butylbenzene, which accounted for 41% of the total VOCs. *tert*-Butylbenzene is an important intermediate in the production of pharmaceuticals and pesticides, such as fungicides and acaricides [24]. This result suggests that the use of pesticides could be a major contributor to VOCs in agricultural soil.

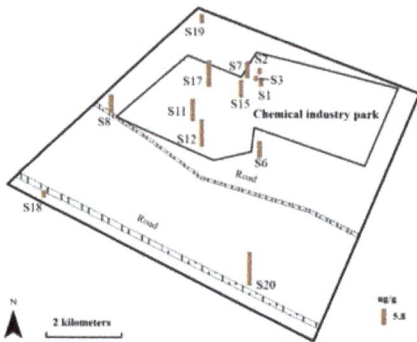

Figure 4. The distribution of total VOCs concentrations in soil at the study area.

Five kinds of VOCs were detected in the soil samples. These were 1,2,4-trimethylbenzene, benzene, *p*-isopropyl benzene, *tert*-butylbenzene, and *sec*-butylbenzene, which are all aromatic hydrocarbons. This was a little different from what was detected at the chemical industrial park in eastern China, which had more alkyl halide in the soil in the park [22]. In this study, benzene was the most frequently detected compound in the soil samples. These results were completely different from those obtained from the water samples, where the main VOCs were halogenated aliphatic hydrocarbons. This difference can be explained by the physical and chemical properties of the VOCs. Compared with aromatic hydrocarbons, aliphatic hydrocarbons have higher water solubility and lower octanol–water distribution coefficients, which lead to different distribution characteristics in water and soil. The concentrations of BTEX in all soil samples ranged from ND to 3.52 ng g^{-1} (mean: 1.55 ng g^{-1}), which was close to the relatively low levels of BTEX in water. Generally, the chemical industrial park had a limited impact on the surrounding soil. The potential sources of VOCs in the soil in the surrounding environment are the use of pesticides, the incomplete combustion of coal and biomass, and automobile exhaust emissions [25,26].

3.3. Environment Risk Assessment

The RQs to fish, *Daphnia*, and green algae for individual VOCs in the InCP and the OuCP water samples were analyzed in this study. An RQ > 1.0 indicates high risk, an RQ of 0.1–1 indicates moderate risk, and an RQ of 0.01–0.1 indicates low risk [3]. The RQs for most of the VOCs, except for hexachlorobutadiene (HCBD), dichloromethane, 1,2-dichloroethane, vinylidene chloride, 1,2,3-trichlorobenzene, *n*-butylbenzene, and naphthalene, detected in eight sampling locations from the InCP and the OuCP were below 0.1. Therefore, these compounds were of low risk to the relevant, sensitive aquatic organisms in the area. Among the sampling locations, W13 (from the OuCP) had the highest RQ, which was 3.32 for HCBD. This result indicated that aquatic organisms were at potentially high risk from HCBD in this waterway. HCBD mainly comes from the production of chlorinated hydrocarbons, paints, plastics, and herbicides [27]. W13 is in an important transportation location, and the potentially high risk of HCBD may be caused by the volatilization and leakage of chemicals during transportation. Therefore, the management of the transportation and storage of chemicals outside the park should be strengthened.

High concentrations of dichloromethane at sampling locations W6 and W8 (from the OuCP) resulted in RQs for dichloromethane that were greater than one. Therefore, dichloromethane is of concern at these sites. The environmental risk of a pollutant is affected not only by its concentration in the environment but also by the pollutant toxicity and the organism sensitivity. Consequently, high pollutant concentrations are not directly related to environmental risk, even though they may induce adverse effects or represent an ecological risk [4,5]. In the present study, dichloromethane did not have the highest RQ, even though it was present in many samples and had the highest concentration among the VOCs in some samples. By contrast, the concentration of hexachlorobutadiene in sample

W13 was far lower than that of dichloromethane in samples W6 and W8, but hexachlorobutadiene showed a higher RQ than dichloromethane for the selected aquatic organisms. Furthermore, the RQs for 1,2-dichloroethane, vinylidene chloride, 1,2,3-trichlorobenzene, n-butylbenzene, and naphthalene at some locations were higher than 0.1, which indicated that these compounds may pose a moderate risk to aquatic organisms.

The RQ_{total} values at the eight locations varied from 3.10×10^{-3} (sample W11) to 4.36 (sample W13). The RQ values for all the detected VOCs in the samples are shown in Table 2. Among the locations, the OuCP water sample collected in the Bingcha Canal had the highest RQ. As mentioned above, the VOCs concentration in this river was higher than that in the water taken from the other OuCP locations. Therefore, the future treatment of water in this river should focus on the removal of VOCs.

Table 2. RQs of water in the rivers from study area.

Location	Sample Site	RQ		
		Min	Max	Sum
InCP	W6	2.02×10^{-3}	1.570	1.881
	W7	1.23×10^{-3}	0.007	0.013
	W8	3.31×10^{-3}	1.142	1.653
	W9	1.53×10^{-3}	0.019	0.034
OuCP	W10	5.19×10^{-3}	0.029	0.081
	W11	3.40×10^{-3}	0.003	0.003
	W12	7.92×10^{-4}	0.003	0.005
	W13	4.39×10^{-3}	3.318	4.359

4. Conclusions

Chemical production had a great impact on the environment inside the chemical industrial park, and the VOCs concentrations of samples taken from inside the park were higher. It is found that the sewage treatment plant has high removal efficiency of VOCs in sewage through the study of the VOCs concentrations and species in the inlet and the outlet of the sewage treatment plant. However, the concentration of the VOCs in the sludge from the sewage treatment plant was higher than that in the soil and the sediment; therefore, careful disposal of the sludge should be considered. Based on the results from the soil and sediment samples, the chemical industrial park had a limited impact on the soil in the surrounding area.

One of the most important measures to control the effect of VOCs in the environment is to control the source emissions of VOCs. In this study, we found that chemical production has an important effect on the distribution of VOCs inside the chemical industrial park, but the emission sources of VOCs in the actual environment are complex. Pesticides, the incomplete combustion of coal and biomass, and automobile exhaust emissions are all potential sources of VOCs in the environment. Therefore, while paying attention to fixed emission sources, such as chemical industrial parks, these scattered, mobile emission sources should also be examined further.

Supplementary Materials: The following are available online: Table S1: PNEC values calculated using chronic values (ChVs) from ECOSAR.

Author Contributions: Conceptualization, W.L., Y.S.; Formal analysis, R.L., T.J., Y.H., J.D. and S.Z.; Validation, R.L., T.J., Y.H. and J.D.; Data curation, R.L.; Investigation, R.L., Y.S., S.Z., T.J., Y.H., J.D. and W.L.; Methodology, W.L. and Y.S.; Project administration, W.L.; Funding acquisition, W.L.; Writing—original draft, R.L.; Writing—review and editing, W.L. and Y.S. All authors have read and agreed to the published version of the manuscript.

Funding: This work was funded by the National Key Research and Development Plan of the Ministry of Science and Technology of China (2018YFC1801602, 2017YFC0213000) and the National Natural Science Foundation of China (22076207).

Institutional Review Board Statement: Not applicable.

Informed Consent Statement: Not applicable.

Data Availability Statement: Not applicable.

Conflicts of Interest: The authors declare no conflict of interest.

References

1. Jia, H.; Gao, S.; Duan, Y.; Fu, Q.; Che, X.; Xu, H.; Wang, Z.; Cheng, J. Investigation of health risk assessment and odor pollution of volatile organic compounds from industrial activities in the Yangtze River Delta region, China. *Ecotoxicol. Environ. Saf.* **2021**, *208*, 111474. [CrossRef]
2. Je, C.-H.; Stone, R.; Oberg, S.G. Development and application of a multi-channel monitoring system for near real-time VOC measurement in a hazardous waste management facility. *Sci. Total Environ.* **2007**, *382*, 364–374. [CrossRef]
3. Cao, F.; Qin, P.; Lu, S.; He, Q.; Wu, F.; Sun, H.; Wang, L.; Li, L. Measurement of volatile organic compounds and associated risk assessments through ingestion and dermal routes in Dongjiang Lake, China. *Ecotoxicol. Environ. Saf.* **2018**, *165*, 645–653. [CrossRef] [PubMed]
4. Chen, X.; Luo, Q.; Wang, D.; Gao, J.; Wei, Z.; Wang, Z.; Zhou, H.; Mazumder, A. Simultaneous assessments of occurrence, ecological, human health, and organoleptic hazards for 77 VOCs in typical drinking water sources from 5 major river basins, China. *Environ. Pollut.* **2015**, *206*, 64–72. [CrossRef] [PubMed]
5. Cho, E.; Khim, J.; Chung, S.; Seo, D.; Son, Y. Occurrence of micropollutants in four major rivers in Korea. *Sci. Total Environ.* **2014**, *491–492*, 138–147. [CrossRef] [PubMed]
6. Insam, H.; Seewald, M.S.A. Volatile organic compounds (VOCs) in soils. *Biol. Fertil. Soils* **2010**, *46*, 199–213. [CrossRef]
7. Ahn, J.; Rao, G.; Mamun, M.; Vejerano, E.P. Soil–air partitioning of volatile organic compounds into soils with high water content. *Environ. Chem.* **2020**, *17*, 545–557. [CrossRef]
8. Shao, M.; Zhang, Y.; Zeng, L.; Tang, X.; Zhang, J.; Zhong, L.; Wang, B. Ground-level ozone in the Pearl River Delta and the roles of VOC and NOx in its production. *J. Environ. Manag.* **2009**, *90*, 512–518. [CrossRef]
9. Shao, M.; Lu, S.; Liu, Y.; Xie, X.; Chang, C.; Huang, S.; Chen, Z. Volatile organic compounds measured in summer in Beijing and their role in ground-level ozone formation. *J. Geophys. Res.: Atmos.* **2009**, *114*, D2. [CrossRef]
10. Guo, H.; Ling, Z.H.; Cheng, H.R.; Simpson, I.J.; Lyu, X.P.; Wang, X.M.; Shao, M.; Lu, H.X.; Ayoko, G.; Zhang, Y.L.; et al. Tropospheric volatile organic compounds in China. *Sci. Total Environ.* **2017**, *574*, 1021–1043. [CrossRef]
11. Li, A.J.; Pal, V.K.; Kannan, K. A review of environmental occurrence, toxicity, biotransformation and biomonitoring of volatile organic compounds. *Environ. Chem. Ecotoxicol.* **2021**, *3*, 91–116. [CrossRef]
12. Kostopoulou, M.N.; Golfinopoulos, S.K.; Nikolaou, A.D.; Xilourgidis, N.K.; Lekkas, T.D. Volatile organic compounds in the surface waters of Northern Greece. *Chemosphere* **2000**, *40*, 527–532. [CrossRef]
13. Liang, X.; Sun, X.; Xu, J.; Ye, D.; Chen, L. Industrial Volatile Organic Compounds (VOCs) Emission Inventory in China. *Environ. Sci.* **2020**, *41*, 4767–4775.
14. Wu, R.; Bo, Y.; Li, J.; Li, L.; Li, Y.; Xie, S. Method to establish the emission inventory of anthropogenic volatile organic compounds in China and its application in the period 2008–2012. *Atmos. Environ.* **2016**, *127*, 244–254. [CrossRef]
15. Ministry of Ecology and Environment of the PRC. *National Bureau of Statistics; Ministry of Agriculture and Rural Affairs of the PRC, Communique of the Second National Survey of Pollution Sources*; Ministry of Ecology and Environment of the PRC: Beijing, China, 2020.
16. Liu, W.; Yao, H.; Xu, W.; Liu, G.; Wang, X.; Tu, Y.; Shi, P.; Yu, N.; Li, A.; Wei, S. Suspect screening and risk assessment of pollutants in the wastewater from a chemical industry park in China. *Environ. Pollut.* **2020**, *263*, 114493. [CrossRef]
17. Jiao, H.; Wang, Q.; Zhao, N.; Jin, B.; Zhuang, X.; Bai, Z. Distributions and Sources of Polycyclic Aromatic Hydrocarbons (PAHs) in Soils around a Chemical Plant in Shanxi, China. *Int. J. Environ. Res. Public Health* **2017**, *14*, 1198. [CrossRef]
18. Shuai, J.; Kim, S.; Ryu, H.; Park, J.; Lee, C.K.; Kim, G.-B.; Ultra, V.U.; Yang, W. Health risk assessment of volatile organic compounds exposure near Daegu dyeing industrial complex in South Korea. *BMC Public Health* **2018**, *18*, 528. [CrossRef] [PubMed]
19. US EPA. *Method 8260D: Volatile Organic Compounds by Gas Chromatography/Mass Spectrometry (GC/MS) 2018*; EPA: Washington, DC, USA, 2018.
20. Santos, J.L.; Aparicio, I.; Alonso, E. Occurrence and risk assessment of pharmaceutically active compounds in wastewater treatment plants. A case study: Seville city (Spain). *Environ. Int* **2007**, *33*, 596–601. [CrossRef]
21. US EPA. Ecological Structure Activity Relationships (ECOSAR) Predictive Model. Available online: https://www.epa.gov/tsca-screening-tools/ecological-structure-activity-relationships-ecosar-predictive-model (accessed on 1 October 2021).
22. Liu, B.; Chen, L.; Huang, L.; Wang, Y.; Li, Y. Distribution of volatile organic compounds (VOCs) in surface water, soil, and groundwater within a chemical industry park in Eastern China. *Water Sci. Technol.* **2014**, *71*, 259–267. [CrossRef] [PubMed]
23. Mackay, D.M.; de Sieyes, N.R.; Einarson, M.D.; Feris, K.P.; Pappas, A.A.; Wood, I.A.; Jacobson, L.; Justice, L.G.; Noske, M.N.; Scow, K.M.; et al. Impact of Ethanol on the Natural Attenuation of Benzene, Toluene, and o-Xylene in a Normally Sulfate-Reducing Aquifer. *Environ. Sci. Technol.* **2006**, *40*, 6123–6130. [CrossRef] [PubMed]
24. Hu, Y.; Wang, H.; Yin, D. Synthesis and application of tert-butylbenzene (in Chinese). *Gansu Sci. Technol.* **2010**, *26*, 58–61.

25. Huang, C.; Chen, C.H.; Li, L.; Cheng, Z.; Wang, H.L.; Huang, H.Y.; Streets, D.G.; Wang, Y.J.; Zhang, G.F.; Chen, Y.R. Emission inventory of anthropogenic air pollutants and VOC species in the Yangtze River Delta region, China. *Atmos. Chem. Phys.* **2011**, *11*, 4105–4120. [CrossRef]
26. Liu, Y.; Shao, M.; Fu, L.; Lu, S.; Zeng, L.; Tang, D. Source profiles of volatile organic compounds (VOCs) measured in China: Part I. *Atmos. Environ.* **2008**, *42*, 6247–6260. [CrossRef]
27. Sun, J.T.; Pan, L.L.; Zhan, Y.; Zhu, L.Z. Spatial distributions of hexachlorobutadiene in agricultural soils from the Yangtze River Delta region of China. *Environ. Sci. Pollut. Res.* **2018**, *25*, 3378–3385. [CrossRef] [PubMed]

Article

Occurrence of Phthalates in Bottled Drinks in the Chinese Market and Its Implications for Dietary Exposure

Xiaohong Xue [1], Yaoming Su [2], Hailei Su [3], Dongping Fan [2], Hongliang Jia [4], Xiaoting Chu [4], Xiaoyang Song [5], Yuxian Liu [6,7,*], Feilong Li [8], Jingchuan Xue [8,*] and Wenbin Liu [9]

[1] College of Science, Dalian Maritime University, 1 Linghai Road, Dalian 116026, China; xuexiaohong@dlmu.edu.cn
[2] South China Institute of Environmental Sciences, Ministry of Ecology and Environment of P.R.C., No. 7, West Street, Yuancun, Tianhe District, Guangzhou 510535, China; suyaoming@scies.org (Y.S.); fandp0104@163.com (D.F.)
[3] State Key Laboratory of Environmental Criteria and Risk Assessment, Chinese Research Academy of Environmental Sciences, Beijing 100012, China; su.hailei@craes.org.cn
[4] International Joint Research Centre for Persistent Toxic Substances (IJRC-PTS), College of Environmental Science and Engineering, Dalian Maritime University, Dalian 116026, China; jiahl@dlmu.edu.cn (H.J.); liuxuesong1998@163.com (X.C.)
[5] Dalian Modern Agricultural Production Development Service Center, 678 Zhongshan Road, Dalian 116026, China; sun_7812@163.com
[6] Key Laboratory of Ministry of Education for Water Quality Security and Protection in Pearl River Delta, School of Environmental Science and Engineering, Guangzhou University, Guangzhou 510006, China
[7] Linköping University-Guangzhou University Research Center on Urban Sustainable Development, Guangzhou University, Guangzhou 510006, China
[8] Key Laboratory for City Cluster Environmental Safety and Green Development of the Ministry of Education, Institute of Environmental and Ecological Engineering, Guangdong University of Technology, Guangzhou 510006, China; fl_li@gdut.edu.cn
[9] Research Center for Eco-Environmental Sciences, Chinese Academy of Sciences, Beijing 100085, China; liuwb@rcees.ac.cn
* Correspondence: liuyuxian@gzhu.edu.cn (Y.L.); xue@gdut.edu.cn (J.X.)

Abstract: Ubiquitous occurrences of phthalic acid esters (PAEs) or phthalates in a variety of consumer products have been demonstrated. Nevertheless, studies on their occurrence in various types of bottled drinks are limited. In this study, fifteen PAEs were analyzed in six categories of bottled drinks (n = 105) collected from the Chinese market, including mineral water, tea drinks, energy drinks, juice drinks, soft drinks, and beer. Among the 15 PAEs measured, DEHP was the most abundant phthalate with concentrations ranging from below the limit of quantification (LOQ) to 41,000 ng/L at a detection rate (DR) of 96%, followed by DIBP (DR: 88%) and DBP (DR: 84%) with respective concentration ranges of below LOQ to 16,000 and to 4900 ng/L. At least one PAE was detected in each drink sample, and the sum concentrations of 15 PAEs ranged from 770 to 48,004 ng/L (median: 6286 ng/L). Significant differences with respect to both PAE concentrations and composition profiles were observed between different types of bottled drinks. The median sum concentration of 15 PAEs in soft drinks was over five times higher than that detected in mineral water; different from other drink types. Besides DEHP, DBIP, and DBP, a high concentration of BMEP was also detected in a tea drink. The estimated daily dietary intake of phthalates (EDI_{drink}) through the consumption of bottled drinks was calculated based on the concentrations measured and the daily ingestion rates of bottled drink items. The EDI_{drink} values for DMP, DEP, DIBP, DBP, BMEP, DAP, BEEP, BBP, DCP, DHP, BMPP, BBEP, DEHP, DOP, and DNP through the consumption of bottled mineral water (based on mean concentrations) were 0.45, 0.33, 12.5, 3.67, 2.10, 0.06, 0.32, 0.16, 0.10, 0.09, 0.05, 0.81, 112, 0.13, and 0.20 ng/kg-bw/d, respectively, for Chinese adults. Overall, the EDI_{drink} values calculated for phthalates through the consumption of bottled drinks were below the oral reference doses suggested by the United States Environmental Protection Agency (U.S. EPA).

Keywords: phthalate; dietary intake; DEHP; DBP; DIBP; bottled drink

Citation: Xue, X.; Su, Y.; Su, H.; Fan, D.; Jia, H.; Chu, X.; Song, X.; Liu, Y.; Li, F.; Xue, J.; et al. Occurrence of Phthalates in Bottled Drinks in the Chinese Market and Its Implications for Dietary Exposure. *Molecules* **2021**, *26*, 6054. https://doi.org/10.3390/molecules26196054

Academic Editor: Carlo Santoro

Received: 14 September 2021
Accepted: 3 October 2021
Published: 6 October 2021

Publisher's Note: MDPI stays neutral with regard to jurisdictional claims in published maps and institutional affiliations.

Copyright: © 2021 by the authors. Licensee MDPI, Basel, Switzerland. This article is an open access article distributed under the terms and conditions of the Creative Commons Attribution (CC BY) license (https://creativecommons.org/licenses/by/4.0/).

1. Introduction

Phthalic acid esters (PAEs) or phthalates, primarily used to make polyvinyl chloride (PVC), or vinyl, flexible and pliant, are a group of chemicals made of aryl esters of phthalic acid and alkyl. PAEs are generally used to soften plastics because of their strong performance, durability, and stability [1,2]. These phthalate plasticizers are used in hundreds of products in our homes, hospitals, cars, and businesses, such as vinyl flooring, plastic packaging, toys, medical tubing, and cosmetics [3–7]. For example, Xu et al. (2020) reported the sum concentrations of dimethyl phthalate (DMP), diethyl phthalate (DEP), and dibutyl phthalate (DBP) ranged from 102 to 710 µg/kg in polyethylene terephthalate (PET) bottles collected from Beijing, China [8]. PAEs are not covalently bound to the plastic [9,10], so they can be easily released into the environment, leading to potential human exposure through ingestion, dermal absorption, and inhalation.

As a group of well-studied endocrine-disrupting chemicals, phthalates exposure has been associated with a variety of health effects, including premature thelarche, endometriosis, low semen quality, diabetes, overweight and obesity, allergy and asthma, and reproductive health [11,12]. Diethylhexyl phthalate (DEHP) is one of the most-studied phthalates, and accumulative evidence showed that DEHP exposure was significantly related to insulin resistance and higher systolic blood pressure as well as reproductive system problems [13,14]. Potential toxicity mechanisms of DEHP exposure include the activation of Kupffer's cells and the nuclear receptor peroxisome proliferator-activated receptor (PPARα) [15–17]. Evidence has shown that phthalates' toxicity heavily depends on their chemical structures [18]. Based on the difference in carbon backbones in the alkyl side chain, phthalates are differentiated into low and high molecular weight categories. Low molecular weight (LMW) phthalate plasticizers have straight carbon backbones of C3-C6 in the alkyl side chains, while high molecular weight (HMW) phthalate plasticizers have straight C7-C13 carbon backbones in the alkyl side chains [18]. Studies have indicated that LMW phthalates can cause adverse reproductive effects, while HMW phthalates and those C1-C2 backbone alkyl phthalates do not show adverse reproductive effects [18]. The US Environmental Protection Agency (EPA) listed DEHP and butyl benzyl phthalate (BBP) as probable and possible human carcinogens, respectively [11]. European authorities have also classified LMW phthalates with C3-C6 backbone alkyl phthalates as presumed human reproductive toxicants.

Parent phthalates and their metabolites have been detected in a variety of human samples, including serum [19], urine [20,21], semen [22], breast milk [23], and breast tumor tissue [24]. Considerable efforts have also been made to characterize the sources of human exposure to PAEs [25,26], and the ubiquitous occurrence of PAEs in both consumer products [25–28] and environmental matrices [29–31] has been reported. The accumulating evidence has shown that the sources and routes of human exposure to individual PAEs can vary depending on their physicochemical properties [26,31–33]. For example, cosmetics and personal care products are the major sources of human exposure to LMW phthalates [25], diet has been a major source of exposure to HMW phthalates, especially DEHP [31,32,34], and inhalation is the predominant exposure route to DMP [32]. Recent studies have indicated that drinking water is also an important source of human exposure to PAEs. For example, Liu et al. (2015) performed a national survey and risk assessment of phthalates in drinking water from waterworks in China and found that DBP and DEHP were the most abundant PAEs among the six PAEs measured at median levels of 0.18 ± 0.47 and 0.18 ± 0.97 µg/L, respectively [35]. Thuy et al. (2021) surveyed the contamination levels and distribution patterns of ten PAEs in various types of water samples, including bottled water and tap water, collected from Hanoi, Vietnam, and suggested widespread occurrence of PAEs in the water samples [36]. However, little is known of the occurrence of phthalates in bottled drinks commercially available in the market, although the consumption of bottled drinks is huge.

This study aims to investigate the occurrence and distribution levels of fifteen typical phthalates in 105 popular branded bottled drinks in the Chinese market, including mineral water (*n* = 19), a tea drink (*n* = 22), an energy drink (*n* = 15), a juice drink (*n* = 15), a soft drink (*n* = 25), and beer (*n* = 9), in order to estimate human exposure to phthalates through the consumption of bottled drinks. We also chemo-metrically investigate if grouping and correlations among PAEs and bottled drinks from the Chinese market exist. To our knowledge, this is the first survey on phthalates in various types of bottled drinks collected from the Chinese market.

2. Materials and Methods

2.1. Standards and Reagents

Fifteen phthalates, including dimethyl phthalate (DMP), diethyl phthalate (DEP), dibutyl phthalate (DBP), dinonyl phthalate (DNP), diamyl phthalate (DAP), dihexyl phthalate (DHP), diisobutyl phthalate (DIBP), butyl benzyl phthalate (BBP), bis(2-Ethoxyethyl) phthalate (BEEP), bis(2-methoxyethyl) phthalate (BMEP), bis(2-*n*-Butoxyethyl) phthalate (BBEP), bis(4-Methyl-2-pentyl)phthalate (BMPP), dinoctyl phthalate (DOP), dicyclohexyl phthalate (DCP), and diethylhexyl phthalate (DEHP), were analyzed in this study. Detailed information regarding the 15 PAEs is shown in Table S1. Nine deuterated internal standards, including d_4-DMP, d_4-DEP, d_4-DBP, d_4-DNP, d_4-DHP, d_4-DIBP, d_4-DOP, d_4-DCP, and d_4-DEHP, were used as surrogate standards in the quantification of phthalates. Both the target and surrogate standards were purchased from AccuStandard, Inc. (New Haven, CT, USA), with a purity of >99%. Analytical-grade acetone and acetonitrile were purchased from Macron Chemicals (Nashville, TN, USA), and hexane and HPLC-grade water were purchased from J. T. Baker (Phillipsburg, NJ, USA).

2.2. Sample Collection and Analysis

A total of 105 bottled drinks were collected from local supermarkets in Dalian, Liaoning Province, China, including mineral water (*n* = 19), a tea drink (*n* = 22), an energy drink (*n* = 15), a juice drink (*n* = 15), a soft drink (*n* = 25), and beer (*n* = 9). The drink samples collected in this study were popular brands that were consumed widely by the Chinese population.

All the drink samples were spiked with surrogate standards prior to extraction, following the extraction protocol described earlier [28]. In brief, 200 ng of each surrogate standard was spiked into 1500 mL of the bottled drink sample. Spiked samples were thoroughly mixed for 5 min and allowed to equilibrate at room temperature for 30 min. Then 10 mL hexane was used for extraction via shaking in a mechanical shaker at 250 oscillations/min for 30 min. After centrifugation, the hexane layer was transferred into a clean glass flask. The extraction was repeated three times, and the hexane extract was combined and concentrated using a rotary evaporator to 1 mL and transferred into a gas chromatography (GC) vial for analysis.

The instrumental analysis protocol of phthalates was described elsewhere in earlier studies [28,31]. Briefly, the analysis was performed using GC (Agilent Technologies 6890, Santa Clara, CA, USA) coupled with mass spectrometry (Agilent Technologies 5973, Santa Clara, CA, USA) in the selection ion monitoring (SIM) mode. The chromatography separation was carried out using a fused silica capillary column (DB-5 ms, 30 m × 0.25 mm × 0.25 µm; Agilent Technologies; Santa Clara, CA, USA). The detailed parameters for the GC-MS condition for PAE analysis are shown in Table S2.

2.3. Quality Assurance/Quality Control

Prior to the analysis of samples, considerable effort was made to reduce the background contamination from the analytical procedures following the earlier studies [7]. Briefly, all glassware was washed with a detergent and Milli-Q water, followed by solvents (i.e., acetone and hexane), baked at 450 °C for overnight, and kept in an oven at 100 °C until use. All solvents were tested for background levels of phthalates and the batches of solvents that contained the lowest levels of phthalates were used throughout the analysis. Prior to each batch of analysis, pure hexane was injected into GC–MS until the background level was stable. Within each batch of ten samples, three solvent blanks and three procedural blanks and a pair of matrix-spike samples were processed together. Trace levels of phthalates found in procedural blanks were subtracted from the measured concentrations in bottled drink samples (Table S2). The quantification of phthalates in the samples was based on the isotope dilution method. The calibration curves were prepared by plotting a concentration−response factor for each target analyte (peak area of analyte divided by peak area of the internal standard) versus the response-dependent concentration factor (the concentration of the analyte divided by the concentration of the internal standard). The regression coefficients (r) were ≥ 0.99 for all calibration curves. The limits of quantification (LOQs) were calculated based on the instrument detection limits (a quantifiable peak must have a signal-to-noise ratio > 10, and a dilution factor in sample preparation (Table S3)). Recoveries of surrogate standards were calculated using matrix spikes of both low (50 ng each PAE) and high (500 ng each PAE) amounts of chemical spikes, and the recoveries ranged from 82% to 113% (Table S3). The relative standard deviation (RSD) was calculated by analyzing a high amount of matrix spike replicates (n = 3) to evaluate the reproducibility and repeatability of the analysis, and the RSDs of PAEs are below 10% (Table S3).

2.4. Statistical Analysis

Basic descriptive statistical analysis was performed using Microsoft Excel (Microsoft Office 2013). For concentrations below the LOQs, a value of half the LOQ was used in the calculation [37,38]. As a major tool for simplifying the large initial datasets, principal component analysis (PCA) has been widely used in the investigation of possible sources of chemical pollutants in the environment [38,39]. Here, Euclidean distance-based constrained analysis of principal coordinates (CAP), a type of principal component analysis, was used to provide information regarding the sources of PAEs in the analyzed bottled drink samples, and to compare the PAEs concentrations among different groups [40]. PRIMER-e (version 7, PRIMER-E, Ivybridge, UK) with PERMANOVA+ add-on software (PRIMER-E Ltd., Ivybridge, UK) was used in the PCA analysis, and the statistical significance level was set at $\alpha < 0.05$.

2.5. Exposure Doses and Health Risk Assessment of PAEs through Consumption of Bottled Drinks

Daily intake of phthalates through the consumption of bottled drinks by the Chinese population was estimated via the following equation [31]:

$$\text{EDIdrink} = \frac{CQ}{bw} \quad (1)$$

where $\text{EDI}_{\text{drink}}$ (ng/kg-bw/d) is the estimated daily intake from drink, C (ng/g) is the phthalate concentration in the drink, Q (g/day) is the average amount of daily intake of the drink, and bw (kg) is body weight. For Chinese adults, 60 kg and 2000 g/day were used as the bw and Q values, respectively.

Both carcinogenic and non-carcinogenic risks of the select PAEs through the consumption of bottled drinks were assessed following the methods described earlier [41]. The selection of PAEs was based on the availability of relevant parameters in the Integrated Risk Information System, prepared and maintained by the U.S. Environmental Protection Agency (U.S. EPA) [42]. The carcinogenic risk (R) from exposure to PAEs via the consumption of bottled drinks was calculated by the following equation:

$$R = SF \times EDI \ (R < 0.01) \quad (2)$$

$$R = 1 - \exp(-EDI \times SF) \ (R \geq 0.01) \quad (3)$$

where *SF* is the carcinogenic slope factor of oral intake. The *SF* value for DEHP is 0.014 (Kg d)/mg [42].

The hazard index (*HI*) was used to assess non-cancer risks, calculated using the following equation:

$$HI = EDI/RfD \quad (4)$$

where *RfD* is the reference dose for the non-carcinogenic health risk of a chemical proposed by the guidelines. The *RfD* of BBP, DBP, DEP, and DEHP is 0.2, 0.1, 0.8, and 0.02 mg/Kg/d, respectively [42]. HI below 1 indicates safety concerns [41].

3. Results and Discussion

3.1. Concentrations of Phthalates in Bottled Drinks

Overall, at least one PAE was detected in each one of the 105 bottled drinks analyzed (Table 1). The sum concentrations of 15 PAEs measured in the 105 bottled drink samples ranged from 770 to 48,004 ng/L at a median level of 6286 ng/L (Table 1). Among the 15 PAEs measured in this study, DEHP was the predominant compound detected in the bottled drinks (detection rate, DR: 96%; median: 2000 ng/L; range: <LOQ–41000 ng/L), followed by DIBP (88%; 2100; <LOQ–16000) and DBP (84%; 820; <LOQ–4900) (Table 1). All other PAEs were also frequently detected in this study with DRs over 59%, but their contributions to the sum mean concentration of the 15 PAEs were low, with a contribution ratio between 0.1% and 4.4% (Table 1).

Table 1. Overall concentrations of phthalates in bottled drinks (n = 105) collected from Dalian, Liaoning Province, China.

Chemical	DR[a] (%)	Mean (ng/L)	SD[b] (ng/L)	GM[c] (ng/L)	Median (ng/L)	Range (ng/L)	Ratio[d] (%)
DMP	75.2	277	837	64.9	65	<LOQ[e]–7300	3.0
DEP	59.0	32.9	44.2	20.3	17	<LOQ–390	0.4
DIBP	87.6	2825	2796	1491	2100	<LOQ–16000	30.7
DBP	83.8	1097	1119	551	820	<LOQ–4900	11.9
BMEP	87.6	404	2074	36.5	39	<LOQ–17000	4.4
DAP	61.0	41.0	174	5.12	3.5	<LOQ–1400	0.5
BEEP	84.8	63.4	61.2	32.8	50	<LOQ–270	0.7
BBP	81.9	97.6	464	13.3	12	<LOQ–3800	1.1
DCP	76.2	14.6	15.8	9.4	9.5	<LOQ–110	0.2
DHP	60.0	6.88	8.46	4.16	3.7	<LOQ–51	0.1
BMPP	75.2	16.4	39.2	6.6	6.2	<LOQ–280	0.2
BBEP	87.6	92.0	111	48.4	61	<LOQ–760	1.0
DEHP	96.2	4193	6949	2025	2000	<LOQ–41000	45.5
DOP	78.1	10.6	13.0	6.28	7	<LOQ–86.0	0.1
DNP	81.0	35.9	137	9.45	9.8	<LOQ–1300	0.4
∑(sum)	100	9207	8952	6127	6286	770–48004	

[a]: DR, detection rate; [b]: SD, standard deviation; [c]: GM, geometric mean; [d]: ratio, concentration ratio (%), calculated as the ratio between the mean concentration of each target analyte versus the mean sum concentration of 15 PAEs; [e]: LOQ, limit of quantification.

Mineral water is the most commonly used bottled drink among the Chinese population. Among the 15 PAEs measured, DEHP was the most frequently detected and most-abundant chemical in the mineral water samples, with a DR of 100% and median concentration of 1600 (range: 500–15,000) ng/L, followed by DIBP (DR: 58%; median: 170; range: <LOQ–940) and DBP (37%; 57.0; <LOQ–320) (Tables 2 and 3). The distribution pattern is similar to that observed when taking other types of bottled drinks into account (Table 1), but different from what was observed in bottled waters collected from other countries such as Vietnam [36]. Differences in packaging material, water source, and other materials used in the production and bottling between the two countries may explain this observation. When compared with the concentration in bottled water or tap water samples collected from other regions, the concentration of DEHP in mineral water was at a higher level (at the same level as that calculated when taking other types of drinks into account), but the concentrations of DIBP and DBP were at lower levels (DIBP and DBP levels were also at higher levels when taking other types of bottled drinks into account), as shown in Table 2. This indicates that bottled mineral water/drinks are an important source of

human exposure to DEHP, similar to other dietary sources such as foodstuffs [28]. The sum concentrations of 15 PAEs measured in 19 bottled mineral water samples ranged from 770 to 16,301 ng/L (median: 1805 ng/L) (Table 3). Median individual PAE concentrations in the bottled drink samples (n = 105) were 1.3–16.1 times higher than that detected in the bottled mineral water samples (n = 19) (Tables 1 and 3).

Table 2. Concentration[a] (ng/L) comparison of DEHP, DIBP, DBP, and BMEP in bottled drink, bottled water, and tap water samples from various studies.

Sample Type	Location	n	DEHP	DIBP	DBP	BMEP	Reference
mineral water	Dalian, China	19	3351 (500–15000)	375 (<LOQ[b]-940)	110 (<LOQ-320)	63 (<LOQ-310)	this study
tea drink	Dalian, China	22	2660 (500–12000)	3770 (<LOQ-9900)	1197 (<LOQ-3600)	1701 (<LOQ-17000)	this study
energy drink	Dalian, China	15	4738 (<LOQ-34000)	1539 (<LOQ-4300)	630 (<LOQ-2900)	63 (<LOQ-220)	this study
juice drink	Dalian, China	15	3682 (440–27000)	4521 (240–16000)	2363 (290–4900)	73 (8.9–190)	this study
soft drink	Dalian, China	25	7198 (<LOQ-41000)	3916 (1000–7200)	1290 (93–3000)	56 (<LOQ-330)	this study
beer	Dalian, China	9	1317 (440–3700)	1974 (330–4100)	1064 (210–3000)	37 ((<LOQ-130)	this study
total	Dalian, China	105	4193 (<LOQ-41000)	2825 (<LOQ-16000)	1097 (<LOQ-4900)	404 (<LOQ-17000)	this study
non-carbonated water	Hanoi, Vietnam	11	873 (227–1950)	1100 (94.0–3930)	1150 (145–3070)	-	Le et al. (2021) [36]
carbonated water	Hanoi, Vietnam	10	1120 (103–2710)	1790 (123–5190)	1740 (93.0–4710)	-	Le et al. (2021) [36]
carbonated soft drinks	Tehran, Iran	4	8423 (6767–14008)	-	-	-	Moazzen et al. (2018) [43]
bottled water	Tianjin, China	6	1074 (880–1257)	-	486 (465–517)	-	Wang et al. (2021) [41]
bottled water	21 global countries	367–379	3420 (nd[c]-9410)	-	5350 (nd-2220)	-	Luo et al. (2018) [44]
bottled water	Tehran, Iran	10	100 (70–120)	-	70 (nd-120)	-	Abtahi et al. (2019) [45]
bottled water	Portugal	7	100 (20–180)	959 (100–1890)	1574 (60–6500)	-	Santana et al. (2013) [46]
tap water	Tehran, Iran	40	150 (nd-380)	-	90 (nd-140)	-	Abtahi et al. (2019) [45]
tap water	Hanoi, Vietnam	7	5340 (1010–14500)	456 (27.0–1390)	796 (14.0–2560)	-	Le et al. (2021) [36]
tap water	China	225	770 (<LOQ-5510)	-	350 (<LOQ-1560)	-	Liu et al. (2015) [35]
tap water	Tianjin, China	6	1338 (1097–1780)	-	541 (380–679)	-	Wang et al. (2021) [41]

[a]: mean concentration and the concentration range were used in this Table; [b]: LOQ, limit of quantification; [c]: nd, non-detected.

Concentrations of PAEs in both bottled mineral water samples and bottled drink samples analyzed in this study were at higher levels compared with those reported in other countries, e.g., Iran [43,45] and Portugal [46]. This may be partly ascribed to the fact that more types of phthalates (15 vs. 6 [43], 6 [45], and 11 [46], respectively) were measured in this study as well as the differences in the sample pretreatment method, analytical technique, data analysis method, etc. Le et al. (2021) [36] recently reported the concentrations of 10 typical PAEs in bottled water collected from Hanoi, Vietnam with the mean concentration being 6400 (range: 1640–15,700) ng/L, which is higher than that detected in mineral water yet lower than that in bottled drinks analyzed in this study. The

PAE concentrations in the bottled drinks detected in this study are lower than that reported by Luo et al. (2018) [44] in bottled waters from 21 countries (mean: 14,900 ng/L; range: n.a.–520,000 ng/L).

Table 3. Concentrations (ng/L) of phthalates in different types of bottled drinks.

	DMP	DEP	DIBP	DBP	BMEP	DAP	BEEP	BBP	DCP	DHP	BMPP	BBEP	DEHP	DOP	DNP	∑(sum)
							mineral water (n = 19)									
DR[a]	26	5	58	37	53	5	32	42	21	11	5	53	100	47	53	100
mean	13.7	9.8	375	110	62.9	1.8	9.5	4.9	3.1	2.6	1.5	24.2	3351	3.8	5.9	3980
SD[b]	6.7	7.8	337	80.8	103	1.0	20.9	8.7	1.0	4.0	0.2	51.2	4321	3.0	5.4	4637
GM[c]	12.6	8.7	249	89.3	13.7	1.7	4.8	3.0	2.9	1.9	1.5	10.8	1989	3.0	3.8	2546
median	86.0	8.0	170	57.0	7.0	1.6	3.1	1.9	2.6	1.6	1.5	6.5	1600	1.6	4.4	1805
range	<LOQ[e]–31.0	<LOQ–42.0	<LOQ–940	<LOQ–320	<LOQ–310	<LOQ–6.00	<LOQ–95.0	<LOQ–40.0	<LOQ–5.80	<LOQ–19.0	<LOQ–2.30	<LOQ–230	500–15,000	<LOQ–12.0	<LOQ–19.0	770–16,301
ratio[d]	0.3	0.3	9.4	2.8	1.6	0.1	0.2	0.1	0.1	0.1	0.04	0.6	84.2	0.1	0.2	
							tea drink (n = 22)									
DR	77	86	82	91	95	77	91	82	77	55	86	95	100	73	86	100
mean	177	38.1	3770	1197	1701	12.8	53.5	26.8	22.6	7.5	12.7	143	2660	16.0	105	9942
SD	291	30.0	2712	1037	4366	21.6	44.4	40.5	23.7	11.1	14.3	193	2380	21.4	274	7045
GM	68.2	28.4	2092	673	100	6.1	31.5	12.7	13.4	3.9	7.0	72.5	2043	7.2	21.6	7553
median	86.0	28.5	3550	1060	79.5	5.6	45.5	15.0	20.0	2.5	6.0	74.5	2050	7.4	25.0	8459
range	<LOQ–1300	<LOQ–110	<LOQ–9900	<LOQ–3600	<LOQ–17,000	<LOQ–100	<LOQ–150	<LOQ–190	<LOQ–110	<LOQ–49	<LOQ–55.0	<LOQ–760	500–12,000	<LOQ–86.0	<LOQ–1300	1277–27,298
ratio	1.8	0.4	37.9	12.0	17.1	0.1	0.5	0.3	0.2	0.1	0.1	1.4	26.8	0.2	1.1	
							energy drink (n = 15)									
DR	87	53	93	80	93	73	100	100	100	100	93	100	80	100	67	100
mean	232	27.5	1539	630	63.3	13.7	94.7	80.8	19.1	10.5	13.7	140	4738	17.4	10.1	7629
SD	304	26.6	955	700	64.5	37.8	62.1	214	10.0	5.6	10.3	56.2	9893	14.2	13.6	10,072
GM	104	18.4	1230	372	37.9	4.4	78.4	26.2	16.9	9.1	10.0	129	1276	14.1	5.0	4688
median	98.0	14.0	1500	510	42.0	3.5	92.0	23.0	17.0	9.2	12.0	130	1100	14.0	3.0	3972
range	<LOQ–1100	<LOQ–98.0	<LOQ–4300	<LOQ–2900	<LOQ–220	<LOQ–150	22.0–270	5.7–850	9.1–37.0	3.7–20	<LOQ–37.0	52–250	<LOQ–34,000	6.2–63.0	<LOQ–53.0	1184–36,505
ratio	3.0	0.4	20.2	8.3	0.8	0.2	1.2	1.1	0.3	0.1	0.2	1.8	62.1	0.2	0.1	
							juice drink (n = 15)									
DR	100	73	100	100	100	80	100	87	87	93	100	100	100	93	93	100
mean	820	38.2	4521	2363	72.9	202	115	51.5	16.2	9.2	16.9	88.6	3682	10.3	48.9	12,057
SD	1875	31.5	4654	1499	57.4	425	76.6	102	14.9	6.6	9.2	64.6	6617	8.2	125	11,012
GM	167	25.8	2631	1827	51.3	18.1	90.6	18.0	11.5	7.1	14.3	69.0	1823	7.7	16.5	8792
median	167	27	3400	2200	57	11	71	17	12	8.3	15	61	1600	8.7	14	10,179
range	18–7300	<LOQ–87	240–16,000	290–4900	8.9–190	<LOQ–1400	26.0–240	<LOQ–410	<LOQ–62	<LOQ–24	2.7–37.0	18–220	440–27,000	<LOQ–32	<LOQ–500	1635–46,541
ratio	6.8	0.3	37.5	19.6	0.6	1.7	1.0	0.4	0.1	0.1	0.1	0.7	30.5	0.1	0.4	
							soft drink (n = 25)									
DR	84	68	100	100	96	72	96	96	92	56	84	88	96	84	96	100
mean	326	44.7	3916	1290	55.7	28.0	64.2	298	15.1	6.8	35.1	74.4	7200	8.6	15.6	13,376
SD	768	76.1	1957	826	71.8	70.9	58.9	917	14.2	10.7	76.0	59.7	9572	7.3	12.4	10,077
GM	92	24.3	3385	991	30.3	5.7	39.3	26.7	11.2	3.8	8.9	46.4	3705	6.0	11.2	10,710
median	110	22	3300	1100	36.0	4.6	39	20	10	3.4	7.0	66	2900	6.2	12.0	10,534
range	<LOQ–3100	<LOQ–390	1000–7200	93–3000	<LOQ–330	<LOQ–310	<LOQ–220	<LOQ–3800	<LOQ–71.0	<LOQ–51.0	<LOQ–280	<LOQ–210	<LOQ–41,000	<LOQ–30.0	<LOQ–52.0	1991–48,004
ratio	2.4	0.3	29.3	9.6	0.4	0.2	0.5	2.2	0.1	0.1	0.3	0.6	53.8	0.1	0.1	
							beer (n = 9)									
DR	89	67	100	100	89	56	100	89	89	67	100	100	100	78	89	100
mean	113	35.8	1974	1064	36.8	5.3	61.0	13.1	8.2	4.7	8.3	84.0	1317	6.0	8.4	4740
SD	105	26.4	1157	861	38.6	4.4	32.8	10.8	4.3	3.1	6.3	97.8	1172	4.9	7.0	2374
GM	66.7	25.5	1572	801	21.9	3.9	54.2	9.7	7.2	3.7	7.0	52.0	989	4.5	6.0	4253
median	64	31	2000	980	30	5.0	52	8.3	7.1	4.4	6.8	59	780	4.9	6.0	4111
range	<LOQ–290	<LOQ–73	330–4100	210–3000	<LOQ–130	<LOQ–14.0	28.0–130	<LOQ–36.0	<LOQ–16.0	<LOQ–9.80	3.0–24.0	7.10–330	440–3700	<LOQ–17.0	<LOQ–20.0	1895–9104
ratio	2.4	0.8	41.7	22.5	0.8	0.1	1.3	0.3	0.2	0.1	0.2	1.8	27.8	0.1	0.2	

[a]: DR, detection rate (%); [b]: SD, standard deviation (ng/L); [c]: GM, geometric mean (ng/L); [d]: ratio, concentration ratio (%), calculated as the ratio between the mean concentration of each target analyte versus the mean sum concentration of 15 PAEs; [e]: LOQ, limit of quantification.

Several studies also investigated the concentrations of PAEs in bottled water samples in China, but the reported concentrations are lower than those found in mineral water in this study. Liu et al. (2015) collected a total of 225 drinking water samples from the waterworks in different regions of China and determined the concentrations of six typical PAEs including DEP, DMP, DBP, BBP, DEHP, and DOP, and the mean sum concentration was 1278 ng/L [35] (mean concentration was 4015 ng/L for mineral water samples in this study). Wang et al. (2021) collected bottled water samples from Tianjin, China, and reported that the mean sum concentration of DBP, BBP, and DEHP was 1960 ng/L [41]. Li et al. (2019) reported the concentrations of seven PAEs (DMP, DEP, DPP, DBP, BzBP,

DEHP, and DnOP) in 60 bottled water samples collected in Beijing, China, and the sum PAE concentrations ranged from 155 to 5200 (mean: 519) ng/L [47].

Overall, the concentration of individual PAE in bottled drinks did not exceed the maximum contaminant levels recommended by national and international authorities (e.g., in China, the guideline values for DEHP, DBP, and DEP are 8, 3, and 300 µg/L, respectively [48]; the guideline value for DEHP in WHO [49] and the U.S. [50] are 8 and 6 µg/L, respectively). However, the concentration of individual PAE in select bottled drinks may exceed the no-observed-adverse-effect level (NOAEL) suggested by U.S. EPA (e.g., NOAELs for BBP and DEHP in water are 0.10 and 0.32 µg/L, respectively [51]). We further calculated the hazard index (HI) for DEHP using the highest concentration of DEHP (41000 ng/L) observed in this study. The results showed that the highest HI of DEHP is 0.07, far less than 1, indicating that DEHP in bottled drinks posed negligible non-carcinogenic health risks to human health by ingestion. However, this still warrants attention when performing health risk assessment of chemical exposure because individuals are exposed to thousands of chemicals simultaneously and they may work synergistically in posing risks to human health.

3.2. Factors Influencing Phthalates Concentrations in Bottled Drinks

The bottled drinks analyzed in this study were grouped into six different types of bottled drinks, including (1) mineral water, (2) tea drink, (3) energy drink, (4) juice drink, (5) soft drink, and (6) beer. Compared with other types of drinks, mineral water samples contain the least phthalates with respect to both DRs and concentrations. Of the 15 PAEs measured in this study, only DEHP was detected in over 60% of mineral water samples (DR: 100%); however, in tea drink, energy drink, juice drink, soft drink, and beer samples, the number of PAEs with DRs over 60% was 14, 14, 15, 14, and 14, respectively (Table 3). With respect to concentrations of PAEs, of the six types of bottled drinks, soft drink had the highest sum concentration of 15 PAEs (range: 1991–48,004 ng/L; median: 10,534 ng/L), followed by juice drink (1635–46,541; 10,179), tea drink (1277–27,298; 8459), beer (1895–9104; 4111), energy drink (1184–36,505; 3972), and mineral water (770–16,301; 1805) (Table 3). The median sum concentration of 15 PAEs detected in soft drink samples is over five times higher than that detected in mineral water samples. Thus, considerable differences between the concentrations of PAEs in different types of bottled drinks were observed in this study. This is the first study showing that drink type can significantly impact the concentrations of PAEs in bottled drinks.

To investigate the contribution of each phthalate to the total phthalate burden, we calculated the ratio of the mean concentration of each phthalate to the mean sum concentration of 15 PAEs (Table 3; Figure 1a). As shown in Table 3, DEHP, DBIP, and DBP are the three major PAEs detected in beer, soft drink, juice drink, and energy drink samples, with a contribution ratio of over 10% (or around 10%). The predominant compounds found in the four types of bottled drinks are DIBP, DEHP, DIBP, and DEHP, respectively, with the respective contribution ratios being 41.7%, 53.8%, 37.5%, and 62.1%. In mineral water samples, DEHP is the predominant PAE with a contribution ratio of 84%, followed by DBIP (ratio: 9.4%), and the contribution of DBP is minor (2.8%). In tea drink samples, besides DEHP, DBIP, and DBP, we also observed a significant contribution of BMEP to the sum PAE concentration with a contribution ratio of 17.1% (Tables 2 and 3). This indicates that tea drink is an important source of human exposure to BMEP.

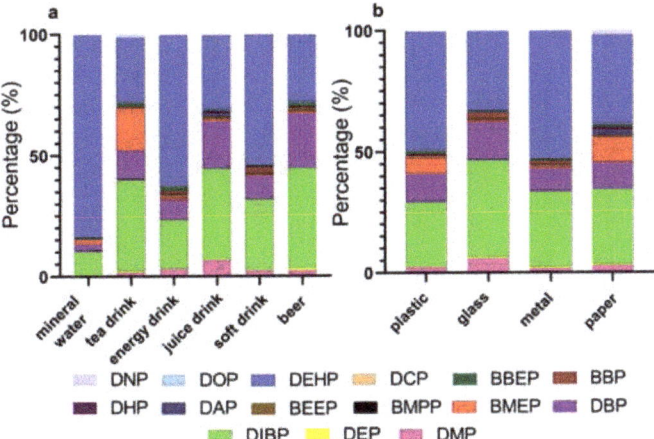

Figure 1. Compositions of total phthalates in different categories of bottled drinks: (**a**) Sorted by bottled drink types; (**b**) sorted by packaging material of bottled drinks.

We further examined the concentrations of PAEs in bottled drinks based on the packaging material, including plastic ($n = 56$), glass ($n = 19$), metal ($n = 22$), and paper ($n = 8$) (Table 4; Figure 1b). As shown in Table 4, of the 15 PAEs measured, the majority of chemicals (12–14) had DRs over 60% in each category. DEHP, DBIP, and DBP are the predominant phthalates found in each category with corresponding contribution ratios of over 10%. Compared with other packaging materials, paper-bottled drinks have a higher concentration of BMEP with a contribution ratio of 10.2%. The highest sum concentration of the 15 PAEs was found in paper-bottled drinks (range: 6418–46,541 ng/L; median: 11119 ng/L), followed by glass-bottled drinks (1635–23,256; 10,190), metal-bottled drinks (3078–48,004; 7501), and plastic-bottled drinks (770–27,298; 4340). This is different from our assumption that plastic may contain higher amounts of PAEs, which might be explained by the following reasons. Firstly, the sample size not large enough to investigate the impact of packaging material on PAEs concentrations within the same drink type. Secondly, even within the same type of packaging material, various sub-types exist. For example, different vendors may use different types of plastic in bottling the drinks. PAE concentrations may vary significantly depending on the specific plastic employed.

Table 4. Concentrations (ng/L) of phthalates in bottled drinks sorted by the packaging material.

	DMP	DEP	DIBP	DBP	BMEP	DAP	BEEP	BBP	DCP	DHP	BMPP	BBEP	DEHP	DOP	DNP	∑(sum)
							plastic ($n = 56$)									
DR[a]	73	57	77	70	82	59	75	75	71	59	64	84	93	71	77	100
mean	217	33.7	1968	920	708	14.9	54.3	36.3	17.0	7.7	13.1	112	2826	12.8	24.9	6964
SD[b]	553	54.9	2205	1240	2816	45.3	51.6	115	18.4	8.8	23.2	136	3689	16.5	42.0	6421
GM[c]	53.7	19.6	861	336	44.1	4.6	25.2	10.5	10.2	4.5	5.7	54.0	1637	6.6	10.0	4559
median	45.0	17.0	1100	280	49.5	3.3	51.5	9.8	11.0	5.2	5.5	76.5	1650	7.4	11.0	4340
range	<LOQ[e]–3100	<LOQ–390	<LOQ–8100	<LOQ–4900	<LOQ–17,000	<LOQ–310	<LOQ–220	<LOQ–850	<LOQ–110	<LOQ–49.0	<LOQ–160	<LOQ–760	<LOQ–17,000	<LOQ–86.0	<LOQ–200	770–27,298
ratio[d]	3.1	0.5	28.3	13.2	10.2	0.2	0.8	0.5	0.2	0.1	0.2	1.6	40.6	0.2	0.4	
							glass ($n = 19$)									
DR	79	53	100	100	95	63	90	84	79	68	79	90	100	74	79	100
mean	524	28.8	3510	1285	63.0	5.6	56.3	222	8.4	7.2	20.7	42.3	6083	5.4	10.2	11,871
SD	1653	26.4	3740	1042	75.9	5.6	75.7	867	5.7	11.2	55.9	31.5	9060	3.8	9.0	9584
GM	94.3	18.8	2085	853	33.1	3.9	29.4	15.9	6.8	4.2	6.7	28.9	2915	4.2	6.7	8418
median	109	12.0	2600	840	43.0	3.4	28.0	10.0	6.2	3.7	6.2	32.0	2600	4.1	8.4	10,190
range	<LOQ–7300	<LOQ–80	240–16,000	<LOQ–3100	<LOQ–310	<LOQ–24.0	<LOQ–270	<LOQ–3800	<LOQ–22.0	<LOQ–51.0	<LOQ–250	<LOQ–100	440–34,000	<LOQ–13.0	<LOQ–36.0	1635–36,505
ratio	4.4	0.2	29.6	10.8	0.5	0.1	0.5	1.9	0.1	0.1	0.2	0.4	51.2	0.1	0.1	

Table 4. Cont.

	DMP	DEP	DIBP	DBP	BMEP	DAP	BEEP	BBP	DCP	DHP	BMPP	BBEP	DEHP	DOP	DNP	∑(sum)
							metal ($n = 22$)									
DR	82	55	100	100	91	59	100	91	86	55	91	91	100	91	91	
mean	248	29.9	3216	1136	42.9	20.9	80.6	155	13.8	4.9	22.7	89.5	5502	10.1	12.3	10,584
SD	486	29.1	1469	804	72.3	45.8	56.8	595	15.0	4.6	58.2	84.7	9848	8.2	11.1	10,841
GM	72.0	19.4	2858	919	21.7	4.94	589	15.8	9.63	3.4	8.6	53.7	2220	7.3	7.6	7782
median	58.0	17.0	3200	970	25.5	3.65	63.0	12.5	9.35	2.8	7.9	59.0	1400	6.3	8.4	7501
range	<LOQ-2100	<LOQ-98.0	750-5800	240-3300	<LOQ-330	<LOQ-170	4.2-220	<LOQ-2800	<LOQ-71.0	<LOQ-16.0	<LOQ-280	<LOQ-330	470-41,000	<LOQ-30.0	<LOQ-36.0	3078-48,004
ratio	2.3	0.3	30.4	10.7	0.4	0.2	0.8	1.5	0.1	0.1	0.2	0.9	52.0	0.1	0.1	
							paper ($n = 8$)									
DR	75	100	100	100	100	63	100	100	75	63	100	100	100	100	88	100
mean	197	44.5	6125	1780	74.5	363	96.4	73.5	14.9	6.0	11.5	78.1	5680	8.7	238	14,791
SD	245	28.8	3925	962	60.8	546	85.3	138	11.2	7.6	9.9	63.6	8714	4.0	462	13,292
GM	76.9	35.2	5252	1531	51.0	23.5	54.7	27.3	9.9	3.8	8.2	57.2	2932	7.8	25.8	11,815
median	109	44.5	4900	1750	66.0	5.6	82.5	21	16	3.2	6.0	57.0	3550	8.2	8.5	11,119
range	<LOQ-700	14-80	2900-14,000	480-3600	11-190	<LOQ-1400	4.80-240	5.90-410	<LOQ-28.0	<LOQ-24.0	2.70-28.0	15.0-180	440-27,000	2.7-14.0	<LOQ-1300	6418-46,541
ratio	1.3	0.3	41.4	12.0	0.5	2.5	0.7	0.5	0.1	0.04	0.1	0.5	38.4	0.1	1.6	

[a]: DR, detection rate (%); [b]: SD, standard deviation (ng/L); [c]: GM, geometric mean (ng/L); [d]: ratio, concentration ratio (%), calculated as the ratio between the mean concentration of each target analyte versus the mean sum concentration of 15 PAEs; [e]: LOQ, limit of quantification.

3.3. Principal Component Analysis (PCA) of Phthalates in Bottled Drinks

PCA was applied to provide information regarding the possible sources of PAEs detected in the bottled drink samples in Dalian, China. Here, we performed a canonical analysis of the principal coordinates (CAP) method to analyze the input dataset after log-transformation and standardization. CAP allows a constrained ordination to be done on the basis of any distance or dissimilarity measure. The analytical results on PAEs present in 105 bottled drink samples showed that the top six principal components, abbreviated as CAP here, explained 78.3% of the total variance in the data, with the top two CAPs explaining 37.3% and 11.0% variance, respectively. The percentages of the total variance explained by other CAPs are all below 10%. This indicates that there is only one major source of PAEs present in the bottled drinks, and a variety of factors are contributing to the PAEs concentrations in the bottled drinks analyzed in this study.

The correlation coefficients between the new abstract principal components and the PAEs were also provided, indicating how well the new abstract principal components correlate with the PAEs (Table S4). The first new abstract principal component, CAP1, correlates positively with all the PAEs measured in this study, implying that higher concentrations of PAEs were linked to higher values of CAP1. This could be explained by the same exposure sources of PAEs present in these bottled drinks.

Permutational multivariate analysis of variance was carried out to compare PAEs concentrations among different types of drinks, and a significant difference ($p < 0.001$) was observed, especially between mineral water and other types of drinks, as shown in Figure 2a. In addition, results of permutational multivariate analysis of variance also indicated the significant difference ($p < 0.001$) of the PAE concentrations among bottled drinks with different packaging materials (Figure 2b). Thus, both drink type and packaging material are associated with the PAEs in the samples. This further corroborated the earlier conclusion that many factors contribute to the PAEs present in the bottled drinks.

3.4. Dietary Exposure to PAEs through Consumption of Bottled Drinks in China

The human exposure doses of 15 PAEs through the ingestion of bottled drinks were estimated based on the mean/maximum concentrations of PAEs measured in different types of bottled drinks, as shown in Table 5. The average daily intake of drink for Chinese adults was estimated as 1 L per day [41]. Mineral water is the most commonly used bottled drink among the Chinese population. Among PAEs, the mean exposure doses of DEHP were the highest from the consumption of mineral water (mean/maximum dose: 112/500 ng/kg-bw/d), followed by DIBP (12.5/31.3) and DBP (3.67/10.7). The mean/maximum human exposure doses from mineral water for other PAEs (DMP, DEP, BMEP, DAP, BEEP, BBP, DCP, DHP, BMPP, BBEP, DOP, and DNP) were 0.45/1.03, 0.33/1.40,

2.10/10.3, 0.06/0.20, 0.32/3.17, 0.16/1.33, 0.10/0.19, 0.09/0.63, 0.05/0.08, 0.81/7.67, 0.13/0.40, and 0.20/0.63 ng/kg-bw/d, respectively. The mean/maximum human exposure doses to the total phthalates were 133/544 ng/kg-bw/d.

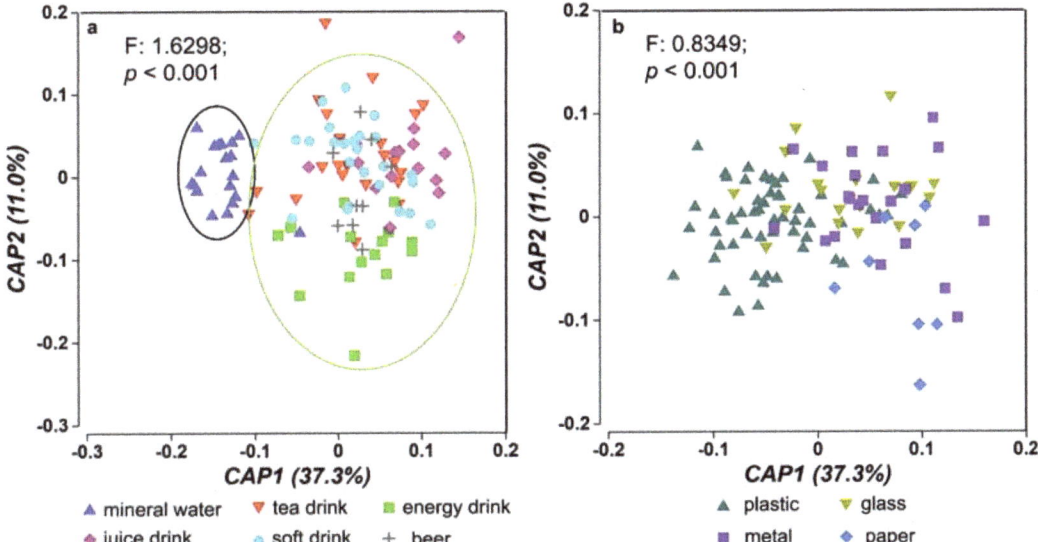

Figure 2. Plots of the canonical analysis of principal (CAP) of PAEs among different types of bottled drinks (**a**) and bottled drinks with different packaging materials (**b**). Permutational multivariate analysis of variance results are also shown in the figure.

Table 5. Estimated daily intake (EDI$_{drink}$, ng/kg-bw/d) of PAEs through ingestion of bottled drinks, based on mean/maximum concentrations.

Chemical	Mineral Water	Energy Drink	Beer	Tea Drink	Juice Drink	Soft Drink
DMP	0.45/1.03	7.72/36.7	3.77/9.67	5.91/43.3	27.3/243	10.9/103
DEP	0.33/1.40	0.92/3.27	1.19/2.43	1.27/3.67	1.27/2.90	1.49/13.0
DIBP	12.5/31.3	51.3/143	65.8/137	126/330	151/533	131/240
DBP	3.67/10.7	21.0/96.7	35.5/100	39.9/120	78.8/163	43.0/100
BMEP	2.10/10.3	2.11/7.33	1.23/4.33	56.7/567	2.43/6.33	1.860/11.0
DAP	0.06/0.20	0.46/5.00	0.18/0.47	0.42/3.33	6.75/46.7	0.93/10.3
BEEP	0.32/3.17	3.16/9.00	2.03/4.33	1.78/5.00	3.82/8.00	2.14/7.33
BBP	0.16/1.33	2.69/28.3	0.44/1.20	0.89/6.33	1.72/13.7	9.95/127
DCP	0.10/0.19	0.64/1.23	0.27/0.53	0.75/3.67	0.54/2.07	0.50/2.37
DHP	0.09/0.63	0.35/0.67	0.16/0.33	0.25/1.63	0.31/0.80	0.23/1.70
BMPP	0.05/0.08	0.46/1.23	0.28/0.80	0.42/1.83	0.56/1.23	1.17/9.33
BBEP	0.81/7.67	4.66/8.33	2.80/11.0	4.78/25.3	2.95/7.33	2.48/7.00
DEHP	112/500	158/1133	43.9/123	88.7/400	123/900	240/1367
DOP	0.13/0.40	0.58/2.10	0.20/0.57	0.53/2.87	0.34/1.07	0.29/1.00
DNP	0.20/0.63	0.34/1.77	0.28/0.67	3.49/43.3	1.63/16.7	0.52/1.73
∑(sum)	133/544	254/1217	158/304	331/910	402/1551	446/1600

Of the six types of bottled drinks analyzed, the highest mean exposure doses of DEHP (240/1367), DIBP (151/533), and DBP (78.8/163) can be obtained through the consumption of soft drinks, juice drinks, and juice drinks, respectively. Other high exposure doses of individual PAEs include DMP though juice drinks (27.3/243) and BMEP through tea drinks (56.7/567). Based on the highest mean exposure doses of each PAE, it can be generalized that the EDI$_{drink}$ values were in the order of 0.10 ng/kg-bw/d for DCP, DHP, BMPP, and

DOP, 1.00 ng/kg-bw/d for DAP, BBP, DNP, DMP, DEP, BEEP, and BBEP, 10.0 ng/kg-bw/d for BMEP and DBP, and 100 ng/kg-bw/d for DIBP and DEHP.

Human exposure doses of PAEs through the consumption of bottled drinks were several orders of magnitude lower than the oral reference doses suggested by the U.S. FDA (20, 100, 200, and 800 µg/kg-bw/d for DEHP, DBP, BBP, and DEP, respectively) [52], even when the highest phthalate concentrations in bottled drinks were used in the estimation. However, humans are exposed to PAEs via multiple pathways including inhalation, diet ingestion, and dermal absorption. The evidence has shown that dietary exposures represent a small fraction of the total exposure doses (e.g., contributed ~10% for DBP, ~10% for DMP, and ~2% for DEP to the total exposures) [28]. Other exposure sources such as personal care products also play crucial roles in human exposure to phthalates. Thus, it is highly likely that the entire human exposure doses to phthalates for individuals might exceed the oral reference doses recommended by the U.S. FDA.

3.5. Health Risk Assessment of Select PAEs through Consumption of Bottled Drinks in China

Human cancer risk caused by DEHP via consumption of different types of bottled drinks was assessed by calculating the carcinogenic risk (R). Based on the mean concentrations of DEHP detected in different types of bottled drinks, the cancer risks of DEHP for mineral water, tea drink, energy drink, juice drink, soft drink, and beer are 1.6×10^{-6}, 1.2×10^{-6}, 2.2×10^{-6}, 1.7×10^{-6}, 3.4×10^{-6}, and 0.6×10^{-6}, respectively. Except for beer, the cancer risks of DEHP for other types of bottled drinks are higher than the maximum acceptable risk level, which is 1.0×10^{-6} [38]. Thus, the potential carcinogenic risk attributable to DEHP present in the bottled drink samples should be of concern for Chinese consumers. Consumption of bottled drinks over a long duration could be harmful to human health.

Non-carcinogenic risks of DEHP, DBP, DEP, and BBP were also evaluated via the calculation of HIs. The results showed that mean HIs for DEHP, DBP, DEP, and BBP were 5.6×10^{-3}, 3.7×10^{-5}, 0.4×10^{-6}, and 0.8×10^{-6}, respectively. These values are far less than 1, indicating that these PAEs in the bottled drinks collected in this study posed negligible non-carcinogenic health risks to human health by ingestion [38]. DEHP is the major chemical contributing to the non-carcinogenic risk of PAEs on average, posing non-carcinogenic risk two orders of magnitude higher than that of DBP. Because non-carcinogenic risk is highly associated with the concentrations of PAEs detected in the samples [38], the risk posed by other non-assessed chemicals (e.g., DIBP) is most likely much lower than DEHP.

It has been known that storage time and temperature can significantly impact the migration of chemicals from packaging material to drinks [38]. When bottled drinks are stored at a high temperature for a long time, human health risks posed by the ingestion of chemicals can be increased significantly, especially the carcinogenic risk [38]. Further, co-exposure of a variety of chemical pollutants under long-term chronic exposure may have a considerable total risk to human health. Therefore, the consumption of bottled drinks could be a non-neglectable risk factor contributing to human health risk.

4. Conclusions

In summary, this is the first study to investigate the occurrence and distribution of fifteen PAEs in various types of bottled drinks in China. Our results indicated the widespread occurrence of PAEs in different types of bottled drinks. Drink type is an important factor determining the concentrations of PAEs in the drinks. Significant differences of PAE concentrations between different types of bottled drinks were observed in this study. For example, the median sum concentration of 15 PAEs in soft drink samples is over five times higher than that detected in mineral water samples. Although human exposure doses of PAEs through the consumption of bottled drinks are much lower than the oral reference doses recommended by U.S. EPA, it is non-neglectable, especially considering the high frequency of the consumption of bottled drinks in daily life. Further, the higher carcinogenic risk posed by DEHP exposure through the consumption of bottled drinks warrants attention from the public.

Our results provide baseline information, for the first time, regarding the occurrence of PAEs in bottled drinks available in the Chinese market, which is helpful for people in choosing appropriate bottled drinks. To minimize PAE exposure, it is recommended to use bottled mineral water, instead of energy drinks, juice drinks, soft drinks, tea drinks, and beer, and avoid the use of bottled drinks with long-term storage at a high temperature. Compared with bottled drinks, tap water is recommended in everyday life. This is especially important for vulnerable members in the community, such as pregnant women, lactating women, infants, and children. Further, it is recommended to develop safer alternatives for DEHP, which is the most frequently observed PAE and can pose a higher carcinogenic risk. Authorities need to take measures to control the content of DEHP present in bottled drinks.

Supplementary Materials: The following are available online, Table S1: title, Detailed information of the 15 PAEs measured in this study; Table S2: title, Instrumental parameters on GC-MS conditions for phthalate analysis; Table S3: title, Concentrations (ng/L) of the target phthalates in procedural blanks and their limits of detection (LOD) and limits of quantification (LOQ); Table S4: title, Correlation coefficients between the new abstract principal components and the PAEs present in the bottled drinks.

Author Contributions: Writing, X.X. and J.X.; data analysis, X.X., Y.S. and H.S.; data collection, Y.S., H.J. and X.C.; sample process, H.S. and D.F.; literature search, D.F.; sample collection, H.J. and X.C.; figures, X.S.; data interpretation, X.S. and Y.L.; study design, Y.L. and J.X.; review, J.X.; revision, J.X., F.L. and W.L. All authors have read and agreed to the published version of the manuscript.

Funding: This research was funded by the National Natural Science Foundation of China, grant number 21707013, 11801052, and 42007374; by the Program for Guangdong Introducing Innovative and Entrepreneurial Teams, grant number 2019ZT08L213; by the Fundamental Research Funds for the Central Universities, grant number 3132019174 and 3132019184; by the Central Public-Interest Scientific Institution Basal Research Fund, grant number PM-zx703-201803-069.

Institutional Review Board Statement: Not applicable.

Informed Consent Statement: Not applicable.

Data Availability Statement: Data is contained within the article or supplementary material.

Conflicts of Interest: The authors declare no conflict of interest.

Sample Availability: Samples analyzed in this study can be purchased from the Chinese market.

References

1. Guo, Y.; Alomirah, H.; Cho, H.S.; Minh, T.B.; Mohd, M.A.; Nakata, H.; Kannan, K. Occurrence of phthalate metabolites in human urine from several Asian countries. *Environ. Sci. Technol.* **2011**, *45*, 3138–3144. [CrossRef]
2. Zhang, Z.M.; Zhang, H.H.; Zhang, J.; Wang, Q.W.; Yang, G.P. Occurrence, distribution, and ecological risks of phthalate esters in the seawater and sediment of Changjiang River Estuary and its adjacent area. *Sci. Total Environ.* **2018**, *619*, 93–102. [CrossRef]
3. Antian, J. Toxicity and health threats of phthalate esters: Review of the literature. *Environ. Health Perspect.* **1973**, *4*, 1–26.
4. Gu, S.; Zheng, H.; Xu, Q.; Sun, C.; Shi, M.; Wang, Z.; Li, F. Comparative toxicity of the plasticizer dibutyl phthalate to two freshwater algae. *Aquat. Toxicol.* **2017**, *191*, 122–130. [CrossRef]
5. Karaconji, I.B.; Jurica, S.A.; Lasic, D.; Jurica, K. Facts about phthalate toxicity in humans and their occurrence in alcoholic beverages. *Arh. Hig. Rada. Toksikol.* **2017**, *68*, 81–92. [CrossRef] [PubMed]
6. Chen, R.Q. SVHCs Their Toxic and Uses. *Dyest. Color.* **2011**, *48*, 47–52.
7. Guo, Y.; Kannan, K. Challenges encountered in the analysis of phthalate esters in foodstuffs and other biological matrices. *Anal. Bioanal. Chem.* **2012**, *404*, 2539–2554. [CrossRef]
8. Xu, X.; Zhou, G.; Lei, K.; LeBlanc, A.G.; An, L. Phthalate esters and their potential Risk in PET bottled water stored under common conditions. *Int. J. Environ. Res. Public Health* **2020**, *17*, 141. [CrossRef] [PubMed]
9. Irvin, E.A.; Calafat, A.M.; Silva, M.J.; Aguilar-Villalobos, M.; Needham, L.L.; Hall, D.B.; Cassidy, B.; Naeher, L.P. An estimate of phthalate exposure among pregnant women living in Trujillo, Peru. *Chemosphere* **2010**, *80*, 1301–1307. [CrossRef]

10. Fromme, H.; Gruber, L.; Seckin, E.; Raab, U.; Zimmermann, S.; Kiranoglu, M.; Schlummer, M.; Schwegler, U.; Smolic, S.; Volkel, W. Phthalates and their metabolites in breast milk–results from the Bavarian Monitoring of Breast Milk (BAMBI). *Environ. Int.* **2011**, *37*, 715–722. [CrossRef] [PubMed]
11. Wang, Y.; Zhu, H.; Kannan, K. A Review of Biomonitoring of Phthalate Exposures. *Toxics* **2019**, *7*, 21. [CrossRef]
12. Benjamin, S.; Masai, E.; Kamimura, N.; Takahashi, K.; Anderson, R.C.; Faisal, P.A. Phthalates impact human health: Epidemiological evidences and plausible mechanism of action. *J. Hazard. Mater.* **2017**, *340*, 360–383. [CrossRef]
13. Grindler, N.M.; Vanderlinden, L.; Karthikraj, R.; Kannan, K.; Teal, S.; Polotsky, A.J.; Powell, T.L.; Yang, I.V.; Jansson, T. Exposure to phthalate, an endocrine disrupting chemical, alters the first trimester placental methylome and transcriptome in women. *Sci. Rep.* **2018**, *8*, 1–9. [CrossRef]
14. Wang, Y.; Qian, H. Phthalates and their impacts on human health. *Healthcare* **2021**, *9*, 603. [CrossRef]
15. Hurst, C.H.; Waxman, D.J. Activation of PPARα and PPARγ by environmental phthalate monoesters. *Toxicol. Sci.* **2003**, *74*, 297–308. [CrossRef]
16. Maloney, E.K.; Waxman, D.J. Trans-activation of PPARalpha and PPARgamma by structurally diverse environmental chemicals. *Toxicol. Appl. Pharmacol.* **1999**, *161*, 209–218. [CrossRef]
17. Praveena, S.M.; Teh, S.W.; Rajendran, R.K.; Kannan, N.; Lin, C.C.; Abdullah, R.; Kumar, S. Recent updates on phthalate exposure and human health: A special focus on liver toxicity and stem cell regeneration. *Environ. Sci. Pollut. Res.* **2018**, *25*, 11333–11342. [CrossRef] [PubMed]
18. European Plasticisers: Comments to the Annex XV-CLH Dossier for DIOP, CAS No. 27554-26-3, Provided by ANSES. Available online: https://echa.europa.eu/documents/10162/e6e57e58-33ce-5e76-22c8-1c9f68842cb9 (accessed on 13 September 2021).
19. Specht, I.O.; Toft, G.; Hougaard, K.S.; Lindh, C.H.; Lenters, V.; Jonsson, B.A.G.; Heederik, D.; Giwercman, A.; Bonde, J.P.E. Associations between serum phthalates and biomarkers of reproductive function in 589 adult men. *Environ. Int.* **2014**, *66*, 146–156. [CrossRef]
20. Suzuki, Y.; Niwa, M.; Yoshinaga, J. Exposure assessment of phthalate esters in Japanese pregnant women by using urinary metabolite analysis. *Environ. Health Prev. Med.* **2009**, *14*, 180–187. [CrossRef] [PubMed]
21. Guo, Y.; Wu, Q.; Kannan, K. Phthalate metabolites in urine from China, and implications for human exposures. *Environ. Int.* **2011**, *37*, 893–898. [CrossRef] [PubMed]
22. Hauser, R.; Meeker, J.D.; Duty, S.; Silva, M.J.; Calafat, A.M. Altered semen quality in relation to urinary concentrations of phthalate monoester and oxidative metabolites. *Epidemiology* **2006**, *17*, 682–691. [CrossRef] [PubMed]
23. Zhu, J.P.; Phillips, S.P.; Feng, Y.L.; Yang, X.F. Phthalate esters in human milk: Concentration variations over a 6-month postpartum time. *Environ. Sci. Technol.* **2006**, *40*, 5276–5281. [CrossRef] [PubMed]
24. Darbre, P.D.; Aljarrah, A.; Miller, W.R.; Coldham, N.G.; Sauer, M.J.; Pope, G.S. Concentrations of parabens in human breast tumours. *J. Appl. Toxicol.* **2004**, *24*, 5–13. [CrossRef]
25. Guo, Y.; Kannan, K. A Survey Phthalates and Parabens in Personal Care Products from the United States and Its Implications for Human Exposure. *Environ. Sci. Technol.* **2013**, *47*, 14442–14449. [CrossRef] [PubMed]
26. Schettler, T. Human exposure to phthalates via consumer products. *Int. J. Androl.* **2006**, *29*, 134–139. [CrossRef] [PubMed]
27. Škrbić, B.D.; Ji, Y.; Živančev, J.R.; Jovanović, G.G.; Zhao, J. Mycotoxins, trace elements, and phthalates in marketed rice of different origin and exposure assessment. *Food Addit. Contam. Part. B Surveill.* **2017**, *10*, 256–267. [CrossRef]
28. Guo, Y.; Zhang, Z.; Liu, L.; Li, Y.; Ren, N.; Kannan, K. Occurrence and profiles of phthalates in foodstuffs from China and their implications for human exposure. *J. Agric. Food Chem.* **2012**, *60*, 6913–6919. [CrossRef]
29. Škrbić, B.D.; Ji, Y.; Đurišić-Mladenović, N.; Zhao, J. Occurrence of the phthalate esters in soil and street dust samples from the Novi Sad city area, Serbia, and the influence on the children's and adults' xposure. *J. Hazard. Mater.* **2016**, *312*, 272–279. [CrossRef]
30. Škrbić, B.D.; Kadokami, K.; Antić, I. Survey on the micro-pollutants presence in surface water system of orthern Serbia and environmental and health risk assessment. *Environ. Res.* **2018**, *166*, 130–140. [CrossRef]
31. Guo, Y.; Kannan, K. Comparative assessment of human exposure to phthalate ethers from house dust in China and the United States. *Environ. Sci. Technol.* **2011**, *45*, 3788–3794. [CrossRef]
32. Wormuth, M.; Scheringer, M.; Vollenweider, M.; Hungerbuhler, K. What are the sources of exposure to eight frequently used phthalic acid esters in Europeans? *Risk Anal.* **2006**, *26*, 803–824. [CrossRef] [PubMed]
33. Clark, K.; Cousins, I.T.; Mackay, D. Assessment of critical exposure pathways. *Handb. Environ. Chem.* **2003**, *3*, 227–262.
34. Itoh, H.; Yoshida, K.; Masunaga, S. Quantitative identification of unknown exposure pathways of phthalates based on measuring their metabolites in human urine. *Environ. Sci. Technol.* **2007**, *41*, 4542–4547. [CrossRef] [PubMed]
35. Liu, X.W.; Shi, J.H.; Bo, T. Occurrence and risk assessment of selected phthalates in drinking water from waterworks in China. *Environ. Sci. Pollut. Res.* **2015**, *22*, 10690–10698. [CrossRef]
36. Le, T.M.; Nguyen, h.m.n.; Nguyen, V.K.; Nguyen, A.V.; Vu, N.D.; Yen, N.T.H.; Hoang, A.Q.; Minh, T.B.; Kannan, K.; Tran, T.M. Profiles of phthalic acid esters (PAEs) in bottled water, tap water, lake water, and wastewater samples collected from Hanoi, Vietnam. *Sci. Total Environ.* **2021**, *788*, 147831. [CrossRef] [PubMed]
37. Škrbic', B.; Živanc̆ev, J.; Mrmoš, N. Concentrations of arsenic, cadmium and lead in selected foodstuffs from Serbian market basket: Estimated intake by the population from the Serbia. *Food Chem. Toxicol.* **2013**, *58*, 440–448. [CrossRef] [PubMed]
38. Škrbić, B.; Đurišić-Mladenović, N.; Cvejanov, J. Principal component analysis of trace elements in Serbian wheat. *J. Agric. Food Chem.* **2005**, *53*, 2171–2175. [CrossRef]

39. Škrbić, B.; Đurišić-Mladenović, N.; Živančev, J.; Tadić, Đ. Seasonal occurrence and cancer risk assessment of polycyclic aromatic hydrocarbons in street dust from the Novi Sad city, Serbia. *Sci. Total Environ.* **2019**, *647*, 191–203. [CrossRef]
40. Anderson, M.J.; Robinson, J. Generalized discriminant analysis based on distances. *Aust. New Zealand J. Stat.* **2003**, *45*, 301–318. [CrossRef]
41. Wang, C.; Huang, P.; Qiu, C.; Li, J.; Hu, S.; Sun, L.; Bai, Y.; Gao, F.; Li, C.; Liu, N.; et al. Occurrence, migration and health risk of phthalates in tap water, barreled water and bottled water in Tianjin, China. *J. Hazard. Mater.* **2021**, *408*, 124891. [CrossRef]
42. U.S. Environmental Protection Agency. Integrated Risk Information System (IRIS). Available online: https://www.epa.gov/iris (accessed on 13 September 2021).
43. Moazzen, M.; Mahvi, H.A.; Shariatifar, N.; Khaniki, J.G.; Nazmara, S.; Alimohammadi, M.; Ahmadkhaniha, R.; Rastkari, N.; Ahmadloo, M.; Akbarzadeh, A.; et al. Determination of phthalate acid esters (PAEs) in carbonated soft drinks with MSPE/GC-MS method. *Toxin Rev.* **2018**, *37*, 319–326. [CrossRef]
44. Luo, Q.; Liu, Z.H.; Yin, H.; Dang, Z.; Wu, P.Z.; Zhu, N.W.; Lin, Z.; Liu, Y. Migration and potential risk of trace phthalates in bottled water: A global situation. *Water Res.* **2018**, *147*, 362–372. [CrossRef] [PubMed]
45. Abtahi, M.; Dobaradaran, S.; Torabbeigi, M.; Jorfi, S.; Gholamnia, R.; Koolivand, A.; Darabi, H.; Kavousi, A.; Saeedi, R. Health risk of phthalates in water environment: Occurrence in water resources, bottled water, and tap water, and burden of disease from exposure through drinking water in Tehran, Iran. *Environ. Res.* **2019**, *173*, 469–479. [CrossRef]
46. Santana, J.; Giraudi, C.; Marengo, E.; Robotti, E.; Pires, S.; Nunes, I.; Gaspar, M.E. Preliminary toxicological assessment of phthalate esters from drinking water consumed in Portugal. *Environ. Sci. Pollut. Res.* **2013**, *21*, 1380–1390. [CrossRef] [PubMed]
47. Li, H.; Li, C.; An, L.; Deng, C.; Su, H.; Wang, L.; Jiang, Z.; Zhou, J.; Wang, J.; Zhang, C.; et al. Phthalate ester in bottled drinking water and their human exposure in Beijing, China. *Food Addit. Contam. Part. B Surveill.* **2019**, *12*, 1–9. [CrossRef] [PubMed]
48. National Standard of the People's Republic of China (NSC) (2006) China's Standard for Drinking Water Quality. Ministry of Health of the People's Republic of China and Standardization Administration of the People's Republic of China, GB 5749–2006. Available online: http://tradechina.dairyaustralia.com.au/wp-content/uploads/2018/08/GB-5749-2006-Standards-for-Drinking-Water-Quality.pdf (accessed on 13 September 2021). (In Chinese)
49. WHO. Guidelines for Drinking-Water Quality-4th ed. WHO Press, Geneva, Switzerland. 2011. Available online: https://apps.who.int/iris/bitstream/handle/10665/44584/9789241548151_eng.pdf (accessed on 13 September 2021).
50. US EPA. 2006 Edition of the drinking water standards and health advisories. Office of Water. US Environmental Protection Agency, Washington, DC, USA. 2006. Available online: https://nepis.epa.gov/Exe/ZyPURL.cgi?Dockey=P1004X78.txt (accessed on 13 September 2021).
51. US EPA (US Environmental Protection Agency). National Recommended Water Quality Criteria–Human Health Criteria Table. US Environmental Protection Agency, Washington, DC, USA. 2015. Available online: https://www.epa.gov/wqc/national-recommended-water-quality-criteria-human-health-criteria-table (accessed on 13 September 2021).
52. US FDA (US Food & Drug Administration). Limiting the Use of Certain Phthalates as Excipients in Center for Drug Evaluation and Research-Regulated Products. US Food & Drug Administration, Washington, DC, USA. 2012. Available online: https://www.fda.gov/regulatory-information/search-fda-guidance-documents/limiting-use-certain-phthalates-excipients-cder-regulated-products (accessed on 13 September 2021).

Article

A QSAR–ICE–SSD Model Prediction of the PNECs for Per- and Polyfluoroalkyl Substances and Their Ecological Risks in an Area of Electroplating Factories

Jiawei Zhang [1,2], Mengtao Zhang [1], Huanyu Tao [1,2], Guanjing Qi [1], Wei Guo [1,3], Hui Ge [1,*] and Jianghong Shi [1,*]

[1] State Environmental Protection Key Laboratory of Integrated Surface Water-Groundwater Pollution Control, School of Environmental Science and Engineering, Southern University of Science and Technology, Shenzhen 518055, China; 11750007@mail.sustech.edu.cn (J.Z.); 11652003@mail.sustech.edu.cn (M.Z.); 11850006@mail.sustech.edu.cn (H.T.); 11849066@mail.sustech.edu.cn (G.Q.); gwfybj@bjut.edu.cn (W.G.)
[2] Environmental Engineering Research Centre, Department of Civil Engineering, The University of Hong Kong, Hong Kong 999077, China
[3] Key Laboratory of Beijing for Water Quality Science and Water Environment Recovery Engineering, Beijing University of Technology, Beijing 100124, China
* Correspondence: geh@sustech.edu.cn (H.G.); shijh@sustech.edu.cn (J.S.)

Abstract: Per- and polyfluoroalkyl substances (PFASs) are a class of highly fluorinated aliphatic compounds that are persistent and bioaccumulate, posing a potential threat to the aquatic environment. The electroplating industry is considered to be an important source of PFASs. Due to emerging PFASs and many alternatives, the acute toxicity data for PFASs and their alternatives are relatively limited. In this study, a QSAR–ICE–SSD composite model was constructed by combining quantitative structure-activity relationship (QSAR), interspecies correlation estimation (ICE), and species sensitivity distribution (SSD) models in order to obtain the predicted no-effect concentrations (PNECs) of selected PFASs. The PNECs for the selected PFASs ranged from 0.254 to 6.27 mg/L. The ΣPFAS concentrations ranged from 177 to 983 ng/L in a river close to an electroplating industry in Shenzhen. The ecological risks associated with PFASs in the river were below 2.97×10^{-4}.

Keywords: PFASs; QSAR–ICE–SSD; electroplating industry; ecological risk assessment

Citation: Zhang, J.; Zhang, M.; Tao, H.; Qi, G.; Guo, W.; Ge, H.; Shi, J. A QSAR–ICE–SSD Model Prediction of the PNECs for Per- and Polyfluoroalkyl Substances and Their Ecological Risks in an Area of Electroplating Factories. *Molecules* **2021**, *26*, 6574. https://doi.org/10.3390/molecules26216574

Academic Editor: Nuno Neng

Received: 8 October 2021
Accepted: 28 October 2021
Published: 30 October 2021

Publisher's Note: MDPI stays neutral with regard to jurisdictional claims in published maps and institutional affiliations.

Copyright: © 2021 by the authors. Licensee MDPI, Basel, Switzerland. This article is an open access article distributed under the terms and conditions of the Creative Commons Attribution (CC BY) license (https://creativecommons.org/licenses/by/4.0/).

1. Introduction

Per- and polyfluoroalkyl substances (PFASs) consist of carbon chains of different lengths where the hydrogen atoms are completely (perfluorinated) or partly (polyfluorinated) substituted by fluorine atoms. PFASs are widely used in the textile/leather treatment industry, manufacture of fluoropolymers, semiconductor industry, and electroplating industry. From 1951 to 2015, an estimated 2610–21,400 t of long-chain perfluoroalkyl carboxylic acids (PFCAs) were produced [1]. Due to the toxic effects, tissue accumulation, long-range transport, and environmental persistence of PFASs, perfluorooctanesulfonic acid (PFOS) was listed under the Stockholm Convention on Persistent Organic Chemicals and perfluorooctanoic acid (PFOA) was being considered for listing by 2017 [2]. As a result, around 3 million companies developed PFAS alternatives, for which they claim intellectual property rights protection [2].

An ecological risk assessment (ERA) aims to qualitatively or quantitatively describe the possibility that adverse ecological effects occur because of exposure to one or more stressors (e.g., chemical substances) [3]. An ERA has been adopted as an important methodology in many studies of typical PFASs such as PFOS [4] and PFOA [5]. The predicted no-effect concentration (PNEC) is expressed as the lowest concentration of adverse effects in an ecosystem of a given chemical substance [6]. The ratio of the PNEC to the measured exposure concentration (MEC) is known as the risk quotient (RQ), which is a screening-level descriptor of the ecological risk. To reduce the uncertainty associated with an ERA, the species sensitivity distribution (SSD) method is widely used to derive the PNEC [7,8]. An SSD is a cumulative probability distribution of the toxicity measurements of a chemical obtained from single-species bioassays of various species that can be used to estimate the ecotoxicological impacts of a chemical [9]. The robustness and accuracy of the SSD method strongly depend on the amount of species toxicity data [6,10,11]. Owing to the wide variety of PFASs and many emerging alternatives, the acute toxicity data for emerging PFASs and their alternatives are relatively limited [12]. The combination of quantitative structure-activity relationship (QSAR) models with interspecies correlation estimation (ICE) models can greatly expand the ability to predict untested chemicals and their potential effects on untested species; this has aroused a wide research interest [13–16]. QSAR models provide opportunities to estimate the ecotoxicity values for certain species (usually standard test species such as zebrafish) based on the knowledge of chemical structures or properties [17]. ICE models use available toxicity data of tested species (i.e., surrogate species) to predict those of untested species (i.e., predicted species) [18]. QSAR–ICE models can fill the data gap to generate SSDs, providing practical applications for the ERA of chemicals with limited data [13].

In aquatic systems, PFAS concentrations are higher in industrialized and urbanized areas than in less populated and remote regions in China [19]. Our previous study investigated the concentrations of PFOA and PFOS in the effluent of a sewage treatment plant in Beijing, which were found to be 29.9–71.5 ng/L and 60.1–233 ng/L, respectively [20]. These results indicated that the activated sludge process could not effectively remove PFOA and PFOS. Another previous study in the Fenhe River in Shanxi Province showed that the PFOA and PFOS concentrations were 2.49–4.79 ng/L and 3.54–16.2 ng/L, respectively [21]. Yamazaki et al. [22] reported that the PFAS concentrations ranged from non-detected to 1.5 ng/L in rivers and lakes on the Qinghai–Tibet Plateau, corresponding with low industrial levels. Based on the estimations of Wang et al. [19], the electroplating industry was the most important source of PFASs discharged into the aquatic environment.

Few studies have been undertaken on the occurrence and ERA of PFASs in the surface waters surrounding areas where electroplating industries operate. Relatively limited toxicity data make it difficult to develop the ERA of PFASs and their alternatives. In addition, QSAR–ICE–SSD models developed for estimating the PNECs of PFASs and their alternatives have been rarely reported in recent studies. Accordingly, the objectives of this study are (1) to construct QSAR–ICE–SSD models to predict the PNECs of PFASs and their alternatives and (2) to assess the ecological risk of PFASs in a river near electroplating factories.

2. Materials and Methods
2.1. Construction of QSAR–ICE Models

Following the procedures recommended in the Technical Guidance Document on Risk Assessment of the European Commission [6], the Guidelines for Ecological Risk Assessment of United States Environmental Protection Agency (US EPA) [3], and the literature [23–25], the process of collecting toxicity data can be summarized briefly as follows: four species (*Pseudokirchneriella subcapitata*, *Chlorella vulgaris*, *Daphnia magna*, and *Danio rerio*) representing three trophic levels in the aquatic environment were selected as model species for QSAR models. The acute toxicity data were mainly obtained from the US EPA ECOTOX database (http://cfpub.epa.gov/ecotox/ (accessed on 4 May 2021)), the literature, and relevant government documents. Structurally similar chemicals in the same group (i.e., PFASs) were used in the QSAR modeling. Chemicals that contained at least one -CF_2- were originally considered as PFASs and further checked against the list of PFASs of the US EPA (https://comptox.epa.gov/dashboard/chemical_lists/pfasmaster (accessed on 4 May 2021)) [26]. Data screening followed the principles of accuracy, relevance, and reliability [27]. The test methods were in accordance with standard test methods (e.g., the methods of the Organization for Economic Cooperation and Development). The toxicity endpoints were the median lethal concentration (LC_{50}) or the median effect concentration (EC_{50}). The 48 h $LC(EC)_{50}$ was preferred for invertebrate species and the 96 h $LC(EC)_{50}$

was preferred for other species. When multiple toxicity values were available for the same species and the same endpoint, the geometric mean was taken as the mean toxicity value for the species.

Molecular structure files were obtained from the ChemSpider database (https://chemspider.com/ (accessed on 10 May 2021)) and the molecular energy was optimized using the GAMESS Interface method in ChemBio3D (https://perkinelmerinformatics.com/ (accessed on 12 May 2021)). A total of 12 semi-empirical molecular descriptors were then calculated using the AM1 method in MOPAC 2016 (http://openmopac.net/ (accessed on 12 May 2021)) and the K_{ow} values (shown in Table 1) were calculated using EPI Suite software (https://www.epa.gov/ (accessed on 12 May 2021)). The Chemical Abstracts Service Registry Number (CAS No.), chain lengths, and molecular descriptor values of the selected PFASs are listed in the Supplementary Excel file. There were 27 PFASs selected, in which chain lengths ranged from 2 to 15 and contained PFCAs, perfluoroalkane sulfonic acids (PFSAs), polyfluoroalkyl ether sulfonic acids (PFESAs), cyclic perfluorinated acids, fluorotelomer-based substances, and perfluoroalkyl acid precursors.

Table 1. Molecular descriptors used in this study.

No.	Molecular Descriptors	Abbreviations	Units
1	Heat of formation	HOF	kcal/mol
2	Total energy	TE	EV
3	Electronic energy	EE	EV
4	Core–core repulsion energy	ECCR	EV
5	COSMO area	CA	Å2
6	COSMO volume	CV	Å3
7	Gradient norm	GN	-
8	Gradient norm per atom	GN p A	-
9	Ionization potential	IP	EV
10	Lowest unoccupied molecule orbital energy	ELUMO	EV
11	Highest occupied molecular orbital energy	EHOMO	EV
12	Molecular weight	MW	-
13	Octanol–water partition coefficient	K_{ow}	-

The stepwise regression method in SPSS (https://www.ibm.com/ (accessed on 4 May 2021)) was used to establish the multiple regression statistical models between the logarithmic values of the toxicity data (i.e., log LC(EC)$_{50}$) and the molecular descriptors (including their logarithmic values)). Four QSAR models were validated using SIMCA software (https://www.sartorius.com/ (accessed on 28 May 2021)), in which the non-cross-validation correlation coefficient (r^2) and leave-one-out cross-validation correlation coefficient (q^2) were used as the evaluation indices.

A total of 227 ICE models of native species in China were established and used to estimate the acute toxicity data of 6 chemicals, including 4-dichlorophenol, triclosan, tetrabromobisphenol A, nitrobenzene, PFOS, and octachlorodiphenyl [28]. These ICE models were used after verifying the application domain.

2.2. Sample Treatment and Analysis of PFASs

Water samples were collected from a river near electroplating factories in Shenzhen. S1 and S5 were located approximately 500 m upstream of the factories and S2, S3, and S4 were located downstream. The locations of the water sampling points around the plant are shown in Figure 1. Each water sample was collected in a polypropylene sample bottle and stored at 4 °C in a sampling box. Upon arrival at the laboratory, 500 mL of each water sample was filtered through a glass microfiber filter (GFF: diameter 150 mm; pore size 0.7 μm). The pH of the water was adjusted to 3.0 with a hydrochloric acid solution. The samples were then stored at 4 °C in the laboratory.

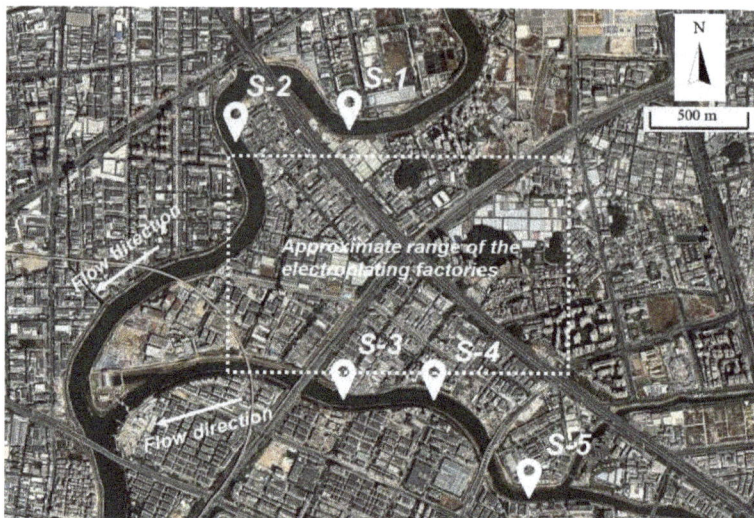

Figure 1. Sampling sites near the electroplating factories.

The water samples were extracted following a previously established method [29] with a few modifications. Briefly, Oasis WAX cartridges were preconditioned with 4 mL of 0.1% $NH_4OH/MeOH$, 4.0 mL of methanol, and 4.0 mL of deionized water. Each filtrate was passed through a preconditioned cartridge at a flow rate of 5–10 mL min^{-1}. The cartridge was then washed with 4 mL of deionized water and 25 mM of an acetic acid–ammonium acetate buffer solution (pH = 4). The WAX cartridges were placed in a centrifuge tube and centrifuged at 3000 rpm for 2 min to remove the excess water. An elution was carried out with 4 mL of methanol and 4 mL of 0.1% $NH_4OH/MeOH$. The eluent was evaporated to dryness under a gentle N_2 stream in a water bath at 40 °C, redissolved in 1.0 mL of methanol, transferred to a liquid chromatography (LC) vial, and evaporated to dryness under a gentle N_2 stream. Each sample was reconstituted with methanol (0.5 mL) and spiked with an internal standard (0.5 ng).

Ultra-high-performance liquid chromatography combined with Q-Exactive Orbitrap Tandem Mass Spectrometry (UPLC-Q-Exactive MS) was applied for the non-target screening of the PFASs and 19 certified standards (Table S1) were applied for the further quantification of the PFASs. An RRHD Extend-C_{18} column (2.1 mm × 50 mm, 1.8 µm, Agilent) was used for separation. Mobile phase A consisted of 2 mM ammonium acetate/water, and mobile phase B consisted of methanol. The elution gradient was set as follows: 5–35% B for 1 min; 35–55% B for 7 min; 55–95% B for 17 min and maintained at 18 min; then back to the initial conditions (95% A) for 18.1 min and maintained at 20 min. The flow rate was set at 0.25 mL/min and the column oven was maintained at 35 °C. A 5 µL aliquot was injected into the LC-Q-Exactive MS system. The mass spectrometer was operated in the negative electrospray ionization in full scan mode (m/z 100–1000) (Table S1). The chromatograms are shown in Figure S1. The exact mass of the PFASs was applied to the screening and quantification of the PFASs.

All target analytes were quantified using an internal standard calibration curve (r > 0.99). The method reproducibility was evaluated based on the relative standard deviation (RSD) of the recovery of the spiked replicates. The limits of detection (LOD) were estimated based on signal-to-noise ratios of 3:1. The mean procedural recovery of the PFASs ranged from 81 to 122% and the LODs of the PFASs were 1–70 ng/L (Table S1). One procedural blank and one procedural recovery sample were also analyzed for each batch of samples to check for laboratory contamination and accuracy.

2.3. Ecological Risk Characterization

RQ methods were used in this study to roughly characterize the estimated ecological risks posed by the PFASs (as shown in Equation (1)). The ecological risks could be divided into four grades: high risk (RQ ≥ 1); medium risk (1 > RQ ≥ 0.1); low risk (0.1 > RQ ≥ 0.01); and no risk (RQ < 0.01) [30].

$$RQ = \frac{MEC}{PNEC}. \quad (1)$$

The PNEC values were extrapolated by SSDs. The log normal parametric fitting method was used for the construction of the SSD curves. The cumulative distribution function (CDF) is shown in Equation (2). The threshold concentration for protecting 95% of the species (i.e., the hazardous concentration for 5% of species, HC_5) was obtained from the constructed SSD curve. The PNEC values were obtained using Equation (3). The model construction and related statistical calculations were completed using R (https://r-project.org (accessed on 2 July 2021)) and related packages such as "ssdtools" (https://bcgov.github.io/ssdtools/ (accessed on 2 July 2021)). The goodness-of-fit test for the normal distribution of the toxicity data was conducted using the Anderson–Darling test, Kolmogorov–Smirnov test, or Cramér–von Mises test.

$$CDF = (x, \mu, \sigma)\frac{1}{2} + \frac{1}{2}\text{erf}[\frac{\ln x - \mu}{\sqrt{2}\sigma}]. \quad (2)$$

$$PNEC = \frac{HC_5}{AF} \quad (3)$$

where AF is the assessment factor, which was set to 5 in this study [6].

3. Results and Discussion

3.1. Predicted Toxicity Data by QSAR–ICE Models

The acute toxicity of PFASs to *Pseudokirchneriella subcapitata*, *Chlorella vulgaris*, *Daphnia magna*, and *Danio rerio* were collected (Table S2). There were 14 EC_{50} values (from 2.1 to 1130 mg/L) for *Pseudokirchneriella subcapitata*, 10 EC_{50} values (from 3.9 to 4030 mg/L) for *Chlorella vulgaris*, 10 LC_{50} values (from 0.06 mg/L to 2.58 × 10^5 mg/L) for *Daphnia magna*, and 12 LC_{50} values (from 8.4 to 1500 mg/L) for *Danio rerio*. Based on the collected data, the calculated molecular descriptors (Supplementary Excel file), and a stepwise multiple linear regression, QSAR models for the four species were constructed (Equations (4)–(7) in Table 2). The four QSAR models were validated using the conventional correlation coefficient (r^2) and the leave-one-out cross-validation correlation coefficient (q^2) (Table 2). Generally, QSAR models with $r^2 > 0.6$ and $q^2 > 0.5$ can be regarded as having a relatively good predictive ability [31]. In this study, although the QSAR models showed only passable fitting degrees (R^2) due to the relatively small datasets (*n*), the r^2 and q^2 values were >0.6 and >0.5, respectively, indicating that the established QSAR models had a good prediction ability and statistical significance ($p < 0.05$).

Table 2. QSAR models and their validation parameters.

Species	Models	Equations	n [a]	R^2 [b]	r^2 [c]	q^2 [d]	p [e]
Pseudokirchneriella subcapitata	log EC_{50} = −log K_{ow} × 8.82 + TE × 47.8 + log E_{LUMO} × 1.47 − E_{CCR} × 39.7 + 50.3	(4)	14	0.770	0.742	0.701	0.006
Chlorella vulgaris	log EC_{50} = −4.18 × K_{ow} − 0.332 × E_{CCR} − 4.29	(5)	10	0.592	0.751	0.673	0.043
Daphnia magna	log LC_{50} = −K_{ow} × 4.09 + log TE × 9.75 − E_{CCR} × 7.03 + log E_{LUMO} × 1.63 + 1.95	(6)	10	0.370	0.605	0.580	0.045
Danio rerio	log LC_{50} = −K_{ow} × 1.03 − E_{CCR} × 1.04 + E_{LUMO} × 0.318 + 2.94	(7)	12	0.558	0.722	0.630	0.046

[a] *n*: number of toxicity data. [b] R^2: coefficient of determination of the multiple regression. [c] r^2: conventional correlation coefficient or non-validation correlation coefficient. [d] q^2: leave-one-out cross-validation correlation coefficient. [e] p: statistical significance.

The molecular descriptors of the established models practically explained the mechanism of acute toxicity (MOA). There was a positive correlation between the log EC_{50} and the total energy (TE), which is a molecular descriptor related to the molecular energies and stabilities of PFASs. These include molecular internal energy, translational kinetic energy, the energy of electrons in a molecule, the vibration energy between atoms in a molecule, and the energy of a molecule rotating around the center of a mass. A higher TE value indicates that the molecule is not easily polarized or absorbed by cells, thus resulting in a lower toxicity [32]. There was a positive correlation between the log EC_{50} and the lowest unoccupied molecule orbital energy (ELUMO). As the electronegativity of the F atom is the strongest, the PFASs reacted with the action site of the target organism as the electron acceptor. According to the frontier orbital theory, the occurrence of the reaction is related to the difference between the highest occupied orbital energy (EHOMO) of the electron donor and the ELUMO of the electron acceptor; that is, EHOMO–ELUMO (also known as the energy band gap). The larger the band gap, the easier the reaction and the stronger the binding force between the electron donor and the electron acceptor. Hence, the larger the band gap, the more obvious the toxicity and the lower the log LC_{50} value [33]. There was a negative correlation between the log LC_{50} and the nuclear–nuclear repulsive energy (ECCR). The electron cloud of atoms in a molecule is deformed more easily with an increase in the ECCR value, which makes PFASs more likely to polarize and enter a cell [33]. The log EC_{50} was negatively correlated with the octanol–water partition coefficient (K_{ow}), which is related to the lipophilicity of PFASs. With an increase in the K_{ow} value, PFASs accumulate more easily in an organism, thus corresponding with a higher toxicity. K_{ow} is a key physico-chemical parameter serving as a classic molecular descriptor in QSAR modeling [34]. In this study, all 4 QSAR models contained K_{ow} (or log K_{ow}), indicating the universality of K_{ow} in predicting aquatic acute toxicity. Moreover, it has been shown that K_{ow} is also important in applying QSAR models to predict toxicity in rodents [35,36] and in vitro toxicity assays [37,38]. In the practice of chemical management, K_{ow} can be used to justify waiving ecotoxicity tests (if log K_{ow} < 3) to assess bioaccumulation (if log K_{ow} < 3, the chemical can be considered to be non-bioaccumulative) [34]. As a result, K_{ow}-based QSAR modeling can be an effective tool for predicting the toxicity of different endpoints in screening levels.

The acute toxicity data of perfluorobutyric acid (PFBA), PFOA, perfluorobutanesulfonic acid (PFBS), perfluorohexanesulfonic acid (PFHxS), PFOS, and 6:2 chlorinated polyfluoroalkyl ether sulfonate (6:2 Cl-PFESA) for the four selected species were predicted using the four QSAR models, as shown in Table 3. The results predicted by the QSAR models showed that the toxicities of novel PFASs or substitutes such as 6:2 Cl-PFESA, PFBA, and PFBS were higher than those of PFOS and PFOA. The insertion of an oxygen atom into the 6:2 Cl-PFESA molecule could increase the activity of the molecule. The experimental results of other studies have also indicated that the presence of oxygen atoms could increase the toxicity of PFASs [39]. Due to their smaller molecular weight, short-chain PFAS substitutes (e.g., PFBA and PFBS) may be more easily polarized and absorbed by cells, thus increasing toxicity.

Table 3. Predicted acute toxicity data (mg/L) of six PFASs by QSAR models.

PFASs	CAS No.	*Pseudokirchneriella subcapitata*	*Chlorella vulgaris*	*Daphnia magna*	*Danio rerio*
PFBA	375-22-4	67.1	112	37.4	1410
PFOA	335-67-1	478	150	570	98.5
PFBS	375-73-5	2840	222	487	1000
PFHxS	355-46-4	1030	258	821	256
PFOS	1763-23-1	53	309	173	61.3
6:2 Cl-PFESA	73606-19-6	1.3	84.9	10.9	32.7

Based on the measured toxicity data collected from databases (Table S3) and the predicted data of the QSAR models, the 13, 34, 17, 13, 13, and 13 ICE models available for PFBA, PFOA, PFBS, PFHxS, PFOS, and 6:2 Cl-PFESA (Table S4) were selected for their toxicity extrapolation, respectively [28]. The acute toxicity data estimated by the QSAR–ICE models constructed for the above six substances are listed in Table S5.

3.2. Calculation and Comparison of the PNEC Values of SSDs Produced Using Predicted and Measured Data

Figure 2 shows the SSD curves based on the predicted data by the QSAR–ICE models (the six PFASs) and measured data, respectively. The results of the goodness-of-fit tests for the acute toxicity data are shown in Table S6. The results of the goodness-of-fit tests for all six PFASs were less than the corresponding thresholds, indicating that the obtained acute toxicity data were consistent with the log normal distribution. The HC_5 and PNEC values are presented in Table 4. The order of the HC_5 values was ranked from low to high, which was 6:2 Cl-PFESA < PFBA < PFOS < PFOA < PFBS < PFHxS. The measured acute toxicity data used in the SSD curves of PFOA and PFOS are shown in Table S7. The HC_5 values obtained by the two methods were compared in order to evaluate the accuracy of the QSAR–ICE–SSD models. As shown in Table 4, the HC_5 values of the QSAR–ICE–SSD models were 1.16 times (PFOA) and 1.20 times (PFOS) higher than the calculated values based on the measured toxicity data. As a result, the QSAR–ICE–SSD models had a certain reliability for predicting PNEC values when limited data were available.

Table 4. The HC_5 and PNEC values based on the predicted data (six PFASs) and measured data (PFOA and PFOS).

	PFBA	PFOA	PFOA (Measured)	PFBS	PFHxS	PFOS	PFOS (Measured)	6:2 Cl-PFESA
HC_5 (mg/L)	4.02	31.4	27	50.5	64.5	10.5	8.72	1.27
PNEC (mg/L)	0.804	6.27	-	10.1	12.9	2.09	-	0.254

3.3. Concentrations of PFASs in the River near the Electroplating Factories

The Σ_{19}PFAS concentrations ranged from 177 to 983 ng/L in the river water samples (Figure 3); the mean values of PFOS, PFBS, and PFHxS were 254, 132, and 9.18 ng/L, respectively. One study on PFASs in 28 rivers in eastern China showed that the PFAS concentration ranges were 39–212 ng/L and 0.68–146 ng/L in Shanghai and Zhejiang Province, respectively [40]. Another study of fluoropolymer facilities showed that PFAS concentrations ranged from 0.96 to 4534.41 ng/L in nearby rivers [41]. Industrial processes involving the use of PFASs are a conspicuous source of PFASs for the environment. Major downstream industrial users, such as electroplating facilities, have started to use alternatives [42].

3.4. Ecological Risks of PFASs

Based on the monitoring data of PFASs in this study, PFBA, PFOA, PFBS, PFHxS, PFOS, and 6:2 Cl-PFESA were the mainly detected PFASs in the river near the electroplating facilities. The PNEC values of the six typical PFASs were calculated using the QSAR–ICE–SSD models. The RQ values of the PFASs in this study and in four other electroplating areas in Guangdong Province in China [43] were then calculated, as listed in Table 5. The results showed that the six PFASs posed no ecological risks to the river although, compared with other electroplating areas, the RQ values of PFOA, PFBS, PFOS, and 6:2 Cl-PFESA were higher in this study. Only ecological risks based on acute PNEC values were calculated due to limited data. However, it has been suggested that PFASs may have reproductive and growth adverse effects on aquatic organisms [12]. PFASs are persistent, bioaccumulative, and can be transported long distances, thus causing lasting damage to the aquatic organism [12]. The ecological risks of PFASs in this study may have been underestimated.

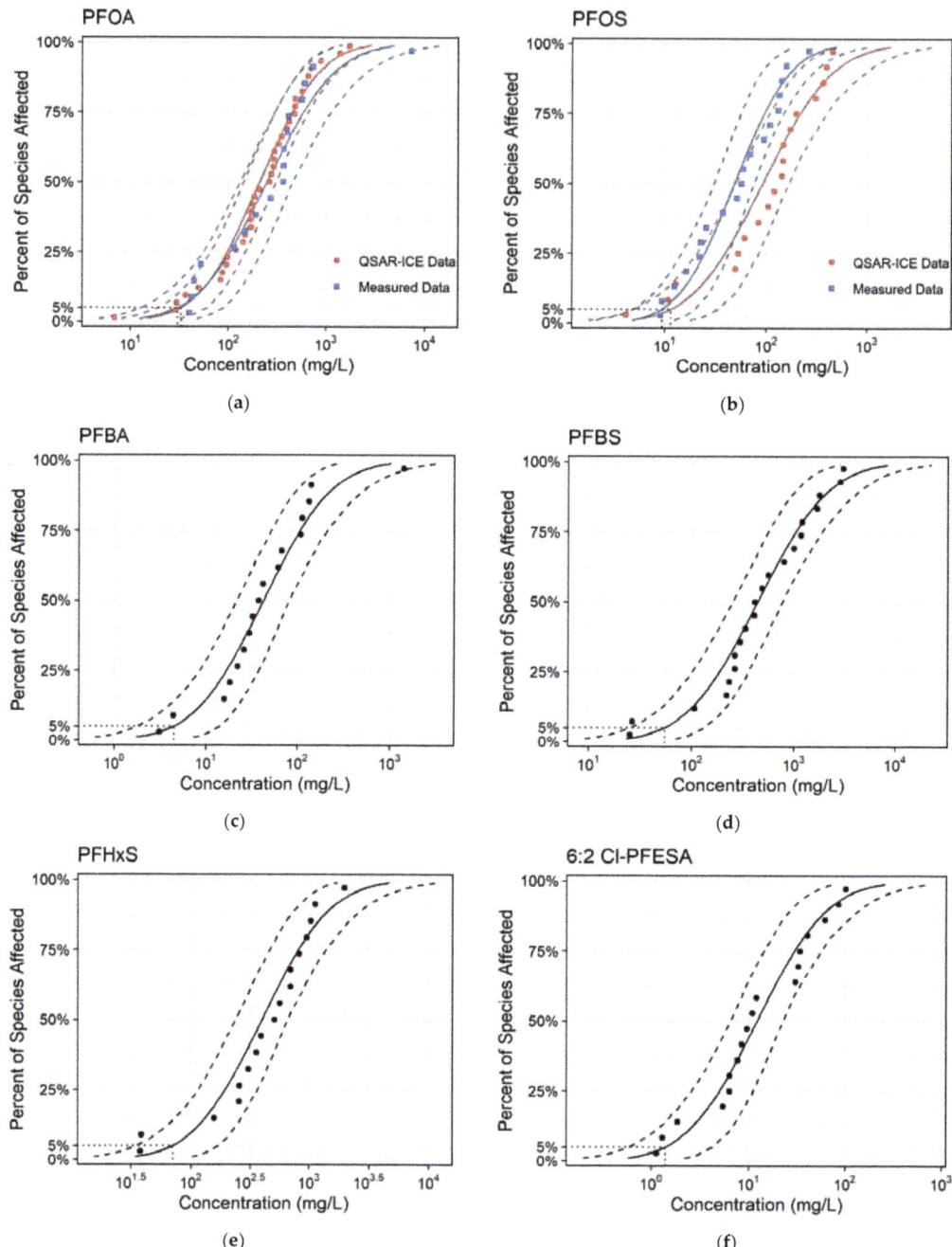

Figure 2. SSD curves based on the predicted data (six PFASs) and measured data (PFOA and PFOS). (**a**) PFOA (**b**) PFOS (**c**) PFBA (**d**) PFBS (**e**) PFHxS (**f**) 6:2 Cl-PFESA.

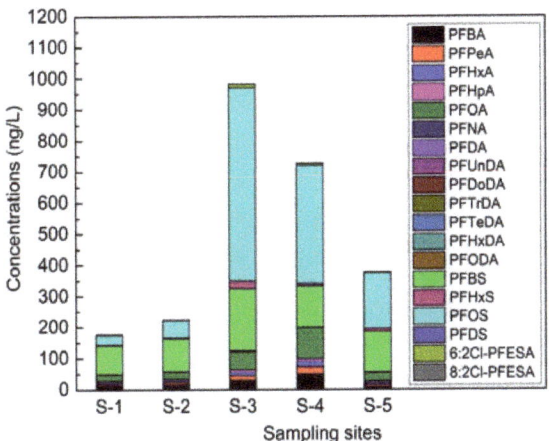

Figure 3. PFAS concentrations in a river near electroplating factories.

Table 5. RQ values ($\times 10^{-6}$) of six PFASs in this study and in the literature.

Sample Sites	RQ Values	PFBA	PFOA	PFBS	PFHxS	PFOS	6:2 Cl-PFESA
This Study	Range	11.5–60.9	3.25–15.8	9–20	0.23–1.83	15.3–297	5.1–49.8
	Mean	29.1	7.26	13.1	0.71	121	19.1
Gaoping [43]	Range	2.44–24.1	0.18–2.81	0–0.6	0.04–0.48	0–7.66	0–1.38
	Mean	11.1	0.57	0.21	0.23	1.07	0.12
Humen [43]	Range	2.44–42.3	0.53–3.24	0.13–1.24	0–27,000	0–4.41	0–3.82
	Mean	24.7	1.42	0.64	9290	1.29	0.39
Boluo [43]	Range	16.2–54.3	0.064–2.99	0–14.9	0–1.87	0–15.6	0–1.26
	Mean	24.3	0.76	4.17	0.31	2.03	0.16
Shatian [43]	Range	22–105	0.99–6.98	0.23–4.7	0–0.3	0–6.44	0
	Mean	38.9	3.4	1.7	0.2	1.57	0

3.5. Implications and Limitations

It has been shown that there are around 4700 PFASs on the global market [26]. The development of rapid in silico methods avoiding time-consuming and laborious animal experiments is necessary. QSAR–ICE–SSD models can be used to derive screening-level PNEC values in both prospective and retrospective assessments for novel PFASs where ecotoxicity data are lacking. The acute toxicity data of at least 15 species covering three trophic levels of an ecosystem can be derived [26]. These data can meet the requirements of the minimum datasets for the construction of SSD models, improving ecological relevance and reducing the uncertainty caused by the limited data quantity [3,6,44]. We selected as much as possible of the acute toxicity data of four model species from a wide range of PFAS groups (e.g., PFCAs, PFSAs, PFESAs) selected for QSAR modeling. The selected ICE models were also developed from data containing PFOS. This improved the adaptability and reliability of the model, reducing the uncertainty caused by the construction of the models [45]. As mentioned in Section 3.1, K_{ow} was found to be a key molecular descriptor in predicting aquatic acute toxicity in QSAR modeling. A possible future research direction could be to identify the role of K_{ow} in QSAR modeling to predict other endpoints (e.g., no observed adverse effect level (NOAEL), benchmark dose (BMD) of acute toxicity in rodents, or in vitro toxicity assays). This can help us understand the MOA of PFASs and integrate the data between an ERA and a human health risk assessment [46].

A major limitation of this study was that the acute toxicity data used to develop the QSAR model for each species was quite limited. Correspondingly, the small data sizes led to a just passable fitting effect of the QSAR models and limited the use of machine learning algorithms such as random forest [47,48]. One possible improvement of this issue is the selection of more acute toxicity data in QSAR modeling from not only PFASs but also other organic chemicals based on the same MOA [49]. However, this method is based on a sufficient understanding of the MOA of PFASs, in which further study is needed [26]. Another limitation of the QSAR–ICE–SSD approach was that only acute toxicity data were used and only acute PNEC values could be derived. Given the current limited availability of chronic (e.g., growth and reproductive effect) no observed effect concentration (NOEC), lowest observed effect concentration (LOEC), and 10% effect concentration (EC_{10}) data, it was not possible to follow our approach to derive chronic PNEC values. Using an acute-to-chronic ratio to extrapolate the chronic toxicity data for each species is a possible way; however, it can increase the uncertainty of the data quality [50].

4. Conclusions

In summary, the QSAR–ICE–SSD models predicted the following HC_5 values for six PFASs: 0.804 mg/L (PFBA), 6.27 mg/L (PFOA), 10.1 mg/L (PFBS), 12.9 mg/L (PFHxS), 2.09 mg/L (PFOS), and 0.254 mg/L (6:2 Cl-PFESA). The Σ_{19}PFAS concentrations were 177–983 ng/L in the nearby river of electroplating factories in Shenzhen. The results indicated that these electroplating factories may not be the source of the PFASs in the local aquatic environment. The RQ values of the six PFASs ranged from 2.29×10^{-7} to 2.97×10^{-4} in the nearby river.

Supplementary Materials: The following are available online: Supplementary Excel file; Table S1: PFAS properties and m/z values for quantification; Table S2: The collected acute toxicity of PFASs to four species for the QSAR models; Table S3: The collected acute toxicity of PFASs for the ICE models; Table S4: The ICE models used in this study; Table S5: The predicted acute toxicity of PFASs by QSAR–ICE models; Table S6: The results of the goodness-of-fit tests of the SSD models; Table S7: The measured acute toxicity of PFOA and PFOS; Figure S1: Liquid chromatography coupled to hybrid quadrupole-Orbitrap mass spectrometer LC-MS (Q-Exactive) (Thermo fisher scientific, USA) chromatograms for (A) standard PFASs and their (B) internal standard.

Author Contributions: Conceptualization, writing—original draft preparation, writing—review and editing, and methodology, J.Z. and M.Z.; investigation and visualization, H.T. and G.Q.; writing—review and editing and funding acquisition, W.G.; writing—review and editing and funding acquisition, H.G.; writing—review and editing, supervision, and funding acquisition, J.S. All authors have read and agreed to the published version of the manuscript.

Funding: This research was funded by the Ministry of Science and Technology of the People's Republic of China (2018YFC1801603 and 2018YFC1801605), the Science, Technology and Innovation Commission of Shenzhen Municipality (JCYJ20170817110953833), and the National Natural Science Foundation of China (No. 21976079 and No. 41977325).

Institutional Review Board Statement: Not applicable.

Informed Consent Statement: Not applicable.

Data Availability Statement: Not applicable.

Acknowledgments: This research was supported by the Center for Computational Science and Engineering at Southern University of Science and Technology.

Conflicts of Interest: The authors declare no conflict of interest.

Sample Availability: Not applicable.

References

1. Wang, Z.; Cousins, I.T.; Scheringer, M.; Buck, R.C.; Hungerbühler, K. Global emission inventories for C4–C14 perfluoroalkyl carboxylic acid (PFCA) homologues from 1951 to 2030, Part I: Production and emissions from quantifiable sources. *Environ. Int.* **2014**, *70*, 62–75. [CrossRef]
2. United Nations Environment Programme. Eighth Meeting of the Conference of the Parties to the Stockholm Convention. Available online: http://chm.pops.int/TheConvention/ConferenceoftheParties/Meetings/COP8/tabid/5309/Default.aspx (accessed on 1 September 2021).
3. United States Environmental Protection Agency. Guidelines for Ecological Risk Assessment. Available online: https://www.epa.gov/sites/production/files/2014-11/documents/eco_risk_assessment1998.pdf (accessed on 3 January 2020).
4. Salice, C.J.; Anderson, T.A.; Anderson, R.H.; Olson, A.D. Ecological risk assessment of perfluorooctane sulfonate to aquatic fauna from a bayou adjacent to former fire training areas at a US Air Force installation. *Environ. Toxicol. Chem.* **2018**, *37*, 2198–2209. [CrossRef] [PubMed]
5. Kwak, J.I.; Lee, T.-Y.; Seo, H.; Kim, D.; Kim, D.; Cui, R.; An, Y.-J. Ecological risk assessment for perfluorooctanoic acid in soil using a species sensitivity approach. *J. Hazard. Mater.* **2020**, *382*, 121150. [CrossRef]
6. European Chemicals Bureau. Technical Guidance Document on Risk Assessment. Available online: https://echa.europa.eu/documents/10162/16960216/tgdpart2_2ed_en.pdf (accessed on 3 January 2020).
7. Grist, E.P.M.; O'Hagan, A.; Crane, M.; Sorokin, N.; Sims, I.; Whitehouse, P. Bayesian and Time-Independent Species Sensitivity Distributions for Risk Assessment of Chemicals. *Environ. Sci. Technol.* **2006**, *40*, 395–401. [CrossRef] [PubMed]
8. Caldwell, D.J.; Hutchinson, T.H.; Heijerick, D.; Anderson, P.D.; Sumpter, J.P. Derivation of an Aquatic Predicted No-Effect Concentration for the Synthetic Hormone, 17α-Ethinyl Estradiol. *Environ. Sci. Technol.* **2008**, *42*, 7046–7054. [CrossRef] [PubMed]
9. Garner, K.L.; Suh, S.; Lenihan, H.S.; Keller, A.A. Species Sensitivity Distributions for Engineered Nanomaterials. *Environ. Sci. Technol.* **2015**, *49*, 5753–5759. [CrossRef] [PubMed]
10. Wheeler, J.R.; Grist, E.P.M.; Leung, K.M.Y.; Morritt, D.; Crane, M. Species sensitivity distributions: Data and model choice. *Mar. Pollut. Bull.* **2002**, *45*, 192–202. [CrossRef]
11. Maltby, L.; Blake, N.; Brock, T.C.M.; Van den Brink, P.J. Insecticide species sensitivity distributions: Importance of test species selection and relevance to aquatic ecosystems. *Environ. Toxicol. Chem.* **2005**, *24*, 379–388. [CrossRef]
12. Ankley, G.T.; Cureton, P.; Hoke, R.A.; Houde, M.; Kumar, A.; Kurias, J.; Lanno, R.; McCarthy, C.; Newsted, J.; Salice, C.J.; et al. Assessing the Ecological Risks of Per- and Polyfluoroalkyl Substances: Current State-of-the Science and a Proposed Path Forward. *Environ. Toxicol. Chem.* **2020**, *40*, 564–605. [CrossRef]
13. He, J.; Tang, Z.; Zhao, Y.; Fan, M.; Dyer, S.D.; Belanger, S.E.; Wu, F. The Combined QSAR-ICE Models: Practical Application in Ecological Risk Assessment and Water Quality Criteria. *Environ. Sci. Technol.* **2017**, *51*, 8877–8878. [CrossRef]
14. Douziech, M.; Ragas, A.M.J.; van Zelm, R.; Oldenkamp, R.; Jan Hendriks, A.; King, H.; Oktivaningrum, R.; Huijbregts, M.A.J. Reliable and representative in silico predictions of freshwater ecotoxicological hazardous concentrations. *Environ. Int.* **2020**, *134*, 105334. [CrossRef]
15. Zhang, S.; Wang, L.; Wang, Z.; Fan, D.; Shi, L.; Liu, J. Derivation of freshwater water quality criteria for dibutyltin dilaurate from measured data and data predicted using interspecies correlation estimate models. *Chemosphere* **2017**, *171*, 142–148. [CrossRef] [PubMed]
16. Raimondo, S.; Barron, M.G. Application of Interspecies Correlation Estimation (ICE) models and QSAR in estimating species sensitivity to pesticides. *SAR QSAR Environ. Res.* **2020**, *31*, 1–18. [CrossRef] [PubMed]
17. Escher, B.I.; Hermens, J.L.M. Modes of action in ecotoxicology: Their role in body burdens, species sensitivity, QSARs, and mixture effects. *Environ. Sci. Technol.* **2002**, *36*, 4201–4217. [CrossRef] [PubMed]
18. Dyer, S.D.; Versteeg, D.J.; Belanger, S.E.; Chaney, J.G.; Raimondo, S.; Barron, M.G. Comparison of species sensitivity distributions derived from interspecies correlation models to distributions used to derive water quality criteria. *Environ. Sci. Technol.* **2008**, *42*, 3076–3083. [CrossRef] [PubMed]
19. Wang, T.; Wang, P.; Meng, J.; Liu, S.; Lu, Y.; Khim, J.S.; Giesy, J.P. A review of sources, multimedia distribution and health risks of perfluoroalkyl acids (PFAAs) in China. *Chemosphere* **2015**, *129*, 87–99. [CrossRef]
20. Zhang, H.; Shi, J.; Higashiguchi, T.; Bo, T.; Niu, J. Quantitative determination and mass flow analysis of perfluorooctanesulfonate (PFOS) and perfluorooctanoate (PFOA) during reversed A^2O wastewater treatment process. *Acta Sci. Circumstantiae* **2014**, *34*, 872–880.
21. Higashiguchi, T.; Shi, J.; Zhang, H.; Liu, X. Distribution of Perfluorooctanesulfonate and Perfluorooctanoate in Water and the Sediment in Fenhe River, Shanxi Province. *Environ. Sci.* **2013**, *34*, 4211–4217.
22. Yamazaki, E.; Falandysz, J.; Taniyasu, S.; Hui, G.; Jurkiewicz, G.; Yamashita, N.; Yang, Y.-L.; Lam, P.K.S. Perfluorinated carboxylic and sulphonic acids in surface water media from the regions of Tibetan Plateau: Indirect evidence on photochemical degradation? *J. Environ. Sci. Health Part A* **2016**, *51*, 63–69. [CrossRef]
23. Jin, X.; Wang, Y.; Jin, W.; Rao, K.; Giesy, J.P.; Hollert, H.; Richardson, K.L.; Wang, Z. Ecological Risk of Nonylphenol in China Surface Waters Based on Reproductive Fitness. *Environ. Sci. Technol.* **2014**, *48*, 1256–1262. [CrossRef]
24. Wang, X.-N.; Liu, Z.-T.; Yan, Z.-G.; Zhang, C.; Wang, W.-L.; Zhou, J.-L.; Pei, S.-W. Development of aquatic life criteria for triclosan and comparison of the sensitivity between native and non-native species. *J. Hazard. Mater.* **2013**, *260*, 1017–1022. [CrossRef]

25. Feng, C.L.; Wu, F.C.; Dyer, S.D.; Chang, H.; Zhao, X.L. Derivation of freshwater quality criteria for zinc using interspecies correlation estimation models to protect aquatic life in China. *Chemosphere* **2013**, *90*, 1177–1183. [CrossRef]
26. Cousins, I.T.; DeWitt, J.C.; Glüge, J.; Goldenman, G.; Herzke, D.; Lohmann, R.; Miller, M.; Ng, C.A.; Scheringer, M.; Vierke, L.; et al. Strategies for grouping per- and polyfluoroalkyl substances (PFAS) to protect human and environmental health. *Environ. Sci. Process. Impacts* **2020**, *22*, 1444–1460. [CrossRef] [PubMed]
27. Klimisch, H.J.; Andreae, M.; Tillmann, U. A Systematic Approach for Evaluating the Quality of Experimental Toxicological and Ecotoxicological Data. *Regul. Toxicol. Pharm.* **1997**, *25*, 1–5. [CrossRef] [PubMed]
28. Wang, X.; Fan, B.; Fan, M.; Belanger, S.; Li, J.; Chen, J.; Gao, X.; Liu, Z. Development and use of interspecies correlation estimation models in China for potential application in water quality criteria. *Chemosphere* **2020**, *240*. [CrossRef] [PubMed]
29. Taniyasu, S.; Kannan, K.; So, M.K.; Gulkowska, A.; Sinclair, E.; Okazawa, T.; Yamashita, N. Analysis of fluorotelomer alcohols, fluorotelomer acids, and short- and long-chain perfluorinated acids in water and biota. *J. Chromatogr. A* **2005**, *1093*, 89–97. [CrossRef]
30. Bhandari, G.; Atreya, K.; Vašíčková, J.; Yang, X.; Geissen, V. Ecological risk assessment of pesticide residues in soils from vegetable production areas: A case study in S-Nepal. *Sci. Total Environ.* **2021**, *788*, 147921. [CrossRef] [PubMed]
31. Cao, H.; Wu, Y.; Ren, C.; Zhou, X.; Jiang, G. 3D-QSAR Studies on biarhibitors for the bromodomains of CBP/P300. *Chemistry* **2018**, *81*, 548–554.
32. Yan, W.; Lin, G.; Zhang, R.; Liang, Z.; Wu, L.; Wu, W. Studies on molecular mechanism between ACE and inhibitory peptides in different bioactivities by 3D-QSAR and MD simulations. *J. Mol. Liq.* **2020**, *304*, 112702. [CrossRef]
33. Tuppurainen, K. Frontier orbital energies, hydrophobicity and steric factors as physical qsar descriptors of molecular mutagenicity. A review with a case study: MX compounds. *Chemosphere* **1999**, *38*, 3015–3030. [CrossRef]
34. Ferrari, T.; Lombardo, A.; Benfenati, E. QSARpy: A new flexible algorithm to generate QSAR models based on dissimilarities. The log Kow case study. *Sci. Total Environ.* **2018**, *637–638*, 1158–1165. [CrossRef] [PubMed]
35. Moon, J.; Lee, B.; Ra, J.-S.; Kim, K.-T. Predicting PBT and CMR properties of substances of very high concern (SVHCs) using QSAR models, and application for K-REACH. *Toxicol. Rep.* **2020**, *7*, 995–1000. [CrossRef] [PubMed]
36. Wang, L.-L.; Ding, J.-J.; Pan, L.; Fu, L.; Tian, J.-H.; Cao, D.-S.; Jiang, H.; Ding, X.-Q. Quantitative structure-toxicity relationship model for acute toxicity of organophosphates via multiple administration routes in rats and mice. *J. Hazard. Mater.* **2021**, *401*, 123724. [CrossRef] [PubMed]
37. Conroy-Ben, O.; Garcia, I.; Teske, S.S. In silico binding of 4,4′-bisphenols predicts in vitro estrogenic and antiandrogenic activity. *Environ. Toxicol.* **2018**, *33*, 569–578. [CrossRef] [PubMed]
38. Kropf, C.; Begnaud, F.; Gimeno, S.; Berthaud, F.; Debonneville, C.; Segner, H. In Vitro Biotransformation Assays Using Liver S9 Fractions and Hepatocytes from Rainbow Trout (*Oncorhynchus mykiss*): Overcoming Challenges with Difficult to Test Fragrance Chemicals. *Environ. Toxicol. Chem.* **2020**, *39*, 2396–2408. [CrossRef]
39. Sheng, N.; Cui, R.; Wang, J.; Guo, Y.; Wang, J.; Dai, J. Cytotoxicity of novel fluorinated alternatives to long-chain perfluoroalkyl substances to human liver cell line and their binding capacity to human liver fatty acid binding protein. *Arch. Toxicol.* **2018**, *92*, 359–369. [CrossRef]
40. Lu, Z.; Song, L.; Zhao, Z.; Ma, Y.; Wang, J.; Yang, H.; Ma, H.; Cai, M.; Codling, G.; Ebinghaus, R.; et al. Occurrence and trends in concentrations of perfluoroalkyl substances (PFASs) in surface waters of eastern China. *Chemosphere* **2015**, *119*, 820–827. [CrossRef]
41. Wang, P.; Lu, Y.; Wang, T.; Fu, Y.; Zhu, Z.; Liu, S.; Xie, S.; Xiao, Y.; Giesy, J.P. Occurrence and transport of 17 perfluoroalkyl acids in 12 coastal rivers in south Bohai coastal region of China with concentrated fluoropolymer facilities. *Environ. Pollut.* **2014**, *190*, 115–122. [CrossRef] [PubMed]
42. Wang, Z.; Cousins, I.T.; Scheringer, M.; Hungerbühler, K. Fluorinated alternatives to long-chain perfluoroalkyl carboxylic acids (PFCAs), perfluoroalkane sulfonic acids (PFSAs) and their potential precursors. *Environ. Int.* **2013**, *60*, 242–248. [CrossRef]
43. Li, C. Contamination Characteristics of F-53B, OBS and other Poly- and per Fluoroalky Substances in Typical Areas. Master's Dissertation, Qingdao Technological University, Qingdao, China, 2016.
44. Hoondert, R.P.J.; Oldenkamp, R.; de Zwart, D.; van de Meent, D.; Posthuma, L. QSAR-Based Estimation of Species Sensitivity Distribution Parameters: An Exploratory Investigation. *Environ. Toxicol. Chem.* **2019**, *38*, 2764–2770. [CrossRef]
45. Kostal, J.; Plugge, H.; Raderman, W. Quantifying Uncertainty in Ecotoxicological Risk Assessment: MUST, a Modular Uncertainty Scoring Tool. *Environ. Sci. Technol.* **2020**, *54*, 12262–12270. [CrossRef] [PubMed]
46. de Knecht, J.; Rorije, E.; Maslankiewicz, L.; Dang, Z. Feasibility of using interspecies relationships for integration of human and environmental hazard assessment. *Hum. Ecol. Risk Assess. An. Int. J.* **2021**, *27*, 1715–1731. [CrossRef]
47. Wang, Z.; Chen, J.; Hong, H. Developing QSAR Models with Defined Applicability Domains on PPARγ Binding Affinity Using Large Data Sets and Machine Learning Algorithms. *Environ. Sci. Technol.* **2021**, *55*, 6857–6866. [CrossRef] [PubMed]
48. Perlich, C.; Provost, F.; Simonoff, J.S. Tree induction vs. logistic regression: A learning-curve analysis. *J. Mach. Learn. Res.* **2003**, *4*, 211–255. [CrossRef]
49. Nendza, M.; Müller, M.; Wenzel, A. Discriminating toxicant classes by mode of action: 4. Baseline and excess toxicity. *SAR QSAR Environ. Res.* **2014**, *25*, 393–405. [CrossRef]
50. Kienzler, A.; Halder, M.; Worth, A. Waiving chronic fish tests: Possible use of acute-to-chronic relationships and interspecies correlations. *Toxicol. Environ. Chem.* **2017**, *99*, 1129–1151. [CrossRef]

Article

A 110 Year Sediment Record of Polycyclic Aromatic Hydrocarbons Related to Economic Development and Energy Consumption in Dongping Lake, North China

Wei Guo [1,*], Junhui Yue [1], Qian Zhao [1], Jun Li [1], Xiangyi Yu [2] and Yan Mao [2,*]

[1] Key Laboratory of Beijing for Water Quality Science and Water Environment Recovery Engineering, Faculty of Architecture, Civil and Transportation Engineering, Beijing University of Technology, Beijing 100124, China; yuejunhui3214@163.com (J.Y.); ZQ1176170762@163.com (Q.Z.); jglijun@bjut.edu.cn (J.L.)
[2] Solid Waste and Chemicals Management Center of MEE, Beijing, 100029, China; yuxiangyi@meescc.cn
* Correspondence: gwfybj@bjut.edu.cn (W.G.); maoyan@meescc.cn (Y.M.)

Abstract: A sedimentary record of the 16 polycyclic aromatic hydrocarbon (PAH) pollutants from Dongping Lake, north China, is presented in this study. The influence of regional energy structure changes for 2–6-ring PAHs was investigated, in order to assess their sources and the impact of socioeconomic developments on the observed changes in concentration over time. The concentration of the ΣPAH_{16} ranged from 77.6 to 628.0 ng/g. Prior to the 1970s, the relatively low concentration of ΣPAH_{16} and the average presence of 44.4% 2,3-ring PAHs indicated that pyrogenic combustion from grass, wood, and coal was the main source of PAHs. The rapid increase in the concentration of 2,3-ring PAHs between the 1970s and 2006 was attributed to the growth of the urban population and the coal consumption, following the implementation of the Reform and Open Policy in 1978. The source apportionment, which was assessed using a positive matrix factorization model, revealed that coal combustion was the most important regional source of PAHs pollution (>51.0%). The PAHs were mainly transported to the site from the surrounding regions by atmospheric deposition rather than direct discharge.

Keywords: PAHs; historical trends; shallow lake; economic parameters; sources; sediment core

1. Introduction

Polycyclic aromatic hydrocarbons (PAHs) are a class of persistent organic pollutants (POPs) which are ubiquitously present in the environment [1]. Because of their potential as carcinogens, mutagens, and teratogens [2], PAHs have elicited serious concern worldwide [3,4]. Sixteen PAH compounds have been included in the priority pollutant list for risk control and management of the US Environmental Protection Agency (US EPA) [5]. PAHs originate mainly from the incomplete combustion and/or pyro-synthesis of organic materials through fossil-fuel usage, biomass burning, industrial processes, waste incineration, and vehicle exhaust emissions [6–8]. PAHs can be transported into aquatic ecosystems through various processes, including atmospheric deposition [9], wastewater discharge from urban sewage treatment plants, as well as industrial sites [10], or surface runoff from urban or industrial areas [11]. The PAHs discharged into the aquatic environment are prone to combining with fine particles, being ultimately deposited into sediments due to their hydrophobic nature and resulting low solubility [12,13]. This makes sediments an important sink for PAHs, which directly affects the dwelling organisms and aquatic environment safety [14–16].

PAHs in dated sediments provided an ideal archive of historical information about the anthropogenic contamination of aquatic ecosystems and emission of pollutants from energy consumption [17–21]. Thus, a dated sediment core analysis can reconstruct very well the chronology of PAHs pollution and clearly identify their fate in any water-related ecosystem [22–24]. China is currently the world's second largest economy and has undergone a rapid growth in population and industrial and agricultural outputs, as well as energy consumption and transportation infrastructures, during recent decades [25]. Earlier studies have reported that PAH contamination is often associated with the economic development and energy consumption in China [26,27]. In 2004, the emissions of PAHs in China (up to 114 Gg) accounted for an estimated 22% of total global PAH emissions [28]. Coal burning was responsible for 60% of PAH emissions in China [29]. The sedimentary records of organic contamination in lakes and marine areas have shown an increase in PAH quantities following the ~1850s, providing information with regard to regional fuel consumption and intensity of anthropogenic activities [14,30,31]. However, studies have mainly focused on developed regions, and limited information is available regarding the temporal trends of PAHs concentration within inland shallow lakes in the north of China. This leads to a significant lack of data about the impact of lifestyle and energy consumption on the concentration of PAHs.

Thus, it could be useful to evaluate the implications of the regional economic development and energy consumption on the temporal variation of PAH pollution to understand the factors related to the historical changes in PAH emissions [32,33]. The present study aims to (1) investigate the residual levels and temporal distribution of PAHs in sediment core from a typical inland shallow lake, Dongping Lake, located in the north of China, (2) speculate on the possible PAH sources in the undisturbed sediment profiles in combination with a detailed chronology study, and (3) reconstruct the historical trends of PAH contamination related to the economic development in the region (population size, gross domestic product, and energy consumption).

2. Materials and Methods

2.1. Study Area and Sampling

This study was carried out at Dongping Lake, a shallow freshwater lake with 627 km^2 surface area and 4×10^9 m^3 storage volume, which is located in Tai'an city in the western part of Shandong province, China (35°30′–36°20′ N, 116°00′–116°30′ E) (Figure 1). Dongping Lake is the second largest freshwater source in Shandong province. Most of the lake area is no more than 3 m in depth. Dongping Lake is mainly affected by a warm and semi-humid continental monsoonal climate. The multi-annual mean temperature is 13.3 °C with an average annual precipitation of 640 mm [34]. Dongping Lake serves as an important flood control project in the lower reaches of the Yellow River and represents the last water reservoir along the Eastern Route of China's South-to-North Water Diversion Project [35]. The water in the lake flows north through the Xiaoqing River, eventually entering the Yellow River. Dawen River is the major inflow to Dongping Lake, whose recharge sources are mainly supplied by surface runoff and rainfall [36]. There has been increasing concern about water pollution by different types of contaminants in recent years, mainly due to the large inputs of industrial, agricultural, and urban sewage activities from the Dawen River. In recent years, the volume of sewage for the Dawen River reached 2.6×10^9 tons, which has led to a severe degradation of the water quality of Dongping Lake [37].

The study was carried out in the autumn of 2016 in the Dongping Lake (Figure 1). Two parallel sediment cores with lengths of 60 cm were collected using a Beeker 04.23 core sampler (100 cm length × 57 mm ID; Eijkelkamp Co., Giesbeek, The Netherlands) at the still water area (DP: 35°57′23.1″ N, 116°11′35.6″ E) of Dongping Lake, where the water depth is approximately 3.0 m. The sediment core was sliced into 2 cm thick sections. Following this step, each section was placed in a precleaned aluminum foil, before being frozen, freeze-dried, ground into a fine homogenized powder, and finally stored at −20 °C until further treatment.

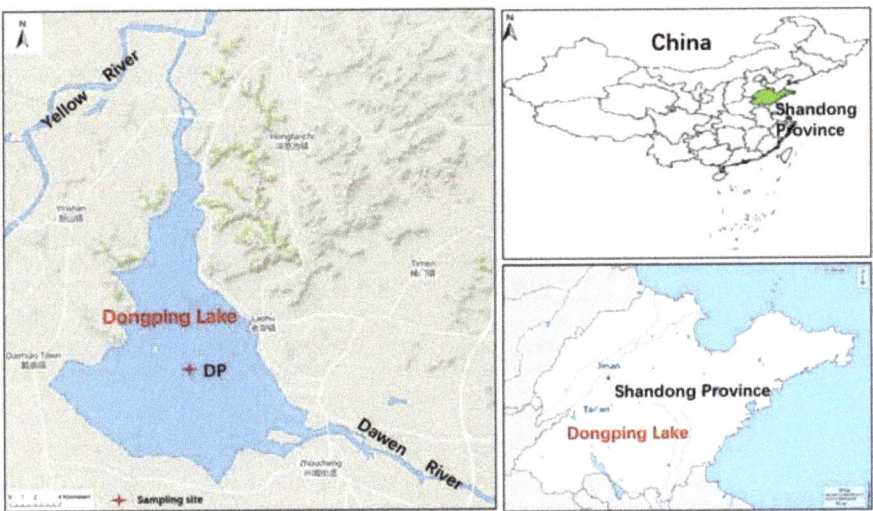

Figure 1. Map of the study area and location of the sampling site.

2.2. Total Organic Carbon Analysis and Sediment Core Dating

Total organic carbon (TOC) of sediment core samples was measured using a Perkin-Elmer PE 2400 Series II elemental analyzer. With the aim of dating the sediment core samples, the radioisotope activity concentrations (e.g., ^{137}Cs, ^{210}Pb, and ^{226}Ra) in samples were determined using a well-type HPGe gamma detector (GCW3523, Canberra Inc., USA) at the Institute of Geology and Geophysics, Chinese Academy of Sciences. Before analyzing the radionuclides, each subsample was stored in sealed centrifuge tubes for 3 weeks, allowing radioactive equilibration [25]. The activities of ^{137}Cs and ^{210}Pb were respectively determined by the γ emissions at 662 keV and 46.5 keV, while the activity of ^{226}Ra was determined by the γ emissions at 352 keV. Detection errors were within 5% for both ^{137}Cs and ^{210}Pb. ^{210}Pbex was obtained by subtracting the ^{226}Ra activity from the total ^{210}Pb activity [38]. The ^{210}Pb geochronology was calculated from a constant rate of supply (CRS) model [39] according to the following equation:

$$t = \lambda^{-1} ln(A_0/A_z), \tag{1}$$

where A_0 and A_z are the ^{210}Pb$_{ex}$ accumulation fluxes at the surface layer of the sediment core and depth z, respectively, while λ is the ^{210}Pb$_{ex}$ radioactive decay constant (0.03114 year^{-1}). The profiles of ^{137}Cs activity in samples were compared to scattering nuclides from nuclear testing, thermonuclear weapons testing in the middle 1960s, and nuclear accidents such as the Chernobyl nuclear site in 1986 [40]. Thus, the ^{137}Cs activity was used as an independent chrono-marker to enhance the dating accuracy from ^{210}Pb [24].

2.3. PAH Extraction and Analysis

The extraction of PAHs was performed according to a previously reported method [14]. Briefly, about 2.0 g (dw) of each sample was extracted with 250 mL of a dichloromethane–hexane mixture (1:1, v/v) for 24 h using a Soxhlet apparatus. A known mixture of surrogates (naphthalene-d8, acenaphthene-d10, phenanthrene-d10, and chrysene-d12) was added to each blank and sample before extraction. The extract passed through a glass column packed with 1:2 alumina–silica gel (v/v) containing 1 g of anhydrous sodium sulfate overlaying the silica gel. The eluents containing PAHs were collected by eluting 70 mL of hexane-dichloromethane (7:3, v/v) and were then concentrated to 1.0 mL. After adding a known quantity of an internal standard (hexamethylbenzene), the PAHs were analyzed by gas chromatography and mass spectrometry (GC/MS).

In the study, 16 PAHs, namely, naphthalene (Naph), acenaphthene (Aceph), acenaphthylene (Ace), fluorene (Fl), phenanthrene (Phen), anthracene (Ant), fluoranthene (Flu), pyrene (Pyr), benz[a]anthracene (BaA), chrysene (Chr), benzo[b]fluoranthene (BbF), benzo[k]fluoranthene (BkF), benzo[a]pyrene (BaP), dibenz[ah]anthracene (DBA), benzo[ghi]perylene (BgP), and indeno[1,2,3-cd]pyrene (InP), were detected. A Varian 4000 mass spectrometer (Varian Inc., Palo Alto, CA) coupled with a Varian CP-3800 gas chromatograph equipped with a Varian VF-5MS column (30 m × 0.25 mm × 0.25 μm) was employed to quantitatively determine PAHs. The column ramp temperature was programmed to rise from 80 °C (dwell time of 3 min) to 230 °C (dwell time of 2 min) with a rate of 15 °C/min, followed by a ramp to 290 °C (dwell time of 8 min) with a rate of 5 °C/min. The injection volume was 1 μL in splitless mode. All data were subject to strict quality control procedures. The spiked recoveries of 16 PAHs in samples were in the range of 75.3–107.4%. The method detection limits (MDLs) for each PAH ranged from 0.12 to 1.07 ng/g.

2.4. Positive Matrix Factorization (PMF) Model for Source Apportionment

The PMF model was used for source apportionment of sedimentary PAHs [33]. The model is based on an advanced multivariate factor analysis method that relies on weighted least squares calculation and was developed in 1994 [41]. The United States Environmental Protection Agency PMF user guide (version 5.0) explains the model in detail [42]. In theory, the PMF model can be described by Equation (2).

$$X_{ij} = \sum_{j=1}^{p} g_{ik} f_{kj} + e_{ij}, \qquad (2)$$

where X_{ij} is the concentration of the *i*-th species that was determined by the *j*-th sample, whereas, g_{ik} is the *i*-th species concentration, which was detected in source *k*; f_{kj} represents the contribution of the *k*-th source to the *j*-th sample, and e_{ij} is the error for species *j* to sample *i* [43]. The objective function $Q(E)$ of the PMF model is defined by Equation (3).

$$Q(E) = \sum_{i=1}^{n} \sum_{j=1}^{m} [(X_{ij} - \sum_{k=1}^{p} g_{ik} k_{kj})/s_{ij}]^2, \qquad (3)$$

where $Q(E)$ is the weighted sum of the squares for the difference in value between the original dataset and the PMF output [44], whereas s_{ij} is the uncertainty in the *j*-th PAH to sample *i* [45].

2.5. Data Analysis

Origin Pro 8.0 was used to plot the experimental data. Statistical analyses were performed using SPSS 13.0 (SPSS Inc., Microsoft Co., USA). The correlation coefficients between the measured parameters were calculated through a two-tailed test and the Pearson correlation coefficient.

3. Results and Discussion

3.1. Sediment Chronology

Figure 2 shows the vertical distribution of excess ^{210}Pb (^{210}Pb$_{ex}$) and ^{137}Cs for the sediment core collected from Dongping Lake. The ^{137}Cs activity was low (<17 Bq/kg) throughout the sediment core; however, a typical peak of ^{137}Cs activity (16.8 Bq/kg) could be identified at 22 cm depth. This corresponded well with the ^{137}Cs atmospheric fallout peak from the nuclear bomb testing in 1963 [46]. The ^{210}Pb$_{ex}$ activity showed a continuous increase from 6.7 Bq/kg (at the core bottom) to 55.5 Bq/kg (at the core top). The ^{210}Pb$_{ex}$ activity profile showed a definite exponential decay together with increasing depth ($R^2 = 0.934$); thus, a CRS dating model was applied to date the sediment core to determine the chronologies of ^{210}Pb [39]. The ^{210}Pb$_{ex}$ CRS model provided an age of 1963 at the depth of 22 cm, which was verified by the age provided by the independent ^{137}Cs dating peak. Mover, the enrichment factor of Pb significantly increased after 1960 due to gasoline-related Pb emission [47], which was consistent with the historical trend of ^{210}Pb$_{ex}$ activity in sediment core. Thus, the chronology based on the ^{210}Pb$_{ex}$ CRS model was considered reliable. According to this age–depth model, the estimated average sedimentation rate was approximately 0.56 cm/year at the site, meaning that the 60 cm core covered ~110 years of sedimentary history between 1907 and 2016.

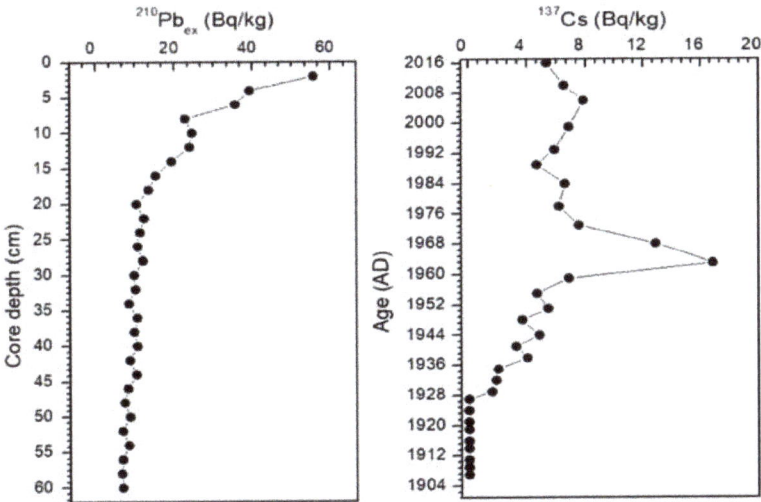

Figure 2. Age–depth profiles of excess ^{210}Pb (^{210}Pb$_{ex}$) and ^{137}Cs in the sediment core.

It is noted that ^{210}Pb$_{ex}$ activity was higher than found in earlier studies from the adjacent areas, for instance, those from Baiyangdian Lake in Hebei Province [48] and Lake Gonghai in Shanxi Province [23]. This is because the ^{210}Pb$_{ex}$ activity in sediment cores is positively correlated with the atmospheric deposition flux [23]. The average sediment accumulation rate in the sediment core from Lake Gonghai (0.17 g/cm^2/year) [23] was lower than that performed at the Lake Chaohu (0.23 g/cm^2/year) [49] and the Dongping Lake (0.22 g/cm^2/year) [50]. In addition, atmospheric deposition of ^{210}Pb is also influenced by local precipitation. The climate between the two sites is also quite different, with an annual average precipitation of 456 mm at Lake Gonghai and 680 mm at Dongping Lake, which could lead to large variance of the ^{210}Pb$_{ex}$ activity in the sediment cores.

3.2. Temporal Variation of PAH Concentration and Composition

The concentration of PAHs at different dates of the lake's sediment core is illustrated in Table S1 and Figure 3a. All 16 priority PAHs were detected in the sediment core back to the 1970s, yet only nine individual PAHs were detected before the 1970s in the totality of the core samples. Increased industrial and agricultural activity after the 1970s may have increased the input of PAHs such as Ace, Fl, and DBA [14]. An overall increasing trend of total PAH concentration (ΣPAHs) was observed from 1907 to 2016, and the ΣPAH concentration ranged from 77.6 to 628.0 ng/g. Three temporal trends for the concentration of ΣPAHs in the sediment core were characterized. In the first stage from 1907 to the 1950s, the concentration of ΣPAHs spread over a relatively narrow range (77.6–122.1 ng/g) showing a similar constant trend before the mid-20th century. In the second stage from the 1950s to 2006, the concentrations of ΣPAHs increased sharply, reaching around 628.0 ng/g in ca. 2006. This increase can be attributed to rapid industrialization and urbanization following the establishment of the People's Republic of China in 1949, as well as the implementation of the Reform and Open Policy in 1978 [26]. Moreover, the construction of the Dongping Industrial Park in 2002–2005 and the consequently increasing development activities around the lake not only further deteriorated the water quality of Dongping Lake [51], but also increased the input of PAHs. In the third stage from 2006 to 2016, the concentrations of ΣPAHs (average value of 337.6 ng/g) decreased compared to the period of 2000–2006. The implementation of pollution control measures, which were carried out to guarantee the water quality and safety of the South-to-North Water Diversion in the catchment since the 2005, may have already reduced the emissions of PAHs and deposition in the lake sediment [36]. Compared with other lakes subject to more frequent industrial and human activities in China, Dongping Lake is a lake mainly developed for agriculture and tourism, and the peak (455.4 ng/g) PAHs in this location were lower than observed in Dianchi Lake (4560.8 ng/g) [52], Chaohu Lake (2500 ng/g) [32], and Taihu Lake (1600 ng/g) [30]. In addition, the peak period (in the 1990s–2000s) of PAHs in Dongping Lake appeared later than that in the aforementioned lakes (in the 1980s–1990s). The different temporal trends of the ΣPAH concentration in the sediment core of lakes in China indicated different histories of industrialization and development intensity in different catchments.

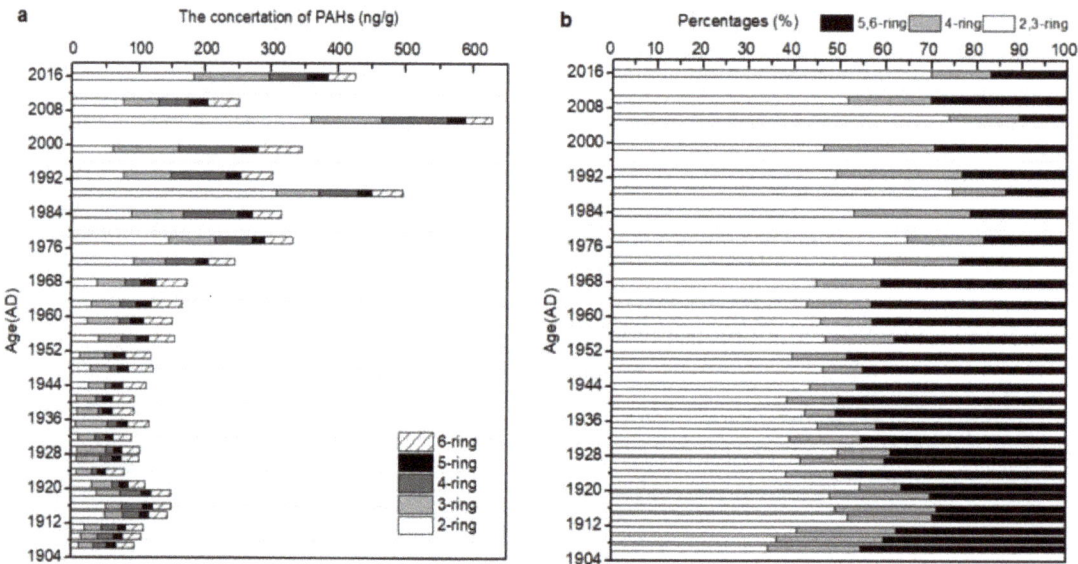

Figure 3. The historical variations of the concentrations (**a**) and percentages (**b**) of polycyclic aromatic hydrocarbons (PAHs) with different rings (2–6 rings) in the sediment core of Dongping Lake. Two-ring PAHs include Naph; three-ring PAHs include Aceph, Ace, Fl, Phen, and Ant; four-ring PAHs include Flu, Pyr, BaA, and Chr; five-ring PAHs include BbF, BkF, BaP, and DBA; six-ring PAHs include InP and BgP.

Different types of PAHs were grouped according to their number of aromatic rings, and the concentration and percentage of PAHs ranging from 2–6 rings were calculated (Figure 3). The concentration of two- and four-ring PAHs slowly increased from the 1900s to the 1920s, and then decreased and maintained a steady value between the 1920s and the 1950s, finally showing a significantly increasing trend after the 1950s. Concentrations of three-, five-, and six-ring PAHs increased from the 1900s to recent years, with the three-ring PAH concentration increasing even more. The 2,3-ring PAHs exhibited the highest ratios to total PAHs compared to other multiring PAHs, considering the period from the 1970s to the recent years, with an average of 60.1% over the last four decades. The two total PAHs peaks in 1989 and 2006 had the highest 2,3-ring PAH contribution to the total PAH amount

(>74%), suggesting that the more frequent and high-intensity occurrence of petrogenic discharge and low–moderate temperature combustions (e.g., incomplete grass and wood burning) allegedly occurred in these periods [32,51]. Compared with the early and late 2000s, the water level of Dongping Lake in 2006 was at a low level, which reduced the water environment capacity and further deteriorated the water quality [53]. In addition, according to the statistical data of Tai'an city, the amount of biomass combustion in 2006 was the highest during the 2000s, which was twice that in 2010 [54], which resulted in more 2,3-ring PAHs entering the lake through atmospheric emissions [32]. These factors resulted in the highest concentration of PAHs and a high proportion of 2,3-ring PAHs in 2006 (Figure 3). This study further confirmed that the deposited PAHs found in the sediment core of the lake are strictly derived from the transportation and precipitation of atmospheric aerosols [55] and direct emissions from petrogenic sources [23], as observed in some lakes of rural areas in Thailand [24,56], as well as in Baiyangdian Lake [14] and Yangzonghai Lake [25] from China.

Furthermore, the relationship between PAH concentration and TOC content in sediment cores was investigated using Pearson's coefficient rank correlation. A significant positive correlation was found between PAH and TOC concentrations at the 0.05 level of significance ($r > 0.7$, $p < 0.01$, $n = 30$), with correlation coefficients of 0.776, 0.905, and 0.859 for 2,3-ring PAHs, four-ring PAHs, and 5,6-ring PAHs, respectively. This indicates that PAH pollution in sediment core was primarily controlled by TOC. Several previous studies have also demonstrated that the TOC in the sediment is an important indicator for determining the fate, sorption dynamics, and sequestration mechanisms of PAHs [19,57,58]. The correlation between TOC and four-ring or 5,6-ring PAH concentration was better compared to that between TOC and 2,3-ring PAH concentration. This may be attributed to the strong hydrophobicity and degradation resistance of four-ring or 5,6-ring PAHs [23,59].

3.3. Sources Analysis

Molecular diagnostic ratios of specific PAHs were applied to identify a more specific origin of PAHs found in the environment [32]. Many former studies used this methodology to confirm the possible emission sources of PAHs [14,60,61]. PAH molecular ratios for Ant/(Ant + Phen) and Fl/(Fl + Pyr) were applied to calculate the possible sources of PAHs. The results for this study are shown in Figure 4. Ant/(Ant + Phen) ratios lower than 0.1 suggest that the PAHs principally originated from petrogenic source. Ratios higher than 0.1 are considered representative of pyrogenic sources including biomass, coal, and petroleum combustion [25,60]. Fl/(Fl + Pyr) ratios less than 0.4 are often considered to be typical of petroleum contamination. Ratios greater than 0.5 imply that PAH compounds were primarily generated from pyrogenic sources, especially grass, wood, and coal burning, while values of Fl/(Fl + Pyr) that fall between 0.4 and 0.5 point to liquid fossil-fuel combustion [58,62]. For the Ant/(Ant + Phen) ratios, all values were >0.1 in the sediment core, which indicates a combustion source. The Fl/(Fl + Pyr) ratios showed the two types of PAHs sources. The values exceeded the threshold of 0.5 from the 1907s to the 1920s and from the 1950s to the recent years, which suggests a strong pyrogenic signal from grass, wood, and coal combustion. For the period ranging from 1907 to the 1920s, the pyrogenic source was mainly from the combustion of grass and wood related to natural or anthropogenic wildfire. For the period from the 1950s to recent years, the pyrogenic source was mainly related to the combustion of biomass and coal due to the increasing industrial activities since the 1950s. Further studies have shown that the contribution of coal and biomass combustion is about 44% and 24% to the PAHs found in the Shandong province, respectively [27]. Thus, the residential indoor wood and crop burning and coal combustion may be the two major emission sources of PAHs in the region. The Fl/(Fl + Pyr) ratios ranged from 0.38 to 0.49, corresponding to the period of time of the 1920s to the 1950s, reflecting PAHs from liquid fossil-fuel combustion. This might show the effect of the Chinese Liberation War (1946–1949), World War II (1937–1945), and the Chinese Civil Revolutionary War (1927–1937) on the temporal distribution of PAHs [14].

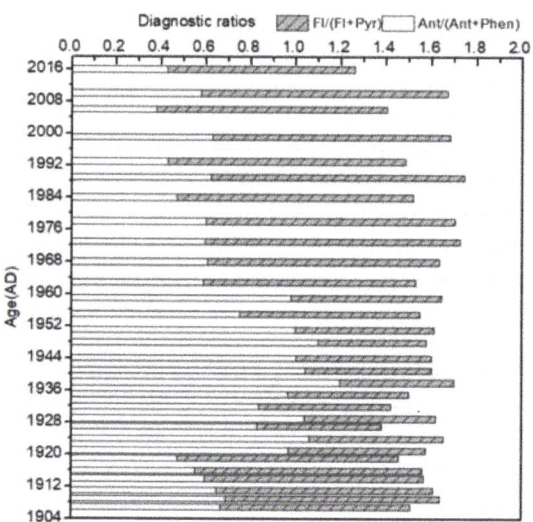

Figure 4. Diagnostic ratios of polycyclic aromatic hydrocarbon (PAH) ratios in the sediment core of Dongping Lake.

3.4. The Impact of Economic Parameters on the Change in PAH Concentrations and Sources

Historical changes in the concentrations of PAHs in this study followed the general temporal trends reported from the socioeconomic development data in Tai'an city in Shandong province. Due to the limitation of historical statistics, the only used data were gross domestic product (GDP), total population, rural population, road freight capacity, and coal consumption data from 1949 to 2016; similarly, petroleum consumption and natural gas consumption data from 1975 to 2016 were used (Figure 5) [54]. According to Figure 3a, the concentration of ΣPAHs increased sharply from the 1950s to 2006, and then decreased from 2006 to recent years. Since the founding of the People's Republic of China in 1949, an increase in population from 2.7×10^6 to 5.7×10^6 people has been observed, along with increases in GDP from 1.0×10^2 to 3.3×10^5 million yuan, coal consumption from 6.4×10^4 to 2.9×10^7 tons standard coal, and road freight capacity from 6.6×10^5 to 1.2×10^8 Tons in Tai'an city; this growth has contributed to an increase and accumulation of PAHs in the catchments, with a good correlation at the 0.05 level of significance ($r > 0.75$, $p < 0.01$, $n = 15$). The intensification of the industrial activities around lake areas such as the Dongping Industrial Park development since 2002 [51], combined with the impact of pollutant transmission phenomena in the area, allowed the concentration of PAHs to reach a peak in 2006. Later, as the Chinese government carried out environmental protection and implemented energy conservation and emissions reduction [14], together with the development of the South-to-North Water Diversion Project in the region [36], the emissions of PAHs were effectively reduced, leading to a decline in the accumulation of PAHs in the lake. For example, the road freight capacity gradually decreased from 1.2×10^8 to 6.0×10^7 tons, with the coal consumption decreasing from 2.9×10^7 to 1.4×10^7 tons of standard coal and the petroleum consumption decreasing from 6.2×10^5 to 3.2×10^5 tons of standard coal (Figure 5).

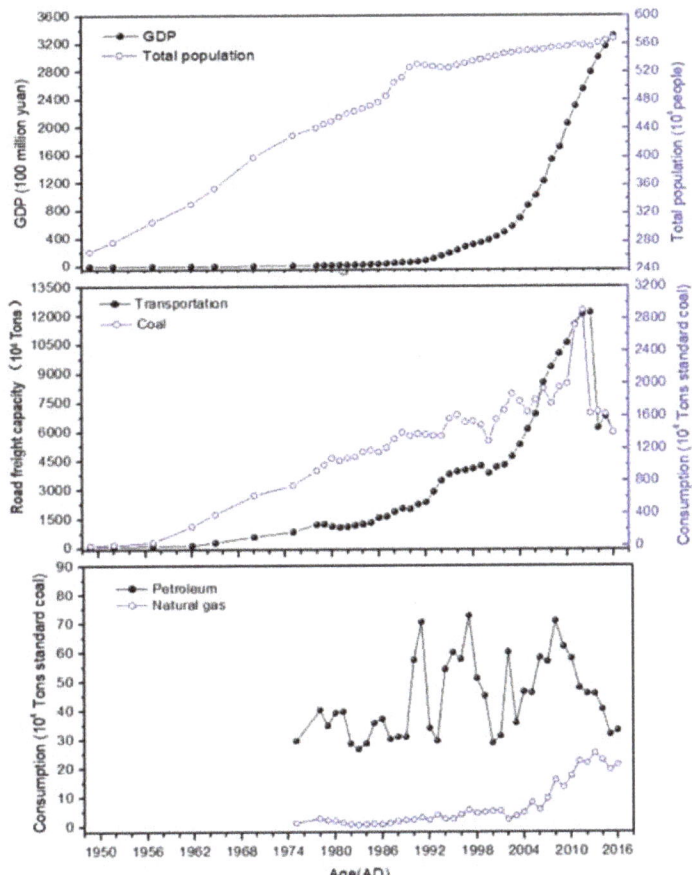

Figure 5. Statistical data on the socioeconomic development and energy consumption in Tai'an city.

The relationship between the vertical distribution of the concentration of different-ring PAHs and energy consumption (Figure 5) in the period of 1975–2016 was further analyzed. With regard to the consumption of coal from 1975 to 2016, the value gradually increased, reaching a peak in 2012 and then gradually faded. On the other hand, petroleum consumption was relatively stable at a value (3.3×10^5 tons standard coal) before the 1990s, and then fluctuated between this latter value and $3.0–7.1 \times 10^5$ tons standard coal. The consumption of natural gas augmented slowly from 0.7 to 8.4×10^4 tons standard coal between 1975 and 2006, and then increased rapidly to 2.5×10^5 tons standard coal. The results shown in Figures 3 and 5 indicate that the increase in concentration for the 2,3-ring and four-ring PAHs was consistent with the surge of coal consumption, thus suggesting that the household energy usage structure was a major factor impacting the concentration of PAHs [33]. Although coal consumption has declined since 2010, it is still the main structural energy source in the region, accounting for 96.7% of the overall energy structure [54]. The correlation coefficients between the concentration of 2,3-ring and four-ring PAHs and coal consumption reached values of 0.75 and 0.91, respectively, between 1975 and 2016. Coal consumption emits higher levels of 2,3 ring and four-ring PAHs compared to the burning of petroleum products or natural gas [33,63]. A comparison of the concentrations of 5,6-ring PAHs with petroleum consumption may reflect a relationship between PAH pollution and vehicles exhaust emissions [33]. The correlation coefficient calculated between the concentration of 5,6-ring PAHs and the petroleum consumption was 0.85 between 1975 and 2016. The positive relationship between 5,6-ring PAHs and the traffic volume was observed to be 0.89 between 1975 and 2016, confirming once again this relationship. Moreover, the concentrations of different-ring PAHs had a good correlation with the natural gas consumption at the 0.05 level of significance ($r > 0.61$, $p < 0.01$, $n = 15$), suggesting that natural gas is gradually becoming a new source of contribution to PAH pollution.

The main source components were classified according to the data of Naph, Phen, Ant, Fl, Pyr, BbF, BaP, InP, and BgP having the largest contribution for the period from the 1970s to 2016; the PMF source file is shown in Figure 6. There are four factors extracted by PMF model; the percentage contributions to PAH sources were 51.4% by Factor 1, 11.5% by Factor 2, 4.3% by Factor 3, and 32.8% by Factor 4. Factor 1 showed high levels of Phen, Fl, and Pyr, suggesting that this factor might be coal combustion [33,63]. Moreover, a high correlation coefficient ($R^2 = 0.81$) between Factor 1 contributions and the total consumption of coal from the 1970s to 2016 was observed. Factor 2 had high levels of BbF, BaP, InP, and BgP (i.e., high-molecular-weight PAHs), indicating this factor as a possible marker of gasoline and diesel emissions [64,65]. The correlation coefficient of Factor 2 contributions and road freight capacity was 0.55, which further confirmed Factor 2 as likely related to traffic emissions. Factor 3 correlated strongly with Ant and Pyr, which is associated with refined petroleum combustion or crude oil leakage [14]. The correlation coefficient between Factor 3 contributions and the total consumption of petroleum was 0.53, indicating that this factor might represent petrogenic sources. Factor 4 showed a high loading of Naph and Fl and was identified as an important indicator of biomass combustion [4,63]. Furthermore, the correlation coefficient between Factor 4 contribution and the rural population reached 0.66, suggesting that the factor might represent the suburban lifestyle of biomass burning for cooking and heating [33]. According to the PMF results, four sources were successfully identified: (1) coal combustion sources (51.4% of total factor contributions), (2) traffic emissions (11.5% of total), (3) petrogenic sources (4.3% of total), and (4) biomass combustion and contribution (32.8% of total). Hence, coal combustion was recognized as the dominant source of PAHs in Dongping Lake in the last four decades.

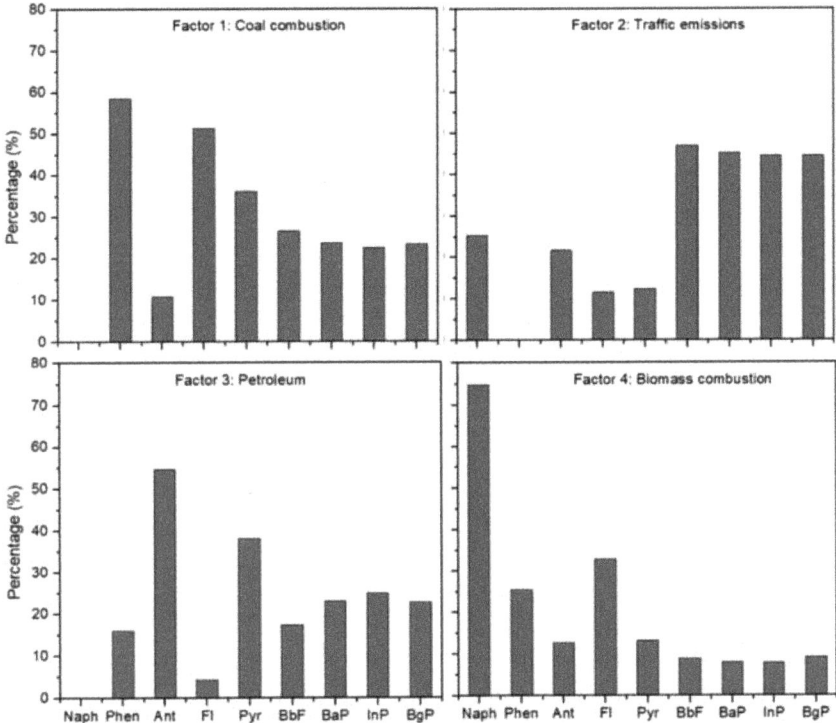

Figure 6. Four main source components of sedimentary polycyclic aromatic hydrocarbons (PAHs) obtained using a positive matrix factorization (PMF) model for Dongping Lake.

4. Conclusions

The historical variation of PAH pollution in the sediment core of Dongping Lake was investigated in this study. The concentration of the ΣPAH_{16} fluctuated from 77.6 to 628.0 ng/g (with a mean value of 198.3 ng/g), which was significantly lower compared to the lakes located in the areas with frequent industrial and human activities. The $\Sigma PAH16$ was mainly composed of 2,3-ring PAHs (48.7%), followed by 5,6-ring PAH (35.4%) and four-ring PAHs (15.9%). The concentration of 2,3-ring, four-ring, and 5,6-ring PAHs varied from 32.3 to 464.5 ng/g, 6.1 to 96.8 ng/g, and 39.6 to 101.8 ng/g, respectively. Since the 1970s, the increase in population and the construction of industrial parks have promoted the accumulation of PAHs in the area. In addition, the main energy structure significantly affects the input and composition of PAHs. The molecular diagnostic ratios of specific PAHs demonstrated that pyrogenic sources were the main PAH sources in the sediment core from Dongping Lake, with coal combustion (51.4% of total sources contributions) and biomass combustion (32.8% of total) being the dominant sources of PAHs since the 1970s, according to analysis of the PMF model. The results indicate that an appropriate adjustment of regional energy structure and encouragement of clean energy use can help reduce the impact of PAHs on lakes, improve air quality, and reduce carbon emissions.

Supplementary Materials: The following are available online: Table S1. Concentration of PAHs and total organic carbon (TOC) in sediment core from the Dongping Lake.

Author Contributions: Conceptualization, W.G. and Y.M.; formal analysis, Q.Z.; validation, J.Y.; data curation, X.Y.; investigation, W.G. and Y.M.; methodology, W.G. and J.L.; project administration, W.G.; funding acquisition, W.G. and Y.M.; writing—original draft, W.G. and J.Y.; writing—review and editing, W.G. and Y.M. All authors read and agreed to the published version of the manuscript.

Funding: This work was financially supported by the National Key R&D Program of China (Nos. 2018YFC1801605). This work was also partially funded by the National Natural Science Foundation of China (Nos. 41977325).

Institutional Review Board Statement: Not applicable.

Informed Consent Statement: Not applicable.

Data Availability Statement: Not applicable.

Conflicts of Interest: The authors declare no conflict of interest.

References

1. Hernandez-Vega, J.C.; Cady, B.; Kayanja, G.; Mauriello, A.; Cervantes, N.; Gillespie, A.; Lavia, L.; Trujillo, J.; Alkio, M.; Colon-Carmona, A. Detoxification of polycyclic aromatic hydrocarbons (PAHs) in Arabidopsis thaliana involves a putative flavonol synthase. *J. Hazard. Mater.* **2017**, *321*, 268–280. [CrossRef]
2. Yan, J.; Wang, L.; Fu, P.P.; Yu, H. Photomutagenicity of 16 polycyclic aromatic hydrocarbons from the US EPA priority pollutant list. *Mutat. Res.* **2004**, *557*, 99–108. [CrossRef] [PubMed]
3. Dushyant, R.D.; Rahul, K.R.; Jwalant, K.B.; Haren, B.G.; Bhumi, K.S.; Bharti, P.D. Distribution, sources and ecological risk assessment of PAHs in historically contaminated surface sediments at Bhavnagar coast, Gujarat, India. *Environ. Pollut.* **2016**, *213*, 338–346.
4. Feng, J.L.; Liu, M.L.; Zhao, J.H.; Hu, P.T.; Zhang, F.; Sun, J.H. Historical trends and spatial distributions of polycyclic aromatic hydrocarbons in the upper reach of the Huai River, China: Evidence from the sedimentary record. *Appl. Geochem.* **2019**, *103*, 59–67. [CrossRef]
5. Manoli, E.; Samara, C.; Konstantinou, I.; Albanis, T. Polycyclic aromatic hydrocarbons in the bulk precipitation and surface waters of Northern Greece. *Chemosphere* **2000**, *41*, 1845–1855. [CrossRef]
6. Ahrens, M.J.; Depree, C.V. A source mixing model to apportion PAHs from coal tar and asphalt binders in street pavements and urban aquatic sediments. *Chemosphere* **2010**, *81*, 1526–1535. [CrossRef] [PubMed]
7. Tobiszewski, M.; Namie snik, J. PAH diagnostic ratios for the identification of pollution emission sources. *Environ. Pollut.* **2012**, *162*, 110–119. [CrossRef]
8. Olkowska, E.; Kudlak, B.; Tsakovski, S.; Ruman, M.; Simeonov, V.; Polkowska, Z. Assessment of the water quality of Kłodnica River catchment using self-organizing maps. *Sci. Total Environ.* **2014**, *476-477*, 477–484. [CrossRef]
9. Inomata, Y.; Kajino, M.; Sato, K.; Ohara, T.; Kurokawa, J.-I.; Ueda, H.; Tang, N.; Hayakawa, K.; Ohizumi, T.; Akimoto, H. Emission and atmospheric transport of particulate PAHs in Northeast Asia. *Environ. Sci. Technol.* **2012**, *46*, 4941–4949. [CrossRef] [PubMed]
10. Hayakawa, K. Oil spills and polycyclic aromatic hydrocarbons. In *Polycyclic Aromatic Hydrocarbons: Environmental Behavior and Toxicity in East Asia*; Hayakawa, K., Ed.; Springer: Singapore, Singapore, 2018; pp. 213–223.
11. Parajulee, A.; Lei, Y.D.; Kananathalingam, A.; McLagan, D.S.; Mitchell, C.P.J.; Wania, F. The transport of polycyclic aromatic hydrocarbons during rainfall and snowmelt in contrasting landscapes. *Water Res.* **2017**, *124*, 407–414. [CrossRef]
12. Warren, N.; Allan, I.J.; Carter, J.E.; House, W.A.; Parker, A. Pesticides and other micro-organic contaminants in fresh water sedimentary environments-a review. *Appl. Geochem.* **2003**, *18*, 159–194. [CrossRef]
13. Mouhri, A.; Motelay-Masei, A.; Masei, N.; Fournier, M.; Laignel, B. Polycyclic aromatic hydrocarbon transport processes on the scale of a flood event in the rural watershed of Le Bebec, France. *Chemosphere* **2008**, *73*, 443–450. [CrossRef] [PubMed]
14. Guo, W.; Pei, Y.; Yang, Z.; Chen, H. Historical changes in polycyclic aromatic hydrocarbons (PAHs) input in Lake Baiyangdian related to regional socioeconomic development. *J. Hazard. Mater.* **2011**, *187*, 441–449. [CrossRef] [PubMed]
15. Lu, M.; Zeng, D.C.; Liao, Y.; Tong, B. Distribution and characterization of organochlorine pesticides and polycyclic aromatic hydrocarbons in surface sediment from Poyang Lake, China. *Sci. Total Environ.* **2012**, *433*, 491–497. [CrossRef] [PubMed]
16. McGrath, J.A.; Joshua, N.; Bess, A.S.; Parkerton, T.F. Review of polycyclic aromatic hydrocarbons (PAHs) sediment quality guidelines for the protection of benthic life. *Integr. Environ. Assess. Manag.* **2019**, *15*, 505–518. [CrossRef]
17. Pietzsch, R.; Patchineelam, S.R.; Torres, J.P.M. Polycyclic aromatic hydrocarbons in recent sediments from a subtropical estuary in Brazil. *Mar. Chem.* **2010**, *118*, 56–66. [CrossRef]
18. Bandowe, B.A.M.; Frankl, L.; Grosjean, M.; Tylmann, W.; Mosquera, P.V.; Hampel, H.; Schneider, T. A 150-year record of polycyclic aromatic compound (PAC) deposition from high Andean Cajas National Park, southern Ecuador. *Sci. Total Environ.* **2018**, *621*, 1652–1663. [CrossRef]
19. Ma, X.H.; Han, X.X.; Jiang, Q.L.; Huang, T.; Yang, H.; Yao, L. Historical records and source apportionment of polycyclic aromatic hydrocarbons over the past 100 years in Dianchi Lake, a plateau lake in Southwest China. *Arch. Environ. Contam. Toxicol.* **2018**, *75*, 187–198. [CrossRef] [PubMed]
20. Gardes, T.; Portet-Koltalo, F.; Debret, M.; Humbert, K.; Levaillant, R.; Simon, M.; Copard, Y. Temporal trends, sources, and relationships between sediment characteristics and polycyclic aromatic hydrocarbons (PAHs) and polychlorinated biphenyls (PCBs) in sediment cores from the major Seine estuary tributary, France. *Appl. Geochem.* **2020**, *122*, 104749. [CrossRef]
21. Du, J.J.; Jing, C.Y. Anthropogenic PAHs in lake sediments: A literature review (2002–2018). *Environ. Sci. Process. Impacts* **2018**, *20*, 1649–1666. [CrossRef] [PubMed]

22. Alonso-Hernández, C.M.; Tolosa, I.; Mesa-Albernas, M.; Díaz-Asencio, M.; Corcho-Alvarado, J.A.; Sánchez-Cabeza, J.A. Historical trends of organochlorine pesticides in a sediment core from the Gulf of Batabanó, Cuba. *Chemosphere* **2015**, *137*, 95–100. [CrossRef] [PubMed]
23. Zhan, C.L.; Wan, D.J.; Han, Y.M.; Zhang, J.Q. Historical variation of black carbon and PAHs over the last ~200 years in central North China: Evidence from lake sediment records. *Sci. Total Environ.* **2019**, *690*, 891–899. [CrossRef] [PubMed]
24. Han, Y.M.; Bandowe, B.A.M.; Schneider, T.; Pongpiachan, S.; Ho, S.S.H.; Wei, C.; Wang, Q.Y.; Xing, L.; Wilcke, W. A 150-year record of black carbon (soot and char) and polycyclic aromatic compounds deposition in Lake Phayao, north Thailand. *Environ. Pollut.* **2021**, *269*, 116148. [CrossRef] [PubMed]
25. Yuan, H.Z.; Liu, E.F.; Zhang, E.L.; Luo, W.L.; Chen, L.; Wang, C.; Lin, Q. Historical records and sources of polycyclic aromatic hydrocarbons (PAHs) and organochlorine pesticides (OCPs) in sediment from a representative plateau lake, China. *Chemosphere* **2017**, *173*, 78–88. [CrossRef] [PubMed]
26. Liu, L.Y.; Wang, J.Z.; Wei, G.L.; Guan, Y.F.; Wong, C.S.; Zeng, E.Y. Sediment records of polycyclic aromatic hydrocarbons (PAHs) in the continental shelf of China: Implications for evolving anthropogenic impacts. *Environ. Sci. Technol.* **2012**, *46*, 6497–6504. [CrossRef] [PubMed]
27. Zhang, Y.J.; Lin, Y.; Cai, J.; Liu, Y.; Hong, L.N.; Qin, M.M.; Zhao, Y.F.; Ma, J.; Wang, X.S.; Zhu, T.; et al. Atmospheric PAHs in north China: Spatial distribution and sources. *Sci. Total Environ.* **2016**, *565*, 994–1000. [CrossRef] [PubMed]
28. Zhang, Y.; Tao, S. Global atmospheric emission inventory of polycyclic aromatic hydrocarbons (PAHs) for 2004. *Atmos. Environ.* **2009**, *43*, 812–819.
29. Xu, S.; Liu, W.; Tao, S. Emission of polycyclic aromatic hydrocarbons in China. *Environ. Sci. Technol.* **2006**, *40*, 702–708. [CrossRef]
30. Liu, G.Q.; Zhang, G.; Jin, Z.D.; Li, J. Sedimentary record of hydrophobic organic compounds in relation to regional economic development: A study of Taihu Lake, East China. *Environ. Pollut.* **2009**, *157*, 2994–3000. [CrossRef]
31. Lin, T.; Qin, Y.; Zheng, B.; Li, Y.; Zhang, L.; Guo, Z. Sedimentary record of polycyclic aromatic hydrocarbons in a reservoir in Northeast China. *Environ. Pollut.* **2011**, *163*, 256–260. [CrossRef] [PubMed]
32. Chang, J.; Zhang, E.L.; Liu, E.F.; Liu, H.J.; Yang, X.Q. A 60-year historical record of polycyclic aromatic hydrocarbons (PAHs) pollution in lake sediment from Guangxi Province, Southern China. *Anthropocene* **2018**, *24*, 51–60. [CrossRef]
33. Ma, X.H.; Wan, H.B.; Zhou, J.; Luo, D.; Huang, T.; Yang, H.; Huang, C.C. Sediment record of polycyclic aromatic hydrocarbons in Dianchi lake, southwest China: Influence of energy structure changes and economic development. *Chemosphere* **2020**, *248*, 126015. [CrossRef] [PubMed]
34. Compiling Team for Annals of Dongping County Shandong Province. Annals of Dongping County (1986–2003). Zhonghua Book Company: Beijing, China, 2006. (In Chinese)
35. Wang, Y.Q.; Yang, L.J.; Kong, L.H.; Liu, E.F.; Wang, L.F.; Zhu, J.R. Spatial distribution, ecological risk assessment and source identification for heavy metals in surface sediments from Dongping Lake, Shandong, East China. *Catena* **2015**, *125*, 200–205. [CrossRef]
36. Yang, L.W.; Chen, S.Y.; Zhang, J.; Yu, S.Y.; Deng, H.G. Environmental factors controlling the spatial distribution of subfossil Chironomidae in surface sediments of Lake Dongping, a warm temperate lake in North China. *Environ. Earth Sci.* **2017**, *76*, 524. [CrossRef]
37. Mei, R.B.; Wang, Y.; Wang, X.Q.; Qin, S. Reason analysis and prevention measure of pollution in Dawen River Basin. *Environment and Development* **2014**, *26*, 125–127. (In Chinese)
38. Liu, G.Q.; Zhang, G.; Li, X.D.; Li, J.; Peng, X.Z.; Qi, S.H. Sedimentary record of polycyclic aromatic hydrocarbons in a sediment core from the Pearl River Estuary, South China. *Mar. Pollut. Bull.* **2005**, *51*, 912–921. [CrossRef] [PubMed]
39. Appleby, P.G.; Oldfield, F. The calculation of Lead-210 dates assuming a constant rate of supply of unsupported ^{210}Pb to the sediment. *Catena* **1978**, *5*, 1–8. [CrossRef]
40. Piazza, R.; Ruiz-Fernández, A.C.; Frignani, M.; Vecchiato, M.; Bellucci, L.G.; Gambaro, A.; Pérez-Bernal, L.H.; Páez-Osuna, F. Historical PCB fluxes in the Mexico City Metropolitan Zone as evidenced by a sedimentary record from the Espejo de los Lirios lake. *Chemosphere* **2009**, *75*, 1252–1258. [CrossRef]
41. Paatero, P.; Tapper, U. Positive matrix factorization: A non-negative factor model with optimal utilization of error estimates of data values. *Environmetrics* **1994**, *5*, 111–126. [CrossRef]
42. US EPA. EPA positive matrix factorization (PMF) 5.0 fundamentals & user guide. 2014. Available online: http://www.epa.gov/heasd/research/pmf.html (accessed on 10 November 2021).
43. Hopke, P.K. Recent developments in receptor modeling. *J. Chemom.* **2003**, *17*, 255–265. [CrossRef]
44. Lin, T.; Hu, L.M.; Guo, Z.G.; Yang, Y.S. Deposition fluxes and fate of polycyclic aromatic hydrocarbons in the Yangtze River estuarine-inner shelf in the East China sea. *Glob. Biogeochem. Cycles* **2013**, *27*, 77–87. [CrossRef]
45. Wang, C.L.; Zou, X.Q.; Gao, J.H.; Zhao, Y.F.; Yu, W.W.; Li, Y.L. Pollution status of polycyclic aromatic hydrocarbons in surface sediments from the Yangtze River Estuary and its adjacent coastal zone. *Chemosphere* **2016**, *162*, 80–90. [CrossRef] [PubMed]
46. Abril, J.M.; Brunskill, G.J. Evidence that excess ^{210}Pb flux varies with sediment accumulation rate and implications for dating recent sediments. *J. Paleolimnol.* **2014**, *52*, 121–137. [CrossRef]
47. Eichler, A.; Gramlich, G.; Kellerhals, T.; Tobler, L.; Schwikowski, M. Pb pollution from leaded gasoline in South America in the context of a 2000-year metallurgical history. *Sci. Adv.* **2015**, *1*, e1400196. [CrossRef] [PubMed]

48. Guo, W.; Huo, S.L.; Ding, W.J. Historical record of human impact in a lake of northern China: Magnetic susceptibility, nutrients, heavy metals and OCPs. *Ecol. Indic.* **2015**, *57*, 74–81. [CrossRef]
49. Li, C.C.; Huo, S.L.; Xi, B.D.; Yu, Z.Q.; Zeng, X.Y.; Zhang, J.T.; Wu, F.C.; Liu, H.L. Historical deposition behaviors of organochlorine pesticides (OCPs)in the sediments of a shallow eutrophic lake in Eastern China: Roles of the sources and sedimentological conditions. *Ecol. Indic.* **2015**, *53*, 1–10. [CrossRef]
50. Chen, Y.Y.; Chen, S.Y.; Ma, C.M.; Yu, S.Y.; Yang, L.W.; Zhang, Z.K.; Yao, M. Palynological evidence of natural and anthropogenic impacts on aquatic environmental changes over the last 150 years in Dongping Lake, North China. *Quaern. Int.* **2014**, *349*, 2–9. [CrossRef]
51. Chen, Y.Y.; Chen, S.Y.; Liu, J.Z.; Yao, M.; Sun, W.B.; Zhang, Q. Environmental evolution and hydrodynamic process of Dongping Lake in Shandong Province, China, over the past 150 years. *Environ. Earth Sci.* **2013**, *68*, 69–75. [CrossRef]
52. Guo, J.Y.; Wu, F.C.; Liao, H.Q.; Zhao, X.L.; Li, W.; Wang, J.; Wang, L.F.; Giesy, J.P. Sedimentary record of polycyclic aromatic hydrocarbons and DDTs in Dianchi Lake, an urban lake in Southwest China. *Environ. Sci. Pollut. R.* **2013**, *20*, 5471–5480. [CrossRef] [PubMed]
53. Li, J.X.; Chen, Y.Y.; Han, F.; Chen, S.Y. Water level change of Dongping Lake and its impact on water quality from 1990 to 2016. *Chin. Agric. Sci. Bull.* **2021**, *37*, 94–100.
54. Tai'an Statistical Bureau. *Tai'an Statistical Yearbook*; China Statistical Publishing House Tai'an: Beijing, China, 2017.
55. Gustafsson, O.; Krusa, M.; Zencak, Z.; Sheesley, R.J.; Granat, L.; Engstrom, E.; Praveen, P.S.; Rao, P.S.P.; Leck, C.; Rodhe, H. Brown clouds over south Asia: Biomass or fossil fuel combustion? *Science* **2009**, *323*, 495–498. [CrossRef]
56. Pongpiachan, S.; Hattayanone, M.; Cao, J. Effect of agricultural waste burning season on PM2.5-bound polycyclic aromatic hydrocarbon (PAH) levels in Northern Thailand. *Atmos. Pollut. Res.* **2017**, *8*, 1069–1080. [CrossRef]
57. Viguri, J.; Verde, J.; Irabien, A. Environmental assessment of polycyclic aromatic hydrocarbons (PAHs) in surface sediments of the Santander Bay, Northern Spain. *Chemosphere* **2002**, *48*, 157–165. [CrossRef]
58. Guo, W.; He, M.C.; Yang, Z.F.; Lin, C.Y.; Quan, X.C.; Men, B. Distribution, partitioning and sources of polycyclic aromatic hydrocarbons in Daliao River water system in dry season, China. *J. Hazard. Mater.* **2009**, *164*, 1379–1385. [CrossRef]
59. Tamamura, S.; Sato, T.; Ota, Y.; Wang, X.; Tang, N.; Hayakawa, K. Longrange transport of polycyclic aromatic hydrocarbons (PAHs) from the eastern Asian continent to Kanazawa, Japan with Asian dust. *Atmos. Environ.* **2007**, *41*, 2580–2593. [CrossRef]
60. Barakat, A.O.; Mostafa, A.; Wade, T.L.; Sweet, S.T.; El Sayed, N.B. Spatial distribution and temporal trends of polycyclic aromatic hydrocarbons (PAHs) in sediments from Lake Maryut, Alexandria, Egypt. *Water Air Soil Pollut.* **2011**, *218*, 63–80. [CrossRef]
61. Keshavarzifard, M.; Zakaria, M.P.; Shau Hwai, T.; Yusuff, F.F.M.; Mustafa, S.; Vaezzadeh, V.; Magam, S.M.; Masood, N.; Alkhadher, S.A.A.; Abootalebi-Jahromi, F. Baseline distributions and sources of polycyclic aromatic hydrocarbons (PAHs) in the surface sediments from the Prai and Malacca Rivers, Peninsular Malaysia. *Mar. Pollut. Bull.* **2014**, *88*, 366–372. [CrossRef] [PubMed]
62. Manneh, R.; Ghanem, C.A.; Khalaf, G.; Najjar, E.; El Khoury, B.; Iaaly, A.; El Zakhem, H. Analysis of polycyclic aromatic hydrocarbons (PAHs) in Lebanese surficial sediments: A focus on the regions of Tripoli, Jounieh, Dora, and Tyre. *Mar. Pollut. Bull.* **2016**, *110*, 578–583. [CrossRef]
63. Ravindra, K.; Sokhi, R.; Van Grieken, R. Atmospheric polycyclic aromatic hydrocarbons: Source attribution, emission factors and regulation. *Atmos. Environ.* **2008**, *42*, 2895–2921. [CrossRef]
64. Nemr, A.E.L.; Said, T.O.; Khaled, A.; Sikaily, A.E.L.; Allah, A.M.A. The distribution and sources of polycyclic aromatic hydrocarbons in surface sediments along the Egyptian Mediterranean coast. *Environ. Monit. Assess.* **2007**, *124*, 343–359. [CrossRef]
65. Sofowote, U.M.; Mccarry, B.E.; Marvin, C.H. Source apportionment of PAH in Hamilton Harbour suspended sediments: Comparison of two factor analysis methods. *Environ. Sci. Technol.* **2008**, *42*, 6007–6014. [CrossRef] [PubMed]

Article

Determination of Hexabromocyclododecane in Expanded Polystyrene and Extruded Polystyrene Foam by Gas Chromatography-Mass Spectrometry

Tianao Mao [1,2], Haoyang Wang [3,*], Zheng Peng [3], Taotao Ni [2], Tianqi Jia [2,4], Rongrong Lei [2,4] and Wenbin Liu [1,2,4,*]

1. Hangzhou Institute for Advanced Study, University of Chinese Academy of Sciences, Hangzhou 310024, China; maotianao@foxmail.com
2. Research Center for Eco-Environmental Sciences, Chinese Academy of Sciences, Beijing 100085, China; ttaoni@yeah.net (T.N.); tqjjia@126.com (T.J.); leirongr@163.com (R.L.)
3. Environmental Protection and Foreign Cooperation and Exchange Center of Ministry of Ecology and Environment, Beijing 100035, China; peng.zheng@fecomee.org.cn
4. University of Chinese Academy of Sciences, Beijing 100049, China
* Correspondence: wang.haoyang@fecomee.org.cn (H.W.); liuwb@rcees.ac.cn (W.L.); Tel.: +86-10-82268590 (H.W.); +86-10-62849356 (W.L.); Fax: +86-10-62849339 (W.L.)

Abstract: A gas chromatography-mass spectrometry (GC/MS) method for the determination of hexabromocyclododecane (HBCD) in expanded polystyrene and extruded polystyrene foam (EPS/XPS) was developed. The EPS/XPS samples were ultrasonically extracted with acetone and the extracts were purified by filtration through a microporous membrane (0.22 μm) and solid-phase extraction. The samples were analyzed using a GC/MS using the selected ion monitoring mode. The ions 157, 319 and 401 were selected as the qualitative ions, while ion 239 was chosen as the quantitative ion. An HBCD standard working solution with a concentration range of 1.0–50.0 mg/L showed good linearity. The detection limit of HBCD was 0.5 mg/kg, meeting the LPC limit (<100 or 1000 mg/kg). Six laboratories were selected to verify the accuracy of the method, and 10 samples were tested. The interlaboratory relative standard deviation range was 3.68–9.80%. This method could play an important role in controlling HBCD contamination in EPS/XPS.

Keywords: gas chromatography-mass spectrometry; HBCD; EPS; XPS; POPs

Citation: Mao, T.; Wang, H.; Peng, Z.; Ni, T.; Jia, T.; Lei, R.; Liu, W. Determination of Hexabromocyclododecane in Expanded Polystyrene and Extruded Polystyrene Foam by Gas Chromatography-Mass Spectrometry. *Molecules* **2021**, *26*, 7143. https://doi.org/10.3390/molecules26237143

Academic Editors: Giorgio Vilardi and Alessandra Gentili

Received: 30 August 2021
Accepted: 22 November 2021
Published: 25 November 2021

Publisher's Note: MDPI stays neutral with regard to jurisdictional claims in published maps and institutional affiliations.

Copyright: © 2021 by the authors. Licensee MDPI, Basel, Switzerland. This article is an open access article distributed under the terms and conditions of the Creative Commons Attribution (CC BY) license (https://creativecommons.org/licenses/by/4.0/).

1. Introduction

Hexabromocyclododecane (HBCD) is flame retardant with high bromine content that has long been used in the manufacture of expanded polystyrene (EPS) and extruded polystyrene (XPS) boards for fire protection and insulation in buildings. In the 1980s, HBCD was detected in air, sludge and sediment in Sweden [1]. Since then, many researchers in other countries have also confirmed the widespread presence of HBCD in soils [2], outdoor air [3,4], seawater [5], and house dust [6]. Also, HBCD has been detected in arctic regions [7] and breast milk [8]. There is global concern about the potential toxic effects of HBCD on humans and ecosystems [3,9] because of its persistence in the environment, bioaccumulation, and bioamplification in fish [10], birds [11], and mammals [12,13]. Recently many studies in animals have shown that HBCD can affect the expression of related genes in rats [14] and zebrafish [10]. Furthermore, HBCD promoted the production and accumulation of fat in vivo and in vitro [15]. HBCD was added to the control list in the Stockholm Convention on Persistent Organic Pollutants in May 2013 and banned from future production and use [16].

The latest information on global production of HBCD indicates that total production in 2011 was estimated at 31,000 metric tons, and it was mainly produced in China, Europe, and the United States. Considering that 90% of HBCD on average is used in the manufacture of EPS/XPS each year [17], large quantities of HBCD-containing EPS/XPS building materials have accumulated globally over the past decade. Because of the long service life of EPS/XPS (20–50 years) [18,19], many of these building materials are still in use. However, a study shows that the use of HBCD in some consumer products is unregulated [20], and there is still a risk of HBCD entering the environment from existing construction materials and waste through wear and tear during product use, weathering, and leachate from landfill [21]. As of 2019, HBCD must be destroyed or irreversibly transformed to prevent its entry into the environment when it is at or above the designated LPC limits (100 or 1000 mg/kg) in accordance with the European Union's Basel Convention General Technical Guidelines for Persistent Organic Pollutants Waste Management [22]. Therefore, a simple and quick method is required to determine whether HBCD in EPS/XPS exceeds the standard (LPC). Many methods have been developed to determine HBCD concentration in EPS or XPS foam like X-ray fluorescence spectroscopy (XRF), flowing atmospheric pressure afterglow mass spectrometry, and liquid chromatography-tandem mass spectrometry (LC-MS/MS) [23–28].

In this study, the GC/MS method was developed to determine the HBCD concentration in EPS/XPS products. The solvent and purification processes were investigated to separate the HBCD and EPS/XPS matrix in the extract dilution. Six laboratories were selected to carry out the verification work, and 10 samples were tested to verify the feasibility of the method. This study aimed to establish a simple, inexpensive and effective analytical method for determination of HBCD in EPS/XPS. It is used to help more countries and regions judge whether HBCD in construction waste exceeds LPC limits, and provides a reference for the recycling or destruction and irreversible transformation of waste containing HBCD.

2. Materials and Methods

2.1. Samples and Chemicals

HBCD standard (purity >97%) was purchased from Dr. Ehrenstorfer GmbH (Augsburg, Germany). Acetone (pesticide residue grade) was obtained from Fisher Chemical (Thermo Fisher Scientific Inc., Waltham, MA, USA). Toluene and n-Hexane (pesticide residue grade) were purchased from J.T. Baker Corporation (Phillipsburg, NJ, USA). Methanol and dichloromethane (analytical pure) were purchased from Sinopharm (Beijing, China). Polytetrafluoroethylene filtration membrane (0.22 µm) was purchased from Jintengyi Technology Co., Ltd. (Beijing, China). Five EPS and five XPS samples were collected from 10 companies in Shandong China. 20 mg HBCD was dissolved with 2 mL acetone and diluted to 100 mL with n-hexane to prepare the HBCD stock solution (200 mg/L). Working solutions with HBCD concentrations of 1.0, 2.0, 5.0, 10.0, 20.0, and 50.0 mg/L were prepared by dilute with n-hexane of the HBCD stock solution.

2.2. Sample Preparation

The EPS/XPS samples were cut into particles smaller than 5 mm. 0.1 g (±0.1 mg) of EPS/XPS sample was weighed from the particles above, then placed in a 10 mL stoppered colorimetric tube, and 5 mL of acetone was added to dissolve the sample. The colorimetric tube was placed on a vortex mixer for 2 min and then extracted by ultrasonication (KQ-100E, Kunshan Ultrasonic Instrument Co., Ltd., Kunshan, China) for 20 min at room temperature. The supernatant was filtered through a 0.22 µm polytetrafluoroethylene filter membrane. The filtrate was collected and n-hexane was added to give a volume of 20 mL. Because the concentration of HBCD in the EPS/XPS samples were usually very high, only part of the filtrate was analyzed. For the EPS sample, the HBCD concentration in typical EPS product was approximately 0.6–0.8%, 2.5% (500 µL) of the filtrate was analyzed. For the XPS sample, the typical HBCD in XPS was approximately 4–6%, 0.25% (50 µL) of the filtrate was analyzed and n-hexane (450 µL) was added to make the volume up to 500 µL. Next, the sample was purified using a solid-phase extraction (SPE) column (LC-Si, 500 mg, 6 mL, Supelco Inc., St. Louis, MO, USA). The SPE column was activated with 6 mL of n-hexane. The EPS/XPS filtrates were eluted once with 6 mL of acetone. All of the eluent was collected and then concentrated under a stream of nitrogen gas. The concentrated samples were reconstituted with n-hexane to 1 mL. Finally, the samples were placed in 2 mL brown vials for GC/MS analysis. An EPS sample without HBCD was selected as the blank sample, extracted, and purified together with other EPS/XPS samples.

2.3. GC/MS Analyses

The samples were analyzed using an Agilent 6890A/5973C GC/MS with a DB-5 MS capillary column (Agilent Technologies, Santa Clara, CA, USA). The length of DB-5 MS capillary column is 30 m, the inner diameter is 0.25 mm and the film thickness is 0.25 m. The following heating procedure of oven temperature was applied: initial temperature is 60 °C, held at 60 °C for 2 min, increased to 270 °C at 15 °C/min and held at this temperature for 5 min, then increased to 290 °C at 5 °C/min and held at this temperature for 5 min. The carrier gas used helium (purity $\geq$99.999%), and the flow rate was 1.0 mL/min. The sampling method was split-less injection and the injection volume was 1 µL. The inlet temperature was 230 °C. The ionization mode was electron ionization, the ionization energy was 70 eV, and the ion source temperature was 230 °C. The quadrupole temperature was 150 °C and the interface temperature was 280 °C. The data acquisition mode used selected ion monitoring. The selected monitoring ions (m/z) are shown in Table 1. The ions 157, 319 and 401 were selected as the qualitative ions, while ion 239 was chosen as the quantitative ion.

Table 1. Selected monitoring ions (m/z) and allowable relative deviations.

Monitoring of the Ion (m/z)	Ion Species	Ion Ratio (%)	Permissible Relative Deviation (%)
157	Qualitative ion	100	-
239	Quantitative ion	89	±15
319	Qualitative ion	55	±20
401	Qualitative ion	19	±20

2.4. Quality Assurance and Quality Control

Three samples of EPS and XPS each were selected to estimate the matrix effect (*ME*) according to the method reported by Caban et al. [29]. The *ME* is calculated according to the following formula.

$$ME = \left(\frac{B-C}{A} - 1\right) \times 100\% \quad (1)$$

where *A* is the peak area of the HBCD standard solution (20 mg/L). *B* is the peak area of the EPS/XPS sample with HBCD standard (20 mg/L) added before injection. *C* is the peak area of the non-spiked EPS/XPS sample. The samples showed acceptable matrix effects ranging from -10.9% to 22.0%.

The HBCD was not detected in the blank sample. The recovery of the blank spiked sample test is between 87% and 113%.

3. Results and Discussion

3.1. Optimization of the Sample Pretreatment Conditions

Because dissolution of polystyrene from EPS/XPS would cause matrix interference in the HBCD analysis, different extraction solvents were screened to optimize the method. Through literature research [23,26,30,31], we chose methanol, toluene, acetone, dichloromethane, n-hexane, n-hexane/isopropanol (1:1, v/v), acetone/methylene chloride (1:1, v/v), toluene/methylene chloride (1:1, v/v), acetone/n-hexane (1:1, v/v) and n-hexane/methylene chloride (1:1, v/v) to conduct experiments on the dissolution of EPS/XPS and the extraction effect of HBCD.

EPS/XPS partially dissolved in acetone, n-hexane, and n-hexane/isopropanol, but did not dissolve in methanol. Toluene and dichloromethane completely dissolved EPS/XPS.

Although acetone emulsified EPS/XPS, minimal dissolution of EPS/XPS occurred, and this solvent gave a very high HBCD extraction efficiency. Therefore, acetone was selected as the extraction solvent. The extract was then purified using SPE. We found that HBCD was adsorbed by the LC-Si filler in the SPE column, and was completely eluted by acetone. The results of the matrix effect also proved that effectiveness of acetone.

3.2. Optimization of the GC/MS Conditions

HBCD can be analyzed using a 15 or 30 m capillary column. We found that a 30 m capillary column gave effective separation of HBCD in EPS/XPS and good peak shapes. The heating procedure was optimized, and the retention time of HBCD was 25.40 min. The injection temperatures are the important variable in the determination of HBCD. When the injection temperature is too low, it will lead to incomplete gasification of HBCD and reduce the amount available for detection. If the injection temperature is too high, it will cause thermal isomerisation and degradation of HBCD [32–34]. In a comparison of the chromatograms obtained with injection temperatures of 190 °C, 230 °C, and 270 °C, the best results were obtained at 230 °C (Figure 1). At 190 °C, the gasification of HBCD was incomplete and the response was low, which was only 20–25% of those at 230 °C. At 270 °C, the HBCD was decomposed, the impurity peak (retention time from 17.40 min to 18.6 min) increased five times of those at 230 °C. Consequently, 230 °C was selected as the optimum injection temperature, and a good response was obtained using optimized GC/MS conditions.

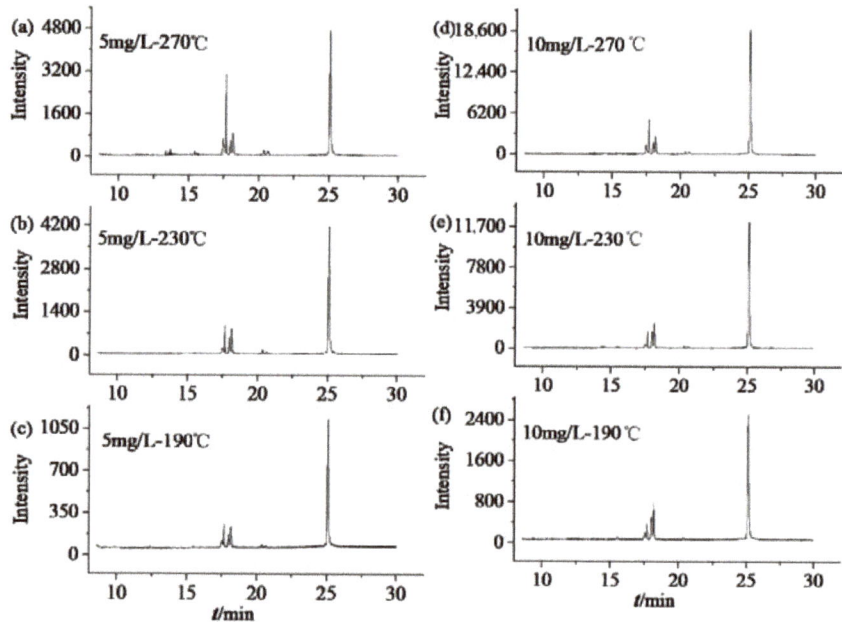

Figure 1. Chromatograms of HBCD at different injection temperatures, (**a**) 5 mg/L-270 °C, (**b**) 5 mg/L-230 °C, (**c**) 5 mg/L-190 °C, (**d**) 10 mg/L-270 °C, (**e**) 10 mg/L-230 °C, and (**f**) 10 mg/L-190 °C.

3.3. Linear Range and Detection Limit

The HBCD working solutions were analyzed using the optimized GC/MS conditions. A 1.0 µL aliquot of each HBCD standard working solution was injected. The retention times and chromatographic peak areas were recorded and a standard curve was plotted (Figure 2).

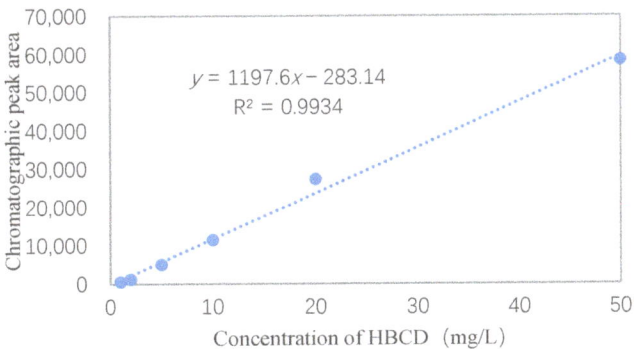

Figure 2. The calibration curve of HBCD.

The linear equation for the standard curve of HBCD was $Y = 1197.6X - 283.1$, and the R^2 was 0.9934. These results met the test requirements for HBCD analysis [35].

HBCD in real samples were accurately quantified by using a calibration curve. According to the HBCD working curve, the concentration of HBCD in the sample is calculated according to the following formula.

$$\rho = \frac{C \times V_1 \times V_3}{V_2 \times M} \times D \qquad (2)$$

where ρ is the concentration of HBCD in EPS or XPS samples (mg/kg). C is concentration of HBCD in the sample solution (mg/L). V_1 is volume of extract (ml), V_2 is volume of extracted solution transferred (ml) and V_3 is volume of injection solution (ml). M is sample mass (g). D is dilution multiple.

The detection limit was determined as described in the references [35]. Seven blank spiked samples with HBCD concentration of 5 mg/kg were continuously tested. The method detection limit is then calculated as 3.143 times the standard deviation. In the present study, the limit of quantitation of detection is 0.5 mg/kg.

3.4. Method Precision and Accuracy

To evaluate the method precision and accuracy [35], three blank samples were spiked with HBCD standard solutions to give concentrations of 5, 10, and 20 mg/kg. The HBCD concentration in each of these samples was determined six times. The 18 analysis results were shown in Table 2. The relative standard deviation (RSD) range for the three blank samples was 6.64–6.92%, and the recovery range was 87–113%.

Table 2. Test data for the method precision and accuracy (n = 6).

HBCD Add the Amount	5 mg/kg	10 mg/kg	20 mg/kg
Measured mean $\bar{x}_i$	4.9	9.9	21.3
The standard deviation S_i	0.34	0.69	1.4
Relative standard deviation RSD (%)	6.90%	6.92%	6.64%
Recovery range	87~106%	89~108%	96~113%

We compared the results from our study with those from previous studies (Table 3). The accuracy of our method met the test requirements.

Table 3. Comparison of experimental results from different studies.

Test Material	Testing Equipment	Concentrations g/kg	RSDs %	Limit of Quantitation mg/kg	The Literature
EPS/XPS	GC/MS	4.5~33.5	6.64~6.92	0.5	In this study
EPS/XPS	XRF	5~12	7~16	50	[23]
EPS	LC-MS/MS	6.849~7.105	0.2~0.6	0.005	[26]
EPS/XPS	XRF, LC-MS/MS, NMR	6.7~11.1	44~104	300	[31]

3.5. Actual Sample Analysis and Interlaboratory Verification

To verify the effectiveness of the method, five EPS and five XPS samples were collected from 10 companies. Five verification laboratories (L1–L5) and our laboratory (L6) were commissioned to carry out verification tests for the developed method. The HBCD analysis results from the six laboratories were shown in Figure 3.

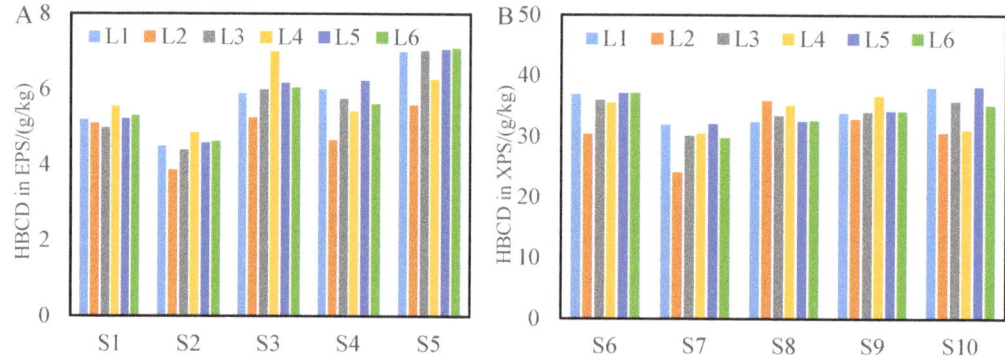

Figure 3. Results from the six laboratories for EPS (**A**) and XPS (**B**).

For the five EPS samples, the HBCD concentration range was 4.5–6.7 g/kg and the RSD range was 3.78–9.76%. For the five XPS samples, the HBCD concentration range was 29.8–35.5 g/kg, and the RSD range was 3.68–9.80%. The overall RSD range of the 10 EPS/XPS samples was 3.68–9.80% and the R^2 of the calibration curve of six laboratories were between 0.991 and 0.998.

Through our study, we found that the content of HBCD in EPS/XPS is approximately 1%, with a range of 0.45–0.67% in EPS and 2.98–3.55% in XPS. Such a high concentration of HBCD may be dangerous to human health [36] and pose a challenge to the treatment of construction waste containing EPS/XPS [27,28]. Improper handling methods may cause HBCD to leak into the environment, causing serious damage to the ecological environment [37]. Production of HBCD-containing EPS/XPS was expected to increase by 2.7 million tons per year until 2013, and then decrease by about 7% (18,000 tons) after the inclusion of HBCD in the Stockholm Convention. Because HBCD-containing EPS/XPS products have been produced for more than 40 years, we can infer that 40 to 70 million tons of these products have been produced globally [38]. Consequently, a method is needed to determine HBCD levels in these plastics to provide a basis for the recycling and disposal of HBCD.

4. Conclusions

A method was established and verified for the determination of HBCD in the product and waste of EPS/XPS by GC/MS. Satisfactory recovery and precision results were obtained for 10 EPS/XPS samples. The advantages of this method are its high precision, low detection limit, simplicity, speed, and suitability for the detection of HBCD in EPS/XPS samples. Although HBCD has recently been banned, EPS/XPS products containing HBCD are still used on the surface of buildings, and will last for 20–50 years until the demolition of these structures. In the future, HBCD in EPS/XPS waste generated from the demolition of these buildings needs to be detected and analyzed. Therefore, the detection of HBCD in EPS/XPS samples by GC/MS still has great application value.

Author Contributions: Conceptualization, T.M., H.W. and W.L.; methodology, T.M. and Z.P.; validation, T.N. and R.L.; formal analysis, T.M. and Z.P.; investigation, T.M., T.J. and T.N.; writing—original draft preparation, T.M.; writing—review and editing, Z.P. and W.L.; supervision, W.L. and R.L. All authors have read and agreed to the published version of the manuscript.

Funding: This work was funded by the National Key Research and Development Plan (2018YFC1801602), and the National Natural Science Foundation of China (22076207).

Institutional Review Board Statement: Not applicable.

Informed Consent Statement: Not applicable.

Data Availability Statement: Not applicable.

Acknowledgments: Not applicable.

Conflicts of Interest: The authors declare no conflict of interest.

References

1. Sellstrom, U.; Kierkegaard, A.; de Wit, C.; Jansson, B. Polybrominated diphenyl ethers and hexabromocyclododecane in sediment and fish from a Swedish river. *Environ. Toxicol. Chem.* **1998**, *17*, 1065–1072. [CrossRef]
2. Huang, L.; Wang, W.; Shah, S.B.; Hu, H.; Xu, P.; Tang, H. The HBCDs biodegradation using a Pseudomonas strain and its application in soil phytoremediation. *J. Hazard. Mater.* **2019**, *380*, 120833. [CrossRef]
3. Ma, S.; Yu, Y.; Yang, Y.; Li, G.; An, T. A new advance in the potential exposure to "old" and "new" halogenated flame retardants in the atmospheric environments and biota: From occurrence to transformation products and metabolites. *Crit. Rev. Environ. Sci. Technol.* **2020**, *50*, 1935–1983. [CrossRef]
4. Zhiqiang, Y.; Laiguo, C.; Bixian, M.; Minghong, W.; Guoying, S.; Jiamo, F.; Ping'an, P. Diastereoisomer- and enantiomer-specific profiles of hexabromocyclododecane in the atmosphere of an urban city in South China. *Environ. Sci. Technol.* **2008**, *42*, 3996–4001.
5. Gong, W.; Wang, J.; Cui, W.; Zhu, L. Distribution characteristics and risk assessment of TBBPA in seawater and zooplankton in northern sea areas, China. *Environ. Geochem. Health* **2021**, *43*, 4759–4769. [CrossRef] [PubMed]
6. Kuribara, I.; Kajiwara, N.; Sakurai, T.; Kuramochi, H.; Motoki, T.; Suzuki, G.; Wada, T.; Sakai, S.; Takigami, H. Time series of hexabromocyclododecane transfers from flame-retarded curtains to attached dust. *Sci. Total Environ.* **2019**, *696*, 133957. [CrossRef] [PubMed]
7. Jorundsdottir, H.; Lofstrand, K.; Svavarsson, J.; Bignert, A.; Bergman, A. Polybrominated diphenyl ethers (PBDEs) and hexabromocyclododecane (HBCD) in seven different marine bird species from Iceland. *Chemosphere* **2013**, *93*, 1526–1532. [CrossRef]
8. Inthavong, C.; Hommet, F.; Bordet, F.; Rigourd, V.; Guerin, T.; Dragacci, S. Simultaneous liquid chromatography-tandem mass spectrometry analysis of brominated flame retardants (tetrabromobisphenol A and hexabromocyclododecane diastereoisomers) in French breast milk. *Chemosphere* **2017**, *186*, 762–769. [CrossRef]
9. Lee, J.-G.; Anh, J.; Kang, G.-J.; Kim, D.; Kang, Y. Development of an analytical method for simultaneously determining TBBPA and HBCDs in various foods. *Food Chem.* **2020**, *313*, 126027. [CrossRef]
10. Guo, Z.; Zhang, L.; Liu, X.; Yu, Y.; Liu, S.; Chen, M.; Huang, C.; Hu, G. The enrichment and purification of hexabromocyclododecanes and its effects on thyroid in zebrafish. *Ecotoxicol. Environ. Saf.* **2019**, *185*, 109690. [CrossRef]
11. Reindl, A.; Falkowska, L.; Grajewska, A. Hexabromocyclododecane contamination of herring gulls in the coastal area of the southern Baltic Sea. *Oceanol. Hydrobiol. Stud.* **2020**, *49*, 147–156. [CrossRef]
12. Shi, J.; Wang, X.; Chen, L.; Deng, H.; Zhang, M. HBCD, TBECH, and BTBPE exhibit cytotoxic effects in human vascular endothelial cells by regulating mitochondria function and ROS production. *Environ. Toxicol.* **2021**, *36*, 1674–1682. [CrossRef]
13. Yi, S.; Liu, J.G.; Jin, J.; Zhu, J. Assessment of the occupational and environmental risks of hexabromocyclododecane (HBCD) in China. *Chemosphere* **2016**, *150*, 431–437. [CrossRef]
14. Farmahin, R.; Gannon, A.M.; Gagne, R.; Rowan-Carroll, A.; Kuo, B.; Williams, A.; Curran, I.; Yauk, C.L. Hepatic transcriptional dose-response analysis of male and female Fischer rats exposed to hexabromocyclododecane. *Food Chem. Toxicol.* **2019**, *133*, 110262. [CrossRef] [PubMed]
15. Xie, X.; Yu, C.; Ren, Q.; Wen, Q.; Zhao, C.; Tang, Y.; Du, Y. Exposure to HBCD promotes adipogenesis both in vitro and in vivo by interfering with Wnt6 expression. *Sci. Total Environ.* **2020**, *705*, 135917. [CrossRef] [PubMed]
16. Nations, U. *Stockholm Convention on Persistent Organic Pollutants. Depositary Notifications: C.N.934.2013; TREATIES-XXVII.15 (Amendment to Annex A)*; UNEP: Stockholm, Sweden, 2013.

17. UNEP. *Guidance for the Inventory of Hexabromocyclododecane (HBCD). Chapter 8. Inventory of Contaminated Sites Containing HBCD*; UNEP: Stockholm, Sweden, 2017.
18. Posner, S.R.S.; Olsson, E. *Exploration of Management Options for HBCDD*; UNECE: Montreal, QC, Canada, 2010; pp. 10–11.
19. UNEP. *Risk Profile on Hexabromocyclododecane; UNEP/POPS/POPRC.6/13/Add.2*; UNEP: Geneva, Switzerland, 2010.
20. Rani, M.; Shim, W.J.; Han, G.M.; Jang, M.; Song, Y.K.; Hong, S.H. Hexabromocyclododecane in polystyrene based consumer products: An evidence of unregulated use. *Chemosphere* **2014**, *110*, 111–119. [CrossRef] [PubMed]
21. Muenhor, D.; Harrad, S.; Ali, N.; Covaci, A. Brominated flame retardants (BFRs) in air and dust from electronic waste storage facilities in Thailand. *Environ. Int.* **2010**, *36*, 690–698. [CrossRef]
22. UNEP. *General Technical Guidelines on the Environmentally Sound Manage Ment of Wastes Consisting of, Containing or Contaminated with Persistent Organic Pollutants*; UNEP: Stockholm, Sweden, 2019.
23. Schlummer, M.; Vogelsang, J.; Fiedler, D.; Gruber, L.; Wolz, G. Rapid identification of polystyrene foam wastes containing hexabromocyclododecane or its alternative polymeric brominated flame retardant by X-ray fluorescence spectroscopy. *Waste Manag. Res.* **2015**, *33*, 662–670. [CrossRef]
24. Smoluch, M.; Silberring, J.; Reszke, E.; Kuc, J.; Grochowalski, A. Determination of hexabromocyclododecane by flowing atmospheric pressure afterglow mass spectrometry. *Talanta* **2014**, *128*, 58–62. [CrossRef]
25. Hu, D.C.; Hu, X.Z.; Li, J.; Wang, P.; Guo, S.F. Determinations of hexabromocyclododecane (HBCD) isomers in plastic products from china by LC-MS/MS. *J. Liq. Chromatogr. Relat. Technol.* **2012**, *35*, 558–572. [CrossRef]
26. Baek, S.Y.; Kim, B.; Lee, S.; Lee, J.; Ahn, S. Accurate determination of hexabromocyclododecane diastereomers in extruded high-impact polystyrene: Development of an analytical method as a candidate reference method. *Chemosphere* **2018**, *210*, 296–303. [CrossRef] [PubMed]
27. Takigami, H.; Watanabe, M.; Kajiwara, N. Destruction behavior of hexabromocyclododecanes during incineration of solid waste containing expanded and extruded polystyrene insulation foams. *Chemosphere* **2014**, *116*, 24–33. [CrossRef]
28. Stubbings, W.A.; Harrad, S. Laboratory studies on leaching of HBCDD from building insulation foams. *Emerg. Contam.* **2019**, *5*, 36–44. [CrossRef]
29. Caban, M.; Migowska, N.; Stepnowski, P.; Kwiatkowski, M.; Kumirska, J. Matrix effects and recovery calculations in analyses of pharmaceuticals based on the determination of β-blockers and β-agonists in environmental samples. *J. Chromatogr. A* **2012**, *1258*, 117–127. [CrossRef]
30. AbouElwafa, A.M.; S., D.D.; Martin, S.; Harald, B.; Stuart, H. A rapid method for the determination of brominated flame retardant concentrations in plastics and textiles entering the waste stream. *J. Sep. Sci.* **2017**, *40*, 3873–3881.
31. Jeannerat, D.; Pupier, M.; Schweizer, S.; Mitrev, Y.N.; Favreau, P.; Kohler, M. Discrimination of hexabromocyclododecane from new polymeric brominated flame retardant in polystyrene foam by nuclear magnetic resonance. *Chemosphere* **2016**, *141*, 1391–1397. [CrossRef] [PubMed]
32. Covaci, A.; Voorspoels, S.; Ramos, L.; Neels, H.; Blust, R. Recent developments in the analysis of brominated flame retardants and brominated natural compounds. *J. Chromatogr. A* **2007**, *1153*, 145–171. [CrossRef] [PubMed]
33. Koppen, R.; Becker, R.; Jung, C.; Nehls, I. On the thermally induced isomerisation of hexabromocyclododecane stereoisomers. *Chemosphere* **2008**, *71*, 656–662. [CrossRef]
34. Morris, S.; Bersuder, P.; Allchin, C.R.; Zegers, B.; Boon, J.P.; Leonards, P.E.G.; de Boer, J. Determination of the brominated flame retardant, hexabromocyclodocane, in sediments and biota by liquid chromatography-electrospray ionisation mass spectrometry. *TrAC Trends Anal. Chem.* **2006**, *25*, 343–349. [CrossRef]
35. Chain. *The Technical Guidance for the Preparation and Revision of the Standard for Environ-Mental Monitoring and Analysis Methods*; MEPC: Beijing, China, 2010.
36. Cao, X.H.; Lu, Y.L.; Zhang, Y.Q.; Khan, K.; Wang, C.C.; Baninla, Y. An overview of hexabromocyclododecane (HBCDs) in environmental media with focus on their potential risk and management in China. *Environ. Pollut.* **2018**, *236*, 283–295. [CrossRef]
37. Wu, Y.; Xiao, Y.; Wang, G.; Shi, W.; Sun, L.; Chen, Y.; Li, D. Research progress on status of environmental pollutions of polybrominated diphenyl ethers, hexabromocyclodocane, and tetrabromobisphenol A: A review. *Huanjing Huaxue Environ. Chem.* **2021**, *40*, 384–403. [CrossRef]
38. Li, L.; Weber, R.; Liu, J.G.; Hu, J.X. Long-term emissions of hexabromocyclododecane as a chemical of concern in products in China. *Environ. Int.* **2016**, *91*, 291–300. [CrossRef] [PubMed]

Article

Tetrabromobisphenol A Disturbs Brain Development in Both Thyroid Hormone-Dependent and -Independent Manners in *Xenopus laevis*

Mengqi Dong [1,2], Yuanyuan Li [1,2], Min Zhu [1,2], Jinbo Li [1,2] and Zhanfen Qin [1,2,*]

1. State Key Laboratory of Environmental Chemistry and Ecotoxicology, Research Center for Eco-Environmental Sciences, Chinese Academy of Sciences, Beijing 100085, China; dongmengqi029@163.com (M.D.); yyli@rcees.ac.cn (Y.L.); minzhu21@163.com (M.Z.); chnlijinbo@163.com (J.L.)
2. University of Chinese Academy of Sciences, Beijing 100049, China
* Correspondence: qinzhanfen@rcees.ac.cn; Tel.: +86-10-62919177

Citation: Dong, M.; Li, Y.; Zhu, M.; Li, J.; Qin, Z. Tetrabromobisphenol A Disturbs Brain Development in Both Thyroid Hormone-Dependent and -Independent Manners in *Xenopus laevis*. *Molecules* 2022, 27, 249. https://doi.org/10.3390/molecules27010249

Academic Editor: Nikolaos S. Thomaidis

Received: 22 November 2021
Accepted: 30 December 2021
Published: 31 December 2021

Publisher's Note: MDPI stays neutral with regard to jurisdictional claims in published maps and institutional affiliations.

Copyright: © 2021 by the authors. Licensee MDPI, Basel, Switzerland. This article is an open access article distributed under the terms and conditions of the Creative Commons Attribution (CC BY) license (https://creativecommons.org/licenses/by/4.0/).

Abstract: Although tetrabromobisphenol A (TBBPA) has been well proven to disturb TH signaling in both in vitro and in vivo assays, it is still unclear whether TBBPA can affect brain development due to TH signaling disruption. Here, we employed the T3-induced *Xenopus* metamorphosis assay (TIXMA) and the spontaneous metamorphosis assay to address this issue. In the TIXMA, 5–500 nmol/L TBBPA affected T3-induced TH-response gene expression and T3-induced brain development (brain morphological changes, cell proliferation, and neurodifferentiation) at premetamorphic stages in a complicated biphasic concentration-response manner. Notably, 500 nmol/L TBBPA treatment alone exerted a stimulatory effect on tadpole growth and brain development at these stages, in parallel with a lack of TH signaling activation, suggesting the involvement of other signaling pathways. As expected, at the metamorphic climax, we observed inhibitory effects of 50–500 nmol/L TBBPA on metamorphic development and brain development, which was in agreement with the antagonistic effects of higher concentrations on T3-induced brain development at premetamorphic stages. Taken together, all results demonstrate that TBBPA can disturb TH signaling and subsequently interfere with TH-dependent brain development in *Xenopus*; meanwhile, other signaling pathways besides TH signaling could be involved in this process. Our study improves the understanding of the effects of TBBPA on vertebrate brain development.

Keywords: tetrabromobisphenol A; *Xenopus laevis*; brain development; thyroid hormone; biphasic concentration-response

1. Introduction

Tetrabromobisphenol A (TBBPA) is usually used as brominated flame retardants in various products, including printed circuit boards, adhesives, coatings, etc. [1]. The extensive production and use of TBBPA have led to wide existence in the environment, especially in China. It was reported that the TBBPA concentration in surface water from Lake Chaohu (China) reached 4.87 μg/L [2]. Monitoring studies show that TBBPA is frequently detected in human samples, such as urine, blood, breast milk, and adipose tissue, including samples from pregnant women [3–5]. Although the European Union (EU) approved TBBPA as no significant adverse effects for human health based on a risk assessment of TBBPA following a series of standard guideline-compliant toxicity tests [6,7], there is growing evidence that TBBPA could exert certain adverse effects, including carcinogenicity, in laboratory animals [8]. Especially, TBBPA has been shown to interfere with thyroid hormone receptor (TR), thereby altering TR-mediated gene transcription in vitro assays, indicating its TH signaling disrupting activity [9,10]. Importantly, several research groups, including ours, reported that TBBPA antagonized TH-induced metamorphic development in amphibians and even inhibited spontaneous metamorphosis at the metamorphic climax, along with changes in the expressional levels of TH-response genes, demonstrating the TBBPA's actions on TH signaling in vivo [11–13].

Given the key roles of TH in brain development, we suspect that TBBPA could disturb TH-dependent brain development in vertebrates. However, few studies have addressed this issue. In fact, whether or not TBBPA can exert adverse effects on the nervous system is still controversial. Based on a series of standard guideline-compliant toxicity tests, the EU concluded that TBBPA is not a neurotoxicant for both adult and developing mammals [6]. Kacew and Hayes [14] declared that TBBPA has little neurotoxicity to vertebrates, including humans, considering limited neurobehavioral effects and its limited presence in the brain. However, there is increasing evidence that TBBPA could exert adverse effects, such as increased anxiety and reduced social behaviors, in laboratory mammals and fish [15]. Nevertheless, no data has addressed whether or not the neurotoxic effects of TBBPA are associated with its TH signaling disruption. Indeed, it is challenging to employ rodents to reveal whether chemicals could exert neurotoxic effects through TH signaling due to the involvement of multiple signaling pathways. In contrast to rodents, amphibians undergo metamorphic development, including brain remodeling, mainly regulated by TR-mediated TH signaling that activates the expression of a series of TH-response genes. Notably, TH addition in water can induce precocious metamorphosis of pre-metamorphic tadpoles, characterized by upregulated expression of TH-response genes and subsequent cellular, histological and morphological changes of various organs [16,17]. Thus, the metamorphic development of amphibians, especially the model species *Xenopus laevis*, acts as an excellent model to dissect TH signaling and TH-dependent development [17,18]. Previously, we developed the T3-induced *Xenopus* metamorphosis assay (TIXMA) for evaluating TH signaling disruption and subsequent effects on TH-dependent development [19,20]. Specifically, TH-response gene expression is used as endpoints for TH signaling disruption, while morphological, histological, and cellular parameters are employed for evaluating the effects on TH-dependent development. Since the TIXMA involves the parameters of brain remodeling, it offers a great chance to evaluate the interference of chemicals on TH signaling and subsequent effects on brain development [21]. In our previous study investigating TH signaling disruption of TBBPA [13], we noticed that TBBPA appeared to have an effect on T3-induced brain morphological alternations in *Xenopus*.

The aim of this study was to determine whether low concentrations of TBBPA could interact with TH signaling in *Xenopus* brains and alter TH-dependent brain development. Following the TIXMA we established previously [13,22], pre-metamorphic tadpoles were treated with TBBPA in the absence or presence of T3. On day two after exposure, TH-response gene expression was measured for evaluating TH-signaling disruption in the brain, while on day four, brain morphological and cellular changes were examined for evaluating the effects on TH-dependent brain development. Additionally, the effects of TBBPA on spontaneous TH-dependent brain development were investigated to verify the results observed in the TIXMA. This study is expected to enhance our understanding of the neurodevelopmental effects of TBBPA in vertebrates.

2. Results

2.1. TBBPA Affects T3-Induced TH-Response Gene Expression in Xenopus laevis Brains

As expected, two days of T3 treatment dramatically upregulated the expression of all TH-response genes (*klf9*, *thrb*, *thibz*, *mmp13*, *tgm2*, and *st3*) in *Xenopus* brains ($p < 0.05$) (Figure 1). TBBPA exposure alone had no effects on the expression levels of these genes. In the presence of T3; however, 5 nmol/L and/or 50 nmol/L TBBPA resulted in higher expression of these TH-response genes, whereas the highest concentration caused lower expression of *thrb*, *thibz*, *klf9*, and *tgm2*, compared with T3 treatment. Altogether, these observations show that 5–500 nmol/L TBBPA affected T3-induced TH-response gene expression in a biphasic concentration-response manner.

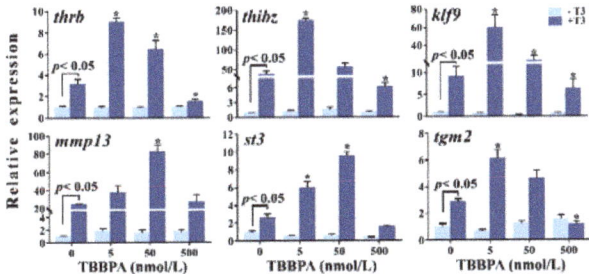

Figure 1. The relative expression of six thyroid hormone-response genes in brains of premetamorphic *Xenopus laevis* tadpoles following two-day exposure to tetrabromobisphenol A (TBBPA) in the absence or presence of 1 nmol/L T3. Data are shown as mean ± SEM. * indicates significant differences between TBBPA + T3 treatment and T3 treatment ($p < 0.05$).

2.2. TBBPA Alters Brain Morphology in the Absence and Presence of T3

After four days of treatment, the 500 nmol/L TBBPA-treated tadpoles had longer hindlimbs with more obvious toes and smaller head area than controls, showing that 500 nmol/L TBBPA promoted pre-metamorphic development (Figure 2A). As expected, T3 treatment induced precocious metamorphosis, characterized by body weight loss, hindlimb growth, head shrinkage (Figure 2B), and brain remodeling. Following co-treatment with T3 and 50–500 nmol/L TBBPA, morphological changes exhibited a concentration-dependent weakening trend compared with T3 treatment, indicating that 50–500 nmol/L TBBPA antagonized T3-induced metamorphosis in a concentration-dependent manner. A slight effect was found at the lowest concentration of TBBPA on T3-induced metamorphosis in *Xenopus* brain.

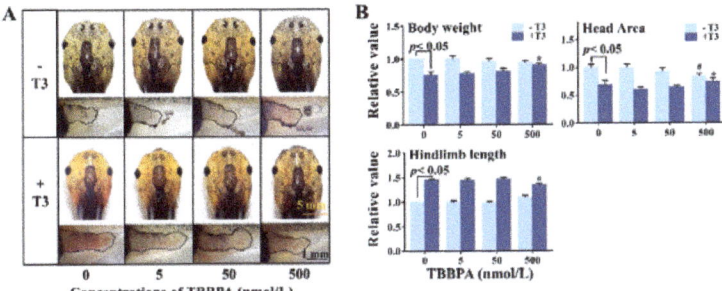

Figure 2. Morphological changes of premetamorphic *Xenopus laevis* following four-day exposure to tetrabromobisphenol A (TBBPA) in the absence or presence of 1 nmol/L T3. (**A**) Representative morphology of the head and the hindlimb of exposed tadpoles. (**B**) Quantitative analysis for body weight, hindlimb length, and head area of exposed tadpoles. Data are shown as mean ± SD. # indicates significant differences between TBBPA treatment and solvent control treatment ($p < 0.05$). * indicates significant differences between TBBPA + T3 treatment and T3 treatment ($p < 0.05$).

In terms of brain morphological changes, mainly longitudinal shortening as well as dorsoventral and lateral broadening, T3 treatment induced dramatic brain remodeling, particularly in the diencephalon. We characterized brain remodeling with ULBW/BL, DL/BL, and DT/BL (Figure 3A). Compared with controls, T3-treatment resulted in increases in ULBW/BL and DT/BL but decreased in DL/BL (Figure 3C). Interestingly, 500 nmol/L TBBPA treatment alone also decreased DL/BL. In 50–500 nmol/L TBPPA and T3 co-treatment groups, tadpoles exhibited lower ULBW/BL and DT/BL values and higher DL/BL values in comparison with T3-treated animals. As a whole, 50–500 nmol/L TBBPA appeared to concentration-dependently antagonize T3-induced increases in ULBW/BL and

DT/BL but decrease in DL/BL. In contrast, 5 nmol/L TBBPA combined with T3 resulted in a significant decrease in DL/BL than T3 treatment, displaying an agonistic effect on T3 action. Collectively, 5–500 nmol/L affected T3-induced brain development in a biphasic concentration-response manner.

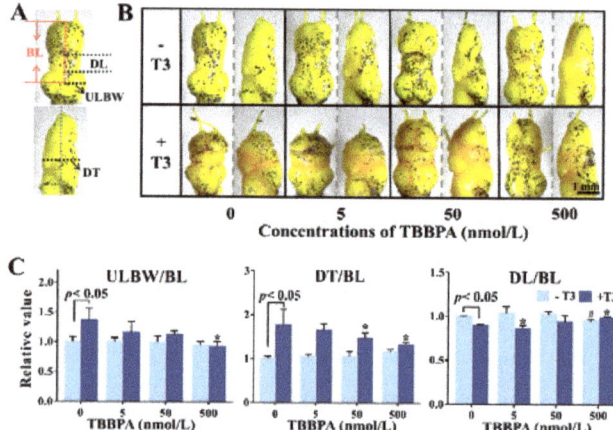

Figure 3. Brain morphological changes of premetamorphic *Xenopus laevis* following four-day exposure to tetrabromobisphenol A (TBBPA) in the absence or presence of 1 nmol/L T3. (**A**) Schematic description of measurement parameters. DL for diencephalon length, ULBW for unilateral brain width, BL for brain length, and DT for diencephalon thickness. (**B**) Representative morphology of the elevation view and side view of the brain. (**C**) Quantitative analysis result of brain parameters. Each parameter was normalized by the mean value of the solvent control tadpoles. Data are shown as mean ± SD. # indicates significant differences between TBBPA treatment and solvent control treatment ($p < 0.05$). * indicates significant differences between TBBPA + T3 treatment and T3 treatment ($p < 0.05$).

2.3. TBBPA Affects Cell Proliferation in the Telencephalon in the Presence of T3

EdU-labeled cells were observed in the subventricular zone (SVZ) in the telencephalon, showing active cell proliferation in this zone (Figure 4A). Some of the EdU-labeled cells were also present in the surrounding of the SVZ, meaning their migration from the SVZ because cell proliferation is known to only occur in the SVZ [23]. Corresponding to brain morphological changes, T3 treatment for four days caused more EdU-labeled cells in the SVZ compared with controls, and some EdU-labeled cells migrated from the SVZ, along with shrunk ventricles, indicating that T3 stimulated cell proliferation and subsequent migration from the zone. Increased cell proliferation was also observed in 500 nmol/L TBBPA treatment group (Figure 4B). When combined with T3, the lowest concentration of TBBPA resulted in more EdU-labeled cells, some of which migrated far away from the SVZ in comparison with T3 treatment. In contrast, decreased EdU-labeled cells and ventricle size in telencephalons were observed in the co-treatment with 500 nmol/L TBBPA and T3 compared to T3 treatment, and even appeared to be comparable with those of controls, showing 500 nmol/L TBBPA antagonized and even counteracted T3 actions. The median concentration (50 nmol/L) appeared to have no pronounced effect on T3-induced cell proliferation. As a whole, TBBPA treatment affected T3-induced cell proliferation in *Xenopus* brains in a biphasic concentration-response manner. In the 500 nmol/L TBBPA treatment alone, there were more EdU-marked cells than the control, showing that 500 nmol/L TBBPA increased cell proliferation, despite being less relative to T3 treatment.

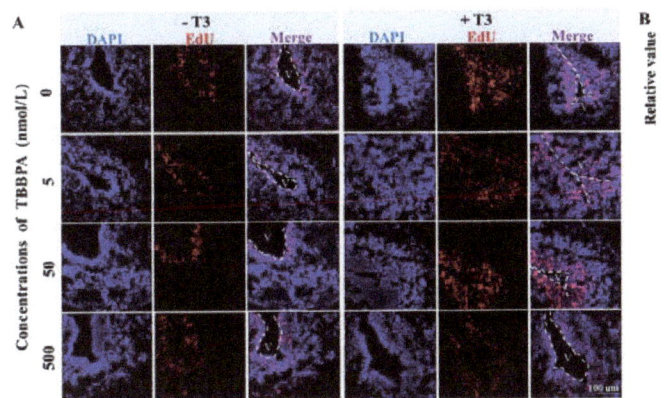

Figure 4. EdU-labeled cell proliferation in telencephalons of pre-metamorphic *Xenopus laevis* following 4-day exposure to tetrabromobisphenol A (TBBPA) in the absence or presence of 1 nmol/L T3. (**A**) Immunofluorescence images of telencephalons at the end of the T3-induced metamorphosis assay. EdU label (red) and DAPI (blue). (**B**) The ratio of EdU positive cells number. Three animals were analyzed for each treatment. V: ventricle. Parameter was normalized by the mean value of the solvent control. Data are shown as mean ± SD. # indicates significant differences between TBBPA treatment and solvent control treatment ($p < 0.05$). * indicates significant differences between TBBPA + T3 treatment and T3 treatment ($p < 0.05$).

2.4. TBBPA Affects Neurodifferentiation in the Telencephalon in the Presence of T3

By immunofluorescence (IF) staining, we detected neuronal marker TUBB2 in the telencephalon to characterize neuronal differentiation in *Xenopus* brain (Figure 5A). Compared with controls, T3-treated brains exhibited significantly higher TUBB2 expression (Figure 5B), with thicker outer neuropil of the telencephalon (Figure 5C), indicating that T3 promoted neurodifferentiation in the brain. Similarly, 50 nmol/L and 500 nmol/L TBBPA-treated brains also exhibit stronger TUBB2 staining and thicker outer neuropil, despite the lack of a statistical difference between the 50 nmol/L TBBPA group and the control, suggesting a stimulatory effect of higher concentrations of TBBPA on neurodifferentiation. In the presence of T3, 5 nmol/L TBBPA led to stronger TUBB2 staining in comparison with T3, along with increased thicknesses of the outer neuropil (Figure 5B,C). However, 500 nmol/L TBBPA antagonized T3 action on neurodifferentiation, which was characterized by thinner outer neuropil and a decreasing trend of TUBB2 expression—combined with T3, while 50 nmol/L TBBPA had no significant effects on T3-induced TUBB2 expression and thickening of the outer neuropil. Overall, 5–500 nmol/L TBBPA affected T3-induced neurodifferentiation in *Xenopus* brains in a biphasic concentration-response manner.

2.5. TBBPA Disturbs Brain Development in Spontaneous Metamorphosis Assay

Following the TIXMA, the effects of TBBPA on spontaneous metamorphosis and brain development were investigated. After six-day treatment since stage 58, all the control tadpoles reached stage 63, whereas some of the TBBPA-treated tadpoles remained at earlier developmental stages. Statistical analysis revealed significant inhibitory effects of 50–500 nmol/L TBBPA on spontaneous metamorphosis (Figure 6A). At stage 63, as SOX2 labeled the cells in the SVZ of the telencephalon, the EdU-labeled proliferating cells in the zone and its surrounding appeared to be concentration-dependently less in TBBPA-treated animals compared with the controls (Figure 6B), with a significant difference between 500 nmol/L TBBPA group and the control group (Figure 6C). Moreover, there were fewer EdU-labeled proliferating cells far away from the SVZ in the TBBPA-treated brains than controls. These observations indicate that TBBPA inhibited brain development at the metamorphic climax.

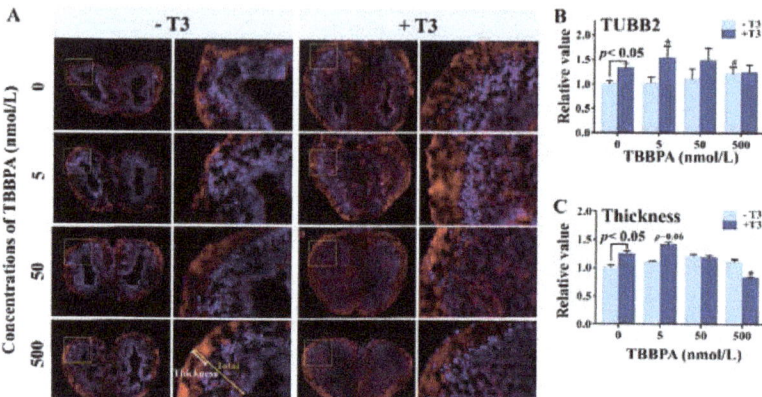

Figure 5. The effect of TBBPA on neurodifferentiation in the brain of NF stage 52 *Xenopus laevis* tadpoles exposed to a series of concentrations of TBBPA in the absence or presence of 1 nmol/L T3. (**A**) the immunofluorescence images of telencephalons retained TUBB2 (red) and DAPI (blue). (**B**) the relative fluorescence intensity of TUBB2, normalized by the control group. (**C**) The relative thickness of the outer neuropil. The data were normalized by the mean value of the solvent control. Data are shown as mean ± SD. # indicates significant differences between TBBPA treatment and solvent control treatment ($p < 0.05$). * indicates significant differences between TBBPA + T3 treatment and T3 treatment ($p < 0.05$). The experiment was repeated three times using tadpoles from different sets of adults. The results were consistent among the three independent experiments.

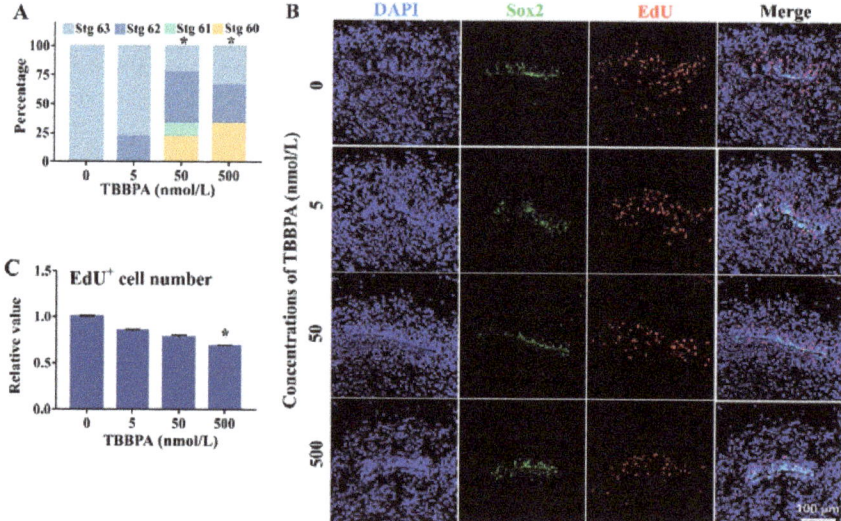

Figure 6. Effects of tetrabromobisphenol A (TBBPA) on spontaneous metamorphosis of *Xenopus laevis* at the metamorphic climax. (**A**) Percentages of tadpoles at different stages after a six-day exposure. * indicates a significant difference between TBBPA treatment and the solvent control group ($p < 0.05$). (**B**) Immunofluorescence images for telencephalons of NF 63 tadpoles. EdU labeled proliferating cells and migrating cells. Sox2 outlined the ventricle zone. (**C**) The ratio of EdU positive cells number. The data were normalized by the mean value of the solvent control tadpoles. * indicates a significant difference between TBBPA treatment and the solvent control group ($p < 0.05$).

3. Discussion

In this study, we employed *X. laevis*, a TH-dependent developmental model, to address the influences of TBBPA on brain development by interfering with TH signaling. In the TIXMA, two days of TBBPA treatment alone did not alter TH-response gene (*thrb*, *thibz*, *klf9*, *mmp13*, *st3*, and *tgm2*) expression in brains of NF stage 52 *Xenopus*, implying the lack of TH signaling disruption at this stage, which differed from potential TH signaling agonism observed in *Xenopus* intestines and hindlimbs [13]. In fact, previous studies generally reported TBBPA as a TH signaling antagonist but not as a TH signaling agonist [24]. However, here, we found that 500 nmol/L TBBPA treatment for four days decreased the DL/BL values and increased cell proliferation and neurodifferentiation in the telencephalon, indicating a stimulatory effect on brain development at pre-metamorphic stages. Given the lack of TH signaling disruption [25,26], it is concluded that the stimulatory effect of TBBPA on brain development at pre-metamorphic stages is not associated with TH signaling. Maybe, there are other signaling pathways, such as Notch/Wnt signaling, that are involved in brain development [27], contributing to the stimulatory effect of TBBPA on tadpole brain development at pre-metamorphic stages. Indeed, TBBPA was reported to affect neural differentiation of embryonic stem cells and neural stem cells by partially disturbing Notch and Wnt signaling [28,29].

In this study, importantly, we found that in the presence of T3, 500 nmol/L TBBPA antagonized T3-induced TH-response gene expression, but 5 nmol/L TBBPA exerted stimulatory effects on T3 actions, showing an apparently biphasic concentration-response manner. These results agree with our previous observations in *Xenopus* intestines treated with TBBPA [13]. Recently, we found that bisphenol F and bisphenol A also affected T3-induced TH-response gene expression in *Xenopus* brains at pre-metamorphic stages in a biphasic concentration-response manner [21]. Given the strong evidence that TBBPA can act as a TH signaling antagonist in vitro [24], the antagonistic effect of 500 nmol/L TBBPA on T3-induced TH-response gene expression implies its TH signaling antagonism in *Xenopus* brains. As for the stimulatory effects of 5–50 nmol/L TBBPA on T3-induced TH-response gene expression, we speculate that some specific TH-dependent signaling pathways could have participated in the regulation of the TH gene expression response by TBBPA, which needs further studies. Overall, the effects of TBBPA on T3-induced TH-response gene expression in *Xenopus* brains at pre-metamorphic stages strongly indicate that TBBPA could interfere with TH-dependent signaling pathways.

Furthermore, in the TIXMA, TBBPA affected T3-induced brain morphological changes, cell proliferation, and neurodifferentiation. These findings reveal that TBBPA disrupted TH-dependent brain development. Notably, the effects of TBBPA on TH-dependent brain development were in a biphasic concentration-response manner, i.e., 5 nmol/L TBBPA exerted agonistic effects, whereas 500 nmol/L exerted antagonistic effects, which correspond to TH-response gene expression in the biphasic concentration-response manner. However, 50 nmol/L TBBPA antagonized T3-induced brain development, but it simulated T3-induced TH-response gene expression. Overall, the biphasic effects of TBBPA at the highest and lowest concentration on T3 actions in TH-response gene expression were in agreement with their effects on phenotype. However, we cannot give a reasonable explication for the inconsistent effects of the median concentration on TH-response gene expression and brain development, possibly suggesting complex mechanisms involving other signaling pathway(s) besides TH signaling.

T3-induced metamorphosis of *Xenopus* at pre-metamorphic stages can simulate TH-regulated development at the metamorphic climax, at which tadpoles contain high endogenous TH levels. Based on the findings that TBBPA affected T3-induced brain development, the effects of TBBPA on TH-dependent brain development at the metamorphic climax were further investigated. As expected, 50–500 nmol/L TBBPA significantly inhibited metamorphic development and brain development, which agrees with the antagonistic effects of higher concentrations on T3-induced brain development. At the metamorphic climax, however, the lowest concentration of TBBPA did not promote brain development but exhibited an inhibitory trend in terms of cell proliferation and migration in brains, which is inconsistent with the results in the TIXMA. Nevertheless, all findings strongly demonstrate that TBBPA has significant effects on TH-dependent and TH-independent brain development. Given the evolutionary conservation of TH signaling and its roles in brain development [30,31], we infer that TBBPA could exert similar effects on brain development in other vertebrates, including mammals. However, the extrapolation of our findings from *Xenopus* to mammal species warrants cautions due to the difference between waterborne exposure and general oral administration [32–34]. Nevertheless, our study supports previous data that TBBPA exerted significant effects on the nervous system in the literature [35].

4. Materials and Methods

4.1. Chemicals

3,3′,5-triiodo-L-thyronine (T3) from Geel Belgium (New Jersey, USA) was dissolved in ultrapure water (with NaOH) to prepare a 10 mmol/L stock solution. TBBPA from Tokyo Chemical Industry Co. Ltd. (Tokyo, Japan) was dissolved in DMSO to prepare 0.5 mol/L TBBPA stock solution. EdU (5-ethynyl-2′-deoxyuridine) from RiboBio Co. (Guangzhou, China) were dissolved in ultrapure water to prepare 2 mg/mL stock solution. The β2-tubulin (7B9) antibody (SC-47751) and anti-SOX2 antibody (ab97959) were from Santa Cruz Biotechnology (California, USA) and Abcam (Cambridge, UK), respectively. Cy3^{TM} goat anti-mouse IgG (H+L) (A10521) secondary antibody and Cy3^{TM} goat anti-rabbit IgG (H+L) (A10520) were from Invitrogen (Carlsbad, California USA). RNA extraction kit was from Bio Teke Corporation (Beijing, China). Quantscript RT Kit and RealMasterMix (SYBR Green) Kit were from Tiangen (Beijing, China). PCR primers were synthesized by the BGI group (Beijing, China).

4.2. Animal Housing

Xenopus frogs were initially obtained from Nasco (Fort Atkins, WI, USA). Frogs and tadpoles were raised as described previously [36]. Fertilized eggs were obtained by injecting HCG into a pair of adult frogs and incubating them in dechlorinated tap water at 21–22 °C. Tadpoles were raised in a flow-through system (Esen, Beijing, China) at NF stage 46, staged according to the Nieuwkoop and Faber system [37].

4.3. T3-Induced Xenopus Metamorphosis Assay (TIXMA)

At the beginning of exposure, NF stage 52 tadpoles received an intracerebral injection of 0.5 µL of EdU (2 mg/mL) to label proliferating cells. Then tadpoles were randomly transformed into 4 L glass tanks for 5, 50, 500 nmol/L TBBPA exposure alone or combined exposure with 1 nmol/L T3, with 0.001% DMSO as the control. Three replicate tanks with six tadpoles per tank were employed for each treatment. Experiment water was renewed daily.

Following two days of exposure, tadpoles were anesthetized in MS-222 (100 mg/L), and each brain was collected for RT-qPCR analysis. After four days, tadpoles were weighed and photographed for gross morphologic analysis. Then each brain was sampled to fix in MEMFA solution (0.1 mol/L MOPS, 2 mmol/L ethyleneglycoltetraacetic acid, 1 mmol/L MgSO$_4$, 3.7% formaldehyde, pH 7.4) for immunofluorescence (IF) staining and EdU labeling. TIXMA was repeated three times using the offspring from three different sets of adults.

4.4. Spontaneous Metamorphosis Assay

To verify the simulation of T3-induced metamorphosis, the spontaneous metamorphosis assay was conducted. NF stage 58 tadpoles at the early metamorphic climax received one intracerebral injection of 1.5 µL EdU (2 mg/mL) before exposure. Then tadpoles were treated with TBBPA (5, 50, and 500 nmol/L) in 10 L glass tanks as described above. Three replicate tanks and 12 tadpoles were employed for each treatment group. Experiment water was renewed every other daily. After six days, the developmental stage of each tadpole was recorded and statistically analyzed by SPSS, and brains were collected for IF and EdU staining.

4.5. RNA Extraction and RT-qPCR Analysis

The expression of six typical TH-response genes (*thibz*, *thrb*, *klf9*, *st3*, *mmp13*, and *tgm2*) was measured by RT-PCR. The *rpl8* was used as a housekeeping gene [19,38], the expression of *rpl8* was not affected by all treatments. Total RNA extraction, RT-qPCR analysis, and primers information were described in the Supplementary Materials [39–41].

4.6. Analysis for Gross Morphology

The morphological parameters for tadpoles were measured by the Image J software (1.52 a), including hindlimb length (HLL), head area, unilateral brain width (ULBW), and brain length (BL) [19], diencephalon length (DL), and diencephalon thickness (DT). Each parameter was normalized by the mean value of the control.

4.7. IF Staining

The neuronal marker β2-tubulin (TUBB2) was chosen to characterize neurodifferentiation by IF staining, with the stem cell marker SOX2 for labeling the ventricle area. The detailed procedures were described in Supporting Information. Confocal images were obtained on the Leica TCS SP5 and analyzed with Leica LAS AF Lite (Leica Microsystems CMS GmbH). The assay ensured no signal in the negative control. Three telencephalon sections per brain were analyzed by Image J; the neuronal differentiation was quantified by the intensity ratio of the TUBB2 to DAPI. The intensity of each picture was measured by spitting the channel, adjusting the threshold, and measuring to record the Mean value in Image J. The specific value of each exposure group was normalized by the control.

The relative thickness of the outer neuropil was defined by the ratio of the neuropil thickness to the length from the SVZ to the brain edge along the neurite direction. The length was measured using Image J. Three images were analyzed for each brain, with at least three brains for every treatment. The relative value was normalized by the control group.

4.8. EdU Proliferation Assay

EdU was injected as described in TIXMA and spontaneous metamorphosis assay. The brains were fixed, dehydrated, and transversely sectioned at 10 μm, with the same procedure for IF staining. Three animals from three replicate tanks were analyzed for each treatment. For EdU labeling, the Cell-Light™ Apollo 567 Stain Kit (Guangzhou RiboBio Co. Ltd. Guangzhou, China) was used according to the manufacturer's technical information. The assay ensured no signal in the negative control. Confocal images were obtained as described previously. The number of EdU positive cells and the total number of cells (marked by DAPI) were counted manually and divided to get the cell proliferation ratio. Three images were counted in every treatment group, and the relative value was normalized by the control group.

4.9. Data Analysis

RT-qPCR data were present as means ± standard error of the mean (SEM), other quantitative data were shown as means ± standard deviation (SD). SPSS software v 18.0 (USA) was used for statistical analysis with the tank as the statistical unit. Two-way analysis of variance (ANOVA) followed by Dunnett's test was employed for all data analysis, except the Chi-square test for the development stage distribution of spontaneous metamorphosis assay among TBBPA treatments.

5. Conclusions

Our study demonstrates that TBBPA could interfere with TH-signaling and subsequently disrupt TH-dependent brain development in *Xenopus*. In addition, TBBPA could also disrupt brain development, possibly via other signaling pathways besides TH signaling at specific developmental stages. Our study enhances the understanding of the neurodevelopmental effects of TBBPA on brain development in vertebrates.

Supplementary Materials: The methods of total RNA extraction and immunofluorescence staining were described in the Supplementary Materials. Details of primers sequences of all tested genes used for qPCR (Table S1). Table S1: Primers sequences of all tested genes used for qPCR and related information.

Author Contributions: Conceptualization, Z.Q. and M.D.; methodology, M.D.; validation, M.D. and Y.L.; resources, J.L.; data curation, M.D.; writing—original draft preparation, M.D.; writing—review and editing, M.D., Y.L., M.Z. and Z.Q.; visualization, M.D.; supervision, Z.Q.; project administration, Z.Q.; funding acquisition, Z.Q. All authors have read and agreed to the published version of the manuscript.

Funding: This work was supported by the National Key Research and Development Program of China (2018YFA0901103), the National Natural Science Foundation of China (21876196), and the Strategic Priority Research Program of the Chinese Academy of Sciences (XDB14040102).

Institutional Review Board Statement: The Animal Ethics Committee of the Research Center for Eco-Environmental Sciences, Chinese Academy of Sciences, approved all animal protocols and procedures (AEWC-RCEES-2019009).

Informed Consent Statement: Not applicable.

Data Availability Statement: Not applicable.

Acknowledgments: All funds are gratefully appreciated, and we are grateful to Gui Chengmei and Gui Chengli for their help in animal feeding.

Conflicts of Interest: The authors declare no conflict of interest.

References

1. Malkoske, T.; Tang, Y.L.; Xu, W.Y.; Yu, S.L.; Wang, H.T. A review of the environmental distribution, fate, and control of tetrabromobisphenol A released from sources. *Sci. Total Environ.* **2016**, *569–570*, 1608–1617. [CrossRef] [PubMed]
2. Yang, S.W.; Wang, S.R.; Liu, H.L.; Yan, Z.G. Tetrabromobisphenol A: Tissue distribution in fish, and seasonal variation in water and sediment of Lake Chaohu, China. *Environ. Sci. Pollut. Res. Int.* **2012**, *19*, 4090–4096. [CrossRef] [PubMed]
3. Cariou, R.; Antignac, J.P.; Zalko, D.; Berrebi, A.; Cravedi, J.P.; Maume, D.; Marchand, P.; Monteau, F.; Riu, A.; Andre, F.; et al. Exposure assessment of French women and their newborns to tetrabromobisphenol-A: Occurrence measurements in maternal adipose tissue, serum, breast milk and cord serum. *Chemosphere* **2008**, *73*, 1036–1041. [CrossRef] [PubMed]
4. Li, A.J.; Zhuang, T.F.; Shi, W.; Liang, Y.; Liao, C.Y.; Song, M.Y.; Jiang, G.B. Serum concentration of bisphenol analogues in pregnant women in China. *Sci. Total Environ.* **2020**, *707*, 136100. [CrossRef] [PubMed]
5. Shi, Z.X.; Jiao, Y.; Hu, Y.; Sun, Z.W.; Zhou, X.Q.; Feng, J.F.; Li, J.G.; Wu, Y.N. Levels of tetrabromobisphenol A, hexabromocyclododecanes and polybrominated diphenyl ethers in human milk from the general population in Beijing, China. *Sci. Total Environ.* **2013**, *452–453*, 10–18. [CrossRef] [PubMed]
6. European Chemicals Bureau. *European Union Risk Assessment Report-2,2',6,6'-Tetrabromo-4,4'-Isopropylidenediphenol (Tetrabromo bisphenol-A or TBBP-A) (CAS: 79-94-7) Part-Human Health, II-Human Health*; European Chemicals Bureau: Luxembourg, 2006.
7. EFSA. Scientific Opinion on Tetrabromobisphenol A (TBBPA) and its derivatives in food. *EFSA J.* **2011**, *9*.
8. Yu, Y.J.; Yu, Z.L.; Chen, H.B.; Han, Y.J.; Xiang, M.D.; Chen, X.C.; Ma, R.X.; Wang, Z.D. Tetrabromobisphenol A: Disposition, kinetics and toxicity in animals and humans. *Environ. Pollut.* **2019**, *253*, 909–917. [CrossRef] [PubMed]
9. Lu, L.; Zhan, T.; Ma, M.; Xu, C.; Wang, J.; Zhang, C.; Liu, W.; Zhuang, S. Correction to "Thyroid Disruption by Bisphenol S Analogues via Thyroid Hormone Receptor β: In Vitro, In Vivo, and Molecular Dynamics Simulation Study". *Environ. Sci. Technol.* **2018**, *52*, 6617–6625. [CrossRef]
10. Mengeling, B.J.; Wei, Y.; Dobrawa, L.N.; Streekstra, M.; Louisse, J.; Singh, V.; Singh, L.; Lein, P.J.; Wulff, H.; Murk, A.J.; et al. A multi-tiered, in vivo, quantitative assay suite for environmental disruptors of thyroid hormone signaling. *Aquat. Toxicol.* **2017**, *190*, 1–10. [CrossRef]
11. Fini, J.B.; Le Mével, S.; Palmier, K.; Darras, V.M.; Punzon, I.; Richardson, S.J.; Clerget-Froidevaux, M.S.; Demeneix, B.A. Thyroid Hormone Signaling in the *Xenopus laevis* Embryo Is Functional and Susceptible to Endocrine Disruption. *Endocrinol.* **2012**, *153*, 5068–5081. [CrossRef]
12. Kitamura, S.; Kato, T.; Iida, M.; Jinno, N.; Suzuki, T.; Ohta, S.; Fujimoto, N.; Hanada, H.; Kashiwagi, K.; Kashiwagi, A. Anti-thyroid hormonal activity of tetrabromobisphenol A, a flame retardant, and related compounds: Affinity to the mammalian thyroid hormone receptor, and effect on tadpole metamorphosis. *Life Sci.* **2005**, *76*, 1589–1601. [CrossRef]

13. Zhang, Y.F.; Xu, W.; Lou, Q.Q.; Li, Y.Y.; Zhao, Y.X.; Wei, W.J.; Qin, Z.F.; Wang, H.L.; Li, J.Z. Tetrabromobisphenol A disrupts vertebrate development via thyroid hormone signaling pathway in a developmental stage-dependent manner. *Environ. Sci. Technol.* **2014**, *48*, 8227–8234. [CrossRef]
14. Kacew, S.; Hayes, A.W. Absence of neurotoxicity and lack of neurobehavioral consequences due to exposure to tetrabromobisphenol A (TBBPA) exposure in humans, animals and zebrafish. *Arch. Toxicol.* **2020**, *94*, 59–66. [CrossRef]
15. Chen, J.F.; Tanguay, R.L.; Simonich, M.; Nie, S.F.; Zhao, Y.X.; Li, L.L.; Bai, C.L.; Dong, Q.X.; Huang, C.J.; Lin, K.F. TBBPA chronic exposure produces sex-specific neurobehavioral and social interaction changes in adult zebrafish. *Neurotoxicol. Teratol.* **2016**, *56*, 9–15. [CrossRef]
16. Crump, D.; Werry, K.; Veldhoen, N.; Van Aggelen, G.; Helbing, C.C. Exposure to the herbicide acetochlor alters thyroid hormone-dependent gene expression and metamorphosis in *Xenopus laevis*. *Environ. Health Perspect.* **2002**, *110*, 1199–1205. [CrossRef]
17. Morvan-Dubois, G.; Demeneix, B.A.; Sachs, L.M. *Xenopus laevis* as a model for studying thyroid hormone signalling: From development to metamorphosis. *Mol. Cell Endocrinol.* **2008**, *293*, 71–79. [CrossRef]
18. Tata, J.R. Amphibian metamorphosis as a model for the developmental actions of thyroid hormone. *Mol. Cell. Endocrinol* **2006**, *246*, 10–20. [CrossRef]
19. Yao, X.F.; Chen, X.Y.; Zhang, Y.F.; Li, Y.Y.; Wang, Y.; Zheng, Z.M.; Qin, Z.F.; Zhang, Q.D. Optimization of the T3-induced *Xenopus* metamorphosis assay for detecting thyroid hormone signaling disruption of chemicals. *J. Environ. Sci.* **2017**, *52*, 314–324. [CrossRef]
20. Zhu, M.; Chen, X.Y.; Li, Y.Y.; Yin, N.Y.; Faiola, F.; Qin, Z.F.; Wei, W.J. Bisphenol F Disrupts Thyroid Hormone Signaling and Postembryonic Development in *Xenopus laevis*. *Environ. Sci. Technol.* **2018**, *52*, 1602–1611. [CrossRef] [PubMed]
21. Niu, Y.; Zhu, M.; Dong, M.Q.; Li, J.B.; Li, Y.Y.; Xiong, Y.M.; Liu, P.Y.; Qin, Z.F. Bisphenols disrupt thyroid hormone (TH) signaling in the brain and affect TH-dependent brain development in *Xenopus laevis*. *Aquat. Toxicol.* **2021**, *237*, 105902. [CrossRef] [PubMed]
22. Wang, Y.; Li, Y.Y.; Qin, Z.F.; Wei, W.J. Re-evaluation of thyroid hormone signaling antagonism of tetrabromobisphenol A for validating the T3-induced *Xenopus* metamorphosis assay. *J. Environ. Sci.* **2017**, *52*, 325–332. [CrossRef] [PubMed]
23. D'Amico, L.A.; Boujard, D.; Coumailleau, P. Proliferation, migration and differentiation in juvenile and adult *Xenopus laevis* brains. *Brain Res.* **2011**, *1405*, 31–48. [CrossRef]
24. Sun, H.; Shen, O.X.; Wang, X.R.; Zhou, L.; Zhen, S.Q.; Chen, X.D. Anti-thyroid hormone activity of bisphenol A, tetrabromobisphenol A and tetrachlorobisphenol A in an improved reporter gene assay. *Toxicol. Vitro* **2009**, *23*, 950–954. [CrossRef]
25. Wong, J.; Shi, Y.-B. Coordinated regulation of and transcriptional activation by *Xenopus* thyroid hormone and retinoid X receptors. *J. Biol. Chem.* **1995**, *270*, 18479–18483. [CrossRef]
26. Shi, Y.-B.; Wong, J.; Puzianowska-Kuznicka, M.; Stolow, M.A. Tadpole competence and tissue-specific temporal regulation of amphibian metamorphosis: Roles of thyroid hormone and its receptors. *Bioessays* **1996**, *18*, 391–399. [CrossRef] [PubMed]
27. Zhang, R.; Engler, A.; Taylor, V. Notch: An interactive player in neurogenesis and disease. *Cell Tissue Res.* **2018**, *371*, 73–89. [CrossRef] [PubMed]
28. Yin, N.Y.; Liang, S.J.; Liang, S.X.; Yang, R.J.; Hu, B.W.; Qin, Z.F.; Liu, A.F.; Faiola, F. TBBPA and Its Alternatives Disturb the Early Stages of Neural Development by Interfering with the NOTCH and WNT Pathways. *Environ. Sci. Technol.* **2018**, *52*, 5459–5468. [CrossRef] [PubMed]
29. Liang, S.J.; Liang, S.X.; Zhou, H.; Yin, N.Y.; Faiola, F. Typical halogenated flame retardants affect human neural stem cell gene expression during proliferation and differentiation via glycogen synthase kinase 3 beta and T3 signaling. *Ecotoxicol. Environ. Saf.* **2019**, *183*, 109498. [CrossRef]
30. Horn, S.; Heuer, H. Thyroid hormone action during brain development: More questions than answers. *Mol. Cell Endocrinol.* **2010**, *315*, 19–26. [CrossRef] [PubMed]
31. Rovet, J.F. The role of thyroid hormones for brain development and cognitive function. *Endocr. Dev.* **2014**, *26*, 26–43. [CrossRef] [PubMed]
32. Fini, J.B.; Anne, R.; Laurent, D.; Anne, H.; Le, M.S.; Sylvie, C. Parallel biotransformation of tetrabromobisphenol A in *Xenopus laevis* and mammals: *Xenopus* as a model for endocrine perturbation studies. *Toxicol. Sci.* **2012**, *125*, 359–367. [CrossRef]
33. Liu, H.L.; Ma, Z.Y.; Tao, Z.; Yu, N.Y.; Su, G.Y.; Giesy, J.P.L.; Yu, H.X. Pharmacokinetics and effects of tetrabromobisphenol a (TBBPA) to early life stages of zebrafish (*Danio rerio*). *Chemosphere* **2018**, *190*, 243–252. [CrossRef]
34. Hakk, H.; Letcher, R.J. Metabolism in the toxicokinetics and fate of brominated flame retardants—A review. *Environ. Int.* **2003**, *29*, 801–828. [CrossRef]
35. Dong, M.Q.; Li, Y.Y.; Zhu, M.; Qin, Z.F. Tetrabromobisphenol A: A neurotoxicant or not? *Environ. Sci. Pollut. Res. Int.* **2021**, *28*, 54466–54476. [CrossRef] [PubMed]
36. Cai, M.; Li, Y.Y.; Zhu, M.; Li, J.B.; Qin, Z.F. Evaluation of the effects of low concentrations of bisphenol AF on gonadal development using the *Xenopus laevis* model: A finding of testicular differentiation inhibition coupled with feminization. *Environ. Pollut.* **2020**, *260*, 113980. [CrossRef] [PubMed]
37. Nieuwkoop, P.; Faber, J. *Normal Table of Xenopus Laevis (Daudin)*; Garland Publishing: New York, NY, USA, 1994; p. 252.
38. Shi, Y.B.; Liang, C.T. Cloning and characterization of the ribosomal protein L8 gene from *Xenopus laevis*. *Biochim. Biophys. Acta* **1994**, *1217*, 227–228. [CrossRef]

39. Gunderson, M.P.; Veldhoen, N.; Skirrow, R.C.; Macnab, M.K.; Ding, W.; van Aggelen, G.; Helbing, C.C. Effect of low dose exposure to the herbicide atrazine and its metabolite on cytochrome P450 aromatase and steroidogenic factor-1 mRNA levels in the brain of premetamorphic bullfrog tadpoles (Rana catesbeiana). *Aquat. Toxicol.* **2011**, *102*, 31–38. [CrossRef]
40. Hogan, N.S.; Crump, K.L.; Duarte, P.; Lean, D.R.; Trudeau, V.L. Hormone cross-regulation in the tadpole brain: Developmental expression profiles and effect of T3 exposure on thyroid hormone- and estrogen-responsive genes in Rana pipiens. *Gen. Comp. Endocrinol.* **2007**, *154*, 5–15. [CrossRef]
41. Livak, K.J.; Schmittgen, T.D. Analysis of relative gene expression data using real-time quantitative PCR and the 2(-Delta Delta C(T)) Method. *Methods* **2001**, *25*, 402–408. [CrossRef]

Article

Effects of Turning Frequency on Ammonia Emission during the Composting of Chicken Manure and Soybean Straw

Qianqian Ma [1,2], Yanli Li [1,2], Jianming Xue [3,4], Dengmiao Cheng [5] and Zhaojun Li [1,2,*]

1. Key Laboratory of Plant Nutrition and Fertilizer, Ministry of Agriculture and Rural Affairs, Institute of Agricultural Resources and Regional Planning, Chinese Academy of Agricultural Sciences, Beijing 100081, China; mqq1027@126.com (Q.M.); liyanli02@caas.cn (Y.L.)
2. China-New Zealand Joint Laboratory for Soil Molecular Ecology, Institute of Agricultural Resources and Regional Planning, Chinese Academy of Agricultural Sciences, Beijing 100081, China
3. SCION, Private Bag 29237, Christchurch 8440, New Zealand; jianming.xue@scionresearch.com
4. College of Biology and the Environment, Nanjing Forestry University, Nanjing 210037, China
5. Research Center for Eco-Environmental Engineering, Dongguan University of Technology, Dongguan 523808, China; chengdm@dgut.edu.cn
* Correspondence: lizhaojun@caas.cn

Abstract: Here, we investigated the impact of different turning frequency (TF) on dynamic changes of N fractions, NH_3 emission and bacterial/archaeal community during chicken manure composting. Compared to higher TF (i.e., turning every 1 or 3 days in CMS1 or CMS3 treatments, respectively), lower TF (i.e., turning every 5 or 7 days in CMS5 or CMS7 treatments, respectively) decreased NH_3 emission by 11.42–18.95%. Compared with CMS1, CMS3 and CMS7 treatments, the total nitrogen loss of CMS5 decreased by 38.03%, 17.06% and 24.76%, respectively. Ammonia oxidizing bacterial/archaeal (AOB/AOA) communities analysis revealed that the relative abundance of *Nitrosospira* and *Nitrososphaera* was higher in lower TF treatment during the thermophilic and cooling stages, which could contribute to the reduction of NH_3 emission. Thus, different TF had a great influence on NH_3 emission and microbial community during composting. It is practically feasible to increase the abundance of AOB/AOA through adjusting TF and reduce NH_3 emission the loss of nitrogen during chicken manure composting.

Keywords: composting; turning frequency; ammonia oxidizing bacterial; ammonia oxidizing archaeal; N fractions; ammonia emission

1. Introduction

As a sustainable, effective and ecofriendly approach to deal with livestock and poultry waste, composting is a dynamic biological process driven by microbial populations, self-heating and the biodegradative process of waste [1–4]. Composting manure has been shown to have a lot of agronomic benefits, such as a reduction in waste material mass and water content, pathogen suppression, weed seeds killing, and reduction of phytotoxic substances and unpleasant odors, eventually turning the manure into a stable nutrient source of organic fertilizer needed for crop production [5,6]. However, the large amount of loss of nitrogen during the composting process is one of the key disadvantages of conventional aerobic composting [6–8]. Studies have shown that about 16–74% of the total nitrogen (TN) at the initial stage is lost and approximately 46.8–77.4% of the TN is unavoidably lost with the release of NH_3 during composting [9,10]. The NH_3 emissions can not only result of reducing the quality of compost products but also cause secondary environmental pollution [11–13]. Therefore, it is necessary to lower the emission of NH_3 during composting in order to minimize environmental impacts.

The differences in transfer rate of heat as well as mass can cause spatial gradient of air humidity, oxygen content, temperature, volatile solid content and so on. These can further lead to less porosity and poor ventilation in the compost [14]. The aforementioned physicochemical properties of the pile including temperature, humidity and porosity could influence nitrogen loss and NH_3 emission, but also the quality of compost products during aerobic composting [12,15]. It has been shown that turning of composts can aerate the composting pile by increasing porosity, further promoting activities of microbial organisms that are responsible for degrading the compost materials and generating the heat. Therefore, turning is needed to achieve a homogeneous fermentation process [16] and ensure the abundant oxygen of the compost [17]. Previous research demonstrated that there was a strong relationship between turning frequency (TF) and some physicochemical indicators of compost or compost maturity [18,19]. For instance, the TF could affect total bacterial abundance, TN, temperature, pH, content of moisture, ratio of C/N and germination index (GI) of composting piles [19–21]. Aeration also strongly relates to composting efficiency and gas emission in the aerobic fermentation process. It seemed that turning of materials is the most convenient and widely used method of aeration for the minimization of TN loss and NH_3 emissions during composting, which can promote the fermentation process [19,22].

So far, many studies have shown that TF can improve the quality of compost [19,22,23]. For example, proper TF (every five days) could improve final product quality (pathogen reduction) and cut down the composting time of garden waste (30~36 days) [24]. In addition, the longest thermophilic stage were obtained with TF of every 7 days compared to every 5, 10 and 15 days when composting Camellia oleifera shell with goat manure [23]. Soto-Paz et al. also found that it took less time to reach the maximum temperature with higher TF when he investigated the effects of TF (1, 2 and 3 turnings/week) on co-composting of biowaste and sugarcane filter-cake. It is thought that the increasing TF may reduce compaction and ensure aeration inside piles, which further results in higher biological activity and consequently larger heat release [25]. However, higher frequency of turning may cause more depletion of degradable materials attributed to a higher heat, water loss and NH_3 emission through evaporation and convection. Turning the compost material more frequently from once weekly to daily could cause more loss of TN, but this varied with different cases [25]. Both excessive and less turning could significantly affect the decomposition of piles through reducing material temperature and humidity, which reduces the loss of nutrients, results in long fermentation period and bad quality of the final product [19,26]. Therefore, choosing appropriate TF is very important for achieving an effective composting process.

The nitrogen loss mainly in NH_3 emission is the most prominent factor limiting the efficient use of manure [27]. The nitrate formation from NH_3 oxidation has been considered as a way to preserve nitrogen in the final product of composting [9]. The conversion of ammonia to nitrate (i.e., nitrification) during the assimilation process is inseparable from the ammonia-oxidizing bacteria or archaea (AOB/AOA) [9]. In recent years, the increasing and widespread attention were paid on impacts of microbial composition on the quality and maturation of compost [28,29]. Oxidation of ammonia driven by AOB and AOA producing ammonia monooxygenase (*amo*A) is the first and rate-limiting step of nitrification [9]. In general, organic matter is decomposed during the composting process consisted of complex bioprocessing, relying on the activity of microorganisms. It is crucial for successful composting to achieve and maintain a favorable composition of microorganisms [30]. Therefore, exploring the dynamic changes of key functional groups (e.g., AOB, AOA) of microbial communities in relation to NH_3 emissions under different turning frequencies of the composting materials will provide better insight into the AOB and AOA regulations of NH_3 emissions for minimizing the nitrogen losses during composting.

In our study, we analyzed the structure and diversity of bacterial communities, also including AOB and AOA abundance to quantify the target microbes responsible for nitrification by high-throughput sequencing of 16S rRNA gene amplification. The main objectives of this study were: (1) to understand the effects of TF on composting and product quality, N transformation and NH_3 emissions, and succession microbial structure and diversity during the composting, (2) to investigate if the contribution of AOB and AOA to ammonia oxidation could vary at different composting stages and be manipulated by TF, (3) to identify key driving factors shaping AOB and AOA communities and influencing NH_3 emission and nitrogen transformation during the composting.

2. Materials and Methods

2.1. Composting Process and Sampling

The composting trial was conducted in an organic fertilizer factory located at Daxing District in Beijing, China. Composting materials applied in this experiment consisted of chicken manure (CM) and soybean straw (S). The fresh chicken manure was gathered from a large farm in Daxing District and was air dried (15–20 °C) reaching a water content of <30%, ground and sieved using a 10 mm mesh. The soybean straw came from farms located near the factory also air dried to a water content <10%, ground to a granule size smaller than 10 mm. The chicken manure and soybean straw were mixed well to get the compost with the initial C:N of 8:1, humidity of 60% [5].

Sixty kilograms of the above mixture were put into each of twelve plastic boxes (150 L) with lids. There were four treatments assigned with four different turning frequencies of the compost: (1) once a day (CMS1), (2) once every 3 days (CMS3), (3) once every 5 days (CMS5) and (4) once every 7 days (CMS7). Each TF treatment was repeated three times, with 12 boxes in total. The turning process of compost was carried out by mixing the treated substrate manually with a garden shovel.

The 12 boxes with composting materials were arranged in a room according to the randomized block design. The temperature of composting piles was monitored twice a day (9:00 am and 18:00 pm) for calculating an average daily temperature. The duration of composting was 66 days and the pile temperature was close to room temperature in the end. The samples were collected after completely mixed on day 1, 3, 5, 7, 10, 12, 15, 17, 22, 29, 35, 43 and 66 during composting. To collect the representative sample from each replication box, five sub-samples were taken at five locations (one in the center and four at the corners) of a box at the depth of 15 cm, bulked together and mixed well, then kept in polyethylene bags. Each sample was split into three parts. The first part was air-dried for the dry-based chemical analysis. The second part was kept fresh at 4 °C to analyze NH_4^+-N, NO_3^--N, etc., and the third was preserved at −80 °C for DNA extraction and 16S rRNA, bacterial *amo*A (AOB) and archaeal *amo*A (AOA) gene sequencing analyses.

2.2. Physicochemical Properties Analyses of NH_3 Emission

All collected samples were analyzed for moisture content, pH, germination index (GI), TN and total carbon (TC), ammonium nitrogen (NH_4^+-N), nitrate nitrogen (NO_3^--N). The moisture content was measured by drying the sample at 105 °C until constant weight was achieved. The fresh samples were used to measure pH, GI, NH_4^+-N and NO_3^--N according to Test Methods for the Examination of Composting and Compost (TMECC, 2002). pH values were detected by MP521 pH meter (Shanghai, China) after extracted in 1:5 (w/v). The contents of TN and TC were measured for air-dried samples using elemental analyzer (Vario EL III, Elementar Analysensysteme GmbH, Langenselbold, Germany). To determine the concentration of NH_4^+-N and NO_3^--N, fresh samples were extracted at 25 °C using 0.5 M K_2SO_4 (1:10 w/v), and the filtrates were analyzed by using indophenol blue technique [31,32]. Released ammonia was collected with the gas collecting device during composting processes. The NH_3 emission was measured by adsorbing the exhaust gas with H_3BO_3 and titrated against HCl [33,34].

2.3. DNA Extraction, PCR Amplification and Sequence Analysis

DNA extraction was performed using the Fast DNA SPIN extraction kits (MP Biomedicals, Santa Ana, CA, USA) and subsequently kept at −80 °C for further analysis. The quantity of extracted DNA was determined using NanoDrop ND-1000 spectrophotometer (Thermo Fisher Scientific, Waltham, MA, USA) and the quality using agarose gel electrophoresis.

Bacterial 16S rRNA gene V3–V4 regions (total bacteria) was amplified using the forward primer 338F (5′-ACTCCTACGGGAGGCAGCA-3′) and the reverse primer 806R (5′-GGACTACH VGGGTWTCTAAT-3′). The specific primer sets for bacterial *amo*A (AOB) and archaeal *amo*A (AOA) gene were F(5′-GGGGTTTCTACTGGTGGT-3′), R(5′-CCCCTCKGS AAAGCCTTCTTC-3′) [35] and Arch-amoA26F (5′-GACTACATMTTCTAYACWGAYTGGG C-3′), Arch-*amo*A417R (5′-GGKGTC ATRTATGGWGGYAAYGTTGG-3′) [36], respectively. In present study, the related high-throughput sequencing analysis was processed using the Illumina MiSeq platform by Shanghai Personal Biotechnology Corporation., Ltd. (Shanghai, China) to have a closer look at the microbial communities.

Sequences were analyzed and quality-filtered using QIIME 2.0 (Quantitative Insights Into Microbial Ecology, Version 2019.4) and Vsearch (v2.13.4) software as described by [37,38]. In a word, raw sequence data were demultiplexed, quality filtered, denoised, merged and chimera removed using the DADA2, including a strict quality control by getting rid of reads with ambiguous bases, singletons and chimera [39]. The reads were selected to amplicon sequence variants (ASV), or unique sequences, using DADA2 and taxonomically identified. ASV taxonomic was classified through BLAST searching the representative sequences set against the Silva 132 [40] or NCBI Database [41].

2.4. Bioinformatics and Statistical Analysis

Microbial functions were analyzed by PICRUSt2 (Gavin M. Douglas, et al., preprint) upon MetaCyc (https://metacyc.org/ (accessed on 17 July 2021)) and KEGG (https://www.kegg.jp/ (accessed on 23 August 2021)) databases. Bioinformatic analyses of sequence data were mainly performed using QIIME 2.0. ASV-level alpha-diversity for microbial community was evaluated with Good's coverage, Chao1, Observed species, Shannon index and Simpson index. Beta diversity analysis was carried out using Bray-Curtis and visualized by non-metric multidimensional scaling (NMDS) to investigate the structural variation and community function of microbial communities. Taxa relative abundances at both phylum and genus levels were statistically compared among treatments and visualized as bar chart. Linear discriminant analysis effect size (LEfSe) was conducted to test differentially abundant taxa between treatments using default parameters [42]. Random forest analysis was applied using "Random Forest" package in the R package [43] to discriminate the samples from various groups. The generalization error was estimated using ten-fold cross-validation for all comparisons from the 16S rRNA, bacterial *amo*A (AOB) and archaeal *amo*A (AOA) data. Microbial functions were predicted by PICRUSt 2, through high-quality sequences of 16S rRNA gene [44] based on MetaCyc (https://metacyc.org/ (accessed on 25 July 2021)).

One-way ANOVA was performed using SPSS19.0 software for testing the treatment effect. Duncan's multiple range test was carried out to compare the significance levels between the means. The relationships between physicochemical parameters and total/ammonia-oxidizing bacterial/archaea community were assessed using Canoco 4.5. Spearman correlation analysis was also performed using SPSS19.0 software to obtain whether there was a significant correlation between environmental factors and species community. All figures for physicochemical parameters were plotted by Graphpad Prism 5.0.

3. Results and Discussion

3.1. Effect of TF on Composting Process, pH and GI Values of the Compost

In this study, the whole composting process could be divided into four phases, namely, mesophilic phase (21–40 °C, days 0–1), thermophilic phase (35–65 °C, days 2–17), cooling phase (65–40 °C, days 18–35) and maturation phase (40–25 °C, days 36–43) according to the temperature variation trend of composting. These are consistent with other researcher's results [45]. For all TF treatments, the temperature rose rapidly at the beginning and entered the thermophilic phase (>55 °C) on day 2 (Figure 1a). In the present study, the pile temperature at the CMS7 treatment was kept above 60 °C for a total of 11 days, which was three days less than the other three treatments. In addition, CMS1 treatment remained above 55 °C for 18 days, with 3 days less than that of CMS5. These results were in accordance with the previous studies. It has been shown that premature turning delayed time to achieve high temperatures and turning too late was not helpful for the temperature rising again and the high temperature keeping during composting [25,46]. Notably, such high TF aeration surplus as CMS1 treatment not only increased the costs of energy but also decreased the temperature of the composting pile quickly, hindering the thermophilic phase achievement [47]. However, too low TF (i.e., CMS7 treatment) resulted in lots of anaerobic zone present in the compost system and therefore caused problems of long fermentation time [48]. An increase in temperature was observed for all treatment after each compost turning as also reported by previous study [49], thereby indicating that proper TF may increase the oxygen content in the compost, which may result in promoting the activities of microbial organism that can degrade the materials in the compost pile and generating heat [17].

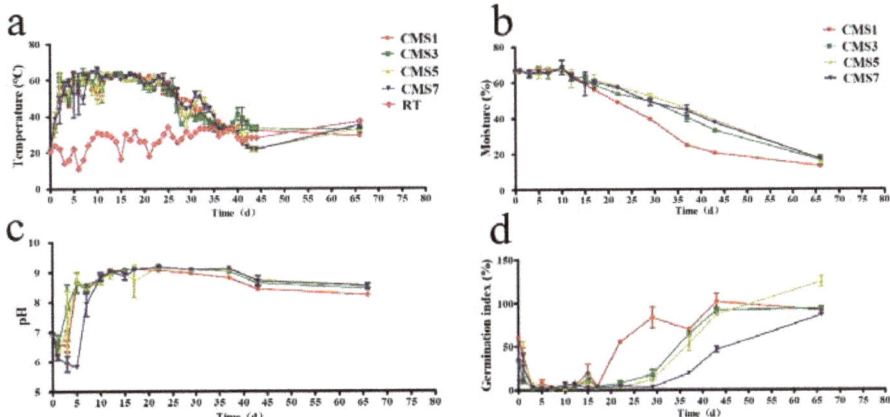

Figure 1. Changes in temperature (**a**), moisture (**b**), pH (**c**) and germination index (**d**) at different treatments during the composting. Every data point represents the mean of three replicates. Bars indicate the SDs of the means.

Moisture content can greatly impact the microbial activity and physicochemical parameters of the compost. As shown in Figure 1b. the moisture content of each treatment decreased from 65–66% at the beginning to 20.41%, 33.16%, 38.71% and 37.35% on day 43 for CMS1, CMS3, CMS5 and CMS7 treatments, respectively. The composting process may be hindered at higher TF, which might cause higher loss of compost moisture. It has been reported that moisture content is the strongest correlative factor with the succession of bacteria and archaea among numerous other factors such as pH, salinity and nutrients [50]. Therefore, TF should be appropriately regulated to obtain an efficient composting process.

A key factor which influences gaseous emissions as well as the microbial activities and structure during composting is pH [33,51]. In the present study, the overall patterns of pH changes during the composting period were similar for all treatments, which increased dramatically from 6.16–6.48 on day 1 to 8.86–9.17 on day 15 and then gradually increased to 9.23 on day 22 and afterwards reduced slightly (Figure 1c). The respiration microorganism and of ammonia (NH_3) emission have significant impacts to the pH changes, as have been reported previously [52,53]. The lower pH obtained under CMS1 treatment during cooling and mature stage ($p < 0.05$), possibly resulted from the release of H^+ ions in the process of nitrification during the transformation of organic nitrogen [54]. At the end of composting, the final pH of materials was also affected by ammonia volatilization [6].

The GI always increases with toxic materials degradation in composting pile [55]. As the composting process progressed, the GI values of CMS1, CMS3 and CMS5 increased, reaching to 91.6%, 93.77% and 124.36% at the end of composting time, respectively (Figure 1d). The compost is recognized basically and sufficiently mature when the value of GI is above 50% and enough well matured when GI value achieves 80% [56,57]. Thus, it can be concluded that the composts of CMS1, CMS3, and CMS5 were stabilized enough at day 43 but the longer time (than 43 days) was required for CMS7 to reach stabilization. TF is commonly thought to be a key factor affecting the rate of composting as well as compost quality [24]. Previous researchers have suggested that compared to every 4-day and every 7-day turning, every 2-day turning can facilitate faster sterilization and maturity during composting [22,58]. In this study, the GI values of the CMS1, CMS3 and CMS5 treatments were 101.52%, 91.64% and 87.24%, respectively, which were significantly higher than that of the CMS7 (46.19%, $p < 0.05$).

3.2. Effect of TF on Nitrogen Transformation and NH_3 Emission

As shown in Figure 2a, the concentration of NH_4^+-N showed an increasing trend among the four treatments with the temperature reaching its highest on day three. These were primarily attributed to organic nitrogen mineralization and ammonification [59]. The contents of NH_4^+-N in CMS1 and CMS3 group were significantly higher than that in CMS5 and CMS7 on day five suggested higher frequency and premature turning at the beginning of the composting process may promote rapid mineralization and the ammonification during composting ($p < 0.05$) [60]. The difference among the four treatments may be due to the combined effects of the degree of NH_3 emissions, organic nitrogen hydrolysis, and nitrification during the composting [9]. During the composting process, NH_4^+-N might be transformed into NO_3^--N by AOB and AOA, which could then result in reduction of NH_3 emission, and decreasing the loss of organic nitrogen [61–64]. However, during thermophilic phase, the excessive NH_3, high temperature and oxygen-deficient of pile can inhibit the proliferation and activity of nitrifying microbial communities [65]. Comprehensively speaking, losses of NH_4^+-N might be the result of the volatilization and nitrification process.

As shown in Figure 2b, the concentration of NO_3^--N was low at 89.60–96.61 mg/kg initially then it increased to 263.63–483.12 mg/kg on day three during the early composting stages. The NO_3^--N concentration decreased gradually during the cooling phase due to a reduction in the concentration of ammonium [66]. The NO_3^--N content of CMS1 is significantly lower than that of CMS3, CMS5 or CMS7 during the cooling and maturity period ($p < 0.05$). This may because that most of the ammonium in CMS1 used as the substrate for generating NO_3^--N was volatilized in the form of NH_3 due to the high TF, reducing the substrate for ammonia-oxidation reaction for the nitrifying microorganisms [67]. Or the moisture content in the compost material was greatly reduced for CMS1, which inhibits the activity of nitrifying microorganisms [50]. In addition, compared with CMS1, CMS3 and CMS7 treatment, the TN loss of CMS5 was decreased by 38.03%, 17.06% and 24.76%, respectively (Figure 2e). However, because nitrogen metabolism is a complex process, synergistic carbon and nitrogen metabolism, for reducing the nitrogen losses, the activity and quantity of enzymes and the genes, should be further assessed during composting. The relatively appropriate TF may reduce the NH_3 emissions through accelerating the proliferation of AOB and AOA. It was also testified by the changes of NH_4^+-N and NO_3-N contents during composting.

As shown in Figure 2c, NH_3 was detected at day 2. With the temperature and pH increased, NH_3 emissions from all of the treatments rapidly increased and reached their highest values. The NH_3 emission from the CMS1 treatment observed on the third, fifth, ninth and thirteenth day was significantly higher than the other three treatments ($p < 0.05$). The high rate of NH_3 emissions for all treatments might be due to quick degradation of organic matter and the fast conversion of NH_4^+-N to NH_3 [13,34,68], associated with the increased temperature and pH during the thermophilic stage [59,69]. Previous study also suggested the NH_3 volatilization had a positive relationship with the amount of aeration or TF [70]. The characteristics of NH_3 emissions from this study are in line with previous research [71,72]. Overall, physicochemical properties are important factors to influence NH_3 volatilization such as temperature, pH, NH_4^+-N concentration, and the microbial community [12,73–75].

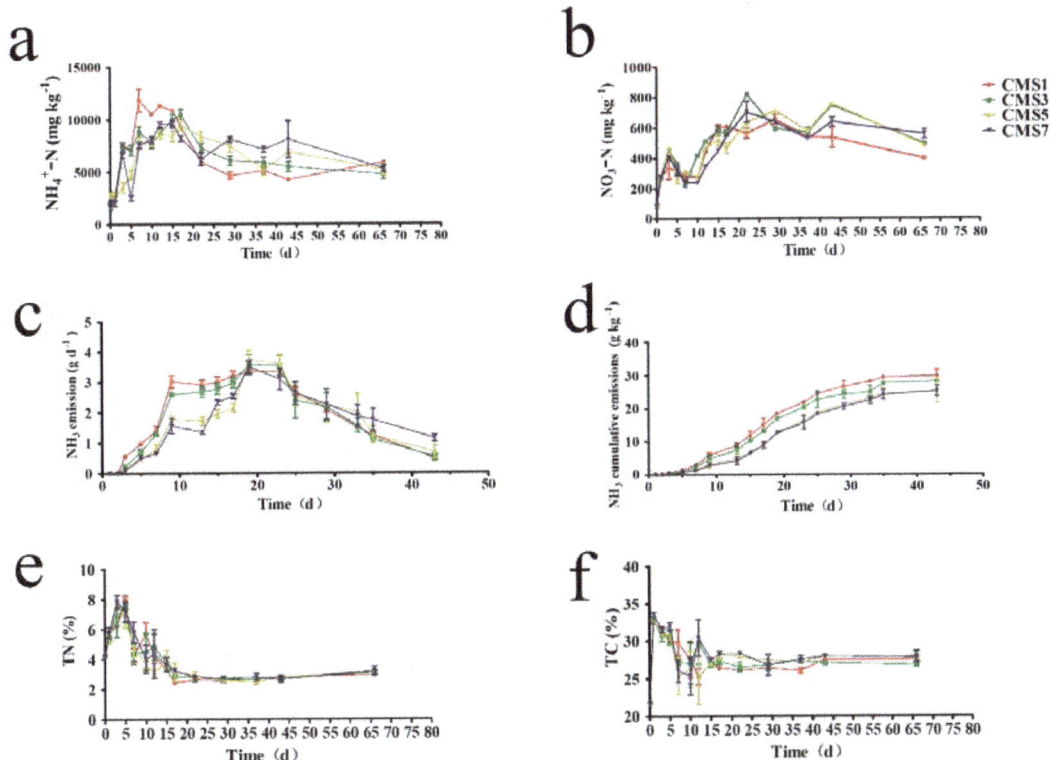

Figure 2. Variation in ammonium nitrogen (**a**), nitrate nitrogen (**b**), NH_3 emission rate (**c**), cumulative NH_3 emission (**d**), TN (**e**) and TC (**f**) in different treatments during the composting. Every data point represents the mean of three replicates. Bars indicate the SDs of the means.

As shown in Figure 2d, the cumulative NH_3 emission profiles indicated that more than 73.67% of NH_3 emissions occurred during the first 23 days in treatments CMS1. However, in the same 23 days, less than 70% of the NH_3 was emitted from the CMS 5, CMS 7 group while more than 70% and less than 73% for CMS3 group. The emission of NH_3 was strongly connected with the temperature of the pile and microorganisms [74,76]. NH_3 is mainly important gas causing nitrogen loss during composting. NH_3 emission is mainly affected by temperature, pH, NH_4^+-N concentration, aeration rate, and moisture content [77]. The differences of NH_3 emissions in all groups were probably due to the interrelationships between pH, temperature, aeration rate and moisture content [78]. The results showed that the higher TF might lead to the emission of relatively larger amount of accumulative NH_3 emission during the early composting stage, but the detected concentration was similar between different TF groups during the late composting period. Therefore, in general, too high TF results in more ammonia emission during composting.

Figure 2e,f showed the variation of TN and TC content with composting time. TN contents for all treatments presented a similar trend. Compared with CMS1, CMS3 and CMS7 treatments, the TN loss of CMS5 decreased by 38.03%, 17.06% and 24.76%, respectively. Many studies have shown that 16~74% of the initial TN is lost during composting [75]. The decrease in TN might be due to the large amount of nitrogen loss caused by the NH_3

volatilization, the degradation and mineralization of complex organic compounds [79]. Nitrogen fixing bacteria might also have contributed to a lesser degree to the increase in TN in the later phase of composting. The TC content of the compost gradually decreased with composting time. This may be attributed to the microorganisms mineralized the organic carbon as a source of energy. The C/N ratio data of compost materials and samples, as shown in the supplementary Table S6 to give more information about the type of composting waste. The ratio of C/N showed a decreasing trend among the four treatments at the first five days and increased on day 7.

3.3. Effects of TF on Bacterial Community Diversity and Composition during Composting

The five alpha diversity indices (a1–a5) for each treatment at day 1, 5, 15, 29 and 43 during composting are shown in Figure 3(a1–a5), respectively. The Good's_coverage index for all samples was over 0.99, indicated that the sequencing depth was enough for this bacterial community analysis. The microbial taxa abundance indices Chao and Observed species were significantly ($p < 0.05$) lower in the CMS5 treatment than other treatments at post-thermophilic stage (i.e., day 15), but higher at cooling stage (i.e., day 29). This suggested a significant effect of different TF on bacterial abundance, especially for the post-thermophilic and cooling stage. The microbial richness and evenness indices Shannon and Simpson were also significantly lower in the CMS5 treatment than other treatments at post-thermophilic stage (i.e., day 15). This indicates a selective effect of different TF against bacterial taxa at different composting stages. The Chao and Observed Species indices rose sharply at post-thermophilic stage of composting, which was attributed to the growth of a range of microbiome. Certain bacteria could proliferate during the thermophilic phase, such as amylolytic microorganisms as previous studies reported [80]. Previous research also suggested higher bacterial abundance and diversity during thermophilic stage of composting for green wastes [81]. However, research has also shown that microbial activities could be inhibited during thermophilic phase, and the diversity fall may be attributed to the dominance of some microorganism taxa [82].

The microbial diversity and phylogenetic distribution might have close relationship with the composting process and the quality of compost. Therefore, the composition and succession of bacterial communities at different stages were analyzed. The composition and relative abundance of bacterial communities at phylum and genus levels are shown in Figure 3(b1,b2), respectively. In total, we detected 27 phyla during composting, with Firmicutes, Proteobacteria, Actinobacteria, Bacteroidetes as the top 4 dominant bacteria, which accounted for 92.32–99.82% of the total sequencing reads. This finding agreed with previous studies [83,84]. These four bacterial phyla are prevalent throughout the whole composting period and have a strong ability to degradation of organic matter [85,86]. Firmicutes often show high enrichment throughout composting because of the ability to form endospores that can help them to keep surviving high temperatures and harsh environment [85,87–89]. In present study, Firmicutes also played a dominant role in the whole composting process, accounting for 44.24–98.20% of the relative abundance. Previous studies also suggested that Proteobacteria are typically the most (or second most) abundant phylum during most aerobic composting [84]. Interestingly, the relative abundance of Firmicutes in high TF treatments (i.e., CMS1 and CMS3) was significantly lower than that in other treatments, but Proteobacteria and Actinobacteria were significantly higher than that in other treatments during cooling and maturation stages. These results might explain a selective effect of higher TF against bacterial taxa during different composting stages and indicate high aeration conditions could stimulate the growth of aerobic bacteria such as Proteobacteria and Actinobacteria [90] but inhibit the activity of anaerobic microorganisms in Firmicutes [91]. Interestingly, the highest abundance of Gemmatimonadetes was found in the CMS5 treatment at the maturation stage. Previous studies also showed that Gemmatimonadetes was probiotics [92], significantly enriched at the maturation stage of vermicomposting with coconut leaf [93], and was predominant in soils amended with alkaline treatments [94]. Thus the beneficial microorganisms may be stimulated by appropriate TF or aeration during aerobic composting [95].

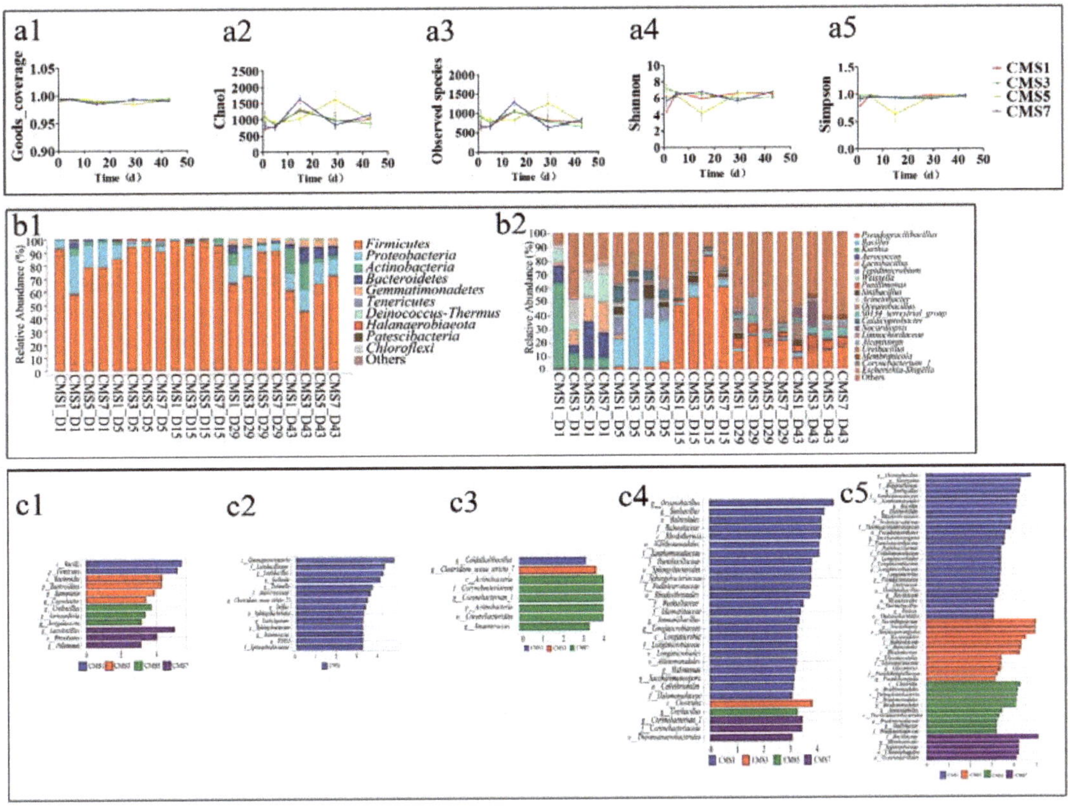

Figure 3. Alterations of bacterial community structure in different treatments during the composting. The α diversity of bacterial community in different treatments during composting process (**a1–a5**). The relative abundance of different bacteria at the (**b1**) phylum and (**b2**) genus levels in different treatments during the composting. LDA (Linear discriminant analysis) Effect Size (Lefse) analysis of the biomarkers ((**c1–c5**) represent day1, 5, 15, 29 and 43, respectively), showing biomarkers of the significant and biological differences from the phylum level to the genus level (LDA score > 3, $p < 0.05$).

In total, we detected 569 genera during composting, with Pseudogracilibacillus, Bacillus, Kurthia, Aerococcus, Lactobacillus, Tepidimicrobium, Weissella, Pusillimonas, Sinibacillus and Acinetobacterwere as the top 10 dominant bacteria, which accounted for 27.57–91.19% of the total sequencing reads. These bacterial genera belong to the phyla of Firmicutes, Proteobacteria, Actinobacteria, Gemmatimonadetes and Bacteroidetes, respectively. Pseudogracilibacillus was the most dominant genus accounting for 1.01–83.53% at whole composting stages. The highest relative abundance of Pseudogracilibacillus was found at day 15 (47.27–83.53%), followed by the cooling and the maturation stages (13.52–25.12%). The ecological function of Pseudogracilibacillus during aerobic composting has seldom been reported. Previous study suggested that Pseudogracilibacillus as neutrophilic aerobes could exist in the high-temperature environments and be associated with the nitrogen cycle [96]. The second dominant genus was Bacillus, accounting for 0.35–50.06% at different composting stages. The relative abundance of Bacillus at the pre-thermophilic stage was 20.01–50.06%, which was significantly higher than that at other stages. This result was in agreement with previous studies, which have reported that the genus Bacillus consists of a large quantity of thermophilic bacteria and can dissimilate and reduce nitrogen compounds [97]. The relative abundance of Kurthia, Aerococcus, Lactobacillus, Weissella, and Acinetobacter accounted for higher than 1% only at the mesophilic phase (day 1). Pusillimonas has been previously identified as the main dominant bacterial community correlated with the heterotrophic nitrification and denitrification of composting and wastewater [98,99]. The abundance of Bacillus, Sinibacillus, Oceanobacillus and Nocardiopsis was significantly higher in CMS1 and CMS3 treatments than other treatments at cooling and mature stages ($p < 0.05$). On the contrary, the abundance of Pseudogracilibacillus, Pusillimonas, S0134_terrestrial_group, Limnochordaceae, Alcanivorax and Membranicola was significantly higher in CMS5 and CMS7 treatments at the maturation stage ($p < 0.05$). This suggested that these bacterial genera might be sensitive to aeration conditions and could be manipulated by TF of composting materials. Sinibacillus, Limnochordaceae and Oceanobacillus genera have been previously identified as the dominant communities responsible for the proteins transportation-related genus of composting [33,100] and wastewater [101]. Interestingly, as a facultative anaerobe with urease activity, the relative abundance of Pseudograciibacillus decreased with decreasing temperature, which may be attributed to the reduction in ammonia emission [102,103].

The significant differences among TF treatments for bacterial communities were identified by LEfSe (Figure 3(c1–c5), Supplementary Figure S2(a1–a5), Table S1). Compared to the mesophilic and thermophilic phases, more taxa were significantly affected by TF during cooling phase and maturation phase. The LEfSe of all taxa showed 12, 14, 8, 29 and 55 bacterial taxa had significant differences (LDA > 3, $p < 0.05$) among the treatments at mesophilic, pre-thermophilic, post-thermophilic, cooling and maturation phases, respectively. The dominant taxa (LDA > 4, $p < 0.05$) were phyla as Firmicutes, Bacteroidetes and genus as *Lactobacillus*, *Ureibacillus* during mesophilic phase, genus as *Lactobacillus* during pre-thermophilic phase, phyla as Actinobacteria and genus as *Corynebacterium_1* during post-thermophilic phase, genus as *Oceanobacillus* and *Snibacillus* during cooling phase, phyla as Bacteroidetes and genus as *Oceanobacillus*, *Georgenia*, *Sinibacillus*, *Bacillus*, *Thermobifida*, *Nocardiopsis*, *Bradymonadales* and Membranicola during maturation phase. It must be noted that microbial diversity is related to the physicochemical properties of compost, which change with the composting time [104]. LEfSe is an accurate and effective method to identify specific microbes (biomarkers) that displayed significant differences in microbial abundance between different treatments [105]. Among all the different taxas above, phyla as Firmicutes, Bacteroidetes, Actinobacteria and only genus as *Ureibacillus*, *Lactobacillus*, *Oceanobacillus*, *Sinibacillus*, *Corynebacterium_1*, *Membranicola* also belonged to the top relative abundance10 phylum and top 20 genus, respectively. This indicated that the dominant taxa were significantly affected by different TF. In addition, it should be noticed that the genus *Ureibacillus* was significantly affected by different TF during both mesophilic and cooling phase. The genus *Lactobacillus* was significantly affected by different TF during both mesophilic and pre-thermophilic phase [106]. Interestingly, *Sinibacillus*, *Corynebacterium_1* and *Oceanobacillus* were all significantly affected by TF for both post-thermophilic and cooling phase. Meanwhile, previous studies emphasized the importance of low-abundance microorganisms to ecosystem function, such as biochemical processes [107], community succession [108] and microbiome function [109]. Therefore, more attention to these different species caused by different TF during different stages may provide a certain amount of theoretical support for optimizing the composting process.

The keystone taxa for the microbial communities of the four groups during composting were determined using a random forest model (Supplementary Figure S3(a1,a2)). At the phylum level, nine taxa were the dominant species occupying the top 10 abundance among the top 10 different important phylum for 16S rRNA while *Cyanobacteria* as less abundant species was out of the top 10 phylum in abundance. At the genus level, nine taxa were the dominant taxa occupying the top 20 different important genus for 16S rRNA, such as *Pseudogracilibacillus*, *Caldicoprobacter*, *Aerococcus*, *S0134_terrestrial_group*, *Pusillimonas*, *Bacillus*, *Membranicola*, *Weissella*, *Ureibacillus* and *Alcanivorax*. It was also shown that these classes were common for dominating the composting process [110]. *Pusillimonas* was widely distributed in environments and can utilize a variety of fatty acids and urea [111]. *Pseudogracilibacillus*, *Ureibacillus* and *Alcanivorax* are affected by the oxygen concentration and related to nitrogen transformation [103,112–114]. So, it was suggested different TF could affect bacteria involved in N cycle, especially ammonium oxidation. Furthermore, TF can also alter the structure of the bacterial community.

3.4. Effects of TF on Ammonia-Oxidizing Bacteria/Ammonia-Oxidizing Archaeal Diversity and Composition during Composting

AOB and AOA are ubiquitous in various environments and play crucial roles in the nitrogen cycling process [115–117]. AOB and AOA also have been found to be common in the composting of various livestock including chicken [118], cow/cattle [4,51,117], sheep [119], pig [120,121]. Based on the Illumina sequencing data, we obtained an average of 56,022 and 113,491 sequence reads per sample ranging from 15,139–140,517 and 40,925–137,869 reads for AOB and AOA, respectively. The alpha diversity indices of AOB and AOA communities in different TF treatments at day1, 15, 29 and 43 of composting were shown in Figure 4(a1–b5), respectively. The Good's_coverage in every sample was over 0.99, suggesting that the sequencing depth was enough for both AOB and AOA community analysis. The Chao and Observed species indices were significantly higher in the CMS1 treatment than in other treatments ($p < 0.05$) for AOB at mature stage (i.e., day 43), while lower in the CMS1 treatment than in other treatments ($p < 0.05$) for AOA at cooling stage (i.e., day 29). The Shannon and Simpson indices for AOA at the post-thermophilic stage were significantly higher in the CMS5 treatment than the other three treatments, suggesting the inhibition proliferation of AOA communities by high TF treatment through moisture loss or excessive aeration [120]. As mentioned above, the increase of NO_3^--N concentration in the post-thermophilic composting stage may because that AOA were able to oxidize ammonium under thermophilic conditions and high pH [71], or caused by high substrate concentration of NH_4^+ to accelerate the nitrification microbial growth [51]. Although it showed relatively higher abundance and diversity at the thermophilic stage of composting in green waste composting [81], microbial activities could be inhibited during thermophilic phase, and the decline in diversity might be attributed to the dominance of some microbial taxa [82]. This could be attributed to the differences in temperature, aeration or moisture caused by different TF during the composting stages.

The composition and relative abundance of AOB community at phylum and genus levels are shown in Figure 4(c1,c2), respectively. In total, we detected 18 phyla during compost, where Proteobacteria, Firmicutes, Actinobacteria and Bacteroidetes, were the top 4 dominant bacteria, which accounted for 56.16–99.59% of the total sequencing reads. In this study, Proteobacteria played the most dominant role in the whole composting process, accounting for 25.42–99.55%. Interestingly, the relative abundance of Actinomycetes (ranged from 0.27–30.47%) and Bacteroidetes (ranged from 0–15.72%) at the post-thermophilic period reached the highest of 30.47% and 15.72%, respectively. This may indicate that certain Actinomyces and Bacteroidetestes of AOB may survive at high temperature. The relative abundance of dominant phylum species at different composting stages can be affected by different TF, for instance, the relative abundance of Actinomycetes in the high TF treatment was significantly higher than that in the other groups at post-thermophilic period. In total, we detected 141 genera with roles for AOB during the compost, where *Nitrosospira*, *Lactobacillus*, *Nitrosomonas*, *Weissella* and so on were the top 10 dominant bacteria, which accounted for 9.45–99.53% of the total sequencing reads. It was reported that *Nitrosomonas* have frequently been investigated in cattle manure or pig slurry amended soils [122,123], wastewater treating [124] and animal manures composting [4,125]. For AOB, *Nitrosospira* and *Nitrosomonas* usually occupied the first or second dominant position in relative abundance during composting [125]. As previously reported, *Nitrosomonas* was typical ammonia-oxidizing bacteria [126] even dominated all the stages whereas *Nitrosospira* dominated the initial of the composting stage and continually decreased in the maturation stage during cow manure composting [4,125]. However, in present study, *Nitrosospira* played a dominant role in the whole process, especially at the post-thermophilic phase, which reached 8.00%~81.01%; however the proportion of *Nitrosospira* sequences increased at higher temperatures [127]. These findings are in agreement with the previous results that *Nitrosospira* can survive at temperatures of up to 75 °C during composting [128,129]. Interesting, *Nitrosomonas* was almost undetectable during the post-thermophilic phase. These results were in inconsistent with the results that *Nitrosomonas* dominated during all the stages of composting [4,125]. Moreover, the relative abundance of *Nitrosospira* in high TF treatment (i.e., CMS1) was significantly lower than that in the other treatments during post-thermophilic, cooling and maturation stage of the compost, while lower relatively abundant *Nitrosomonas* sequences was detected in too high TF (i.e., CMS1) or too low TF (i.e., CMS7) treatment. These findings suggested that proper availability of oxygen, facilitated by turn or aeration, may be an important regulatory factor for AOB in composts [128].

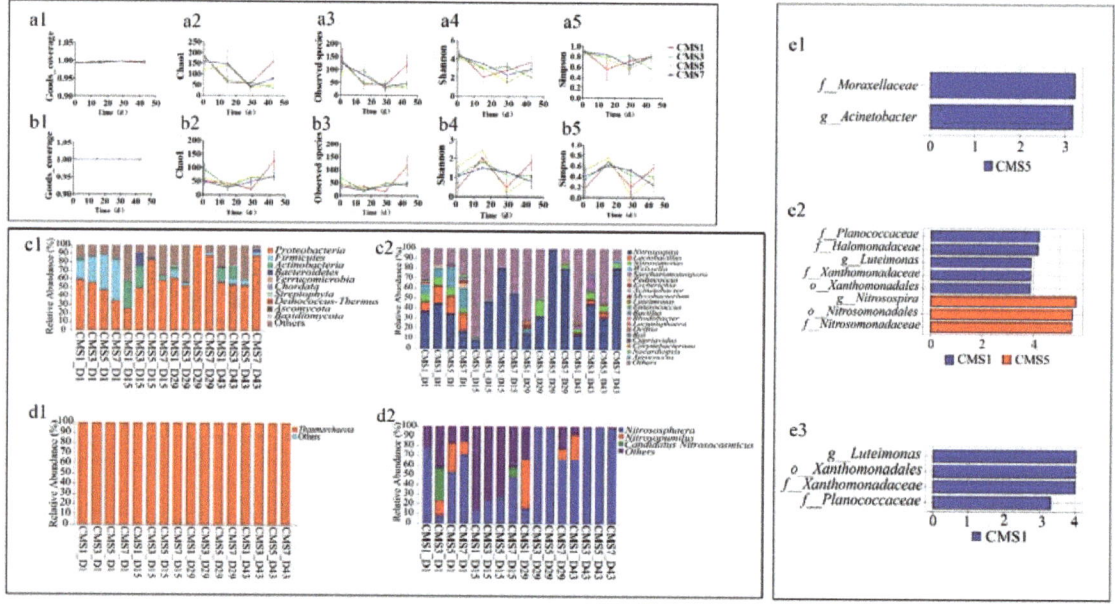

Figure 4. Changes of AOB and AOA community structure in different treatments during the composting. The α diversity of AOB community (**a1–a5**) and AOA community (**b1–b5**) in different treatments during the composting. The relative abundance of different bacteria at the phylum levels of AOB (**c1**) and AOA (**d1**) and genus levels of AOB (**c2**) and AOA (**d2**) in different treatments during the composting. LDA Effect Size (Lefse) analysis of the biomarkers ((**e1–e3**) represent day 15, 29 and 43 for AOB, respectively,) with significant and biological differences from the phylum level to the genus level (LDA score > 3, $p < 0.05$).

Microorganisms including bacteria, archaea and fungi played important roles on chicken manure degradation during the composting process [87,130,131]. However, in present study only a total of 3 genera were detected and identified including *Nitrosophaera*, *Nitrosopumilus* and *Candidatus Nitrosocosmicus*, which accounted for 12.18–99.99% of the total sequencing reads (Figure 4(d1,d2) for AOA at phyhum or genus level). These bacteria all belong to Thaumarchaeota phylum. *Nitrosopumilus* and *Nitrosophaera* were the dominant AOA species in animal manure composting [118,120]. In the present study, *Ni-*

trososphaera played the most dominant role in the whole composting process, accounting for 8.50–99.94%, which was 12.17–47.21% during post-thermophilic phase, and 15.73–99.74% at the cooling and maturation stage. These findings were in accordance with previously studies that Nitrososphaera belong to Thaumarchaeota phylum dominate in compost and can resist high temperatures [118,120]. Nitrosopumilus can be detected in the mesophilic, cooling and maturation phase but not in the post-thermophilic stage suggested Nitrosopumilus may prefer to survive or maintain activity at mesophilic stage [132]. Interestingly, the relative abundance of Nitrososphaera increased with the decrease of TF at post-thermophilic stage furthermore the relative abundance of Nitrososphaera in the high TF treatment was significantly lower than that in the other groups during the post-thermophilic, cooling and maturity periods. These results suggested that the abundance or activity of AOA may be affected by factors such as differing porosity, aeration, or even moisture content [120,133,134].

The significant differences among different TF treatments for the AOB community were identified by LEfSe (Figure 4(e1–e3), Supplementary Figure S5(a1–a3), Table S2). The LEfSe of all taxa showed two, eight and four bacterial taxa of AOB were significant different (LDA > 3, $p < 0.05$) among the treatments at the post-thermophilic (i.e., day 15), cooling (i.e., day 29) and maturation (i.e., day 43) phases, respectively. Specifically, the dominant taxa were genus as Acinetobacter, Nitrosospira and Luteimona during post-thermophilic, cooling and maturation phase, respectively. However, LEfSe analysis showed no significant difference among the treatments for AOA community composition. As mentioned above, these results also indicated that the abundance or activity of AOB might be affected by factors such as differing porosity, aeration, or even moisture content [120,133,134].

Furthermore, keystone taxa for the microbial communities of the four groups during composting were determined using a random forest model (Supplementary Figure S6(a1,a2,b1,b2)). At the phylum level, Seven taxa were the dominant species occupying the top 10 abundance among the top 10 different important genus for AOB such as Lactobacillus, Nitrosospira, Weissella, Nitrosomonas, Acinetobacter, Escherichia and Corynebacterium, while Aerococcus, Muribaculum and Pseudomonas as less abundant species were out of the top 10 abundance genus. In addition, the four groups were consistent with the dominant flora in abundance among species with different importance at the genus level for AOA.

3.5. Key Environmental Factors Shaping Microbial Communities

RDA (Redundancy analysis) analysis was applied to study the relationship between microbial succession and physicochemical parameters. Using the 16S rRNA and AOB sequencing data, RDA analysis results at the genus level showed that RDA1 and RDA2 jointly explained 61.87% (Figure 5a) 40.34% (Figure 5b) and 28.14% (Figure 5c) of the total variance, respectively. Further analysis revealed that the temperature, NH_4^+-N, NH_4^+-N, GI, NH_3 emission and NH_3 cumulative emission were the main factors affecting the 16sRNA taxa. Pusillimonas and Pseudogracilibacillus were positively correlated to NH_3 emission [103,111]. The pH, moisture, NH_3 emission and NH_3 cumulative emission were the main factors affecting the AOB taxa. Interestingly, the moisture content was the only major factor affecting the AOA taxa. This was in agreement with previous study that moisture may affect the amount of dissolved oxygen in the composting [51]. The temperature, pH, NH_4^+-N and NO_3^--N had positive effects on the release of NH_3, while moisture had negative feedback effects on the release of NH_3. These findings suggested that main environmental variables driving the diversity and structure of AOA and AOB communities were different [135]. In this study the NH_4^+-N and NO_3^--N concentration had positive effects on Nitrososphaera and Candidatus Nitrosocosmicus. This agreed with previous studies, which reported high substrate concentration of NH_4^+-N may promote the AOA growth [51,136]. In addition, AOA had a positive relationship with moisture, which supported a previous observation that too high TF accelerated water evaporation and decreased the abundance of ammonia-oxidizing archaea [120].

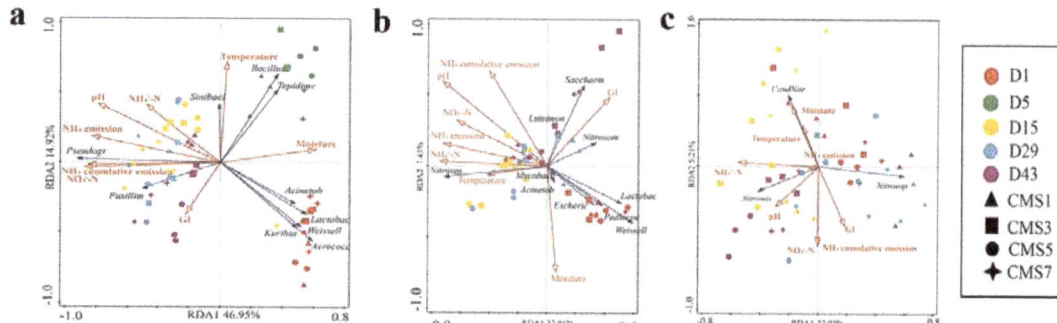

Figure 5. Ordination plots of redundancy analysis for the relationships between environmental factors and bacterial communities at the genus level (**a**), at and between environmental factors and AOB community at the genus level (**b**) and the AOA community at the genus level (**c**) in different treatments during the composting. *Pseudogr* refers to *Pseudogracilibacillus*, *Aerococc* refers to *Aerococcus*, *Lactobac* refers to *Lactobacillus*, *Tepidimi* refers to *Tepidimicrobium*, *Weissell* refers to *Weissella*, *Pusillim* refers to *Pusillimonas*, *Sinibaci* refers to *Sinibacillus*, *Acinetob* refers to *Acinetobacter*, *Nitrosos* refers to *Nitrosospira*, *Nitrosom* refers to *Nitrosomonas*, *Saccharo* refers to *Saccharomonospora*, *Pediococ* refers to *Pediococcus*, *Escheric* refers to *Escherichia*, *Mycobact* refers to *Mycobacterium*, *Luteimon* refers to *Luteimonas*, *Nitrosop* refers to *Nitrosopumilales*, *Nitrosos* refers to *Nitrososphaerales*.

Further analysis combined with spearman correlation was used to test the correlation between pH, moisture, ammonia release and out numbers of AOA and AOB (Supplementary Tables S3–S5). It was found that pH and NH_3 release were significantly correlated with the abundance of AOB and AOA ($p < 0.05$). In addition, NH_3 release was significantly positively correlated with pH and negatively correlated with moisture content ($p < 0.001$) [120]. Specifically, NH_3 release was significantly negatively correlated with *Lactobacillus*, *Weissella*, *Acinetobacter* and positively correlated with *Nitrososphaera* as shown in Supplementary Table S3 during the whole composting ($p < 0.05$). While the NH_3 release was significantly negatively correlated with *Lactobacillus*, *Nitrosomonas*, *Weissella*, *Acinetobacter* and *Nitrososphaera* on day1 and day15 (shown as Supplementary Table S4), the NH_3 release was significantly negatively correlated with *Nitrosospira* and positively correlated with *Acinetobacter* on day 29 and day 43 as shown in Supplementary Table S5 ($p < 0.05$). Therefore, in this study AOA and AOB have an important influence on change of NH_3 emission during composting. *Nitrososphaera* and *Nitrosospira* were significantly negatively correlated with cumulative NH_3 emission ($p < 0.05$) during the early (on day 1 and day 15) and late (on day29 and day 43) stage of composting, respectively. In addition, it was found that pH, moisture, structure and abundance of microbial community (AOB/AOA) would all affect NH_3 emission during the composting.

In this study we found that both AOA and AOB have an important influence on the change of NH_3 emission during composting. *Nitrosospira* and *Nitrososphaera* with high abundance significantly reduced the ammonia emission under turn with suitable frequency during composting of chicken manure. It indicated that we can control the release of NH_3 through increasing the abundance of ammonia oxidizing bacteria and archaea. The results can provide more novel theoretical support for efficient utilization of livestock and poultry waste.

4. Conclusions

This study demonstrated that turning composting materials once every five days (CMS5) had the best effect on the reduction of NH_3 emission and the compost product quality. *Nitrosospira* and *Nitrososphaera* can convert NH_4^+ or NH_3 to NO_2^- and then NO_3^-, resulting in less ammonia emission. The higher relative abundance of ammonia-oxidizing bacteria (*Nitrosospira*) and archaea (*Nitrososphaera*) in the CMS5 treatment could facilitate the aerobic ammonia oxidation and therefore reduce the NH_3 emission. Too high TF promoted NH_3 emission and total N loss. Different TF significantly affected the richness and diversity of the bacterial communities during the whole composting stages. Therefore, though only chicken manure and soybean straw were used in the current study, we can choose appropriate aeration to adapt different application options during the process of livestock and poultry waste composting, according to different material composition. Finally, not only the practical use of composting products be achieved but also the economic value of composting products can be improved.

Supplementary Materials: The following supporting information can be downloaded online. Figure S1: Non-metric multidimensional scaling (NMDS) analysis of bacterial community structure in different turning frequency treatments at various composting stages (a1–a5: represent day1, 5, 15, 29 and 43, respectively). Figure S2: LDA Effect Size (Lefse) analysis of the biomarkers for bacterial community in different treatments. Cladogram showing the biomarkers with significant biological differences from the phylum to genus levels at different composting stages (a1–a5: represent day1, 5, 15, 29 and 43, respectively). Figure S3: Heat map showing the random forest analysis results at the phylum (a1) and genus (a2) levels of bacterial community in different treatments during composting. Figure S4: Non-metric multidimensional scaling (NMDS) analysis of AOB and AOA communities in different turning frequency treatments during various composting stages (a1–a4: represent AOB at day1, 15, 29 and 43, respectively, b1–b4: represent AOA at day1, 15, 29 and 43, respectively). Figure S5: LDA Effect Size (Lefse) analysis of the biomarkers for AOB community in different treatments. Cladogram showing the biomarkers with significant biological differences from the phylum to genus levels at different composting stages (a1–a3: represent day 15, 29 and 43, respectively). Figure S6: Heat map showing the random forest analysis results at the phylum (a1, AOB and b1, AOA) and genus (a2, AOB and b2, AOA) levels of these communities in different treatments during composting. Figure S7: Predicted metabolic pathways involved in composting based on 16S rRNA gene sequencing. Table S1: Numbers of significantly changed phylum/class/order/family/genus identified by the LEfSe analysis based on 16S rRNA gene sequencing. Table S2: Numbers of significantly changed phylum/class/order/family/genus identified by the LEfSe analysis based on AOB gene sequencing. Table S3: Relationships between pH, H_2O, NH_3 emission, NH_3 cumulative emissons and ammonia oxidizing bacteria/archaea of the whole composting (Day 1, 15, 29 and 43). Table S4: Relationships between pH, H_2O, NH_3 emission, NH3 cumulative emissons and ammonia oxidizing bacteria/archaea of the mesophilic and thermophilic stage during composting (Day1,15). Table S5: Relationships between pH, H_2O, NH_3 emission, NH_3 cumulative emissons and ammonia oxidizing bacteria/archaea of the cooling and mature stage during composting (Day 29, 43). Table S6: C/N ratio of all treatments during composting.

Author Contributions: Q.M. was responsible for writing the manuscript. Y.L. was responsible for data curation. J.X. was responsible for writing—review & editing. D.C. was responsible for analyzing the data. Z.L. supervised the study. All authors have read and agreed to the published version of the manuscript.

Funding: This research was funded by the National Natural Science Foundation of China (31972943), the Natural Key Technologies R&D Program (2018YFD0500206), and the Special Fund for Ago-scientific Research in the Public Interest (201503124).

Institutional Review Board Statement: Not applicable.

Informed Consent Statement: Not applicable.

Data Availability Statement: All data generated or analyzed during this study are included in this published article.

Acknowledgments: All the authors gratefully acknowledge the National Natural Science Foundation of China (31972943), the Natural Key Technologies R&D Program (2018YFD0500206), and the Special Fund for Ago-scientific Research in the Public Interest (201503124).

Conflicts of Interest: The authors declare no conflict of interest.

References

1. Akdeniz, N. A systematic review of biochar use in animal waste composting. *Waste Manag.* **2019**, *88*, 291–300. [CrossRef]
2. Zhou, S.; Wen, X.; Cao, Z.; Cheng, R.; Qian, Y.; Mi, J.; Wang, Y.; Liao, X.; Ma, B.; Zou, Y.; et al. Modified cornstalk biochar can reduce ammonia emissions from compost by increasing the number of ammonia-oxidizing bacteria and decreasing urease activity. *Bioresour. Technol.* **2021**, *319*, 124120. [CrossRef]
3. Zouari, I.; Masmoudi, F.; Medhioub, K.; Tounsi, S.; Trigui, M. Biocontrol and plant growth-promoting potentiality of bacteria isolated from compost extract. *Antonie Van Leeuwenhoek* **2020**, *113*, 2107–2122. [CrossRef]
4. Meng, Q.; Han, Y.; Zhu, H.; Yang, W.; Bello, A.; Deng, L.; Jiang, X.; Wu, X.; Sheng, S.; Xu, Y.; et al. Differences in distribution of functional microorganism at DNA and cDNA levels in cow manure composting. *Ecotoxicol. Environ. Saf.* **2020**, *191*, 110161. [CrossRef] [PubMed]
5. Bernal, M.P.; Alburquerque, J.A.; Moral, R. Composting of animal manures and chemical criteria for compost maturity assessment. A review. *Bioresour. Technol.* **2009**, *100*, 5444–5453. [CrossRef] [PubMed]
6. Ogunwande, G.A.; Osunade, J.A.; Adekalu, K.O.; Ogunjimi, L.A. Nitrogen loss in chicken litter compost as affected by carbon to nitrogen ratio and turning frequency. *Bioresour. Technol.* **2008**, *99*, 7495–7503. [CrossRef] [PubMed]
7. Qu, J.; Zhang, L.; Zhang, X.; Gao, L.; Tian, Y. Biochar combined with gypsum reduces both nitrogen and carbon losses during agricultural waste composting and enhances overall compost quality by regulating microbial activities and functions. *Bioresour. Technol.* **2020**, *314*, 123781. [CrossRef]
8. Jiang, J.; Kang, K.; Chen, D.; Liu, N. Impacts of delayed addition of N-rich and acidic substrates on nitrogen loss and compost quality during pig manure composting. *Waste Manag.* **2018**, *72*, 161–167. [CrossRef] [PubMed]
9. Cáceres, R.; Malińska, K.; Marfà, O. Nitrification within composting: A review. *Waste Manag.* **2018**, *72*, 119–137. [CrossRef]
10. Tsutsui, H.; Fujiwara, T.; Inoue, D.; Ito, R.; Matsukawa, K.; Funamizu, N. Relationship between respiratory quotient, nitrification, and nitrous oxide emissions in a forced aerated composting process. *Waste Manag.* **2015**, *42*, 10–16. [CrossRef]
11. Wang, R.; Zhao, Y.; Xie, X.; Mohamed, T.A.; Zhu, L.; Tang, Y.; Chen, Y.; Wei, Z. Role of NH_3 recycling on nitrogen fractions during sludge composting. *Bioresour. Technol.* **2020**, *295*, 122175. [CrossRef]
12. Jiang, T.; Schuchardt, F.; Li, G.; Guo, R.; Zhao, Y. Effect of C/N ratio, aeration rate and moisture content on ammonia and greenhouse gas emission during the composting. *J. Environ. Sci.* **2011**, *23*, 1754–1760. [CrossRef]
13. Chen, J.; Hou, D.; Pang, W.; Nowar, E.E.; Tomberlin, J.K.; Hu, R.; Chen, H.; Xie, J.; Zhang, J.; Yu, Z.; et al. Effect of moisture content on greenhouse gas and NH_3 emissions from pig manure converted by black soldier fly. *Sci. Total Environ.* **2019**, *697*, 133840. [CrossRef] [PubMed]
14. He, X.; Han, L.; Ge, J.; Huang, G. Modelling for reactor-style aerobic composting based on coupling theory of mass-heat-momentum transport and Contois equation. *Bioresour. Technol.* **2018**, *253*, 165–174. [CrossRef]
15. Chowdhury, M.A.; de Neergaard, A.; Jensen, L.S. Potential of aeration flow rate and bio-char addition to reduce greenhouse gas and ammonia emissions during manure composting. *Chemosphere* **2014**, *97*, 16–25. [CrossRef]
16. Amlinger, F.; Peyr, S.; Cuhls, C. Green house gas emissions from composting and mechanical biological treatment. *Waste Manag. Res. J. Int. Solid Wastes Public Clean. Assoc. ISWA* **2008**, *26*, 47–60. [CrossRef]
17. Rincón, C.A.; De Guardia, A.; Couvert, A.; Soutrel, I.; Guezel, S.; Le Serrec, C. Odor generation patterns during different operational composting stages of anaerobically digested sewage sludge. *Waste Manag.* **2019**, *95*, 661–673. [CrossRef]
18. Soto-Paz, J.; Oviedo-Ocaña, E.R.; Manyoma, P.C.; Marmolejo-Rebellón, L.F.; Torres-Lozada, P.; Barrena, R.; Sánchez, A.; Komilis, D. Influence of mixing ratio and turning frequency on the co-composting of biowaste with sugarcane filter cake: A mixture experimental design. *Waste Biomass Valorization* **2019**, *11*, 2475–2489. [CrossRef]
19. Kalamdhad, A.S.; Kazmi, A.A. Effects of turning frequency on compost stability and some chemical characteristics in a rotary drum composter. *Chemosphere* **2009**, *74*, 1327–1334. [CrossRef] [PubMed]
20. Tiquia, S.M. Microbiological parameters as indicators of compost maturity. *J. Appl. Microbiol.* **2005**, *99*, 816–828. [CrossRef]
21. Tiquia, S.M. Reduction of compost phytotoxicity during the process of decomposition. *Chemosphere* **2010**, *79*, 506–512. [CrossRef]
22. Nguyen, V.-T.; Le, T.-H.; Bui, X.-T.; Nguyen, T.-N.; Lin, C.; Vo, T.-D.-H.; Nguyen, H.-H.; Nguyen, D.D.; Senoro, D.B.; Dang, B.-T.; et al. Effects of C/N ratios and turning frequencies on the composting process of food waste and dry leaves. *Bioresour. Technol. Rep.* **2020**, *11*, 100527. [CrossRef]
23. Zhang, J.; Ying, Y.; Yao, X. Effects of turning frequency on the nutrients of Camellia oleifera shell co-compost with goat dung and evaluation of co-compost maturity. *PLoS ONE* **2019**, *14*, e0222841. [CrossRef]
24. Manu, M.K.; Kumar, R.; Garg, A. Decentralized composting of household wet biodegradable waste in plastic drums: Effect of waste turning, microbial inoculum and bulking agent on product quality. *J. Clean. Prod.* **2019**, *226*, 233–241. [CrossRef]
25. Soto-Paz, J.; Oviedo-Ocaña, E.R.; Manyoma-Velásquez, P.C.; Torres-Lozada, P.; Gea, T. Evaluation of mixing ratio and frequency of turning in the co-composting of biowaste with sugarcane filter cake and star grass. *Waste Manag.* **2019**, *96*, 86–95. [CrossRef] [PubMed]
26. Onwosi, C.O.; Igbokwe, V.C.; Odimba, J.N.; Eke, I.E.; Nwankwoala, M.O.; Iroh, I.N.; Ezeogu, L.I. Composting technology in waste stabilization: On the methods, challenges and future prospects. *J. Environ. Manag.* **2017**, *190*, 140–157. [CrossRef]
27. Yang, X.C.; Han, Z.Z.; Ruan, X.Y.; Chai, J.; Jiang, S.W.; Zheng, R. Composting swine carcasses with nitrogen transformation microbial strains: Succession of microbial community and nitrogen functional genes. *Sci. Total Environ.* **2019**, *688*, 555–566. [CrossRef] [PubMed]

28. Wang, J.; Liu, Z.; Xia, J.; Chen, Y. Effect of microbial inoculation on physicochemical properties and bacterial community structure of citrus peel composting. *Bioresour. Technol.* **2019**, *291*, 121843. [CrossRef]
29. Jiang, J.; Wang, Y.; Guo, F.; Zhang, X.; Dong, W.; Zhang, X.; Zhang, X.; Zhang, C.; Cheng, K.; Li, Y.; et al. Composting pig manure and sawdust with urease inhibitor: Succession of nitrogen functional genes and bacterial community. *Environ. Sci. Pollut. Res. Int.* **2020**, *27*, 36160–36171. [CrossRef]
30. Xiao, R.; Awasthi, M.K.; Li, R.; Park, J.; Pensky, S.M.; Wang, Q.; Wang, J.J.; Zhang, Z. Recent developments in biochar utilization as an additive in organic solid waste composting: A review. *Bioresour. Technol.* **2017**, *246*, 203–213. [CrossRef]
31. Sims, G.K.; Ellsworth, T.R.; Mulvaney, R.L. Microscale determination of inorganic nitrogen in water and soil extracts. *Commun. Soil Sci. Plant Anal.* **1995**, *26*, 303–316. [CrossRef]
32. Lazcano, C.; Gómez-Brandón, M.; Domínguez, J. Comparison of the effectiveness of composting and vermicomposting for the biological stabilization of cattle manure. *Chemosphere* **2008**, *72*, 1013–1019. [CrossRef]
33. Mao, H.; Lv, Z.; Sun, H.; Li, R.; Zhai, B.; Wang, Z.; Awasthi, M.K.; Wang, Q.; Zhou, L. Improvement of biochar and bacterial powder addition on gaseous emission and bacterial community in pig manure compost. *Bioresour. Technol.* **2018**, *258*, 195–202. [CrossRef] [PubMed]
34. Yang, F.; Li, G.; Shi, H.; Wang, Y. Effects of phosphogypsum and superphosphate on compost maturity and gaseous emissions during kitchen waste composting. *Waste Manag.* **2015**, *36*, 70–76. [CrossRef] [PubMed]
35. Rotthauwe, J.H.; Witzel, K.P.; Liesack, W. The ammonia monooxygenase structural gene *amo*A as a functional marker: Molecular fine-scale analysis of natural ammonia-oxidizing populations. *Appl. Environ. Microbiol.* **1997**, *63*, 4704–4712. [CrossRef] [PubMed]
36. Park, S.J.; Park, B.J.; Rhee, S.K. Comparative analysis of archaeal 16S rRNA and *amo*A genes to estimate the abundance and diversity of ammonia-oxidizing archaea in marine sediments. *Extrem. Life Under Extrem. Cond.* **2008**, *12*, 605–615. [CrossRef] [PubMed]
37. Rognes, T.; Flouri, T.; Nichols, B.; Quince, C.; Mahé, F. VSEARCH: A versatile open source tool for metagenomics. *PeerJ* **2016**, *4*, e2584. [CrossRef] [PubMed]
38. Bolyen, E.; Rideout, J.R.; Dillon, M.R.; Bokulich, N.A.; Abnet, C.C.; Al-Ghalith, G.A.; Alexander, H.; Alm, E.J.; Arumugam, M.; Asnicar, F.; et al. Reproducible, interactive, scalable and extensible microbiome data science using QIIME 2. *Nat. Biotechnol.* **2019**, *37*, 852–857. [CrossRef]
39. Callahan, B.J.; McMurdie, P.J.; Rosen, M.J.; Han, A.W.; Johnson, A.J.; Holmes, S.P. DADA2: High-resolution sample inference from Illumina amplicon data. *Nat. Methods* **2016**, *13*, 581–583. [CrossRef]
40. Quast, C.; Pruesse, E.; Yilmaz, P.; Gerken, J.; Schweer, T.; Yarza, P.; Peplies, J.; Glöckner, F.O. The SILVA ribosomal RNA gene database project: Improved data processing and web-based tools. *Nucleic Acids Res.* **2013**, *41*, D590–D596. [CrossRef]
41. Altschul, S.F.; Madden, T.L.; Schaffer, A.A.; Zhang, J.H.; Zhang, Z.; Miller, W.; Lipman, D.J. Gapped BLAST and PSI-BLAST: A new generation of protein database search programs. *Nucleic Acids Res.* **1997**, *25*, 3389–3402. [CrossRef]
42. Segata, N.; Izard, J.; Waldron, L.; Gevers, D.; Miropolsky, L.; Garrett, W.S.; Huttenhower, C. Metagenomic biomarker discovery and explanation. *Genome Biol.* **2011**, *12*, R60. [CrossRef]
43. Bellino, A.; Baldantoni, D.; Picariello, E.; Morelli, R.; Alfani, A.; De Nicola, F. Role of different microorganisms in remediating PAH-contaminated soils treated with compost or fungi. *J. Environ. Manag.* **2019**, *252*, 109675. [CrossRef]
44. Langille, M.G.; Zaneveld, J.; Caporaso, J.G.; McDonald, D.; Knights, D.; Reyes, J.A.; Clemente, J.C.; Burkepile, D.E.; Vega Thurber, R.L.; Knight, R.; et al. Predictive functional profiling of microbial communities using 16S rRNA marker gene sequences. *Nat. Biotechnol.* **2013**, *31*, 814–821. [CrossRef]
45. Cheng, D.; Feng, Y.; Liu, Y.; Xue, J.; Li, Z. Dynamics of oxytetracycline, sulfamerazine, and ciprofloxacin and related antibiotic resistance genes during swine manure composting. *J. Environ. Manag.* **2019**, *230*, 102–109. [CrossRef]
46. Kelleher, B.P.; Leahy, J.J.; Henihan, A.M.; O'Dwyer, T.F.; Sutton, D.; Leahy, M.J. Advances in poultry litter disposal technology–A review. *Bioresour. Technol.* **2002**, *83*, 27–36. [CrossRef]
47. Huan, C.; Fang, J.; Tong, X.; Zeng, Y.; Liu, Y.; Jiang, X.; Ji, G.; Xu, L.; Lyu, Q.; Yan, Z. Simultaneous elimination of H_2S and NH_3 in a biotrickling filter packed with polyhedral spheres and best efficiency in compost deodorization. *J. Clean. Prod.* **2020**, *284*, 124708. [CrossRef]
48. Xu, Z.; Zhao, B.; Wang, Y.; Xiao, J.; Wang, X. Composting process and odor emission varied in windrow and trough composting system under different air humidity conditions. *Bioresour. Technol.* **2020**, *297*, 122482. [CrossRef] [PubMed]
49. He, X.; Yin, H.; Fang, C.; Xiong, J.; Han, L.; Yang, Z.; Huang, G. Metagenomic and q-PCR analysis reveals the effect of powder bamboo biochar on nitrous oxide and ammonia emissions during aerobic composting. *Bioresour. Technol.* **2021**, *323*, 124567. [CrossRef] [PubMed]
50. Angel, R.; Soares, M.I.; Ungar, E.D.; Gillor, O. Biogeography of soil archaea and bacteria along a steep precipitation gradient. *ISME J.* **2010**, *4*, 553–563. [CrossRef] [PubMed]
51. Ren, L.; Cai, C.; Zhang, J.; Yang, Y.; Wu, G.; Luo, L.; Huang, H.; Zhou, Y.; Qin, P.; Yu, M. Key environmental factors to variation of ammonia-oxidizing archaea community and potential ammonia oxidation rate during agricultural waste composting. *Bioresour. Technol.* **2018**, *270*, 278–285. [CrossRef]
52. Li, H.; Duan, M.; Gu, J.; Zhang, Y.; Qian, X.; Ma, J.; Zhang, R.; Wang, X. Effects of bamboo charcoal on antibiotic resistance genes during chicken manure composting. *Ecotoxicol. Environ. Saf.* **2017**, *140*, 1–6. [CrossRef] [PubMed]

53. Liu, T.; Kumar Awasthi, M.; Kumar Awasthi, S.; Ren, X.; Liu, X.; Zhang, Z. Influence of fine coal gasification slag on greenhouse gases emission and volatile fatty acids during pig manure composting. *Bioresour. Technol.* **2020**, *316*, 123915. [CrossRef]
54. Nigussie, A.; Bruun, S.; Kuyper, T.W.; de Neergaard, A. Delayed addition of nitrogen-rich substrates during composting of municipal waste: Effects on nitrogen loss, greenhouse gas emissions and compost stability. *Chemosphere* **2017**, *166*, 352–362. [CrossRef]
55. Yang, F.; Li, G.X.; Yang, Q.Y.; Luo, W.H. Effect of bulking agents on maturity and gaseous emissions during kitchen waste composting. *Chemosphere* **2013**, *93*, 1393–1399. [CrossRef]
56. Saidi, N.; Kouki, S.; M'Hiri, F.; Jedidi, N.; Mahrouk, M.; Hassen, A.; Ouzari, H. Microbiological parameters and maturity degree during composting of Posidonia oceanica residues mixed with vegetable wastes in semi-arid pedo-climatic condition. *J. Environ. Sci.* **2009**, *21*, 1452–1458. [CrossRef]
57. Ravindran, B.; Nguyen, D.D.; Chaudhary, D.K.; Chang, S.W.; Kim, J.; Lee, S.R.; Shin, J.; Jeon, B.H.; Chung, S.; Lee, J. Influence of biochar on physico-chemical and microbial community during swine manure composting process. *J. Environ. Manag.* **2019**, *232*, 592–599. [CrossRef] [PubMed]
58. Tiquia, S.M.; Tam, N.F.Y.; Hodgkiss, I.J. Effects of turning frequency on composting of spent pig-manure sawdust litter. *Bioresour. Technol.* **1997**, *62*, 37–42. [CrossRef]
59. Li, R.; Wang, J.J.; Zhang, Z.; Shen, F.; Zhang, G.; Qin, R.; Li, X.; Xiao, R. Nutrient transformations during composting of pig manure with bentonite. *Bioresour. Technol.* **2012**, *121*, 362–368. [CrossRef] [PubMed]
60. Awasthi, M.K.; Wang, Q.; Awasthi, S.K.; Wang, M.; Chen, H.; Ren, X.; Zhao, J.; Zhang, Z. Influence of medical stone amendment on gaseous emissions, microbial biomass and abundance of ammonia oxidizing bacteria genes during biosolids composting. *Bioresour. Technol.* **2018**, *247*, 970–979. [CrossRef]
61. Zeng, G.; Zhang, L.; Dong, H.; Chen, Y.; Zhang, J.; Zhu, Y.; Yuan, Y.; Xie, Y.; Fang, W. Pathway and mechanism of nitrogen transformation during composting: Functional enzymes and genes under different concentrations of PVP-AgNPs. *Bioresour. Technol.* **2018**, *253*, 112–120. [CrossRef]
62. Wang, K.; Wu, Y.; Li, W.; Wu, C.; Chen, Z. Insight into effects of mature compost recycling on N_2O emission and denitrification genes in sludge composting. *Bioresour. Technol.* **2018**, *251*, 320–326. [CrossRef] [PubMed]
63. Chen, X.; Zhao, X.; Ge, J.; Zhao, Y.; Wei, Z.; Yao, C.; Meng, Q.; Zhao, R. Recognition of the neutral sugars conversion induced by bacterial community during lignocellulose wastes composting. *Bioresour. Technol.* **2019**, *294*, 122153. [CrossRef]
64. An, X.; Cheng, Y.; Miao, L.; Chen, X.; Zang, H.; Li, C. Characterization and genome functional analysis of an efficient nitrile-degrading bacterium, *Rhodococcus rhodochrous* BX2, to lay the foundation for potential bioaugmentation for remediation of nitrile-contaminated environments. *J. Hazard. Mater.* **2020**, *389*, 121906. [CrossRef] [PubMed]
65. Zeng, G.; Zhang, J.; Chen, Y.; Yu, Z.; Yu, M.; Li, H.; Liu, Z.; Chen, M.; Lu, L.; Hu, C. Relative contributions of archaea and bacteria to microbial ammonia oxidation differ under different conditions during agricultural waste composting. *Bioresour. Technol.* **2011**, *102*, 9026–9032. [CrossRef] [PubMed]
66. Wu, J.; Wei, Z.; Zhu, Z.; Zhao, Y.; Jia, L.; Lv, P. Humus formation driven by ammonia-oxidizing bacteria during mixed materials composting. *Bioresour. Technol.* **2020**, *311*, 123500. [CrossRef]
67. Agyarko-Mintah, E.; Cowie, A.; Van Zwieten, L.; Singh, B.P.; Smillie, R.; Harden, S.; Fornasier, F. Biochar lowers ammonia emission and improves nitrogen retention in poultry litter composting. *Waste Manag.* **2017**, *61*, 129–137. [CrossRef]
68. Jiang, T.; Ma, X.; Tang, Q.; Yang, J.; Li, G.; Schuchardt, F. Combined use of nitrification inhibitor and struvite crystallization to reduce the NH_3 and N_2O emissions during composting. *Bioresour. Technol.* **2016**, *217*, 210–218. [CrossRef]
69. Guo, H.; Gu, J.; Wang, X.; Nasir, M.; Yu, J.; Lei, L.; Wang, J.; Zhao, W.; Dai, X. Beneficial effects of bacterial agent/bentonite on nitrogen transformation and microbial community dynamics during aerobic composting of pig manure. *Bioresour. Technol.* **2020**, *298*, 122384. [CrossRef]
70. Wang, X.; Bai, Z.H.; Yao, Y.; Gao, B.B.; Chadwick, D.; Chen, Q.; Hu, C.S.; Ma, L. Composting with negative pressure aeration for the mitigation of ammonia emissions and global warming potential. *J. Clean. Prod.* **2018**, *195*, 448–457. [CrossRef]
71. Szanto, G.L.; Hamelers, H.V.; Rulkens, W.H.; Veeken, A.H. NH_3, N_2O and CH_4 emissions during passively aerated composting of straw-rich pig manure. *Bioresour. Technol.* **2007**, *98*, 2659–2670. [CrossRef]
72. Jiang, T.; Schuchardt, F.; Li, G.X.; Guo, R.; Luo, Y.M. Gaseous emission during the composting of pig feces from Chinese Ganqinfen system. *Chemosphere* **2013**, *90*, 1545–1551. [CrossRef]
73. Koyama, M.; Nagao, N.; Syukri, F.; Rahim, A.A.; Kamarudin, M.S.; Toda, T.; Mitsuhashi, T.; Nakasaki, K. Effect of temperature on thermophilic composting of aquaculture sludge: NH_3 recovery, nitrogen mass balance, and microbial community dynamics. *Bioresour. Technol.* **2018**, *265*, 207–213. [CrossRef] [PubMed]
74. Awasthi, M.K.; Wang, Q.; Huang, H.; Ren, X.; Lahori, A.H.; Mahar, A.; Ali, A.; Shen, F.; Li, R.; Zhang, Z. Influence of zeolite and lime as additives on greenhouse gas emissions and maturity evolution during sewage sludge composting. *Bioresour. Technol.* **2016**, *216*, 172–181. [CrossRef] [PubMed]
75. Guo, R.; Li, G.; Jiang, T.; Schuchardt, F.; Chen, T.; Zhao, Y.; Shen, Y. Effect of aeration rate, C/N ratio and moisture content on the stability and maturity of compost. *Bioresour. Technol.* **2012**, *112*, 171–178. [CrossRef]
76. Luo, W.H.; Yuan, J.; Luo, Y.M.; Li, G.X.; Nghiem, L.D.; Price, W.E. Effects of mixing and covering with mature compost on gaseous emissions during composting. *Chemosphere* **2014**, *117*, 14–19. [CrossRef] [PubMed]

77. Liang, Y.; Leonard, J.J.; Feddes, J.J.; McGill, W.B. Influence of carbon and buffer amendment on ammonia volatilization in composting. *Bioresour. Technol.* **2006**, *97*, 748–761. [CrossRef]
78. Petric, I.; Sestan, A.; Sestan, I. Influence of initial moisture content on the composting of poultry manure with wheat straw. *Biosyst. Eng.* **2009**, *104*, 125–134. [CrossRef]
79. He, Z.; Lin, H.; Hao, J.; Kong, X.; Tian, K.; Bei, Z.; Tian, X. Impact of vermiculite on ammonia emissions and organic matter decomposition of food waste during composting. *Bioresour. Technol.* **2018**, *263*, 548–554. [CrossRef]
80. Simujide, H.; Aorigele, C.; Wang, C.-J.; Lina, M.; Manda, B. Microbial activities during mesophilic composting of manure and effect of calcium cyanamide addition. *Int. Biodeterior. Biodegrad.* **2013**, *83*, 139–144. [CrossRef]
81. Liu, L.; Wang, S.; Guo, X.; Zhao, T.; Zhang, B. Succession and diversity of microorganisms and their association with physico-chemical properties during green waste thermophilic composting. *Waste Manag.* **2018**, *73*, 101–112. [CrossRef] [PubMed]
82. Chroni, C.; Kyriacou, A.; Fau-Manios, T.; Manios, T.; Fau-Lasaridi, K.-E.; Lasaridi, K.E. Investigation of the microbial community structure and activity as indicators of compost stability and composting process evolution. *Bioresour. Technol.* **2009**, *100*, 3745–3750. [CrossRef]
83. Ren, G.; Xu, X.; Qu, J.; Zhu, L.; Wang, T. Evaluation of microbial population dynamics in the co-composting of cow manure and rice straw using high throughput sequencing analysis. *World J. Microbiol. Biotechnol.* **2016**, *32*, 016–2059. [CrossRef] [PubMed]
84. Liu, Y.; Cheng, D.; Xue, J.; Weaver, L.; Wakelin, S.A.; Feng, Y.; Li, Z. Changes in microbial community structure during pig manure composting and its relationship to the fate of antibiotics and antibiotic resistance genes. *J. Hazard. Mater.* **2020**, *389*, 23. [CrossRef] [PubMed]
85. Wang, K.; Mao, H.; Wang, Z.; Tian, Y. Succession of organics metabolic function of bacterial community in swine manure composting. *J. Hazard. Mater.* **2018**, *360*, 471–480. [CrossRef]
86. Cai, L.; Cao, M.K.; Chen, T.B.; Guo, H.T.; Zheng, G.D. Microbial degradation in the co-composting of pig manure and biogas residue using a recyclable cement-based synthetic amendment. *Waste Manag.* **2021**, *126*, 30–40. [CrossRef]
87. Wan, J.; Wang, X.; Yang, T.; Wei, Z.; Banerjee, S.; Friman, V.P.; Mei, X.; Xu, Y.; Shen, Q. Livestock manure type affects microbial community composition and assembly during composting. *Front. Microbiol.* **2021**, *12*, 621126. [CrossRef]
88. Liu, Y.; Feng, Y.; Cheng, D.; Xue, J.; Wakelin, S.; Li, Z. Dynamics of bacterial composition and the fate of antibiotic resistance genes and mobile genetic elements during the co-composting with gentamicin fermentation residue and lovastatin fermentation residue. *Bioresour. Technol.* **2018**, *261*, 249–256. [CrossRef]
89. Bello, A.; Han, Y.; Zhu, H.; Deng, L.; Yang, W.; Meng, Q.; Sun, Y.; Egbeagu, U.U.; Sheng, S.; Wu, X.; et al. Microbial community composition, co-occurrence network pattern and nitrogen transformation genera response to biochar addition in cattle manure-maize straw composting. *Sci. Total Environ.* **2020**, *721*, 137759. [CrossRef]
90. Yin, J.; Yu, X.; Zhang, Y.; Shen, D.; Wang, M.; Long, Y.; Chen, T. Enhancement of acidogenic fermentation for volatile fatty acid production from food waste: Effect of redox potential and inoculum. *Bioresour. Technol.* **2016**, *216*, 996–1003. [CrossRef]
91. Tang, Y.; Shigematsu, T.; Morimura, S.; Kida, K. The effects of micro-aeration on the phylogenetic diversity of microorganisms in a thermophilic anaerobic municipal solid-waste digester. *Water Res.* **2004**, *38*, 2537–2550. [CrossRef] [PubMed]
92. Xie, J.J.; Liu, Q.Q.; Liao, S.; Fang, H.H.; Yin, P.; Xie, S.W.; Tian, L.X.; Liu, Y.J.; Niu, J. Effects of dietary mixed probiotics on growth, non-specific immunity, intestinal morphology and microbiota of juvenile pacific white shrimp, Litopenaeus vannamei. *Fish Shellfish Immunol.* **2019**, *90*, 456–465. [CrossRef]
93. Gopal, M.; Bhute, S.S.; Gupta, A.; Prabhu, S.R.; Thomas, G.V.; Whitman, W.B.; Jangid, K. Changes in structure and function of bacterial communities during coconut leaf vermicomposting. *Antonie Van Leeuwenhoek* **2017**, *110*, 1339–1355. [CrossRef]
94. Lu, H.; Wu, Y.; Liang, P.; Song, Q.; Zhang, H.; Wu, J.; Wu, W.; Liu, X.; Dong, C. Alkaline amendments improve the health of soils degraded by metal contamination and acidification: Crop performance and soil bacterial community responses. *Chemosphere* **2020**, *257*, 127309. [CrossRef]
95. Chen, G.; Bin, L.; Tang, B.; Huang, S.; Li, P.; Fu, F.; Wu, L.; Yang, Z. Rapid reformation of larger aerobic granular sludge in an internal-circulation membrane bioreactor after long-term operation: Effect of short-time aeration. *Bioresour. Technol.* **2019**, *273*, 462–467. [CrossRef]
96. Ma, S.; Xiong, J.; Cui, R.; Sun, X.; Huang, G.; Han, L.; Xu, Y.; Kan, Z.; Gong, X. Effects of intermittent aeration on greenhouse gas emissions and bacterial community succession during large-scale membrane-covered aerobic composting. *J. Clean. Prod.* **2020**, *266*, 121551. [CrossRef]
97. Verbaendert, I.; Boon, N.; De Vos, P.; Heylen, K. Denitrification is a common feature among members of the genus Bacillus. *Syst. Appl. Microbiol.* **2011**, *34*, 385–391. [CrossRef] [PubMed]
98. Zainudin, M.H.; Mustapha, N.A.; Maeda, T.; Ramli, N.; Sakai, K.; Hassan, M. Biochar enhanced the nitrifying and denitrifying bacterial communities during the composting of poultry manure and rice straw. *Waste Manag.* **2020**, *106*, 240–249. [CrossRef]
99. Alcántara, C.; Domínguez, J.M.; García, D.; Blanco, S.; Pérez, R.; García-Encina, P.A.; Muñoz, R. Evaluation of wastewater treatment in a novel anoxic-aerobic algal-bacterial photobioreactor with biomass recycling through carbon and nitrogen mass balances. *Bioresour. Technol.* **2015**, *191*, 173–186. [CrossRef]
100. Zhang, L.; Li, L.; Pan, X.; Shi, Z.; Feng, X.; Gong, B.; Li, J.; Wang, L. Enhanced growth and activities of the dominant functional microbiota of chicken manure composts in the presence of maize straw. *Front. Microbiol.* **2018**, *9*, 1131. [CrossRef] [PubMed]
101. Bian, X.; Gong, H.; Wang, K. Pilot-scale hydrolysis-aerobic treatment for actual municipal wastewater: Performance and microbial community analysis. *Int. J. Environ. Res. Public Health* **2018**, *15*, 477. [CrossRef] [PubMed]

102. Ma, S.; Fang, C.; Sun, X.; Han, L.; He, X.; Huang, G. Bacterial community succession during pig manure and wheat straw aerobic composting covered with a semi-permeable membrane under slight positive pressure. *Bioresour. Technol.* **2018**, *259*, 221–227. [CrossRef]
103. Glaeser, S.P.; McInroy, J.A.; Busse, H.J.; Kämpfer, P. *Pseudogracilibacillus auburnensis* gen. nov., sp. nov., isolated from the rhizosphere of Zea mays. *Int. J. Syst. Evol. Microbiol.* **2014**, *64*, 2442–2448. [CrossRef]
104. He, P.; Wei, S.; Shao, L.; Lü, F. Aerosolization behavior of prokaryotes and fungi during composting of vegetable waste. *Waste Manag.* **2019**, *89*, 103–113. [CrossRef] [PubMed]
105. Goecks, J.; Nekrutenko, A.; Taylor, J. Galaxy: A comprehensive approach for supporting accessible, reproducible, and transparent computational research in the life sciences. *Genome Biol.* **2010**, *11*, 2010–2011. [CrossRef]
106. Zaitsev, G.M.; Tsitko, I.V.; Rainey, F.A.; Trotsenko, Y.A.; Uotila, J.S.; Stackebrandt, E.; Salkinoja-Salonen, M.S. New aerobic ammonium-dependent obligately oxalotrophic bacteria: Description of *Ammoniphilus oxalaticus* gen. nov., sp. nov. and *Ammoniphilus oxalivorans* gen. nov., sp. nov. *Int. J. Syst. Bacteriol.* **1998**, *1*, 151–163. [CrossRef] [PubMed]
107. Vuono, D.C.; Regnery, J.; Li, D.; Jones, Z.L.; Holloway, R.W.; Drewes, J.E. rRNA Gene expression of abundant and rare activated-sludge microorganisms and growth rate induced micropollutant removal. *Environ. Sci. Technol.* **2016**, *50*, 6299–6309. [CrossRef]
108. van Elsas, J.D.; Chiurazzi, M.; Mallon, C.A.; Elhottova, D.; Kristufek, V.; Salles, J.F. Microbial diversity determines the invasion of soil by a bacterial pathogen. *Proc. Natl. Acad. Sci. USA* **2012**, *109*, 1159–1164. [CrossRef]
109. Hol, W.H.; de Boer, W.; Termorshuizen, A.J.; Meyer, K.M.; Schneider, J.H.; van Dam, N.M.; van Veen, J.A.; van der Putten, W.H. Reduction of rare soil microbes modifies plant-herbivore interactions. *Ecol. Lett.* **2010**, *13*, 292–301. [CrossRef]
110. Partanen, P.; Hultman, J.; Paulin, L.; Auvinen, P.; Romantschuk, M. Bacterial diversity at different stages of the composting process. *BMC Microbiol.* **2010**, *10*, 94. [CrossRef]
111. Park, M.S.; Park, Y.J.; Jung, J.Y.; Lee, S.H.; Park, W.; Lee, K.; Jeon, C.O. *Pusillimonas harenae* sp. nov., isolated from a sandy beach, and emended description of the genus *Pusillimonas*. *Int. J. Syst. Evol. Microbiol.* **2011**, *61*, 2901–2906. [CrossRef]
112. Tohno, M.; Kitahara, M.; Matsuyama, S.; Kimura, K.; Ohkuma, M.; Tajima, K. *Aerococcus vaginalis* sp. nov., isolated from the vaginal mucosa of a beef cow, and emended descriptions of *Aerococcus suis*, *Aerococcus viridans*, *Aerococcus urinaeequi*, *Aerococcus urinaehominis*, *Aerococcus urinae*, *Aerococcus christensenii* and *Aerococcus sanguinicola*. *Int. J. Syst. Evol. Microbiol.* **2014**, *64*, 1229–1236.
113. Song, L.; Liu, H.; Cai, S.; Huang, Y.; Dai, X.; Zhou, Y. *Alcanivorax indicus* sp. nov., isolated from seawater. *Int. J. Syst. Evol. Microbiol.* **2018**, *68*, 3785–3789. [CrossRef]
114. Weon, H.Y.; Lee, S.Y.; Kim, B.Y.; Noh, H.J.; Schumann, P.; Kim, J.S.; Kwon, S.W. *Ureibacillus composti* sp. nov. and *Ureibacillus thermophilus* sp. nov., isolated from livestock-manure composts. *Int. J. Syst. Evol. Microbiol.* **2007**, *57*, 2908–2911. [CrossRef] [PubMed]
115. Wang, W.; Su, Y.; Wang, B.; Wang, Y.; Zhuang, L.; Zhu, G. Spatiotemporal shifts of ammonia-oxidizing archaea abundance and structure during the restoration of a multiple pond and plant-bed/ditch wetland. *Sci. Total Environ.* **2019**, *684*, 629–640. [CrossRef] [PubMed]
116. Alves, R.J.E.; Minh, B.Q.; Urich, T.; von Haeseler, A.; Schleper, C. Unifying the global phylogeny and environmental distribution of ammonia-oxidising archaea based on amoA genes. *Nat. Commun.* **2018**, *9*, 1517. [CrossRef]
117. Yamamoto, N.; Otawa, K.; Nakai, Y. Diversity and abundance of ammonia-oxidizing bacteria and ammonia-oxidizing archaea during cattle manure composting. *Microb. Ecol.* **2010**, *60*, 807–815. [CrossRef] [PubMed]
118. Yamamoto, N.; Oishi, R.; Suyama, Y.; Tada, C.; Nakai, Y. Ammonia-oxidizing bacteria rather than ammonia-oxidizing archaea were widely distributed in animal manure composts from field-scale facilities. *Microbes Environ.* **2012**, *27*, 519–524. [CrossRef] [PubMed]
119. de Gannes, V.; Eudoxie, G.; Dyer, D.H.; Hickey, W.J. Diversity and abundance of ammonia oxidizing archaea in tropical compost systems. *Front. Microbiol.* **2012**, *3*, 244. [CrossRef]
120. Chen, Q.; Wang, J.; Zhang, H.; Shi, H.; Liu, G.; Che, J.; Liu, B. Microbial community and function in nitrogen transformation of ectopic fermentation bed system for pig manure composting. *Bioresour. Technol.* **2021**, *319*, 124155. [CrossRef]
121. Ge, J.; Huang, G.; Li, J.; Sun, X.; Han, L. Multivariate and Multiscale Approaches for Interpreting the Mechanisms of Nitrous Oxide Emission during Pig Manure-Wheat Straw Aerobic Composting. *Environ. Sci. Technol.* **2018**, *52*, 8408–8418. [CrossRef]
122. Fan, F.; Yang, Q.; Li, Z.; Wei, D.; Cui, X.; Liang, Y. Impacts of organic and inorganic fertilizers on nitrification in a cold climate soil are linked to the bacterial ammonia oxidizer community. *Microb. Ecol.* **2011**, *62*, 982–990. [CrossRef] [PubMed]
123. Yang, Z.; Guan, Y.; Bello, A.; Wu, Y.; Ding, J.; Wang, L.; Ren, Y.; Chen, G.; Yang, W. Dynamics of ammonia oxidizers and denitrifiers in response to compost addition in black soil, Northeast China. *PeerJ* **2020**, *21*, e8844. [CrossRef]
124. Ge, C.H.; Dong, Y.; Li, H.; Li, Q.; Ni, S.Q.; Gao, B.; Xu, S.; Qiao, Z.; Ding, S. Nitritation-anammox process-A realizable and satisfactory way to remove nitrogen from high saline wastewater. *Bioresour. Technol.* **2019**, *275*, 86–93. [CrossRef] [PubMed]
125. Zhong, X.Z.; Zeng, Y.; Wang, S.P.; Sun, Z.Y.; Tang, Y.Q.; Kida, K. Insight into the microbiology of nitrogen cycle in the dairy manure composting process revealed by combining high-throughput sequencing and quantitative PCR. *Bioresour. Technol.* **2020**, *301*, 122760. [CrossRef]
126. Siripong, S.; Rittmann, B.E. Diversity study of nitrifying bacteria in full-scale municipal wastewater treatment plants. *Water Res.* **2007**, *41*, 1110–1120. [CrossRef]

127. Avrahami, S.; Jia, Z.; Neufeld, J.D.; Murrell, J.C.; Conrad, R.; Küsel, K. Active autotrophic ammonia-oxidizing bacteria in biofilm enrichments from simulated creek ecosystems at two ammonium concentrations respond to temperature manipulation. *Appl. Environ. Microbiol.* **2011**, *77*, 7329–7338. [CrossRef]
128. Kowalchuk, G.A.; Naoumenko, Z.S.; Derikx, P.J.; Felske, A.; Stephen, J.R.; Arkhipchenko, I.A. Molecular analysis of ammonia-oxidizing bacteria of the beta subdivision of the class Proteobacteria in compost and composted materials. *Appl. Environ. Microbiol.* **1999**, *65*, 396–403. [CrossRef]
129. Guo, H.; Gu, J.; Wang, X.; Yu, J.; Nasir, M.; Zhang, K.; Sun, W. Microbial driven reduction of N_2O and NH_3 emissions during composting: Effects of bamboo charcoal and bamboo vinegar. *J. Hazard. Mater.* **2020**, *390*, 121292. [CrossRef]
130. Shi, M.; Zhao, Y.; Zhang, A.; Zhao, M.; Zhai, W.; Wei, Z.; Song, Y.; Tang, X.; He, P. Factoring distinct materials and nitrogen-related microbes into assessments of nitrogen pollution risks during composting. *Bioresour. Technol.* **2021**, *329*, 124896. [CrossRef] [PubMed]
131. Wei, Y.; Li, Z.; Ran, W.; Yuan, H.; Li, X. Performance and microbial community dynamics in anaerobic co-digestion of chicken manure and corn stover with different modification methods and trace element supplementation strategy. *Bioresour. Technol.* **2021**, *325*, 124713. [CrossRef] [PubMed]
132. Brochier-Armanet, C.; Boussau, B.; Gribaldo, S.; Forterre, P. Mesophilic Crenarchaeota: Proposal for a third archaeal phylum, the Thaumarchaeota. *Nat. Rev. Microbiol.* **2008**, *6*, 245–252. [CrossRef]
133. Chandler, L.; Harford, A.J.; Hose, G.C.; Humphrey, C.L.; Chariton, A.; Greenfield, P.; Davis, J. Saline mine-water alters the structure and function of prokaryote communities in shallow groundwater below a tropical stream. *Environ. Pollut.* **2021**, *284*, 117318. [CrossRef] [PubMed]
134. Stieglmeier, M.; Klingl, A.; Alves, R.J.E.; Rittmann, S.K.R.; Melcher, M.; Leisch, N.; Schleper, C. *Nitrososphaera viennensis* gen. nov., sp. nov., an aerobic and mesophilic, ammonia-oxidizing archaeon from soil and a member of the archaeal phylum Thaumarchaeota. *Int. J. Syst. Evol. Microbiol.* **2014**, *64*, 2738–2752. [CrossRef] [PubMed]
135. Li, X.; Zhang, M.; Liu, F.; Chen, L.; Li, Y.; Li, Y.; Xiao, R.; Wu, J. Seasonality distribution of the abundance and activity of nitrification and denitrification microorganisms in sediments of surface flow constructed wetlands planted with *Myriophyllum elatinoides* during swine wastewater treatment. *Bioresour. Technol.* **2018**, *248*, 89–97. [CrossRef]
136. Erguder, T.H.; Boon, N.; Wittebolle, L.; Marzorati, M.; Verstraete, W. Environmental factors shaping the ecological niches of ammonia-oxidizing archaea. *FEMS Microbiol. Rev.* **2009**, *33*, 855–869. [CrossRef] [PubMed]

Article

A Multiwell-Based Assay for Screening Thyroid Hormone Signaling Disruptors Using *thibz* Expression as a Sensitive Endpoint in *Xenopus laevis*

Jinbo Li [1,2], Yuanyuan Li [1,2], Min Zhu [1,2], Shilin Song [1,2] and Zhanfen Qin [1,2,*]

1. State Key Laboratory of Environmental Chemistry and Ecotoxicology, Research Center for Eco-Environmental Sciences, Chinese Academy of Sciences, Beijing 100085, China; jbli2016_st@rcees.ac.cn (J.L.); yyli@rcees.ac.cn (Y.L.); zhumin2@jshb.gov.cn (M.Z.); songshilin21@mails.ucas.ac.cn (S.S.)
2. University of Chinese Academy of Sciences, Beijing 100049, China
* Correspondence: qinzhanfen@rcees.ac.cn; Tel.: +86-10-6291-9177

Abstract: There is a need for rapidly screening thyroid hormone (TH) signaling disruptors in vivo considering the essential role of TH signaling in vertebrates. We aimed to establish a rapid in vivo screening assay using *Xenopus laevis* based on the T3-induced *Xenopus* metamorphosis assay we established previously, as well as the *Xenopus* Eleutheroembryonic Thyroid Assay (XETA). Stage 48 tadpoles were treated with a series of concentrations of T3 in 6-well plates for 24 h and the expression of six TH-response genes was analyzed for choosing a proper T3 concentration. Next, bisphenol A (BPA) and tetrabromobisphenol A (TBBPA), two known TH signaling disruptors, were tested for determining the most sensitive TH-response gene, followed by the detection of several suspected TH signaling disruptors. We determined 1 nM as the induction concentration of T3 and *thibz* expression as the sensitive endpoint for detecting TH signaling disruptors given its highest response to T3, BPA, and TBBPA. And we identified betamipron as a TH signaling agonist, and 2,2′,4,4′-tetrabromodiphenyl ether (BDE-47) as a TH signaling antagonist. Overall, we developed a multiwell-based assay for rapidly screening TH signaling disruptors using *thibz* expression as a sensitive endpoint in *X. laevis*.

Keywords: thyroid hormone; screening assay; *Xenopus laevis*; *thibz* gene; multiwell plate

1. Introduction

Thyroid hormones (THs, T3 and T4) produced by the thyroid gland are critical for growth and development in vertebrates [1]. THs exert their genomic actions principally through TH signaling, in which THs bind to their receptors (TR), resulting in the recruitment of coregulatory proteins to form complexes, thereby regulating target gene expression by binding to DNA sequences in the promoter of target genes [2,3]. Unfortunately, some endocrine-disrupting chemicals (EDCs) could interfere with TH signaling as well as TH production, metabolism, and transportation. TH signaling disruptors have been demonstrated to disrupt TH signaling through interacting with TR or other components in TH signaling [4], possibly leading to adverse effects on the growth and development in vertebrates [5–7]. For instance, bisphenol A (BPA) and tetrabromobisphenol A (TBBPA) are known TH signaling antagonists that disrupt the binding of THs to TR and affect TH-dependent development in *Xenopus* [8–11], but they also exhibit TH signaling agonism in certain situations [12,13].

Several in vitro binding assays [14] and in vitro transfection assays based on reporter genes [15] have been developed to detect TH signaling disruptors. In vitro TR binding assays can detect whether chemicals bind to TR, but cannot distinguish whether they are agonists or antagonists of TH signaling. In contrast, in vitro transfection assays can screen TH signaling agonists or antagonists but not completely reflect what actually happens

in vivo due to the lack of the capacity to metabolize test chemicals in cells. Nevertheless, the reporter-enzymes-based GH3-TRE-Luciferase assay established by Freitas et al. (2011) [12] has been modified into a high throughput screening assay in the Tox21 project, aiming to screen TH signaling agonists or antagonists. However, it is still necessary to develop rapid in vivo screening assays for detecting TH signaling disruptors.

TH-dependent amphibian metamorphosis offers an opportunity to investigate thyroid disruption including TH signaling disruption in vivo [16,17]. Specifically, the amphibian metamorphosis assay (AMA) [18] has been designed to identify substances with thyroid activity and the interpretation of the AMA may be complicated due to multiple mechanisms of thyroid disruption and the possible involvement of other pathways besides the thyroid system. Furthermore, Fini et al. (2007) [19] established the *Xenopus* Eleutheroembryonic Thyroid Assay (XETA) [20], which has been validated and issued by the Organization for Economic Co-operation and Development (OECD). In the XETA, stage 45 fluorescent transgenic *X. laevis* bearing a TH/bZIP-GFP construct are exposed to chemicals in 6-well plates for 72 h with or without 5 nM T3, and fluorescent intensity is detected to assess the thyroid disrupting activity of the test chemicals. However, the TH/bZIP-GFP transgenic line is unavailable for many laboratories, which restricts the application of the XETA. Previously, we developed a T3-induced *Xenopus* metamorphosis assay (TIXMA) for the detection of TH signaling disruptors [21,22]. In the TIXMA, stage 52 tadpoles are exposed to chemicals in the absence or presence of 1 nM T3 in glass tanks containing 4 L, and TH-response gene expression after 24-h exposure and gross morphology after 96-h exposure are measured. Relative to the XETA employing 6-well plates, the TIXMA requires a large amount of water, especially due to daily renewal, resulting in heavy labor and discharge of lots of chemical-contaminated experimental water. To sum up, a simpler and more rapid universally available in vivo screening assay is still needed for detecting TH signaling disruptors.

The present study aimed to develop a simple and rapid screening assay for detecting TH signaling disruptors using wild-type *Xenopus* tadpoles, referencing the XETA and the TIXMA. Considering their high sensitivity to exogenous TH due to the high expression of the *thyroid hormone receptor (thr)* gene [23,24] and little capability to respond to chemicals that directly or indirectly disrupt TH synthesis due to undeveloped thyroid glands [25,26], tadpoles at stage 48 were employed as test organisms. Tadpoles were exposed to T3 or chemicals or their combination in 6-well plates for 24 h, thereby facilitating rapid and easy screening of TH signaling disruption and reducing possible interferences resulting from unknown mechanisms. Based on the effects of a series of concentrations of T3 on TH-response gene expression, we determined an appropriate concentration of T3 to induce expression of TH-response genes, among which the most sensitive TH-response gene was chosen as the endpoint. Subsequently, we assessed the TH signaling disrupting activity of several suspected TH signaling disruptors, including betamipron, 2,2',4,4'-tetrabromodiphenyl ether (BDE-47), triclocarban (TCC), triclosan (TCS), benzophenone (BP), and benzophenone-3 (BP-3). Betamipron, a drug to reduce nephrotoxicity [27], was reported as a TR agonist in the Tox21 project [28] and requires testing in vivo to confirm the TR agonist. TCC [29] and TCS [30] are two bacteriostatic agents in personal care products, and BP [31] and BP-3 [32] are two UV filters widely used, with BDE-47 [33] as a brominated retardant. These chemicals were reported to have the potential to interfere with the thyroid system, but whether they affect TH signaling is not yet well-documented. Overall, our study is expected to provide a simple and rapid in vivo assay for screening TH signaling disruptors.

2. Results

2.1. Determination of T3 Induction Concentration

The effects of a series of concentrations of T3 (0.5, 1, 2, and 5 nM) on the expression of six TH-response genes were investigated. As shown in Figure 1, all concentrations of T3, including the lowest concentration, significantly up-regulated the expression of six

TH-response genes, *kruppel-like factor 9* (*klf9*), *thyroid hormone induced bZip protein* (*thibz*), *thyroid hormone receptor beta* (*thrb*), *stromelysin-3* (*st3*), *type 3 deiodinase* (*dio3*), and *matrix metallopeptidase 13* (*mmp13*), in a concentration-dependent manner. The changes of *klf9*, *thibz*, *thrb*, *st3*, *dio3*, and *mmp13* in the highest group of T3 (5 nM) were about 5-, 280-, 21-, 6-, 6-, and 4-fold of the control, respectively. Moreover, among the six TH-response genes, *thibz* was the most sensitive, with a 200-fold change following 1 nM T3 treatment. Therefore, 1 nM was selected as the induction concentration of T3 in the following tests.

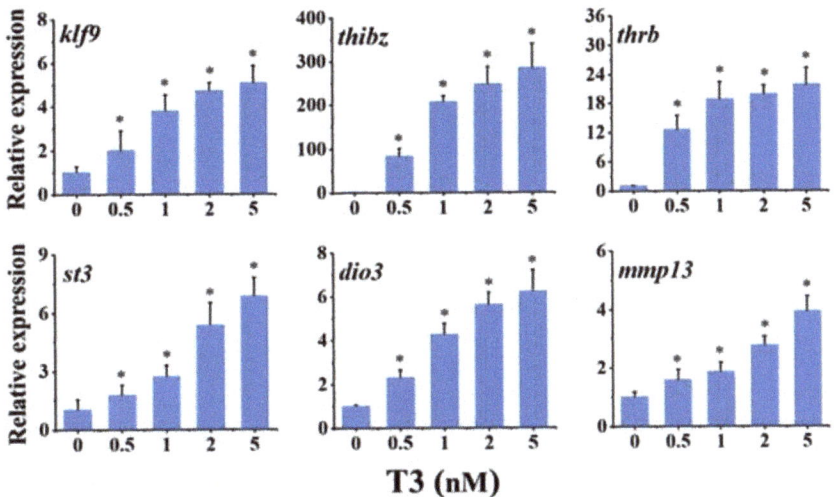

Figure 1. Relative expression of thyroid hormone (TH)-response genes in stage 48 *Xenopus* tadpoles following 24-h exposure to T3. Data are shown as mean ± SEM ($n = 3$). * indicates significant differences between T3 exposure and the control ($p < 0.05$). *klf9*: Kruppel-like factor 9; *thibz*: thyroid hormone induced bZip protein; *thrb*: thyroid hormone receptor beta; *st3*: stromelysin-3; *dio3*: type 3 deiodinase; *mmp13*: matrix metallopeptidase 13.

2.2. Antagonistic Actions of BPA and TBBPA on T3-Induced Gene Expression

The expression of six TH-response genes was measured in tadpoles with BPA or TBBPA exposure in the absence or presence of 1 nM T3. As shown in Figure 2, in the absence of T3, 100 and 1000 nM BPA and TBBPA exposure significantly promoted *thibz* expression. Then, 1000 nM BPA and TBBPA significantly upregulated the expression of *st3*. However, there were no obvious effects on other genes, *klf9*, *thrb*, *dio3*, and *mmp13*. *thibz* expression was significantly induced by as low as 100 nM BPA and TBBPA, indicating that *thibz* was the most sensitive gene among six TH-response genes. Furthermore, T3-induced gene expression of *thibz*, *thrb*, and *dio3* were significantly inhibited by 100 nM BPA, and 1000 nM BPA significantly downregulated all six T3-induced TH-response genes expression. Similar to BPA, 100 nM TBBPA antagonized the T3-induced gene expression of *klf9*, *thibz*, and *thrb*, and 1000 nM TBBPA significantly antagonized all T3-induced genes we tested. In the co-exposure groups, only *thibz* and *thrb* were significantly antagonized by both 100 nM BPA and TBBPA, and T3-induced expression of the other four TH-response genes was antagonized by the higher concentrations of BPA and TBBPA. Taken together, *thibz* expression was chosen as a sensitive endpoint for screening TH signaling disruptors.

2.3. TH Signaling Disrupting Activity of Several Suspected TH Signaling Disruptors

We analyzed the TH signaling disruption activity of betamipron, BDE-47, TCC, TCS, BP, and BP-3 using *thibz* expression as an endpoint. As shown in Figure 3, 1 nM T3 significantly increased the expression of *thibz* expression compared with the control. In the absence of T3, 10–1000 nM betamipron significantly promoted the expression of *thibz* in a concentration-dependent manner, and in the presence of T3, 100–1000 nM betamipron increased the expression of *thibz* compared with the T3 group. BDE-47 inhibited the T3-induced *thibz* expression but did not affect *thibz* expression in the absence of T3. TCC, TCS, BP, and BP-3 had a similar mode of action on *thibz* expression both in the absence and presence of T3. These four chemicals at high concentrations significantly resulted in down-regulation of T3-induced expression of *thibz* expression and high concentrations of these four chemicals alone increased *thibz* expression.

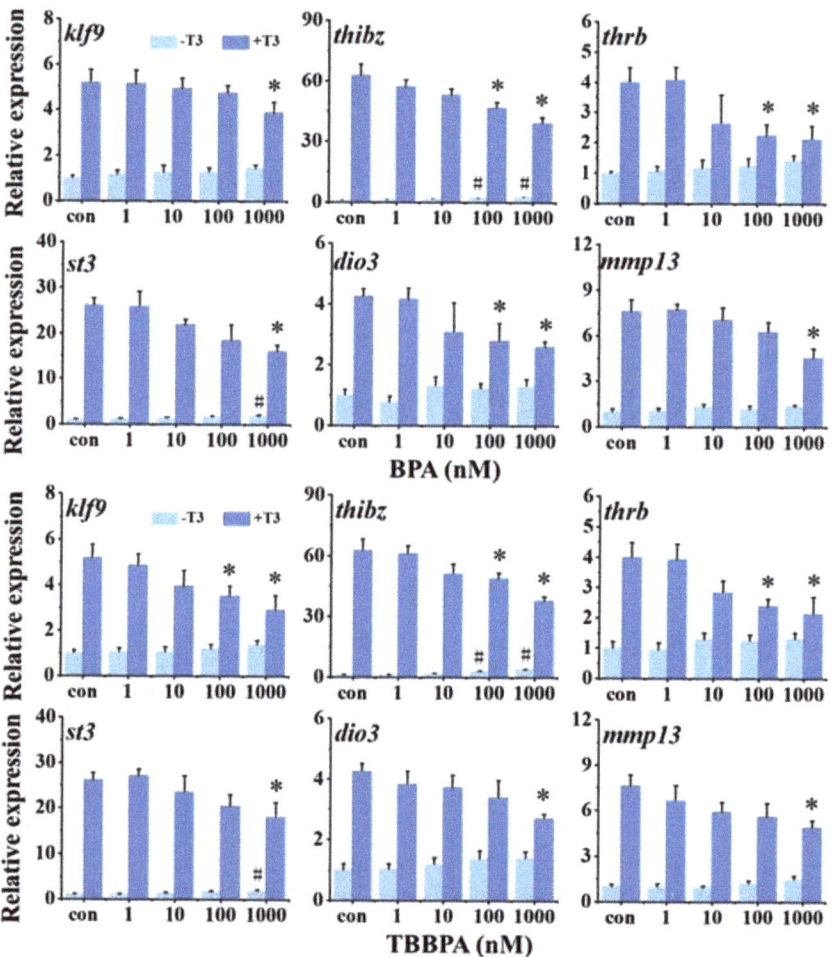

Figure 2. Relative expression of thyroid hormone (TH)-response genes in stage 48 *Xenopus* tadpoles following 24-h exposure to bisphenol A (BPA) and tetrabromobisphenol A (TBBPA) in the absence or presence of 1 nM T3. Data are shown as mean ± SEM (n = 3). * and # indicate significant differences between BPA or TBBPA exposure and the control and between BPA or TBBPA + T3 and T3 treatment, respectively ($p < 0.05$). klf9: Kruppel-like factor 9; thibz: thyroid hormone induced bZip protein; thrb: thyroid hormone receptor beta; st3: stromelysin-3; dio3: type 3 deiodinase; mmp13: matrix metallopeptidase 13.

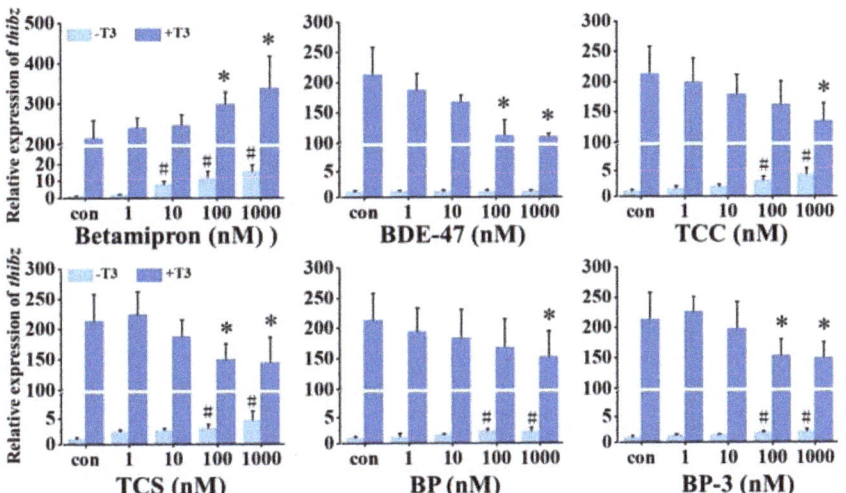

Figure 3. Relative expression of *thibz* (*thyroid hormone induced bZip protein*) in stage 48 *Xenopus* tadpoles following 24-h exposure to betamipron, 2,2′,4,4′-tetrabromodiphenyl ether (BDE-47), triclocarban (TCC), triclosan (TCS), benzophenone (BP), and benzophenone-3 (BP-3) in the absence or presence of 1 nM T3. Data are shown as mean ± SEM (n = 3). * and # indicate significant differences between the chemical exposure and the control and between chemical + T3 and T3 treatment, respectively ($p < 0.05$).

3. Discussion

We developed a multiwell-based screening assay to detect TH signaling disruptors based on *thibz* expression analysis using *Xenopus* tadpoles at stage 48. The concentration-dependent upregulation of all tested TH-response gene expressions by T3 treatment has demonstrated the sensitivity of stage 48 tadpoles to T3 within 24 h. Given that 1 nM T3 dramatically induced upregulation of TH-response gene expression but did not reach the expression climax, we determined 1 nM as the induction concentration of T3, avoiding the possibility that the higher concentrations of T3 would cover up the antagonistic actions of tested chemicals. When 1 nM T3 upregulated the expression of all the test TH-response genes, 1000 and/or 100 nM both BPA and TBBPA antagonized T3 actions, showing that this screening assay is effective in detecting TH signaling antagonists. As mentioned above, the XETA employs transgenic TH/bZIP-GFP tadpoles to indicate TH signaling disruption by measuring GFP fluorescence. Considering the finding that *thibz* expression was the most sensitive to T3 and highly responsive to BPA and TBBPA, we chose *thibz* expression as a sensitive endpoint to detect TH signaling disruptors. To our current understanding, *thibz* is believed to be a gene that specifically responds to TH signaling [34]. Moreover, thyroid glands of stage 48 tadpoles remain undeveloped and thereby tadpoles theoretically have little capability to respond to chemicals that directly or indirectly disrupt TH synthesis, especially within 24 h [25,26]. Therefore, we assure that the assay is relatively specific for screening TH signaling disruptors. Certainly, the results of the screening assay need further confirmation by other assays such as TIXMA. Importantly, the screening assay is more rapid in terms of the 24-h exposure duration relative to the 72-h exposure duration in XETA. In addition, the screening assay, like the XETA, is simpler and easier to implement relative to the TIXMA due to chemical exposure in 6-well plates. Moreover, the exposure duration of 24 h in our assay ensures more rapid detection for TH signaling disruption compared with the 72-h exposure duration in XETA. Additionally, Fini et al. (2007) [19] reported that 1000 nM TBBPA inhibited T3-induced GFP signaling in transgenic TH/bZIP-GFP tadpoles using the XETA assay, but 500 nM failed. Here, we found that both 1000 nM and 100 nM

TBBPA significantly antagonized T3-induced *thibz* expression. Therefore, the screening assay we developed is more sensitive or comparable with the XETA assay for detecting antagonistic actions of chemicals on TH signaling.

Using this screening assay, we examined several chemicals for TH signaling disrupting activity. In the Tox21 project, betamipron was for the first time reported as a TR agonist in the in vitro reporter gene assay [28], with the lack of in vivo data. Here, we found that betamipron significantly upregulated *thibz* expression even at 10 nM in our screening assay, providing the first in vivo evidence that betamipron is a TH signaling agonist. In contrast to betamipron, the screening assay revealed that BDE-47 inhibited T3-induced *thibz* expression as BDE-47 exposure alone had no effect, indicating TH signaling antagonism. Previous studies reported that BDE-47 inhibited TH-dependent development in *X. laevis* [35,36], and decreased the expression of the TH-response genes *thrb* and *klf9* in *X. laevis* [35]. Together with these data, our findings support that BDE-47 could have the potential to antagonize TH signaling, warranting further investigations.

In the screening assay, TCC, TCS, BP, and BP-3 exerted similar effects on *thibz* expression in either the absence or presence of T3, i.e., they stimulated *thibz* gene expression in the absence of T3 but antagonized T3-induced expression in the presence of T3. The effects of these chemicals are similar to those of BPA and TBBPA, implying that they, like BPA and TBBPA, exerted TH signaling antagonistic action in the presence of T3, but could have agonistic action in the absence of T3. Previously, several studies reported that TCS altered TH-response gene expression and TH-dependent growth in amphibians, despite seemly inconsistent effects [37–39]. Similarly, BP-3 was reported to inhibit T3-induced tail resorption in *Rana rugose* and suppressed the T3-induced EGFP activity in transgenic *X. laevis* tadpoles [32], but behaved as TR agonists in HepG2 cells on the activation of TR-mediated transcription [40]. Given these data combined with our results from the screening assay, it is concluded that TCC, TCS, BP, and BP-3 could be TH signaling disruptors, which warrants further studies, including the investigation of the mechanisms for TH signaling disruption.

4. Materials and Methods

4.1. Chemicals

3,3′,5-triiodo-L-thyronine (T3, CAS No. 6893-02-3) was obtained from Geel Belgium (New Jersey, USA) and a stock solution of T3 was prepared by dissolving into ultrapure water. Dimethyl sulfoxide (DMSO, CAS No. 67-68-5) and 3-aminobenzoic acid ethyl ester (MS-222, CAS No. 886-86-2) were purchased from Sigma-Aldrich (St. Louis, MO, USA). Bisphenol A (BPA, CAS No. 80-05-7, Acros Organics, Geel, Belgium), tetrabromobisphenol A (TBBPA, CAS No. 79-94-7, Acros Organics, Geel, Belgium), betamipron (CAS No. 3440-28-6, Aladdin, Shanghai, China), 2,2′,4,4′-tetrabromodiphenyl ether (BDE-47, CAS No. 5436-43-1, Bidepharm, Shanghai, China), triclocarban (TCC, CAS No. 101-20-2, Tokyo Chemical Industry, Tokyo, Japan), triclosan (TCS, CAS No. 3380-34-5, Tokyo Chemical Industry, Tokyo, Japan), benzophenone (BP, CAS No. 119-61-9, Macklin, Shanghai, China) and benzophenone-3 (BP-3, CAS No. 131-57-7, Tokyo Chemical Industry, Tokyo, Japan) were dissolved into DMSO to prepare 10 mM stock solutions. The stock solutions for chemicals were stored at $-20\ °C$ for future exposure experiments. An RNA Extraction kit (AU1201) was obtained from BioTeke Corporation (Wuxi, China). PCR primers were synthesized by BGI Tech Solutions (Beijing, China). DNase/RNase-free water (RT121), Quantscript RT Kit (KR116), and a Real Master Mix (SYBR Green) Kit (FP205) were obtained from TIANGEN Biotech (Beijing, China). Human chorionic gonadotropin (HCG) was purchased from the Ningbo Second Hormone Factory (Ningbo, China) and dissolved into 0.6% sodium chloride.

4.2. Animals and Housing Conditions

Housing and breeding conditions for *X. laevis* were described in our previous study with some adjustments [41]. In brief, sexually mature wild-type adult *X. laevis* were raised at 20 °C with a light/dark cycle of 12 h:12 h in charcoal-filtered tap water. Breeding was induced by injection HCG into a pair of adult frogs (800 IU for the female and 400 IU for the male). After spawning, fertilized eggs were incubated in glass aquariums containing charcoal-filtered tap water with a 12 h light/12 h dark cycle at 22 ± 1 °C. Developmental stages of *Xenopus* tadpoles were identified according to the Nieuwkoop and Faber table [26]. This study was approved by Animal Ethics and Welfare Committee of Research Center for Eco-Environmental Sciences, Chinese Academy of Sciences (AEWC-RCEES-2021040).

4.3. Chemical Exposure and Sampling

To reduce the test size, we employed stage 48 tadpoles in 6-well plates in place of stage 52 tadpoles in glass tanks in the TIXMA. Our previous study has shown that stage 48 tadpoles are highly sensitive to T3 [42]. To determine an appropriate concentration of T3 that can induce TH-response gene expression, a concentration-response experiment (0, 0.5, 1, 2, and 5 nM) was performed. Three replicate wells were set for each treatment, with two tadpoles per well containing 10 mL of the test solution. After 24 h, tadpoles were anesthetized with 100 mg/L MS-222 and then rinsed with water. Tadpoles (tails removed) in each well were pooled for RNA extraction and subsequent analysis for TH-response gene expression. Following the TIXMA, six TH-response genes were chosen, including *klf9*, *thibz*, *thrb*, *st3*, *dio3*, and *mmp13* [21]. Then, stage 48 tadpoles were exposed to BPA and TBBPA (1, 10, 100, and 1000 nM) in the presence or absence of T3 in order to verify the response of this assay to TH signaling antagonists. The most sensitive gene to T3, BPA, and TBBPA was chosen as the endpoint for screening TH signaling disruptors.

Finally, stage 48 tadpoles were exposed to betamipron, BDE-47, TCC, TCS, BP, and BP-3 (1, 10, 100, and 1000 nM) in the absence or presence of T3 to screen their TH signaling disrupting activity. Exposures and sampling were conducted as described above. Each experiment was independently repeated three times using offspring from different pairs of *X. laevis*.

4.4. RNA Extraction and Quantitative Real-Time PCR

For gene expression analysis, the total RNA of tadpoles was extracted by an Automatic Nucleic Acid Extraction Apparatus (BioTeke, Wuxi, China) following the manufacturer's instructions. RNA concentration was measured using a NanoDrop 2000 (Thermo Scientific, Waltham, MA, USA), and RNA quality was examined by A260/A280 ratios and agar-gel electrophoresis. The first-strand cDNA was synthesized from 1000 ng total RNA using the Fast RT Kit according to the manufacturer's instructions (TIANGEN Biotech, Beijing, China). Following this, the first-strand cDNA was stored at −20 °C for gene expression analysis. Expression of TH-response genes was analyzed using SYBR Green I with the Real-time Polymerase Chain Reaction system (Light Cycler 480, Roche, Basel, Switzerland), with *Ribosomal protein L8* (*rpl8*) as a reference gene [43–45] and the expression of *rpl8* was not affected by treatments (Supplementary Materials, Figures S1 and S2). In 10 μL of PCR reaction system, 1 μL of cDNA template, 0.3 μL of forward primer, 0.3 μL of reverse primer, 3.4 μL of DNase/RNase-free water, and 5 μL of 2 × SuperReal PreMix Plus were mixed. All primers and conditions for PCR are listed in Table 1. The relative fold changes of targeted gene expression data using real-time PCR were calculated using the $2^{-\Delta\Delta Ct}$ methods [46].

4.5. Statistical Analysis

The statistical analysis of the experimental data was performed using SPSS software version 16.0 (IBM, Armonk, NY, USA). The data were checked for normal distribution (Kolmogorov–Smirnov test) and homogeneity of variance (Levene test). Data for relative expression of genes are presented as mean ± standard error of the mean (SEM). Statistical differences in relative expression of genes were analyzed by two-way analysis of variance (ANOVA) followed by Tukey's HSD test. Statistical significance was defined as p value < 0.05.

Table 1. Primer sequences of all tested *Xenopus laevis* genes and conditions for quantitative polymerase chain reaction (qPCR).

Gene	Primer Sequences (5′–3′)	Annealing Temperature (°C)	GeneBank ID
rpl8	F: CCGTGGTGTGGCTATGAATC R: TACGACGAGCAGCAATAAGAC	58	NM_001086996.1
klf9	F: GTGGCCACTTGATTTCCCCT R: AAAGACACAAAACAGCGGCG	64	NM_001085597.1
thibz	F: CCACCTCCACAGAATCAGCAG R: AGAAGTGTTCCGACAGCCAAG	62	NM_001085805.1
thrb	F: GAGATGGCAGTGACAAGG R: CAAGGCGACTTCGGTATC	58	NM_001087781.1
st3	F: CCTCTGTCATACACTTACCTT R: TGAACCGTGAGCATTGAG	62	NM_001086342.1
dio3	F: GATGCTGTGGCTGCTGGAT R: ATTCGGTTGGAGTCGGACAC	62	NM_001087863.1
mmp13	F: CCTTGTCAGTGCTTGTCCTATC R: TCCTGGTGTCAGTTCAGAGTC	62	NM_001100931.1

F: forward; R: reverse; *rpl8: ribosomal protein L8; klf9: Kruppel-like factor 9; thibz: thyroid hormone induced bZip protein; thrb: thyroid hormone receptor beta; st3: stromelysin-3; dio3: type 3 deiodinase; mmp13: matrix metallopeptidase 13.*

5. Conclusions

We developed a multiwell-based assay for rapidly screening TH signaling disruptors using thibz expression as a sensitive endpoint in X. laevis which effectively detected the agonistic effect of T3 and the antagonistic effect of BPA and TBBPA on TH signaling within 24 h. Using this screening assay, we identified betamipron as a TH signaling agonist and BDE-47 as a TH signaling antagonist, while TCC, TCS, BP, and BP-3 appeared to exert complex TH signaling disrupting actions in a TH-dependent manner. Overall, all results indicate that this assay is suitable for screening the TH signaling disrupting activity of chemicals.

Supplementary Materials: The following are available online, Figure S1: Relative expression of ribosomal protein L8 (rpl8) in stage 48 *Xenopus* tadpoles following 24-h exposure to T3. Figure S2: Relative expression of rpl8 in stage 48 *Xenopus* tadpoles following 24-h exposure to betamipron, 2,2′,4,4′-tetrabromodiphenyl ether (BDE-47), triclocarban (TCC), triclosan (TCS), benzophenone (BP), and benzophenone-3 (BP-3) in the absence or presence of 1 nM T3.

Author Contributions: Conceptualization, J.L., Z.Q.; Data curation, J.L.; Funding acquisition, Z.Q.; Methodology, J.L.; Project administration, Z.Q.; Resources, J.L., Y.L., M.Z.; Supervision, Z.Q.; Validation, J.L.; Visualization, J.L.; Writing—original draft, J.L., S.S.; Writing—review & editing, J.L., Y.L., M.Z., S.S., Z.Q. All authors have read and agreed to the published version of the manuscript.

Funding: This research was funded by the National Key Research and Development Program of China (2018YFA0901103) and the National Natural Science Foundation of China (21876196).

Institutional Review Board Statement: The Animal Ethics and Welfare Committee of Research Center for Eco-Environmental Sciences, Chinese Academy of Sciences, approved all animal protocols and procedures (AEWC-RCEES-2021040).

Informed Consent Statement: Not applicable.

Data Availability Statement: The authors declare that all data generated or analyzed during this study are included in the published article.

Acknowledgments: All funds are gratefully appreciated. We would like to thank Xiaoxing Liang and Zhaojing Chen for providing language help. We also would like to thank the editor and two anonymous reviewers who kindly reviewed the earlier version of this manuscript and provided valuable suggestions and comments.

Conflicts of Interest: The authors declare that they have no conflict of interest.

References

1. Sachs, L.M.; Campinho, M.A. Editorial: The Role of Thyroid Hormones in Vertebrate Development. *Front. Endocrinol.* **2019**, *10*, 863. [CrossRef] [PubMed]
2. Shi, Y.B. Dual functions of thyroid hormone receptors in vertebrate development: The roles of histone-modifying cofactor complexes. *Thyroid* **2009**, *19*, 987–999. [CrossRef] [PubMed]
3. Yen, P.M.; Ando, S.; Feng, X.; Liu, Y.; Maruvada, P.; Xia, X.M. Thyroid hormone action at the cellular, genomic and target gene levels. *Mol. Cell. Endocrinol.* **2006**, *246*, 121–127. [CrossRef]
4. DeVito, M.; Biegel, L.; Brouwer, A.; Brown, S.; Brucker-Davis, F.; Cheek, A.O.; Christensen, R.; Colborn, T.; Cooke, P.; Crissman, J.; et al. Screening methods for thyroid hormone disruptors. *Environ. Health Perspect.* **1999**, *107*, 407–415. [CrossRef]
5. Zhang, J.; Li, Y.Z.; Gupta, A.A.; Nam, K.; Andersson, P.L. Identification and Molecular Interaction Studies of Thyroid Hormone Receptor Disruptors among Household Dust Contaminants. *Chem. Res. Toxicol.* **2016**, *29*, 1345–1354. [CrossRef] [PubMed]
6. Leemans, M.; Couderq, S.; Demeneix, B.; Fini, J.B. Pesticides with Potential Thyroid Hormone-Disrupting Effects: A Review of Recent Data. *Front. Endocrinol.* **2019**, *10*, 743. [CrossRef]
7. Gilbert, M.E.; O'Shaughnessy, K.L.; Axelstad, M. Regulation of Thyroid-disrupting Chemicals to Protect the Developing Brain. *Endocrinology* **2020**, *161*. [CrossRef]
8. Moriyama, K.; Tagami, T.; Akamizu, T.; Usui, T.; Saijo, M.; Kanamoto, N.; Hataya, Y.; Shimatsu, A.; Kuzuya, H.; Nakao, K. Thyroid hormone action is disrupted by bisphenol A as an antagonist. *J. Clin. Endocr. Metab.* **2002**, *87*, 5185–5190. [CrossRef]
9. Sun, H.; Shen, O.X.; Wang, X.R.; Zhou, L.; Zhen, S.Q.; Chen, X.D. Anti-thyroid hormone activity of bisphenol A, tetrabromobisphenol A and tetrachlorobisphenol A in an improved reporter gene assay. *Toxicol. In Vitro* **2009**, *23*, 950–954. [CrossRef] [PubMed]
10. Niu, Y.; Zhu, M.; Dong, M.; Li, J.; Li, Y.; Xiong, Y.; Liu, P.; Qin, Z. Bisphenols disrupt thyroid hormone (TH) signaling in the brain and affect TH-dependent brain development in *Xenopus laevis*. *Aquat. Toxicol.* **2021**, *237*, 105902. [CrossRef] [PubMed]
11. Zhang, Y.F.; Xu, W.; Lou, Q.Q.; Li, Y.Y.; Zhao, Y.X.; Wei, W.J.; Qin, Z.F.; Wang, H.L.; Li, J.Z. Tetrabromobisphenol A disrupts vertebrate development via thyroid hormone signaling pathway in a developmental stage-dependent manner. *Environ. Sci. Technol.* **2014**, *48*, 8227–8234. [CrossRef]
12. Freitas, J.; Cano, P.; Craig-Veit, C.; Goodson, M.L.; Furlow, J.D.; Murk, A.J. Detection of thyroid hormone receptor disruptors by a novel stable in vitro reporter gene assay. *Toxicol. In Vitro* **2011**, *25*, 257–266. [CrossRef] [PubMed]
13. Hofmann, P.J.; Schomburg, L.; Kohrle, J. Interference of Endocrine Disrupters with Thyroid Hormone Receptor-Dependent Transactivation. *Toxicol. Sci.* **2009**, *110*, 125–137. [CrossRef] [PubMed]
14. Zoeller, R.T.; Tyl, R.W.; Tan, S.W. Current and potential rodent screens and tests for thyroid toxicants. *Crit. Rev. Toxicol.* **2007**, *37*, 55–95. [CrossRef] [PubMed]
15. Grimaldi, M.; Boulahtouf, A.; Delfosse, V.; Thouennon, E.; Bourguet, W.; Balaguer, P. Reporter cell lines for the characterization of the interactions between human nuclear receptors and endocrine disruptors. *Front. Endocrinol.* **2015**, *6*. [CrossRef] [PubMed]
16. Thambirajah, A.A.; Koide, E.M.; Imbery, J.J.; Helbing, C.C. Contaminant and Environmental Influences on Thyroid Hormone Action in Amphibian Metamorphosis. *Front. Endocrinol.* **2019**, *10*, 276. [CrossRef] [PubMed]
17. Sachs, L.M.; Buchholz, D.R. Frogs model man: In vivo thyroid hormone signaling during development. *Genesis* **2017**, *55*. [CrossRef]
18. OECD/OCDE. Test No. 231: Amphibian Metamorphosis Assay. In *OECD Guidelines for the Testing of Chemicals, Section 2*; OECD Publishing: Paris, France, 2009. [CrossRef]
19. Fini, J.B.; Le Mevel, S.; Turque, N.; Palmier, K.; Zalko, D.; Cravedi, J.P.; Demeneix, B.A. An in vivo multiwell-based fluorescent screen for monitoring vertebrate thyroid hormone disruption. *Environ. Sci. Technol.* **2007**, *41*, 5908–5914. [CrossRef]
20. OECD/OCDE. Test No. 248: Xenopus Eleutheroembryonic Thyroid Assay (XETA). In *OECD Guidelines for the Testing of Chemicals, Section 2*; OECD Publishing: Paris, France, 2019. [CrossRef]
21. Yao, X.; Chen, X.; Zhang, Y.; Li, Y.; Wang, Y.; Zheng, Z.; Qin, Z.; Zhang, Q. Optimization of the T3-induced *Xenopus* metamorphosis assay for detecting thyroid hormone signaling disruption of chemicals. *J. Environ. Sci. (China)* **2017**, *52*, 314–324. [CrossRef]
22. Wang, Y.; Li, Y.; Qin, Z.; Wei, W. Re-evaluation of thyroid hormone signaling antagonism of tetrabromobisphenol A for validating the T3-induced *Xenopus* metamorphosis assay. *J. Environ. Sci. (China)* **2017**, *52*, 325–332. [CrossRef] [PubMed]
23. Yaoita, Y.; Brown, D.D. A correlation of thyroid hormone receptor gene expression with amphibian metamorphosis. *Genes Dev.* **1990**, *4*, 1917–1924. [CrossRef] [PubMed]
24. Morvan-Dubois, G.; Demeneix, B.A.; Sachs, L.M. *Xenopus laevis* as a model for studying thyroid hormone signalling: From development to metamorphosis. *Mol. Cell. Endocrinol.* **2008**, *293*, 71–79. [CrossRef] [PubMed]
25. Saxen, L.; Saxen, E.; Toivonen, S.; Salimaki, K. Quantitative investigation on the anterior pituitary-thyroid mechanism during frog metamorphosis. *Endocrinology* **1957**, *61*, 35–44. [CrossRef] [PubMed]
26. Faber, J.; Nieuwkoop, P.D. *Normal Table of Xenopus Laevis (Daudin). A Systematical and Chronological Survey of the Development from the Fertilized Egg till the End of Metamorphosis*; Garland Publishing: New York, NY, USA, 1994; p. 252.
27. Hirouchi, Y.; Naganuma, H.; Kawahara, Y.; Okada, R.; Kamiya, A.; Inui, K.; Hori, R. Preventive Effect of Betamipron on Nephrotoxicity and Uptake of Carbapenems in Rabbit Renal-Cortex. *Jpn. J. Pharmacol.* **1994**, *66*, 1–6. [CrossRef]

28. Paul-Friedman, K.; Martin, M.; Crofton, K.M.; Hsu, C.W.; Sakamuru, S.; Zhao, J.; Xia, M.; Huang, R.; Stavreva, D.A.; Soni, V.; et al. Limited Chemical Structural Diversity Found to Modulate Thyroid Hormone Receptor in the Tox21 Chemical Library. *Environ. Health Perspect.* **2019**, *127*, 97009. [CrossRef] [PubMed]
29. Wu, Y.; Beland, F.A.; Fang, J.-L. Effect of triclosan, triclocarban, 2,2′,4,4′-tetrabromodiphenyl ether, and bisphenol A on the iodide uptake, thyroid peroxidase activity, and expression of genes involved in thyroid hormone synthesis. *Toxicol. In Vitro* **2016**, *32*, 310–319. [CrossRef]
30. Mihaich, E.; Capdevielle, M.; Urbach-Ross, D.; Slezak, B. Hypothesis-driven weight-of-evidence analysis of endocrine disruption potential: A case study with triclosan. *Crit. Rev. Toxicol.* **2017**, *47*, 263–285. [CrossRef] [PubMed]
31. Lee, J.; Kim, S.; Park, Y.J.; Moon, H.-B.; Choi, K. Thyroid Hormone-Disrupting Potentials of Major Benzophenones in Two Cell Lines (GH3 and FRTL-5) and Embryo-Larval Zebrafish. *Environ. Sci. Technol.* **2018**, *52*, 8858–8865. [CrossRef] [PubMed]
32. Kashiwagi, K.; Hanada, H.; Yamamoto, T.; Goto, Y.; Furuno, N.; Kitamura, S.; Ohta, S.; Sugihara, K.; Taniguci, K.; Tooi, O.; et al. 2-Hydroxy-4-methoxybenzophenone (HMB) and 2,4,4′-trihydroxybenzophenone (THB) Suppress Amphibian Metamorphosis. In *Progress in Safety Science and Technology Series*; International Symposium on Safety Science and Technology: Beijing, China, 2008; Volume 7, pp. 970–976.
33. Jiang, Y.; Yuan, L.; Lin, Q.; Ma, S.; Yu, Y. Polybrominated diphenyl ethers in the environment and human external and internal exposure in China: A review. *Sci. Total Environ.* **2019**, *696*, 133902. [CrossRef] [PubMed]
34. Furlow, J.D.; Brown, D.D. In vitro and in vivo analysis of the regulation of a transcription factor gene by thyroid hormone during *Xenopus laevis* metamorphosis. *Mol. Endocrinol.* **1999**, *13*, 2076–2089. [CrossRef] [PubMed]
35. Yost, A.T.; Thornton, L.M.; Venables, B.J.; Jeffries, M.K.S. Dietary exposure to polybrominated diphenyl ether 47 (BDE-47) inhibits development and alters thyroid hormone-related gene expression in the brain of *Xenopus laevis* tadpoles. *Environ. Toxicol. Phar.* **2016**, *48*, 237–244. [CrossRef] [PubMed]
36. Balch, G.C.; Velez-Espino, L.A.; Sweet, C.; Alaee, M.; Metcalfe, C.D. Inhibition of metamorphosis in tadpoles of *Xenopus laevis* exposed to polybrominated diphenyl ethers (PBDEs). *Chemosphere* **2006**, *64*, 328–338. [CrossRef] [PubMed]
37. Fort, D.J.; Mathis, M.B.; Pawlowski, S.; Wolf, J.C.; Peter, R.; Champ, S. Effect of triclosan on anuran development and growth in a larval amphibian growth and development assay. *J. Appl. Toxicol.* **2017**, *37*, 1182–1194. [CrossRef] [PubMed]
38. Marlatt, V.L.; Veldhoen, N.; Lo, B.P.; Bakker, D.; Rehaume, V.; Vallee, K.; Haberl, M.; Shang, D.; van Aggelen, G.C.; Skirrow, R.C.; et al. Triclosan exposure alters postembryonic development in a Pacific tree frog (*Pseudacris regilla*) Amphibian Metamorphosis Assay (TREEMA). *Aquat. Toxicol.* **2013**, *126*, 85–94. [CrossRef] [PubMed]
39. Veldhoen, N.; Skirrow, R.C.; Osachoff, H.; Wigmore, H.; Clapson, D.J.; Gunderson, M.P.; Van Aggelen, G.; Helbing, C.C. The bactericidal agent triclosan modulates thyroid hormone-associated gene expression and disrupts postembryonic anuran development. *Aquat. Toxicol.* **2006**, *80*, 217–227. [CrossRef]
40. Schmutzler, C.; Bacinski, A.; Gotthardt, I.; Huhne, K.; Ambrugger, P.; Klammer, H.; Schlecht, C.; Hoang-Vu, C.; Grueters, A.; Wuttke, W.; et al. The ultraviolet filter benzophenone 2 interferes with the thyroid hormone axis in rats and is a potent in vitro inhibitor of human recombinant thyroid peroxidase. *Endocrinology* **2007**, *148*, 2835–2844. [CrossRef]
41. Lou, Q.Q.; Zhang, Y.F.; Zhou, Z.; Shi, Y.L.; Ge, Y.N.; Ren, D.K.; Xu, H.M.; Zhao, Y.X.; Wei, W.J.; Qin, Z.F. Effects of perfluorooctanesulfonate and perfluorobutanesulfonate on the growth and sexual development of *Xenopus laevis*. *Ecotoxicology* **2013**, *22*, 1133–1144. [CrossRef] [PubMed]
42. Zhang, Y. Methods for Evaluating Thyroid Disruption by Chemicals Using Amphibians and Their Application. Ph.D. Thesis, University of Chinese Academy of Sciences, Beijing, China, 2014.
43. Shi, Y.B.; Liang, V.C.-T. Cloning and characterization of the ribosomal protein L8 gene from *Xenopus laevis*. *Biochim. Biophys. Acta* **1994**, *1217*, 227–228. [CrossRef]
44. Zhu, M.; Chen, X.Y.; Li, Y.Y.; Yin, N.Y.; Faiola, F.; Qin, Z.F.; Wei, W.J. Bisphenol F Disrupts Thyroid Hormone Signaling and Postembryonic Development in *Xenopus laevis*. *Environ. Sci. Technol.* **2018**, *52*, 1602–1611. [CrossRef]
45. Crespi, E.J.; Denver, R.J. Leptin (ob gene) of the South African clawed frog *Xenopus laevis*. *Proc. Natl. Acad. Sci. USA* **2006**, *103*, 10092–10097. [CrossRef]
46. Livak, K.J.; Schmittgen, T.D. Analysis of relative gene expression data using real-time quantitative PCR and the $2^{-\Delta\Delta Ct}$ method. *Methods* **2001**, *25*, 402–408. [CrossRef] [PubMed]

Article

Evaluation of VOCs Emitted from Biomass Combustion in a Small CHP Plant: Difference between Dry and Wet Poplar Woodchips

Enrico Paris [1], Monica Carnevale [1], Beatrice Vincenti [1], Adriano Palma [1], Ettore Guerriero [2], Domenico Borello [3] and Francesco Gallucci [1,*]

1. Council for Agricultural Research and Economics (CREA), Center of Engineering and Agro-Food Processing, Monterotondo (Roma), Via della Pascolare 16, 00015 Monterotondo, Italy; enrico.paris@crea.gov.it (E.P.); monica.carnevale@crea.gov.it (M.C.); beatrice.vincenti@crea.gov.it (B.V.); adriano.palma@crea.gov.it (A.P.)
2. National Research Council of Italy, Institute of Atmospheric Pollution Research (CNR-IIA), Via Salaria km 29,300, 00015 Monterotondo, Italy; ettore.guerriero@iia.cnr.it
3. Department of Mechanical and Aerospace Engineering (DIMA), La Sapienza University of Rome, Via Eudossiana, 18, 00184 Rome, Italy; domenico.borello@uniroma1.it
* Correspondence: francesco.gallucci@crea.gov.it; Tel.: +39-0690675238

Abstract: The combustion of biomass is a process that is increasingly used for the generation of heat and energy through different types of wood and agricultural waste. The emissions generated by the combustion of biomass include different kinds of macro- and micropollutants whose formation and concentration varies according to the physical and chemical characteristics of the biomass, the combustion conditions, the plants, and the operational parameters of the process. The aim of this work is to evaluate the effect of biomass moisture content on the formation of volatile organic compounds (VOCs) during the combustion process. Wet and dry poplar chips, with a moisture content of 43.30% and 15.00%, respectively, were used in a cogeneration plant based on a mobile grate furnace. Stack's emissions were sampled through adsorbent tubes and subsequently analyzed by thermal desorption coupled with the GC/MS. The data obtained showed that, depending on the moisture content of the starting matrix, which inevitably influences the quality of combustion, there is significant variation in the production of VOCs.

Keywords: biomass combustion; POPs; CHP plant; VOCs; renewable energies; emission; organic pollutants

1. Introduction

The use of renewable energy in the European Union has been estimated to increase to a share between 55% and 75% of total gross energy consumption by 2050. In the current energy situation, bioenergies represent the largest renewable and CO_2-neutral energy source for the production of heat, electricity, and transport fuels [1]. Different biochemical or thermochemical conversion processes of biomass can be applied in order to produce power and heat, reduce the consumption of conventional fossil fuel sources, and represent a realistic threat to environmental sustainability [2].

The most appropriate biomass conversion process depends on biomass physical and chemical characteristics, such as moisture content, fixed carbon, volatile solids, C/N ratio, calorific value, ash and cellulose, hemicellulose, and lignin content, which are recognized worldwide as the main factors that affect the conversion processes' efficiency [3].

Several studies have shown the feasibility of using residual biomass, such as shredded prunings or woodchips [4].

Biomass combustion represents the easiest way to generate energy from biomass [5], and it is currently defined as CO_2-neutral because the carbon dioxide generated and emitted during the combustion phase is compensated by that which the biomass has absorbed during the growth phase [6].

In this context, biomass combustion in small combined heat and power plants (CHPs) represents a useful system for distributed renewable energy production, but, at the same time, close attention ought to be paid to the environmental aspects, with particular reference to the atmospheric pollution emitted from the biomass combustion process. The combustion conditions and the physical and chemical parameters influence the process and the related emissions [7–12].

In any combustion phenomenon, incomplete combustion products are generated. In particular, products of pyrolysis remain as unburnt generating particles and gases, such as carbon monoxide (CO), volatile organic compounds (VOCs), particulate matter (PM), and polycyclic aromatic hydrocarbons (PAHs), which contribute to environmental pollution and become a potential issue for the environment and human health.

From this point of view, it is important to evaluate the relationship between the biomass characteristics; the operating condition of the combustion plant and the flue gases produced in terms of particulate matter (PM_{10} and $PM_{2.5}$); oxidized species, such as CO_2, CO, SO_2, and NO_x; inorganic micropollutants, including heavy metals; persistent organic pollutants (POPs), including polycyclic aromatic hydrocarbons (PAHs); dioxins and furans (PCDD/Fs); and volatile organic compounds (VOCs) [13,14]. VOCs represent a very large amount of different organic compounds, whose emission strongly depends on the type of biomass used and the conditions of the combustion process [15]. These chemical compounds (CFCs, alkanes, alkenes, aldehydes, ketones, aromatic compounds, etc.) are defined by an "initial boiling point equal to or less than 250 °C measured at a standard pressure of 101.3 kPa", as expressed in the European Directive on Air Quality 2008/50/CE.

VOCs can have different effects on humans and the environment depending on their chemical characteristics. They can be precursors of photochemical smog under sunlight radiation in the presence of nitrogen oxides (NO_x) [16] (such as alkenes and alkanes from C3 to C8), they can have a high ozone depletion potential (such as CFCs and halons), they can be greenhouse gases (such as CFCs), or they can be directly toxic to humans (such as chlorinated compounds, benzene, etc.). In particular, it has been widely studied how some VOCs emitted by biomass combustion play an important role in tropospheric ozone and photo-oxidant production. For instance, during the smoke plume transport process, VOCs combined with NO_x can be oxidized to generate secondary organic aerosol (SOA) [17,18]. In the function of the described effects, Ciccioli et al. [19] proposed to classify the aforementioned classes as VOC-OX, VOC-STRAT, VOC-TOX, and VOC-CLIM, respectively, which are precursors of photochemical smog, harmful to the stratospheric ozone layer, toxic for humans, and climate-altering.

Several authors have previously investigated the emissions generated from domestic woodstoves and fireplaces of the same hardwood as European beech, Pyrenean oak, and black poplar, demonstrating that the production of such compounds depends on the type of combustion process, the type of plants, and the biomass characterization [20–24]. However, there are few works on the determination of VOCs emitted by Mediterranean vegetation species [19], and they are generally limited to the determination of PM and the main compounds in flue gases [23,25].

This work aims to investigate the emissions of VOCs by the combustion of woodchips from wet and dry poplar, which is one of the most important plantation trees in the European region [26,27]. Biomass moisture content is a parameter that is greatly affected by the type of biomass and by climatic and storage conditions, which negatively affects the combustion process, worsening the uniformity of the burning process. In this work, the VOCs generated by the poplar combustion process in the boiler were evaluated by modifying only the biomass moisture content. To understand the correlation between the physical and chemical biomass characteristics (lignin content, cellulose and hemicellulose content, ash content, carbon contents, hydrogen, nitrogen, sulfur, oxygen, heating value), combustion parameters, and emission of VOCs, a statistical evaluation was carried out.

2. Materials and Methods

2.1. Biomass Characterization

The biomass involved in combustion tests was poplar woody biomass harvested in the research area of CREA-IT in Monterotondo, Italy. The two poplar woodchips used were obtained from the same starting biomass but were differently treated: the first one (dry poplar, DP) underwent an open-air drying process, while the second (wet poplar, WP) was used directly after the harvesting. The biomass combustion test and related analysis were carried out by the Laboratory for Experimental Activities on Renewable Energy from Biomass (LASER-B) of CREA-IT of Monterotondo (Rome). The characterization concerned the determination of moisture content; higher heating value; ash content; and carbon, hydrogen, and nitrogen content. These determinations were made in triplicate. The moisture content, on a wet basis, was measured according to UNI EN ISO 18134-1:2015 and using a Memmert UFP800 drying oven. The higher heating value (HHV) was determined according to UNI EN ISO 18125:2018 and using a Paar 6400 isoperibol calorimeter, and the lower heating value (LHV) was calculated from the HHV and the hydrogen content. The total content of carbon (C), hydrogen (H), and nitrogen (N) was measured according to UNI EN ISO 16948:2015 and using a Costech ECS 4010 CHNS-O elemental analyzer. Ash content was measured according to UNI EN ISO 18122:2016 and using a Lenton EF11/8B muffle furnace.

The evaluation of heating value; ash content; and C, H, and N contents was carried out on a dry basis. In particular, with regard to the three analyses mentioned above, the dried sample obtained from the moisture determination procedure was ground first with the Retsch SM 100 cutting mill for a preliminary size reduction and then with the Retsch ZM 200 rotor mill. The lignin (Lign), cellulose (Cell), hemicellulose (Hem), and chlorine (Cl) content were obtained using the Phyllis database [28]. In order to obtain preliminary data about the thermal behavior of dry and wet poplar, the biomass was studied by thermogravimetric and differential scanning calorimetry analysis (TGA/DSC).

2.2. Experimental TGA Analysis

A thermoanalytical test is the most commonly used method to estimate the thermal kinetics of biomass during a thermochemical conversion process. TGA analysis provides data about the phase variation of the sample, mass loss, and emission production depending on the nature of the sample [29,30]. In this study, thermogravimetric curve and its derivative (DTG) allow us to investigate combustion process dynamics of dry and wet poplar by means of a Mettler Toledo TGA/DSC1 STAR^e System in the following operating conditions: temperature between 25 °C and 1000 °C; a heating rate of 80 °C/min; and an air flow rate of 60 mL/min.

2.3. Experimental Combustion Tests

The combustion tests were carried out through a demonstrative cogeneration plant based on a moving grate furnace (350 kW$_{th}$) and equipped with a steam generator (500 kg/h at 1.2 MPa). The facility (Figure 1) was characterized by a cross-current combustion chamber and a secondary chamber for post-combustion. The biomass loading into the furnace occurred through a combined double auger loading system (DUPLO®) that allowed the use of even unconventional biomasses, and the mobile-grate system allowed the burning of biomasses with different particle sizes and with a moisture content up to 55% on a wet basis. The exhaust gases produced from biomass combustion were treated with a cyclone and a baghouse filter as PM abatement systems before the passage through the chimney. Along the chimney, several sampling points were established in order to evaluate the emissions in relation to the operating conditions of the furnace and the type of biomass.

Figure 1. Details of the CHP plant (double screw mechanism, furnace, baghouse filter, and chimney).

For the PM sampling, a probe (HP5 Dadolab, Cinisello Balsamo MI, Italy) and an isokinetic sampler (ST5 Dadolab, Cinisello Balsamo MI, Italy) were used, respecting the European method [31]. PM was sampled by means of glass microfiber filters and quantified with gravimetric analysis using a Mettler Toledo AL104 Analytical Balance placed in a conditioned room at 20 °C and 50% humidity. The management of the biomass combustion process occurred by means of the control of parameters, such as biomass feeding, grate movement, and air distribution through blowers both inside the furnace and downstream of the filters.

2.4. Sampling and Analysis of VOCs

During the combustion test, the monitoring of VOCs was carried out according to UNI EN 13649, which represents the procedure for the characterization of single VOCs. In particular, such compounds were sampled by thermal desorption multilayer tubes through a dynamic dilution procedure using a "DDS-Tecora" system. This method is recommended when the concentration of water is high enough to cause the risk of condensation. According to this method, the flue gases are diluted with purified air in a mixing chamber before being sampled onto a thermal desorption tube. In this work, a 1:5 dilution factor and a flow of 50 mL/min were used for sampling. The sampling end of an identical, secondary back-up tube was connected to the outlet of the primary sampling tube as a check on breakthrough. Volatile organic compounds from poplar wood combustion were determined using a thermal desorption system (Markes TD 100 xr) coupled with a gas chromatographic mass spectrometer (Agilent 7000—7890A). The tubes were thermo-desorbed with a flow of 50 mL/min up to a temperature of 380 °C for 10 min with a 1:10 split. The analytes were collected on a focusing trap at the temperature of -10 °C and then desorbed for 1 min with a cold trap high temperature at 380 °C with a 40 °C/s rate. The capillary column used for the analysis was DB-502 (Agilent—60 m, 0.32 mm, 1.8 µm). After the determination, VOCs were classified into three groups: chlorofluorocarbons (CFC), chlorinated compounds (Chl), and aromatic compounds (Aro).

2.5. Statistical Analysis

In order to describe the biomass and the relative emission during combustion, principal component analysis (PCA) was performed through the PAST software (PAleontological STatistics) to investigate the correlation between physical and chemical parameters of biomass and the VOCs emission during the combustion process. PCA is a classical multivariate method widely used to interpret variation in a high-dimensional interrelated dataset with a large number of variables. It is a mathematical methodology that uses orthogonal transformation to convert a set of cases of possibly correlated variables into a set of values of uncorrelated variables which are known as principal components (PCs), thus reducing the number of variables. The two samples tested DP and WP were evaluated and compared through PCA, considering 12 variables: moisture; ash; carbon, hydrogen and nitrogen content (C, H, N); cellulose (Cell); hemicellulose (Hem); lignine (Lign); chlorofluorocarbons

(CFC); chlorinated compounds (Chl); aromatic compounds; and chlorine (Cl). Since there were two materials studied, the only PC1 explained 100% of the total variability.

PCA loadings are the coefficients of the linear combination of the original variables from which the principal components (PCs) are constructed. In order to identify which variables have the greatest effect on sample variability, PC1 loadings were calculated. In this way, from the correlation between PC1 and the 12 variables, it was possible to identify correlations between the variables themselves.

3. Results and Discussion

3.1. Biomass Characterization

The chemical–physical characterization of biomass was carried out for the two different types of samples (dry and wet poplar), providing results on parameters that can affect the combustion quality and the related emissions. Biomass characterization results are reported in the following table (Table 1).

Table 1. Chemical and physical biomass parameters.

Parameters	Dry Poplar	Wet Poplar
Moisture %	15.00	43.30
Ash %	1.30	3.12
C %	41.20	47.36
H %	4.05	5.58
N %	1.10	0.31
HHV [MJ/kg]	18.27	17.94
LHV [MJ/kg]	17.43	16.77

The results show comparable values in terms of elemental composition and calorific values content that meet the requirements for poplar chips used in the combustion process. On the contrary, moisture and ash content showed substantially different results and were mostly responsible for incomplete combustion and potentially harmful emissions. In the case of wet poplar, higher ash and moisture content results in a higher amount of unburnt compounds and bad combustion conditions. The sulfur content was below the limit of quantification (LOQ) and, hence, was considered negligible.

3.2. TGA Analysis

A preliminary analysis in TGA was conducted to evaluate the behavior of the matrices subjected to the thermal stress of combustion in a lab-scale test.

In Figure 2, the black curve (TGA) has two main steps that represent, respectively, up to about 100 °C due to the loss of water (the first step) and from about 200 °C up to about 720 °C due to the loss of volatile substances (the second step). The red curve (DTG) highlights these two stages with two peaks corresponding to the main thermal phenomena of mass loss.

In the graph of Figure 3, the black curve (TGA) has three distinct steps that represent the loss of moisture under 100 °C; the second step, which starts at 200 °C and finishes at about 350 °C, corresponds to the active zone and represents the hemicellulose and cellulose degradation. The last decomposition step, named the passive zone, from 350 °C to 750 °C, represents the slow lignin degradation, corresponding to a potential amount of VOC production as reported by the literature data [32]. The DTG curve reflects weight variation shown with the TGA highlight the mass loss in the steps corresponding to the composition of the sample. From the comparison of the curves in Figures 2 and 3, it is observed that there are differences in the thermal behavior of the two samples due mainly to the higher moisture content in the wet poplar; consequently, it is more likely that the use of this matrix will lead to the formation of emission compounds from incomplete combustion. By comparing Figures 2 and 3, it is evident that there is a strong difference in the thermal trend of the same biomass poplar sample with different moisture levels. In fact, dry poplar has a thermogravimetric curve whose course is much more linear than the curve of wet poplar. This indicates that wet poplar subjected to thermal stress will lead to a less homogeneous combustion phenomenon than dry biomass.

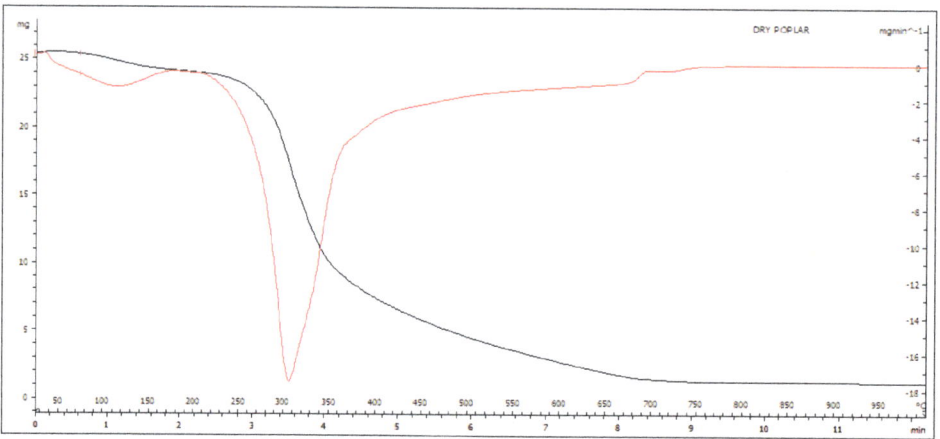

Figure 2. TGA (black curve) and DTG (red curve) of Dry poplar.

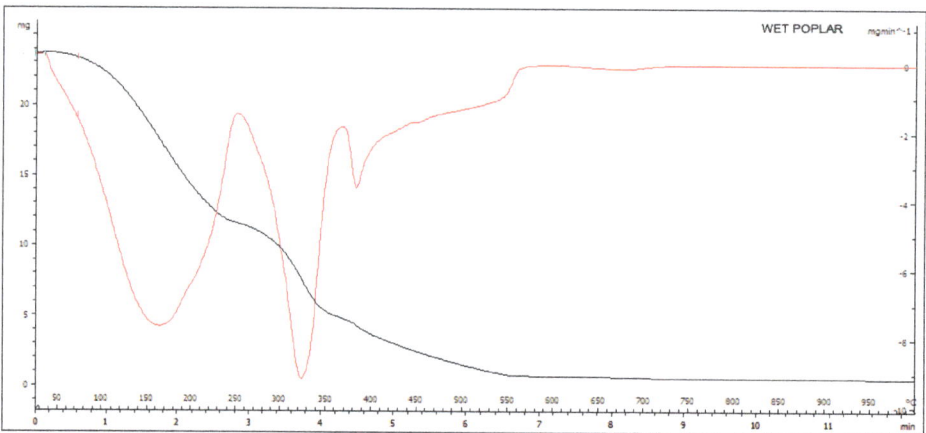

Figure 3. TGA (black curve) and DTG (red curve) of wet poplar.

Figure 4 shows the comparison carried out by means of the DSC analysis related to Figures 2 and 3. It can be observed that as the chamber temperature increases, there is a large exothermic peak of the matrices in the temperature range between 250 and 800 °C. By integrating these heat exchange curves, it can be seen that the dry poplar generates a heat of 163.29 J against the 121.36 J generated by the wet poplar. This result also confirms that the higher moisture negatively affects the combustion conditions.

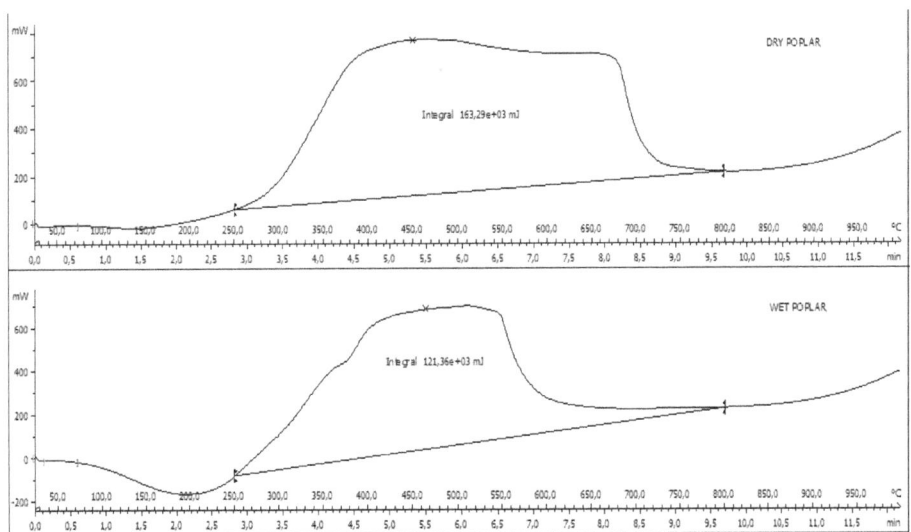

Figure 4. DSC curves, dry poplar, and wet poplar.

3.3. VOCs in Emission

The assessment of VOCs in flue gases was carried out for two different monitoring campaigns related to two different poplar moisture conditions: dry and wet.

During these tests, the temperatures reached by the furnace bed (467 °C) and by the post-combustion chamber (793 °C) were monitored. Several different types of VOCs were found in terms of molecular weight (from freons to 1,2,4-trimethylbenzene) and chemical class (Table 2).

Table 2. Concentration of VOCs in dry and wet poplar combustion emissions.

Compound	Dry Poplar ($\mu g/Nm^3$)	Wet Poplar ($\mu g/Nm^3$)
Dichlorodifluoromethane	0.10	0.38
Chloromethane	4.25	51.15
Bromomethane	<LOQ	14.55
Chloroethane	<LOQ	6.90
1,1-Dichloroethene	1.23	5.85
Dichloromethane	0.95	1.44
Benzene	30.97	60.51
Toluene	6.9	15.46
Chlorobenzene	1.33	5.32
p,m-Xylene	1.10	6.18
o-Xylene	0.45	2.83
Styrene	0.47	1.87
1,2,4-Trimethylbenzene	<LOQ	1.66

From the comparison of dry and wet poplar, it is possible to observe that the highest concentrations are obtained from wet poplar. Backup tubes were analyzed for the assessment of breakthrough volumes, and recovery values above 10% were not found for any analyte.

The analysis of the flue gases emitted by the combustion of two same-origin biomasses with different moisture content shows that the wet biomass produced a higher quantity of VOCs. This phenomenon is due to a higher moisture content that leads to a lower-quality combustion process. In fact, a higher quantity of VOCs is originated from an incomplete combustion phenomenon [33]. Once the VOC concentrations relative to two different moisture levels of the biomass were obtained, the authors thought of constructing the lines that can express an ideal linear trend in the increase in the concentrations of the VOC classes as a function of the increasing moisture content (Figure 5). It is interesting to note that the increase in CFCs and aromatic compounds follows the same type of trend, while the line relating to chlorinated compounds has a less marked increase. This is probably due to the fact that the formation of chlorinated compounds is essentially limited by the presence of chlorine in the matrix, which is the limiting reagent in the formation of the compounds. Moreover, as moisture increases, and, therefore, as combustion conditions worsen, it is likely that organic micropollutants that are heavier than VOCs are formed, such as PAH, PCB, and PCDD/F, whose individual molecules contain more chlorine atoms, thus reducing the formation of chlorinated VOCs. Although the graph is the ideal trend based on only two real points, it is interesting to note how the trend of VOCs formation is positive in all cases as a function of the increase in moisture. This ideal chart forms a basis for comparisons with other works that will be conducted under the same conditions. The authors conducted an in-depth bibliographic search to find other works on VOC emissions from biomass combustion on similar boilers in which the moisture degree of the incoming biomass was specified, but it was found that there is still a lack of studies on this subject.

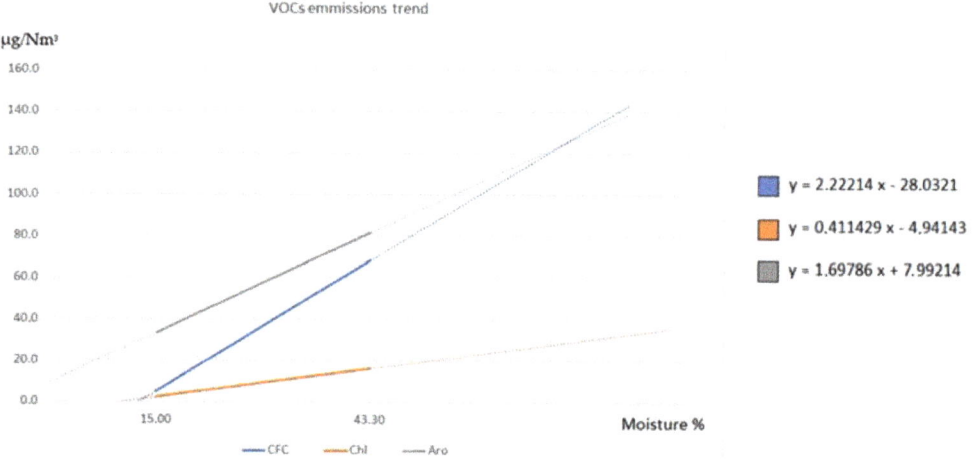

Figure 5. The ideal linear trend in the increase in the concentrations of the VOC classes as a function of the increasing moisture content.

3.4. Principal Component Analysis

The statistical evaluation shows that, in agreement with the bibliographic sources, a correlation can be established between biomass characteristics and VOCs production. Specifically, this study shows how the main variable represented by moisture, under constant burning conditions (type of plant, burning equipment, biomass, air supply, etc.), influences the amount of some classes of VOCs in emissions.

Some studies have evaluated the correlation between the emissions of some classes of volatile organic compounds from Mediterranean species, focusing on the relationship between the content of cellulose, hemicellulose, and lignin and the emission composition, showing a positive correlation between benzene and lignin [34–39]. Other studies show that hemicelluloses decompose during the thermal process and produce large amounts of volatiles at 200 °C [40]. It is well known that drying biomass fuel improves combustion efficiency, increases steam production, reduces air emissions, and improves boiler operation. Except for suspension-firing furnaces, wood-fired boilers and furnaces require a fuel moisture content of below 55% to 65% in order to sustain combustion. For wood-fired incineration, the optimal moisture content is generally much lower, between about 10% and 15% [41]. The water content influences the combustion and the volume of flue gas produced per energy unit. The heating value of the fuel decreases with increasing moisture content. For biofuels, which have a very high moisture content, some problems may occur during firing. High moisture content can cause ignition problems and reduce the combustion temperature, which in turn hinders the combustion of reaction products and consequently affects the quality of combustion [42]. Thermal property values, such as specific heat, thermal conductivity, and emissivity, vary with moisture content [43]. To better understand the relations between biomass moisture and combustion VOCs emissions, an analysis of flue gas and physical and chemical properties of biomass was conducted. The twelve variables coming from these analyses were used to perform a PCA analysis capable of describing the two samples WP and DP. The only principal component 1 (PC1) describes 100% of the sample's variability.

In Figure 6, the PC1 loading for each variable can be observed. Large loadings (positive or negative) indicate that a particular variable has a strong relationship to PC1. The sign of a loading indicates whether a variable and a principal component are positively or negatively correlated. The variables that mostly influence PC1 behavior are CFC, Chl, aromatics, and moisture. These variables are all positively related to PC1 and, consequently, are positively correlated with each other. Positive loadings indicate that variables are positively correlated; an increase in one results in an increase in the other. So, higher moisture values of biomass correspond to higher CFC, Chl, and Aro emissions.

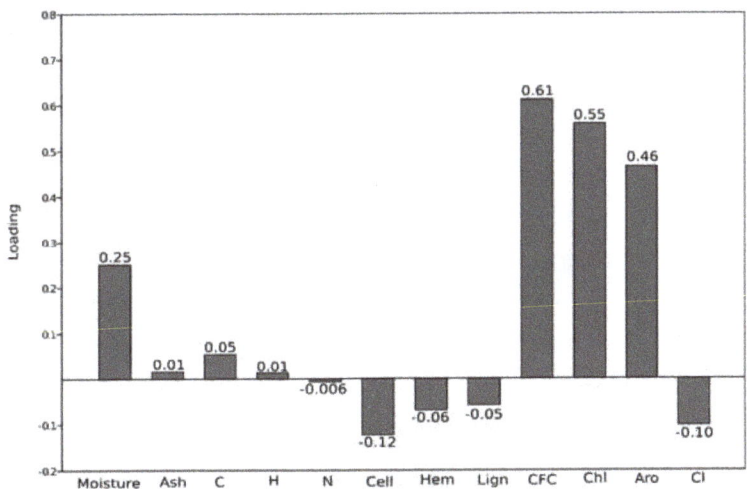

Figure 6. Loadings of PC1 relative to all the variables used for PCA analysis.

Since moisture content was the only parameter that differentiated between WP and DP, and since the combustion parameters were a set constant, it is evident that bad combustion due to biomass moisture content is responsible for the significantly higher production of VOCs.

4. Conclusions

The work proposed aims to highlight how the biomass combustion process is negatively affected by the moisture content of the biomass itself. Compared to other chemical-physical characteristics (e.g., content in ashes, metals present, etc.), moisture content can be reduced quite easily by intervening in the storage conditions or by drying the biomass in the open field under the sun, if seasonality and latitude allow it. The study showed that biomass moisture content leads to a general increase in all classes of VOCs emitted during combustion processes and, thus, inevitably affects the quality of the surrounding air where the phenomenon occurs. It should be noted that such concentrations are even lower than those that would be obtained in the case of uncontrolled combustion or in the phenomena of open burning (e.g., forest fires, domestic fireplaces, combustion of pruning on the sidelines, etc.). Several articles in the literature show the characterization of emissions in terms of emitted organic compounds, even if the majority [20,44] identify almost exclusively aromatic or semi-volatile compounds, neglecting freon and C short-chain chlorinated compounds. However, there are no works in which the formation of volatile organic pollutants is related to the moisture content of the biomass burned. PCA analysis was able to clearly compare two samples with single measurements of many variables and give a preliminary quantification of the relation between biomass moisture and VOCs production from combustion. In general, the work demonstrates how, in a real plant, the moisture of the starting biomass enormously influences the formation of all the classes of VOCs considered (aromatic, CFC, and chlorinated). This is because an increase in the moisture content of the matrix hinders the optimal combustion process and, therefore, leads to the formation of a greater concentration of incomplete combustion compounds (including VOCs).

Author Contributions: Conceptualization: E.P., F.G., M.C., D.B. and E.G.; methodology: B.V., M.C. and A.P.; validation: A.P. and E.G.; formal analysis: M.C.; data curation: A.P., M.C. and E.P.; writing—original draft preparation: M.C. and E.P.; project administration: F.G. All authors have read and agreed to the published version of the manuscript.

Funding: This research was funded by the Italian Ministry of Agricultural, Food, Forestry and Tourism Policies (MiPAAF) under the AGROENER (D.D. n. 26329, 1 April 2016) project.

Institutional Review Board Statement: Not applicable.

Informed Consent Statement: Not applicable.

Data Availability Statement: Not applicable.

Acknowledgments: This study was supported by the Italian Ministry of Agricultural, Food, Forestry and Tourism Policies (MiPAAF) under BTT (D.M. n. 2420, 20 February 2008) and STIMA (D.M. 30681, 10 June 2020) projects.

Conflicts of Interest: The authors declare no conflict of interest.

Sample Availability: Samples of the compounds are available from the authors.

References

1. Pastorello, C.; Caserini, S.; Galante, S.; Dilara, P.; Galletti, F. Importance of activity data for improving the residential wood combustion emission inventory at regional level. *Atmos. Environ.* **2011**, *45*, 2869–2876. [CrossRef]
2. Shan, F.; Lin, Q.; Zhou, K.; Wu, Y.; Fu, W.; Zhang, P.; Song, L.; Shao, C.; Yi, B. An experimental study of ignition and combustion of single biomass pellets in air and oxy-fuel. *Fuel* **2017**, *188*, 277–284. [CrossRef]
3. McKendry, P. Energy production from biomass (part 1): Overview of biomass. *Bioresour. Technol.* **2002**, *83*, 37–46. [CrossRef]
4. Gallucci, F.; Salerno, M.; Guerriero, E.; Amalfi, M.; Palmieri, F. Research facility assessment for biomass combustion in moving grate furnaces. *WASET Int. J. Chem. Mol. Eng* **2017**, *4*, 603–610.
5. Tao, J.; Hou, L.; Li, J.; Yan, B.; Chen, G.; Cheng, Z.; Lin, F.; Ma, W.; Crittenden, J.C. Biomass combustion: Environmental impact of various precombustion processes. *J. Clean. Prod.* **2020**, *261*, 121217. [CrossRef]
6. Proto, A.R.; Palma, A.; Paris, E.; Papandrea, S.F.; Vincenti, B.; Carnevale, M.; Guerriero, E.; Bonofiglio, R.; Gallucci, F. Assessment of wood chip combustion and emission behavior of different agricultural biomasses. *Fuel* **2021**, *289*, 119758. [CrossRef]

7. Johansson, L.S.; Tullin, C.; Leckner, B.; Sjövall, P. Particle emissions from biomass combustion in small combustors. *Biomass Bioenergy* **2003**, *25*, 435–446. [CrossRef]
8. Nussbaumer, T. Combustion and Co-combustion of Biomass: Fundamentals, Technologies, and Primary Measures for Emission Reduction. *Energy Fuels* **2003**, *17*, 1510–1521. [CrossRef]
9. Mcdonald, J.D.; Zielinska, B.; Fujita, E.M.; Sagebiel, J.C.; Chow, J.C.; Watson, J.G. Fine particle and gaseous emission rates from residential wood combustion. *Environ. Sci. Technol.* **2000**, *34*, 2080–2091. [CrossRef]
10. Tissari, J.; Hytönen, K.; Lyyränen, J.; Jokiniemi, J. A novel field measurement method for determining fine particle and gas emissions from residential wood combustion. *Atmos. Environ.* **2007**, *41*, 8330–8344. [CrossRef]
11. Tissari, J.; Lyyränen, J.; Hytönen, K.; Sippula, O.; Tapper, U.; Frey, A.; Saarnio, K.; Pennanen, A.S.; Hillamo, R.; Salonen, R.O.; et al. Fine particle and gaseous emissions from normal and smouldering wood combustion in a conventional masonry heater. *Atmos. Environ.* **2008**, *42*, 7862–7873. [CrossRef]
12. Chen, L.W.A.; Verburg, P.; Shackelford, A.; Zhu, D.; Susfalk, R.; Chow, J.C.; Watson, J.G. Moisture effects on carbon and nitrogen emission from burning of wildland biomass. *Atmos. Chem. Phys.* **2010**, *10*, 6617–6625. [CrossRef]
13. Gallucci, F.; Pari, L.; Longo, L.; Carnevale, M.; Santangelo, E.; Colantoni, A.; Paolini, V.; Guerriero, E.; Tonolo, A. Assessment of Organic Micropollutants (Pcdd/fs and Pcbs) from Biomass Combustion in a Small Chp Facility. Available online: https://dspace.unitus.it/handle/2067/30488?&locale=it (accessed on 29 December 2021).
14. Demirbas, A. Hazardous emissions from combustion of biomass. *Energy Sources Part A Recover. Util. Environ. Eff.* **2008**, *30*, 170–178. [CrossRef]
15. Yan, Y.; Yang, C.; Peng, L.; Li, R.; Bai, H. Emission characteristics of volatile organic compounds from coal-, coal gangue-, and biomass-fired power plants in China. *Atmos. Environ.* **2016**, *143*, 261–269. [CrossRef]
16. Tsai, W.T. Fate of chloromethanes in the atmospheric environment: Implications for human health, ozone formation and depletion, and global warming impacts. *Toxics* **2017**, *5*, 23. [CrossRef] [PubMed]
17. Chen, J.; Li, C.; Ristovski, Z.; Milic, A.; Gu, Y.; Islam, M.S.; Wang, S.; Hao, J.; Zhang, H.; He, C.; et al. A review of biomass burning: Emissions and impacts on air quality, health and climate in China. *Sci. Total Environ.* **2017**, *579*, 1000–1034. [CrossRef]
18. Kroll, J.H.; Seinfeld, J.H. Chemistry of secondary organic aerosol: Formation and evolution of low-volatility organics in the atmosphere. *Atmos. Environ.* **2008**, *42*, 3593–3624. [CrossRef]
19. Ciccioli, P.; Brancaleoni, E.; Frattoni, M.; Cecinato, A.; Pinciarelli, L. Determination of volatile organic compounds (VOC) emitted from biomass burning of Mediterranean vegetation species by GC-MS. *Anal. Lett.* **2001**, *34*, 937–955. [CrossRef]
20. Evtyugina, M.; Alves, C.; Calvo, A.; Nunes, T.; Tarelho, L.; Duarte, M.; Prozil, S.O.; Evtuguin, D.V.; Pio, C. VOC emissions from residential combustion of Southern and mid-European woods. *Atmos. Environ.* **2014**, *83*, 90–98. [CrossRef]
21. Engling, G.; Carrico, C.M.; Kreidenweis, S.M.; Collett, J.L.; Day, D.E.; Malm, W.C.; Lincoln, E.; Min Hao, W.; Iinuma, Y.; Herrmann, H. Determination of levoglucosan in biomass combustion aerosol by high-performance anion-exchange chromatography with pulsed amperometric detection. *Atmos. Environ.* **2006**, *40*, 299–311. [CrossRef]
22. Geng, C.; Yang, W.; Sun, X.; Wang, X.; Bai, Z.; Zhang, X. Emission factors, ozone and secondary organic aerosol formation potential of volatile organic compounds emitted from industrial biomass boilers. *J. Environ. Sci. (China)* **2019**, *83*, 64–72. [CrossRef]
23. Gonçalves, C.; Alves, C.; Fernandes, A.P.; Monteiro, C.; Tarelho, L.; Evtyugina, M.; Pio, C. Organic compounds in PM2.5 emitted from fireplace and woodstove combustion of typical Portuguese wood species. *Atmos. Environ.* **2011**, *45*, 4533–4545. [CrossRef]
24. Alves, C.; Gonçalves, C.; Fernandes, A.P.; Tarelho, L.; Pio, C. Fireplace and woodstove fine particle emissions from combustion of western Mediterranean wood types. *Atmos. Res.* **2011**, *101*, 692–700. [CrossRef]
25. Alves, C.A.; Vicente, A.; Monteiro, C.; Gonçalves, C.; Evtyugina, M.; Pio, C. Emission of trace gases and organic components in smoke particles from a wildfire in a mixed-evergreen forest in Portugal. *Sci. Total Environ.* **2011**, *409*, 1466–1475. [CrossRef]
26. Jia, L.; Liu, S.; Zhu, L.; Hu, J.; Wang, X. Carbon storage and density of poplars in China. *J. Nanjing For. Univ. (Nat. Sci. Ed.)* **2013**, *56*, 1.
27. Chu, D.; Xue, L.; Zhang, Y.; Kang, L.; Mu, J. Surface characteristics of poplar wood with high-temperature heat treatment: Wettability and surface brittleness. *BioResources* **2016**, *11*. [CrossRef]
28. Energy Research Centre of The Netherlands Phyllis2, Database for Biomass and Waste #2907. Available online: https://phyllis.nl/ (accessed on 10 October 2021).
29. Yaman, S. Pyrolysis of biomass to produce fuels and chemical feedstocks. *Energy Convers. Manag.* **2004**, *45*, 651–671. [CrossRef]
30. Strezov, V.; Moghtaderi, B.; Lucas, J.A. Computational calorimetric investigation of the reactions during thermal conversion of wood biomass. *Biomass Bioenergy* **2004**, *27*, 459–465. [CrossRef]
31. EN 13284-1:2017; Stationary Source Emissions. Determination of Low Range Mass Concentration Of Dust—Part 1: Manual Gravimetric Method. 2017. Available online: https://shop.bsigroup.com/products/stationary-source-emissions-determination-of-low-range-mass-concentration-of-dust-manual-gravimetric-method-1/tracked-changes (accessed on 29 December 2021).
32. Jeguirim, M.; Trouvé, G. Pyrolysis characteristics and kinetics of Arundo donax using thermogravimetric analysis. *Bioresour. Technol.* **2009**, *100*, 4026–4031. [CrossRef]
33. Chen, C.; Wang, H.; Qin, Y.; Zhang, Y.; Zheng, S.; Yang, Y.; Jin, S.; Yang, X. Voc characteristics and their source apportionment in the yangtze river delta region during the g20 summit. *Atmosphere* **2021**, *12*, 928. [CrossRef]
34. Garcia, D.P.; Caraschi, J.C.; Ventorim, G.; Vieira, F.H.A.; de Paula Protásio, T. Assessment of plant biomass for pellet production using multivariate statistics (PCA and HCA). *Renew. Energy* **2019**, *139*, 796–805. [CrossRef]

35. Giungato, P.; Barbieri, P.; Cozzutto, S.; Licen, S. Sustainable domestic burning of residual biomasses from the Friuli Venezia Giulia region. *J. Clean. Prod.* **2018**, *172*, 3841–3850. [CrossRef]
36. Barboni, T.; Pellizzaro, G.; Arca, B.; Chiaramonti, N.; Duce, P. Analysis and origins of volatile organic compounds smoke from ligno-cellulosic fuels. *J. Anal. Appl. Pyrolysis* **2010**, *89*, 60–65. [CrossRef]
37. Greenberg, J.P.; Friedli, H.; Guenther, A.B.; Hanson, D.; Harley, P.; Karl, T. Volatile organic emissions from the distillation and pyrolysis of vegetation. *Atmos. Chem. Phys.* **2006**, *6*, 81–91. [CrossRef]
38. Alén, R.; Kuoppala, E.; Oesch, P. Formation of the main degradation compound groups from wood and its components during pyrolysis. *J. Anal. Appl. Pyrolysis* **1996**, *36*, 137–148. [CrossRef]
39. Faix, O.; Fortmann, I.; Bremer, J.; Meier, D. Thermal degradation products of wood: Gas chromatographic separation and mass spectrometric characterization of polysaccharide derived products. *Eur. J. Wood Wood Prod.* **1991**, *48*, 281–285. [CrossRef]
40. Peters, J.; Fischer, K.; Fischer, S. Characterization of emissions from thermally modified wood and their reduction by chemical treatment. *BioResources* **2008**, *3*, 491–502.
41. Roos, C.J. *Biomass Drying and Dewatering for Clean Heat & Power*; U.S. Departement of Energy, CHP Technical Assistance Partnerships, 2008. Available online: http://northwestchptap.org/nwchpdocs/biomassdryinganddewateringforcleanheatandpower.pdf (accessed on 29 December 2021).
42. Khan, A.A.; de Jong, W.; Jansens, P.J.; Spliethoff, H. Biomass combustion in fluidized bed boilers: Potential problems and remedies. *Fuel Process. Technol.* **2009**, *90*, 21–50. [CrossRef]
43. Demirbas, A. Combustion characteristics of different biomass fuels. *Prog. Energy Combust. Sci.* **2004**, *30*, 219–230. [CrossRef]
44. Zhang, Y.; Shao, M.; Lin, Y.; Luan, S.; Mao, N.; Chen, W.; Wang, M. Emission inventory of carbonaceous pollutants from biomass burning in the Pearl River Delta Region, China. *Atmos. Environ.* **2013**, *76*, 189–199. [CrossRef]

Article

Effects of BPZ and BPC on Oxidative Stress of Zebrafish under Different pH Conditions

Ying Han [1,2,*], Yumeng Fei [1,2], Mingxin Wang [1,*], Yingang Xue [1] and Yuxuan Liu [1]

1. School of Environmental & Safety Engineering, Changzhou University, Changzhou 213164, China; feiyumeng1997@163.com (Y.F.); yzxyg@126.com (Y.X.); dinoice0401@163.com (Y.L.)
2. Jiangsu Engineering Research Center of Petrochemical Safety and Environmental Protection, Changzhou 213164, China
* Correspondence: hanying@cczu.edu.cn (Y.H.); wmx@cczu.edu.cn (M.W.)

Abstract: To further understand the toxic effects of bisphenol Z (BPZ) and bisphenol C (BPC) on aquatic organisms, zebrafish (*Danio rerio*) were exposed to 0.02 mg/L BPZ and BPC mixed solution in the laboratory for 28 days. The impacts of BPZ and BPC on the activity of the antioxidant enzymes, expression of antioxidant genes, and estrogen receptor genes in zebrafish under different pH conditions were studied. The changes of glutathione peroxidase (GSH-Px), reduced glutathione (GSH), total superoxide dismutase (T-SOD), catalase (POD), and malondialdehyde (MDA) in the zebrafish were detected by spectrophotometry. The mRNA relative expression levels of CAT, GSH, SOD, ER*a*, and ER*b*1 in the experimental group were determined by fluorescence quantitative PCR. The results showed that SOD activity and MDA content were inhibited under different pH conditions, and the activities of GSH, GSH-Px, and POD were induced. The activities of POD and GSH induced in the neutral environment were stronger than those in an acidic and alkaline environment. The mRNA relative expression levels of SOD and GSH were consistent with the activities of SOD and GSH. The mRNA relative expression levels of CAT were induced more strongly in the neutral environment than in acidic and alkaline conditions, the mRNA relative expression levels of ER*a* were induced most weakly in a neutral environment, and the mRNA relative expression levels of ER*b*1 were inhibited the most in a neutral environment.

Keywords: bisphenol analogs; zebrafish; oxidative stress

1. Introduction

Bisphenol analogs (BPs) are a class of environmental endocrine disrupts with a similar structure to bisphenol A (BPA), which are widely found in various environmental media [1]. On 18 October 2008, Canada was the first country to declare BPA a toxic substance, banning its use in baby bottles and restricting its use in all food packaging and containers in 2010. This was followed by policies to restrict the use of BPA around the world [2]. To meet the needs of the industrial market, the production and use of BPA substitutes have gradually increased, such as bisphenol C (BPC) used in flame retardant preparation and bisphenol Z (BPZ) applied in chemical compounds manufacture [3]. The new double phenol compounds in the process of production, use and waste treatment, etc., also inevitably enter the environment especially through water (more than 90%, water solubility is related to pH values), causing harm to the aquatic environment, ecological situations, and human health [4–7].

Zebrafish (*Danio rerio*) is an important vertebrate model organism. It is named because of the zebra-like stripes on its side and the length of adult fish is 3–5 cm [8]. Zebrafish have organs and systems similar to those of mammals, such as nerves, digestion, reproduction, immunity, endocrinology, and cardiovascular diseases. Therefore, zebrafish were mainly used in studies related to neurodevelopment and genetics at first [9]. Similarly, zebrafish is widely used in environmental toxicology studies and can be used as an important tool to detect the toxicological effects of bisphenol compounds [10].

Oxidative stress refers to the physiological and pathological reactions of cells and tissues caused by the production of reactive oxygen species (ROS) and reactive nitrogen species (RNS) in the body under the harmful stimulation of the internal and external environment. At present, it has become one of the important topics in environmental toxicology research [11]. Although the number of studies on BPA analogs is limited, according to existing research reports, BPA analogs can cause cytotoxicity, reproductive toxicity, neurotoxicity, and endocrine disruption [12,13]. Ullah et al. conducted in vivo experiments on rats and found that exposure to BPA, bisphenol B (BPB), bisphenol F (BPF), and bisphenol S (BPS) for 28 days would cause ROS production and LPO activity in rat sperm and lead to DNA damage in rat sperm [14]. Park et al. conducted experiments on the acute toxicity, vital parameters, and defense bodies of marine rotifers and found that exposure to BPA, BPF, and BPS significantly increased intracellular ROS levels and glutathione S-transferase (GST) activity [15]. Wu et al. reported short-term exposure to BPA and nonylphenol inhibited the contents of total glutathione (TG), reduced glutathione (GSH), oxidized glutathione (GSSH), catalase (CAT), superoxide dismutase (SOD), glutathione peroxidase (GSH-Px), glutathione reductase (GR), glutathione S-transferase, and other antioxidant enzymes in serum of zebrafish embryos [16].

In this study, the toxic mechanism of BPC and BPZ on zebrafish was explored by detecting the expression levels of oxidative stress-related genes and enzyme activities under different pH conditions. The aim of this study was to provide toxicological data for the toxicity of BPC and BPZ on zebrafish and possible impacts on aquatic organisms.

2. Results and Discussion

2.1. Effects of a Mixed Solution of Bisphenol C (BPC) and Bisphenol Z (BPZ) on the Expression of Oxidative Stress-Related Genes and Estrogen Receptor Genes in Zebrafish

2.1.1. Effects of Phase I Mixed Solution on the Expression of Oxidative Stress Gene and Estrogen Receptor Gene in Zebrafish

BPC and BPZ can change the expression levels of related genes in the body's antioxidant defense system. Therefore, this study selected 3 antioxidant genes and 2 estrogen receptor genes as references to reveal the mechanism of the oxidative stress effect of bisphenol estrogen on zebrafish from the transcriptional level. It can be seen from Figure 1 that the expression levels of antioxidant-related genes and estrogen receptor genes in zebrafish are disturbed to varying degrees. Compared with the blank control group, the relative expression level of the CAT gene was significantly upregulated to 1.9 times on day 1 ($p = 0.0004$; $p < 0.01$). In the following 12 days, although the relative expression level decreased, it was still upregulated compared with the blank control group. The relative expression level of the GPX gene was upregulated to 1.3 times at day 4, and compared with the blank control group at other periods except day 4, the relative expression level of the GPX gene was downregulated, showing great overall fluctuation. The relative expression level of the SOD gene was reduced to 0.7 times on the 10th day, which showed an overall downregulation state compared with the blank control group. The relative expression level of the ERb1 gene decreased to 0.2 times on day 13, showing a time-effect relationship. The relative expression level of the ER*a* gene was upregulated to 1.5 times at day 7 and down to 0.4 times at day 1, indicating a wide fluctuation.

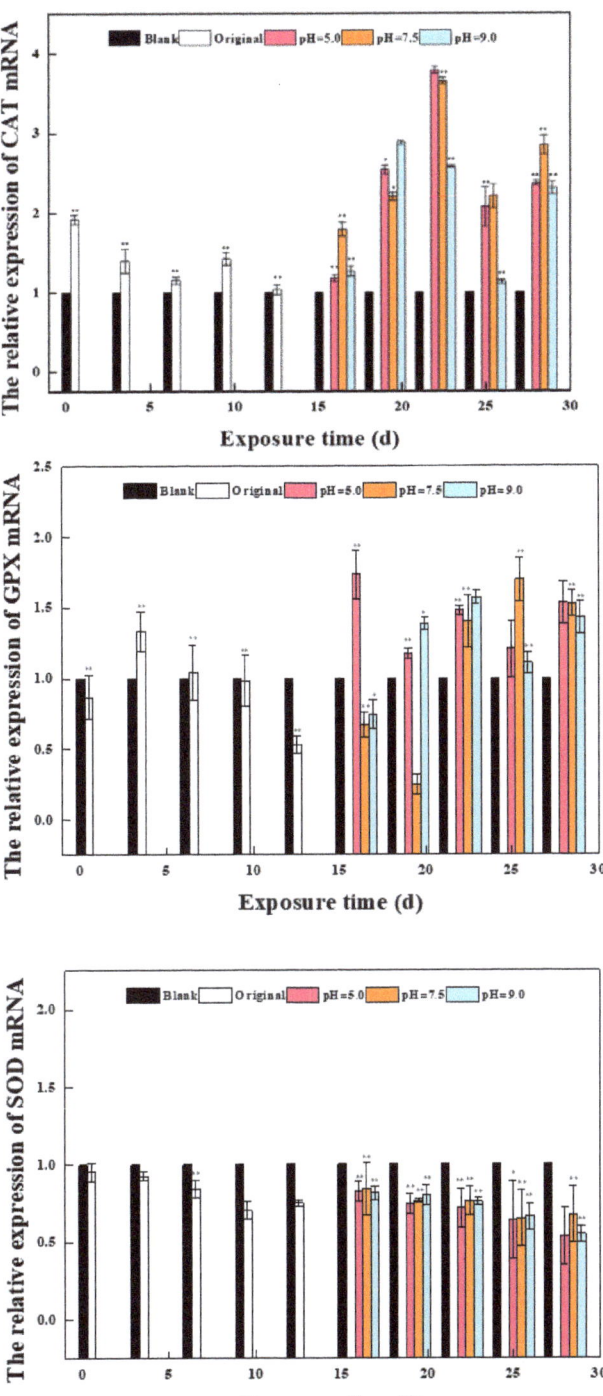

Figure 1. Cont.

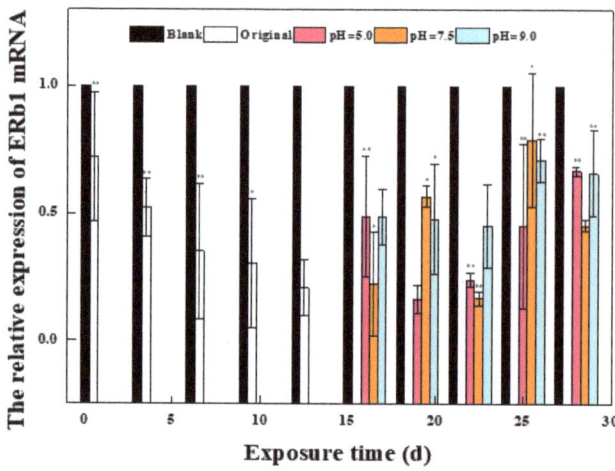

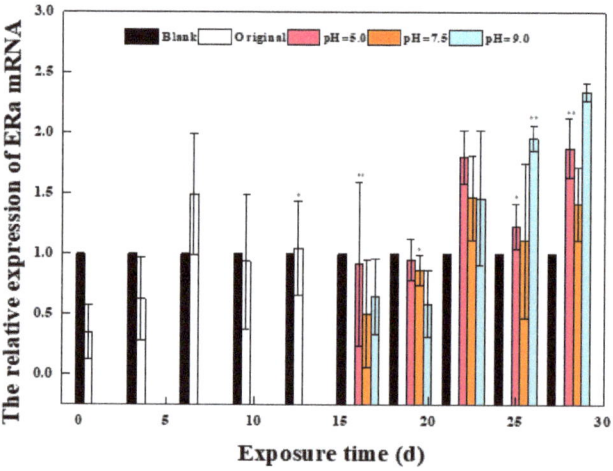

Figure 1. Effects of BPC and BPZ on mRNA expression levels of oxidative stress-related genes in zebrafish ($n = 3$) (Note: $0.01 < * p < 0.05$; $** p < 0.01$, compared with the control group).

2.1.2. Effects of a Mixed Solution of the Second Stage on the Expression of Oxidative Stress Gene and Estrogen Receptor Gene in Zebrafish

In the second stage of exposure, the nine parallel groups in the first stage were divided into three groups with different pH values, and the related gene detection method was the same as in the first stage. It can be seen from Figure 1 that the expression levels of antioxidant genes and the expression of estrogen receptor genes in zebrafish of the three groups with different pH were different. Compared with the blank control group, the relative expression levels of CAT genes in the three groups were upregulated in the first 14 days, which was consistent with the development trend of the first 13 days. The relative expression of the CAT gene was upregulated to the maximum value on day 22 in the low pH group, to 3.8 times on day 22 in the low pH group, and to 3.7 times on day 22 in the middle pH group. The relative expression of the CAT gene in the high pH group reached the maximum value on day 19, which was 2.8 times higher than that in the blank control group. Compared with the blank control group, the relative expression of the GPX gene in the low pH group was upregulated to the maximum value on day 1 and then decreased, but it was still upregulated compared with the control group. The relative expression of the GPX gene in the middle pH group increased to the highest value on day 25, which was 1.7 times that in the control group, and decreased to the lowest value on day 19, which was 0.2 times that in the control group. The relative expression of the CAT gene in the high pH group reached the highest value on day 22, which was 1.6 times that in the control group. Although the relative expression levels of the GPX gene in the three groups fluctuated differently, the final results tended to be consistent. The relative expression of the SOD gene in the second stage was decreased compared with that in the blank control group, which was consistent with that in the first stage, and the fluctuation of the three groups was similar. Wu et al. exposed zebrafish embryos to different concentrations of BPA and nonylphenol, finding that GSH, SOD, and other genes' activities were significantly suppressed. The oxidative stress was onset, which is in accordance with the results of this study [16].

The relative expression of estrogen receptor gene ER*b1* was downregulated in all three groups compared with the blank control group, which was consistent with the results of the first stage. The relative expression level of the ER*b1* gene in the low pH group decreased to the lowest value on day 19, and the relative expression level of the ER*b1* gene in the middle and high pH group decreased to the lowest value on day 22. Compared with the blank control group, the relative expression level of ER*a* in all three groups was upregulated, which was contrary to the results of the first stage. The relative expression of the ER*a* gene in the low and high pH groups was upregulated to the highest value on day 28, which was 1.9 and 2.3 times that in the control group, respectively. The relative expression of the ER*a* gene in the middle pH group was upregulated to the maximum value at day 22, which was 1.4 times that of the control group.

2.2. Effects of a Mixed Solution of BPC and BPZ on the Expression of Oxidative Stress-Related Genes and Estrogen Receptor Genes in Zebrafish

2.2.1. Effects of Mixed Solution at the First Stage on Antioxidant Enzyme Activity and MDA Content in Zebrafish

BPC and BPZ can disbalance the oxidation system and antioxidant system of the body, resulting in excessive production of reactive oxygen species (ROS) and other free radicals in the body. The oxidation degree exceeds the ability of cells to remove oxides themselves, and a large number of oxidation intermediates are produced, leading to oxidative damage of the body.

Figure 2 shows the effects of BPC and BPZ on antioxidant enzyme activity and MDA content in zebrafish at the first stage. Compared with the blank group, SOD activity was inhibited within 13 days, which may be because ROS produced in zebrafish gradually increased with the increase of exposure time, and SOD decreased greatly to eliminate ROS, showing an inhibitory effect. Compared with the blank group, POD, GSH, and GSH-Px activities were induced within 13 days. POD activity was significantly induced at day 9 ($p = 0.034$) and day 12 ($p = 0.046$) ($0.01 < p < 0.05$), GSH activity was also significantly induced at day 9 ($p = 0.006$) and day 12 ($p = 0.001$) ($p < 0.01$), indicating that zebrafish can eliminate excessive free radicals by producing POD, GSH, GSH-Px, and other antioxidant enzymes so that the system can avoid oxidative damage. With the increase of exposure time, MDA content in zebrafish changed. The MDA content in the first phase reached the highest at day 7 (4.6 nmol/mL), which was higher than the control group. Higher MDA content was a stress response mechanism of oxidation that fish tissue showed due to an exogenous pollutants electrophilic group. Exogenous pollutants induced zebrafish to produce large amounts of oxygen free radicals, which in time will combine with the unsaturated fatty acid in biofilm, causing lipid peroxidation reaction and indirectly reflecting the degree of tissue cell damage.

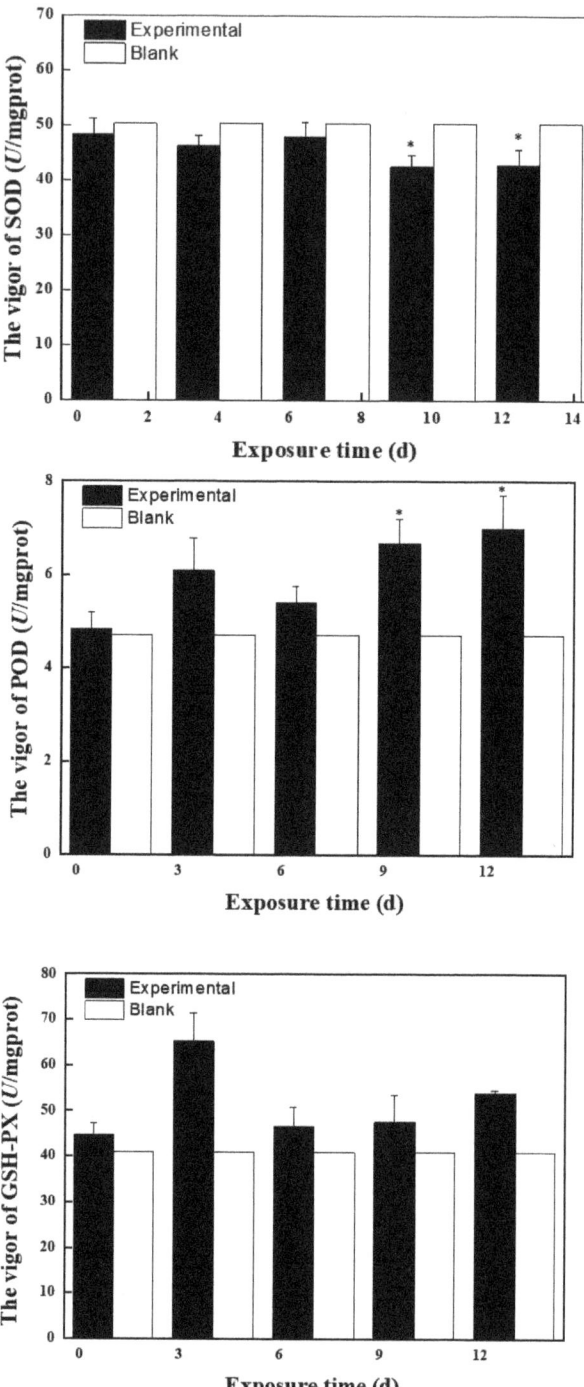

Figure 2. *Cont.*

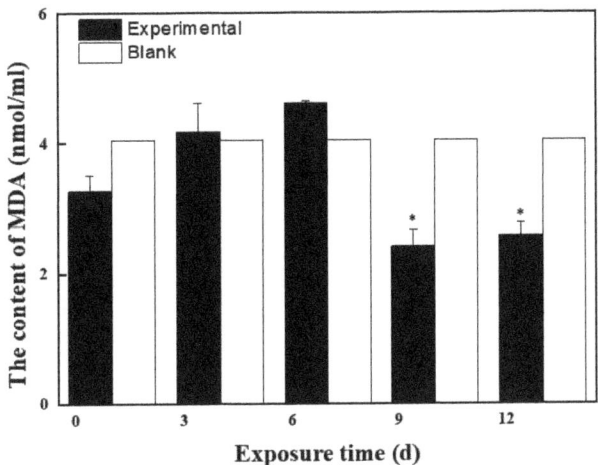

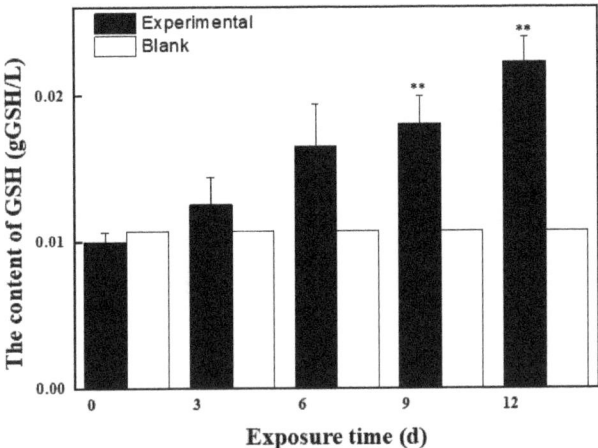

Figure 2. Effects of BPC and BPZ on antioxidant enzyme activities and MDA content in zebrafish ($n = 3$) within 14 days (Note: $0.01 < * p < 0.05$; $** p < 0.01$, compared with the control group).

2.2.2. Effects of Mixed Solution at the Second Stage on Antioxidant Enzyme Activity and MDA Content in Zebrafish

Figure 3 shows the effects of treatment solution with different pH at different exposure times on antioxidant enzyme activity and MDA content in zebrafish in the second stage. Compared with the control group, SOD was mainly inhibited in the middle and high pH groups and reached the lowest value on day 28. On day 28, SOD activity of the three groups, with p values of 0.004, 0.0002, and 0.006, was significantly inhibited ($p < 0.01$) and SOD activity was similar, indicating that exposure solution caused certain oxidative damage to zebrafish, and there was no correlation with the pH value of exposure solution. Compared with the control group, the POD activity of the three groups was induced in the second stage and reached the highest value at day 28. POD activity of the low pH group was 11.6 U/mg Prot, the middle pH group was 13.2 U/mg Prot, and the high pH group was 12.7 U/mg Prot. Compared with the control group, the activity of GSH-Px in the three groups was induced in the second stage. The maximum value of GSH-Px in the middle and low pH groups was 79.7 U/mg Prot and 86.6 U/mg Prot on day 25, and the maximum value of GSH-Px in the high pH group was 69.8 U/mg Prot on day 16. It can be seen that the POD activity in zebrafish was induced similarly in the middle and low pH groups, and the induced level was higher than that in the high pH group. Compared with the control group, GSH activity in the second stage in three groups received induction, in which the pH group of zebrafish in vivo activity of GSH entrainment dropped to close to the control level after rising first, presenting normal distribution, as zebrafish oxidative stress in the body system can clear excess harmful free radicals, protecting zebrafish from oxidative damage. GSH activity was significantly induced in high ($p = 0.043$) and low pH ($p = 0.024$) groups after 22 days ($0.01 < p < 0.05$), the maximum value was reached at day 22 in the low pH group and day 28 in the high pH group. Compared with the first stage, MDA content in the three groups in the second stage changed to varying degrees, and MDA content in the middle and low pH groups increased first and then decreased, indicating that exogenous pollutants induced zebrafish to produce a large number of oxygen free radicals and timely removal. MDA content in the high pH group decreased first and then increased, but it was still lower than that in the control group, indicating that the body cells in zebrafish can still resist the attack of free radicals.

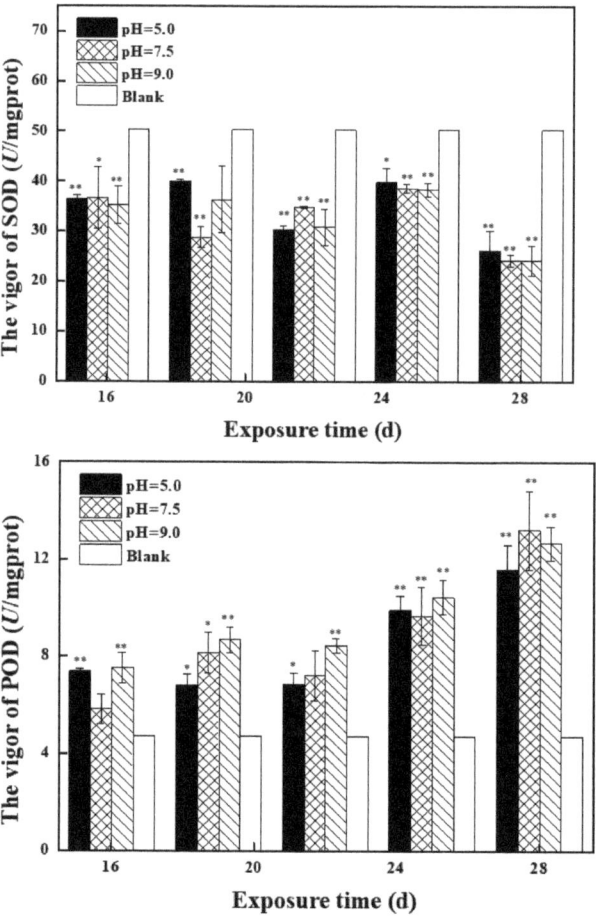

Figure 3. *Cont.*

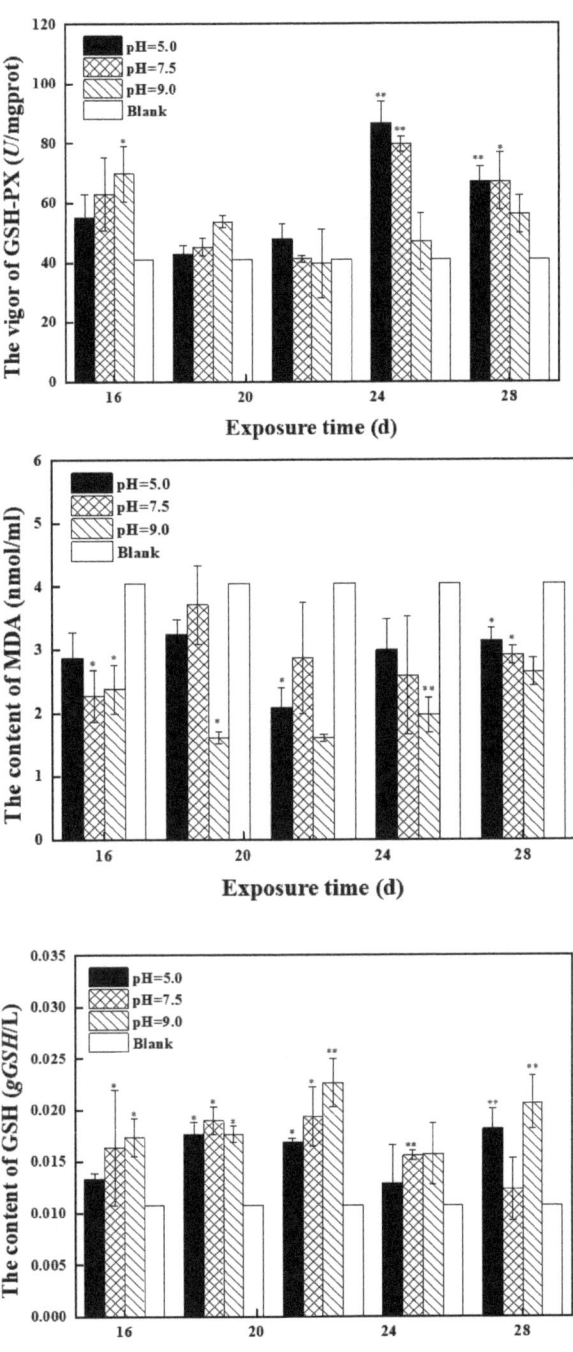

Figure 3. Effects of BPC and BPZ on antioxidant enzyme activity and MDA content in zebrafish (n = 3) under different pH conditions (Note: $0.01 < * p < 0.05$; ** $p < 0.01$, compared with the control group).

3. Materials and Methods

3.1. Chemicals and Materials

Zebrafish (type AB) (protocol number: SCXK 2016-0010) used in this experiment were males and three months old, purchased from Shanghai Jiayu Aquarium in China, with an average body length of 2.5 ± 0.5 cm and an average weight of 0.17 g. After body surface disinfection with 5% sodium chloride solution, the zebrafish were domesticated in tap water which had been dechlorinated after 72 h of aeration. The pH of the test water was 7.74–7.83. The hardness of water was 91–108 mg/L (based on $CaCO_3$); the concentration of dissolved oxygen was 7.45–7.60 mg/L; the temperature was controlled at 25 ± 0.5 °C; and the time distribution was a 14:10 h day-night cycle. Zebrafish were domesticated in the laboratory for more than 7 days, during which they were fed with commercial feed twice a day, and they could not be used in the following experiments until the mortality rate within 7 days was less than 5%.

Bisphenol C (BPC) (>98.0%) and bisphenol Z (BPZ) (≥98.0%) were purchased from Shanghai Aldin Reagent Co., Ltd., Shanghai, China. Dimethyl sulfoxide (DMSO) was purchased from Shanghai Lingfeng Chemical Reagent Co., Ltd., Shanghai, China. Isotope internal standard BPA-$^{13}C_{12}$ ($^{13}C_{12}H_{16}O_2$) (99.0%) was purchased from A ChemTek, Inc., MA, USA. Trizol reagent, dATP, dTTP, dCTP, and dGTP were purchased from Thermo Scientific, Waltham, MA, USA. DNase I was purchased from New England Biotechnology (Suzhou) Co., Ltd., Suzhou, China. Ethyl m-aminobenzoate was purchased from Macklin Co., Ltd., Shanghai, China.

Glutathione peroxidase (GSH-PX) test box, glutathione (GSH) test kit, catalase (POD) test kit, malondialdehyde (MDA) test kit, total superoxide dismutase (T-SOD) test box, and total protein (TP) test kit were purchased from Nanjing Jianceng Institute of Biology, Nanjing, China. SYBR® Premix Ex Taq™ II (Perfect Real Time) Kit and TaKaRa AMV Kit were purchased from TaKaRa Bio Inc., Tokyo, Japan. Ready-to-eat no-hatching harvest shrimp egg feed was purchased from Ching Yee Company, Hong Kong, China.

3.2. Experimental Instruments

Hardness tester (model 16900, Hach Company, Loveland, CO, USA), dissolved oxygen tester (HQ30d, Hach Company, Loveland, CO, USA), electronic analysis balance (ATY124, Shimazu Company of Japan), ultrasonic cleaning machine (UC-4600, Shenzhen Lanjie Ultrasonic Electric Co., Ltd., Shenzhen, China), ultra-pure water machine (UPR table type, Sichuan Youpu Ultra-pure Technology Co., Ltd., Chengdu, China), hand-held homogenizer (S10, Shanghai Jingxin Industrial Development Co., Ltd., Shanghai, China), UV spectrophotometer (UV-3000PC, Shanghai Meipuda Instrument Co., Ltd., Shanghai, China), six-link magnetic heating stirrer (HJ-6A, Jintan District Shuibei Science Experimental Instrument Factory, Changzhou, China), digital display constant temperature water bath (HH4, Shanghai Lichen Bangxi Instrument Technology Co., Ltd., Shanghai, China), low temperature centrifuge (model 5417R, Eppendorf, Hamburg, Germany), real-time PCR System (ABI 7500, Thermo Scientific, Waltham, MA, USA), thermal cycle (MODEL ETC-811, Beijing Dongsheng Innovation Co., Ltd., Beijing, China), gel electrophoresis and imaging system (Bio-Rad Laboratories, Inc., Hercules, CA, USA), and micro UV nucleic acid quantification system (NanoDrop ND-1000, Thermo Scientific, Waltham, MA, USA) were used.

3.3. Experiment Design

The experimental design explaining the experiments throughout the time is shown in Figure 4.

3.3.1. Zebrafish Farming Method

Wild AB strain zebrafish were cultured in an independent breeding system with tap water filtered by activated carbon, UV sterilized, and automatically circulated in the system with a temperature controlled at 25 ± 0.5 °C, while maintaining a photoperiod of 14 (light): 10 (dark). They were fed twice a day with commercially available food.

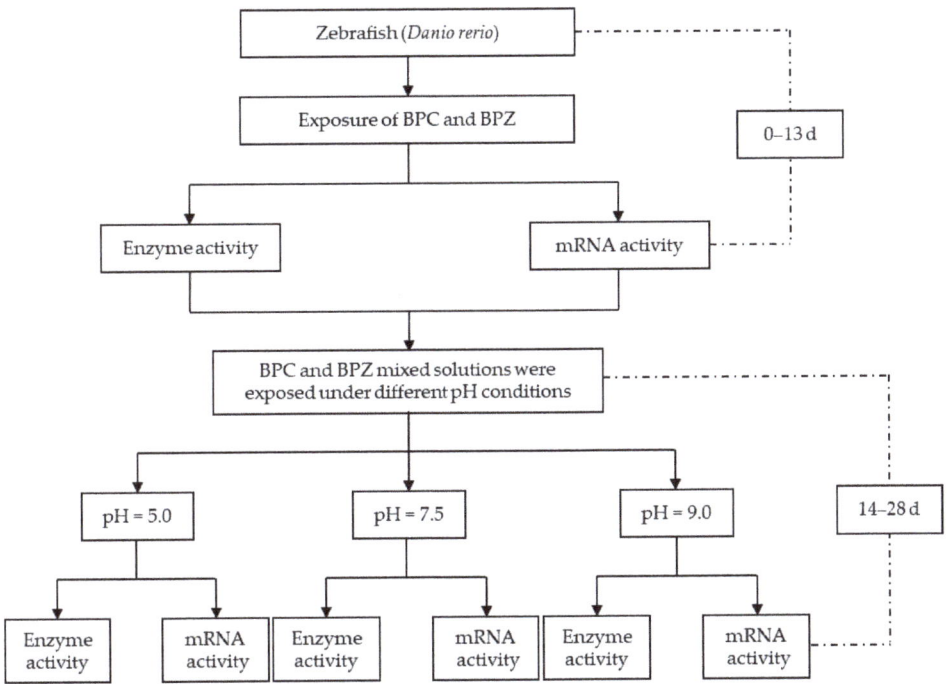

Figure 4. Flow chart of the experimental design in this study.

3.3.2. Method of Exposure

According to the authors' previous experimental results, the semi-lethal concentrations of BPC and BPZ were 2.76 mg/L and 2.57 mg/L, respectively, and the experimental exposure concentration was 1/100 of the semi-lethal concentration [17]. In this paper, DMSO was selected as the cosolvent, and the exposure concentrations of BPC and BPZ were 0.02 mg/L. The exposure stage was divided into two stages, the first stage was 13 days, the second stage was 14–28 days. In the first stage, only 0.02 mg/L of BPC and BPZ mixed solution was added to the natural aerated water. Nine parallel experiments were repeated nine times in each group. In the second stage, by adjusting the pH value of the treatment solution, nine groups of parallel exposed solution were divided into three groups of medium and high pH, with three parallel groups in each group, and the experiment was repeated three times. Three representative pH values were selected in the experiment, namely 5.0, 7.5, and 9.0. During the experiment, 1.2 mol/L HCL solution and 2.0 mol/L NaOH solution were used to adjust the pH of the treatment solution to 5.0 and 9.0, and pH 7.5 was the same natural aeration water as the first 13 days. The exposure method was semi-static, and the treatment solution was replaced every 24 h. The pH, dissolved oxygen, and hardness values of the treatment solution were tested 2 h before and 2 h after the water change. During the experiment, they were fed once a day and fasted for 24 h before analysis and determination. During the experiment, 10 L of the corresponding concentration of experimental liquid was added to the glass aquarium, and the domesticated zebrafish were put into each exposed group with 40 fish in each group ($n = 40$). The physiological indexes in zebrafish were measured 1 d, 4 d, 7 d, 10 d, 13 d, 16 d, 19 d, 22 d, 25 d, and 28 d after exposure. As blank control, 0 d was used for comparison.

3.3.3. Determination of Physiological Indicators in Zebrafish

Three tail zebrafish ($n = 3$) were randomly selected from each group at the sampling time. Zebrafish were situated in 50 mg/L of ethyl m-aminobenzoate solution for 15 min.

Euthanasia of zebrafish was carried out following the Guidelines for the Euthanasia of Animals (2013) published by the America Veterinary Medical Association (AVMA), which was in accordance with the relevant requirements and principles in the animal protection and animal welfare. Zebrafish were frozen in the ice pack immediately; their surface was cleaned with cold saline, filter paper blot moisture was accurately weighed; adding the precooling 0.90% saline (group wet weight: physiological saline = 1:9), using handheld homogenate in the ice water bath machine tissue homogenate in full, made from 10% of the tissue and serum. Centrifuged at 2000 r/min at 4 °C for 10 min, the supernatant was measured for GSH level, MDA content, GSH-Px, SOD, and POD activities. The specific operation was carried out according to the instructions of Nanjing Jian-cheng Kit, and the protein content was determined by Coomassie bright blue method.

3.3.4. Expression of Oxidative Stress Gene and Estrogen Receptor Gene in Zebrafish

The frozen tissue of zebrafish (n = 3) randomly selected from each group was quickly ground in liquid nitrogen and the powder after grinding was added with an appropriate amount of Trizol reagent. The total RNA of the sample was reversely transcribed into cDNA. Primers for each target gene were obtained from Jiangsu Hongzhong Biotech Co., Ltd. (Table 1). The PCR reaction mixture (25 µL) consisted of SYBR®Premix Ex TaqTM II 12.5 µL, nuclease-free water 8.5 µL, forward primer 1 µL, reverse primer 1 µL. and cDNA template 2 µL. Amplification conditions were 95 °C for 2 min, 96 °C denaturation for 10 s, 60 °C denaturation for 30 s, a total of 40 cycles, each RNA repeated 3 times, to form biological replication.

Table 1. Primer sequences of candidate reference genes.

Gene Symbol	Accession NO.	Primer Sequence (5'-3')	Product Length (bp)
β-actin	AF025305.1	F: CGAGCTGTCTTCCCATCCA R: TCACCAACGTAGCTGTCTTTCTG	86
rp17	NM_213213644.2	F: CAGAGGTATCAATGGTGTCAGCCC R: TTCGGAGCATGTTGATGGAGGC	119
CAT	AF170069.1	F: CTCCTGATGTGGCCCGATAC R: TCAGATGCCCGGCCATATTC	126
SOD	BX055516	F: GTCCGCACTTCAACCCTCA R: TCCTCATTGCCACCCTTCC	217
GPX	AW232474	F: AGATGTCATTCCTGCACACG R: AAGGAGAAGCTTCCTCAGCC	94
ERα	AF268283	F: CCC ACA GGA CAA GAG GAA GA R: CCT GGT CAT GCA GAG ACA GA	250
ERβ1	AJ414566	F: GGG GAG AGT TCA ACC ACG GAG R: GCT TTC GGA CAC AGG AGG ACG	89

3.4. Statistical Analysis

Microsoft Excel 2019 software was used to analyze the obtained data, and Origin 8.5 was used for experimental mapping. The experimental results were expressed as mean ± standard deviation (mean ± SD), and the t-test was used to test the statistical difference between the two groups. A * p value < 0.05 means significant difference, ** p < 0.01 means extremely significant difference. Using RP17 as the reference gene, the relative expression of the target gene was calculated by $2^{-\Delta\Delta Ct}$ [18].

4. Conclusions

In the present study, it was found that the bisphenol C (BPC) and bisphenol Z (BPZ) mixture could induce an oxidative stress response and affect the expression of oxidative stress-related genes in zebrafish. Antioxidant enzymes SOD, POD, GSH, and GSH-Px as well as antioxidation-related genes CAT, SOD, and GPX could be potential biological indicators for oxidative stress detection of zebrafish. Under different pH conditions, both SOD in enzymes and genes were inhibited. GSH-Px and GPX were induced in all groups.

POD and GSH were induced in a neutral environment and more strongly than in the acidic and the alkaline environments. The MDA content in the exposed group was lower than that in the blank control group, indicating that free radicals in zebrafish were effectively eliminated. Thus, BPC and BPZ could not only affect the antioxidation system but also cause an estrogen effect on zebrafish. Further studies need to be performed to assess the toxicities of bisphenol analogs, helping to provide evidence for toxicity mechanisms of bisphenol analogs to aquatic organisms.

Author Contributions: Y.H. was responsible for project administration, funding acquisition, supervision, visualization, review, and editing; Y.F. was responsible for data curation, formal analysis, visualization, and writing original draft; M.W. provided help for resources, supervision, visualization, review, and editing; Y.X. provided help for methodology and software; Y.L. was responsible for the investigation. All authors have read and agreed to the published version of the manuscript.

Funding: This research was funded by the National Natural Science Foundation of China (No. 21906009), Jiangsu High-level Innovative and Entrepreneurial Talent Introduction Program Special Fund (No. SCZ2010300004), and The Science and Technology Research Award Package of Changzhou University (No. KYP2102098C).

Institutional Review Board Statement: Not applicable.

Informed Consent Statement: Not applicable.

Data Availability Statement: The authors declare that all data generated or analyzed during this study are included in the published article.

Acknowledgments: The authors acknowledge the zebrafish used for experiments.

Conflicts of Interest: The authors declare no conflict of interest.

Sample Availability: Samples of the compounds are available from the authors.

References

1. Mustieles, V.; D'Cruz, S.C.; Couderq, S.; Rodríguez-Carrillo, A.; David, A. Bisphenol A and its analogues: A comprehensive review to identify and prioritize effect biomarkers for human biomonitoring. *Environ. Int.* **2020**, *144*, 105811. [CrossRef] [PubMed]
2. Sheng, Z.G.; Tang, Y.; Liu, Y.X.; Yuan, Y.; Zhao, B.Q.; Chao, X.J.; Zhu, B.Z. Low concentrations of bisphenol a suppress thyroid hormone receptor transcription through a nongenomic mechanism. *Toxicol. Appl. Pharmacol.* **2012**, *259*, 133–142. [CrossRef] [PubMed]
3. Liu, B.; LehmLer, H.J.; Sun, Y.; Xu, G.; Bao, W. Association of bisphenol A and its substitutes, bisphenol F and bisphenol S, with obesity in United States children and adolescents. *Diabetes Metab. J.* **2019**, *43*, 59. [CrossRef] [PubMed]
4. Braver-Sewradj, S.P.D.; Spronsen, R.V.; Hessel, E.V.S. Substitution of bisphenol A: A review of the carcinogenicity, reproductive toxicity, and endocrine disruption potential of alternative substances. *Crit. Rev. Toxicol.* **2020**, *50*, 128–147. [CrossRef] [PubMed]
5. Mok, S.; Jeong, Y.; Park, M.; Kim, S.; Moon, H.B. Exposure to phthalates and bisphenol analogues among childbearing-aged women in Korea: Influencing factors and potential health risks. *Chemosphere* **2021**, *264*, 128425. [CrossRef] [PubMed]
6. Hc-Wydro, K.; Poe, K.; Broniatowski, M. The comparative analysis of the effect of environmental toxicants: Bisphenol A, S and F on model plant, fungi and bacteria membranes. The studies on multicomponent systems. *J. Mol. Liq.* **2019**, *289*, 111136. [CrossRef]
7. Gavrilescu, M.; Demnerova, K.; Aamand, J.; Agathos, S.; Fava, F. Emerging pollutants in the environment: Present and future challenges in biomonitoring, ecological risks and bioremediation. *New Biotechnol.* **2015**, *32*, 147–156. [CrossRef] [PubMed]
8. Bambino, K.; Chu, J. Zebrafish in Toxicology and Environmental Health. *Curr. Top. Dev. Biol.* **2017**, *124*, 331–367. [PubMed]
9. Nadia, T.; Tai, H.; Da-Woon, J.; Williams, D.R. Fishing for nature's hits: Establishment of the zebrafish as a model for screening antidiabetic natural products. *Evid.-Based Complementary Altern. Med.* **2015**, *2015*, 287847.
10. Stegeman, J.J.; Goldstone, J.V.; Hahn, M.E. Perspectives on zebrafish as a model in environmental toxicology. *Fish Physiol.* **2010**, *29*, 367–439.
11. Zhang, F.; Han, L.; Wang, J.; Shu, M.; Liu, K.; Zhang, Y.; Hsiao, C.D.; Tian, Q.; He, Q. Clozapine induced developmental and cardiac toxicity on zebrafish embryos by elevating oxidative stress. *Cardiovasc. Toxicol.* **2021**, *21*, 399–409. [CrossRef] [PubMed]
12. Chen, D.; Kannan, K.; Tan, H.; Zheng, Z.G.; Widelka, M. Bisphenol analogues other than BPA: Environmental occurrence, human exposure, and toxicity—A review. *Environ. Sci. Technol.* **2016**, *50*, 5438–5453. [CrossRef] [PubMed]
13. Mcdonough, C.M.; Xu, H.S.; Guo, T.L. Toxicity of bisphenol analogues on the reproductive, nervous, and immune systems, and their relationships to gut microbiome and metabolism: Insights from a multi-species comparison. *Crit. Rev. Toxicol.* **2021**, *51*, 283–300. [CrossRef] [PubMed]

14. Ullah, A.; Pirzada, M.; Jahan, S.; Ullah, H.; Khan, M.J. Bisphenol A analogues bisphenol B, bisphenol F, and bisphenol S induce oxidative stress, disrupt daily sperm production, and damage DNA in rat spermatozoa: A comparative in vitro and in vivo study. *Toxicol. Ind. Health* **2019**, *35*, 294–303. [CrossRef] [PubMed]
15. Park, J.C.; Lee, M.C.; Yoon, D.S.; Han, J.; Kim, M.; Hwang, U.K.; Jung, J.H.; Lee, J.S. Effects of bisphenol A and its analogs bisphenol F and S on life parameters, antioxidant system, and response of defensome in the marine rotifer *Brachionus koreanus*. *Aquat. Toxicol.* **2018**, *199*, 21–29. [CrossRef] [PubMed]
16. Wu, M.; Xu, H.; Shen, Y.; Qiu, W.; Yang, M. Oxidative stress in zebrafish embryos induced by short-term exposure to bisphenol A, nonylphenol, and their mixture. *Environ. Toxicol. Chem.* **2011**, *30*, 2335–2341. [CrossRef] [PubMed]
17. Han, Y.; Fei, Y.; Wang, M.; Xue, Y.; Chen, H.; Liu, Y. Study on the joint toxicity of BPZ, BPS, BPC and BPF to zebrafish. *Molecules* **2021**, *26*, 4180. [CrossRef] [PubMed]
18. Schmittgen, T.D.; Livak, K.J. Analyzing real-time PCR data by the comparative CT method. *Nat. Protoc.* **2008**, *3*, 1101–1108. [CrossRef] [PubMed]

Article

Comparison of Dechlorane Plus Concentrations in Sequential Blood Samples of Pregnant Women in Taizhou, China

Ji-Fang-Tong Li [1,2], Xing-Hong Li [1,2,*], Yao-Yuan Wan [1,2], Yuan-Yuan Li [1,2] and Zhan-Fen Qin [1,2]

1 State Key Laboratory of Environmental Chemistry and Ecotoxicology, Research Center for Eco-Environmental Sciences, Chinese Academy of Sciences, 18 Shuangqing Road, Haidian District, Beijing 100085, China; m17853529621@163.com (J.-F.-T.L.); wanyaoyuan21@mails.ucas.ac.cn (Y.-Y.W.); yyli@rcees.ac.cn (Y.-Y.L.); qinzhanfen@rcees.ac.cn (Z.-F.Q.)
2 University of Chinese Academy of Sciences, 19A Yuquan Road, Shijingshan District, Beijing 100049, China
* Correspondence: lxhzpb@rcees.ac.cn; Tel.: +86-10-6291-9177; Fax: +86-10-6292-3563

Abstract: To develop an appropriate sampling strategy to assess the intrauterine exposure to dechlorane plus (DP), we investigated DP levels in sequential maternal blood samples collected in three trimesters of pregnancy, respectively, from women living in Taizhou. The median concentration of DPs (sum of *syn*-DP and *anti*-DP) in all samples was 30.5 pg g^{-1} wet-weight and 5.01 ng g^{-1} lipid-adjusted weight, respectively. The trimester-related DP concentrations were consistently strongly correlated ($p < 0.01$), indicating that a single measurement of DP levels could represent intrauterine exposure without sampling from the same female repeatedly; however, the wet-weight levels significantly increased across trimesters ($p < 0.05$), while the lipid-adjusted levels did not significantly vary. Notably, whether lipid-adjusted weight or wet-weight levels, the variation extent of DP across trimesters was found to be less than 41%, and those for other persistent organic pollutants (POPs) reported in the literature were also limited to 100%. The limitation in variation extents indicated that, regardless of the time of blood collection during pregnancy and how the levels were expressed, a single measurement could be extended to screen for exposure risk if necessary. Our study provides different strategies for sampling the maternal blood to serve the requirement for assessment of in utero exposure to DP.

Keywords: dechlorane plus; maternal blood; sequential samples; variation; correlation

1. Introduction

Dechlorane plus (DP) is a high-chlorinated organic flame retardant. For decades, various products containing DP have been widely used in the electric/electrical industry, such as commercial wire and cable, and connectors for computers and televisions. [1]. Since DP is considered an alternative to decabromodiphenyl ether (BDE-209), which has been phased out [2,3], its application may be expanded in the future. DP exhibits some POP-like properties, one of which is its ability to accumulate in animals and humans [2,4,5]. Studies on animals [6,7] (e.g., fish and poultry) have given evidence on DP transfer from parents to offspring, and Ben's study [8] further confirmed the transplacental transfer of DP in humans. Although its health risk to humans has not yet been better understood until now, animal experiments have shown that exposure to DP in early life could impact axonal growth, musculature, and motor behavior in embryo–larval zebrafish [9], and regulate mRNA expression in chicken embryos [10]. Due to the significant linkage between prenatal exposure and adverse health outcomes at birth and/or in later life, exposure of pregnant women to persistent organic contaminants is an ongoing public health concern [11]. The transplacental transfer and potential toxicity to animals [6,7], together with POP-like properties [2,4,5], implies that we ought to pay specific concern to in utero exposure to DP in humans.

Epidemiological studies generally used a single measurement of maternal blood samples to assess in utero exposure to persistent chemicals [12]. This strategy was cost-efficient, easy to collect and store samples, and also reduced the burden on participants in large epidemiological studies [12,13]; however, by measuring the levels of persistent pollutants in sequential maternal blood samples, some studies had shown that levels of certain POPs might change during pregnancy [14–18]. This was not surprising since important physiological changes would occur over the course of pregnancy [19,20]. Accompanied by these physiological changes, some internal POPs may be redistributed among human compartments, further leading to a change in maternal blood concentrations across gestation [13,18,21,22]. In addition, it was found that changing patterns of levels (based on wet-weight and/or lipid-adjusted weight) in sera of pregnant women for these POPs might differ from compound to compound. For example, the wet-weight levels of per- and polyfluorinated compounds (PFCs) [16–18] showed a general decrease over the course of pregnancy, while those of polychlorinated biphenyls (PCBs)/organic chlorinated pesticides (OCPs) showed a general increase [15,23,24]. In any case, doubts have been raised about the feasibility of using a single measurement to represent in utero exposure to POPs.

The maternal serum that could be a proxy of cord serum for assessing in utero exposure to DP has been documented [8]; however, DP showed different tissue distribution patterns in humans from lipid-related POPs (e.g., PCBs and PBDEs) [8,25–27] and protein-binding PFCs [28,29]. Pan et al. [26] believed that the specific distribution patterns of DP in humans might be regulated together by both lipids and non-lipid factors in the circulatory system. As a result, the prior experiences from POPs reported (e.g., PCBs and PFCs) on targeting one collection of maternal blood during pregnancy to assess in utero exposure [30,31] might not apply to DP. In order to develop an appropriate sampling strategy to obtain knowledge of in utero exposure to DP, we recognized the need to investigate DP levels in different time windows during gestation.

In this study, we determined DP levels in sequential maternal blood samples taken during three trimesters of pregnancy. The main purpose was to examine the inter-period correlations and to investigate the variation patterns/extent of DP concentrations during pregnancy. To the best of our knowledge, this is the first thorough examination of variation in maternal DP levels during different stages of pregnancy.

2. Materials and Methods

2.1. Sample Collection

Since the 1970s, many disposal e-waste products have been transported to Taizhou for recycling purposes. These old devices were generally recycled in informal family-run workshops with primitive methods, such as burning piles of wires and melting circuit boards over coal grills, which allowed the complex chemicals to be easily transferred to the surrounding environment. Consequently, the Taizhou region had serious environmental problems and local residents were at high health risk of exposure to various e-waste-related contaminants, including DP [8,32]. This study was undertaken from 2016 to 2017 in Taizhou, China. Pregnant women living in the Taizhou area with higher exposure levels of DP [25–27] were an ideal group to study fetal health risks associated with DP; therefore, we recruited forty pregnant women in this region as our study participants, who agreed to provide us with serum samples leftover from gestational routine blood monitoring. All pregnant women were residents living near e-waste recycling sites, but they were not engaged in e-waste recycling work. All the donors signed informed consent forms and provided basic personal information in the form of a questionnaire at the first pregnancy medical examination. Demographic characteristics of the participants included age (mean: 27 years; range 19–35 years), pre-pregnancy body mass index (BMI, mean: 21.2 kg m^{-2}; range: 17.6–28.2 kg m^{-2}), and parities (20 primiparas and 6 multiparas). The sampling time of maternal blood was arranged at three trimesters, which were classified as follows: the first trimester, <14 weeks; the second trimester, 14 to 28 weeks; the third trimester, >28 weeks. Blood samples from the 40 women who completed the study might not have all been obtained across three trimesters of pregnancy. As long as two trimesters samples could be taken from the same pregnant woman, the sample data of this woman was included in the study statistics. Finally, a total of 75 samples were obtained, including 26 in the first trimester, 25 in the second trimester, and 24 in the third trimester. A valid sample pair consisted of at least two sequential samples taken in different trimesters. Finally, we obtained 21 pairs between 1st and 2nd trimesters, 20 pairs between 1st and 3rd trimesters, and 19 pairs between 2nd and 3rd trimesters; the valid paired samples were from 30 participants. The average time interval (range) of samples from the same participant taken was 88 days (range: 48–126 days) between the 1st and 2nd trimester, 63 days (range: 35–118 days) between the 2nd and 3rd trimester, and 147 days (range: 91–196 days) between 1st and 3rd trimester, respectively (Table S1). We commissioned hospital staff responsible for collecting these trimester-dependent matched samples during pregnancy. The institutional review board of the Research Center for Eco-Environmental Sciences and the First People's Hospital of Wenling approved the study protocol prior to the collection of samples.

2.2. Sample Pretreatment and Chemical Analysis

The extraction and purification of samples were mainly carried out according to the procedures reported by Ben et al. [25]. In brief, the serum sample was spiked with $^{13}C_{10}$-labelled *syn*-DP and $^{13}C_{10}$-labelled *anti*-DP (4 ng) as surrogate standards, then denatured with hydrochloric acid and isopropanol, and ultrasonically extracted with a mixture of methyl tert-butyl ether and hexane with three times repeats. The extracts were purified by a multi-layer chromatography column, and were washed by a mixture of hexane and dichloromethane. The eluent was concentrated to 20 µL for further analysis, and $^{13}C_{12}$-labelled CB-208 as the injection standard was added before injection.

The target compounds were determined by Agilent 6890 gas chromatography coupled (Agilent Technologies Inc, Santa Clara, CA, USA) with 5973 low-resolution mass spectrometry (Agilent Technologies Inc, Santa Clara, CA, USA) in negative chemical ionization mode. The detailed information on chromatographic separation and the monitored ion fragments of target chemicals could be found in the reports of Ben et al. [25,28]. The method detection limits (MDLs) were defined as three times the standard deviation (SD) of the concentration (4 pg g^{-1}) of the target compounds spiked into matrix blank samples (bovine serum). MDLs were 0.8 pg g^{-1} ww and 0.14 ng g^{-1} lw for *syn*-DP, respectively. MDLs were 0.7 pg g^{-1} ww and 0.12 ng g^{-1} lw for *anti*-DP, respectively. The recoveries of surrogate standards were 73–111% for $^{13}C_{10}$-*syn*-DP and 65–108% for $^{13}C_{10}$-*anti*-DP. The solvent blank, matrix blank, and procedural blank were performed with each batch of samples, and no interferences were found in these quality control samples.

2.3. Serum Lipid Contents and Blood Biochemical Parameters

Serum lipid contents could be obtained generally by two assays. One was calculated according to blood lipid biochemical parameters (e.g., total cholesterol and total triglyceride) [21], and the other was directly determined by the gravimetric method [33]. In the present study, as TC and TG values in some blood samples were not available, a gravimetric method was used to obtain the information for all participants. Levels of blood biochemical parameters were obtained from the local hospital.

2.4. Data Analysis

All analyses were conducted using SPSS 23.0 (SPSS Inc., Chicago, IL, USA). The Kolmogorov–Smirnov test was used to test the normality of the distribution of continuous variables. The paired-sample *t*-test was used to compare the difference between log-transformed inter-trimester levels of target compounds, as well as inter-trimester levels of serum lipid contents and blood biochemical parameters. One-sample *t*-test was used to compare the difference in blood lipid contents between data from our study and Barr's study [12]. Pearson correlation analysis was employed to examine the relationships among log-transformed inter-period concentrations. Partial correlation analysis was further used to examine the influence of age and sampling day intervals on the correlation. p-values < 0.05 were considered statistically significant.

The change in serum concentrations and serum lipid contents were evaluated from paired observations by trimesters. The change (ΔC_i) were defined as the difference in concentration between the later sample and the matched earlier sample, according to Equation (1). In Equation (1), C_{it} represents value measured in a sample taken in trimester t (where $t = 2$ or 3) for the ith participant. $C_{it'}$ represents value measured in a sample taken in trimester t' (where when $t = 2$, $t' = 1$; and when $t = 3$, $t' = 2$ or 1) for the ith participant.

$$\Delta C_i = C_{it} - C_{it'} \qquad (1)$$

The variation extent (P, %) was defined as C_i divided by $C_{it'}$, then multiplied by 100, as expressed in Equation (2).

$$P = 100 \times C_i / C_{it'} \qquad (2)$$

3. Results

3.1. DP Concentrations in All Samples

Table 1 listed DP levels in all samples throughout pregnancy, based on wet-weight (ww) and lipid-adjusted weight (lw). It was found that DP could be detected in all samples, and the whole data set showed abnormal distribution. DPs in Table 1 referred to the sum of *syn*-DP and *anti*-DP, and the median level was 5.01 ng g^{-1} lw and 30.5 pg g^{-1} ww, respectively.

Table 1. Descriptive statistics for dechlorane plus (DP) and serum lipid contents measured in all serum samples of pregnant women.

Compound	All Serum Samples				
	N	Median	Mean	SD	Range
Wet-weight (pg g^{-1})					
syn-DP	75	7.56	17.7	33.1	1.26–199
anti-DP	75	21.4	54.6	109	6.76–602
∑DPs	75	30.5	72.3	142	8.12–801
Lipid-adjusted weight (ng g^{-1})					
syn-DP	75	1.10	2.70	5.20	0.190–33.9
anti-DP	75	3.45	8.15	16.5	0.740–102
∑DPs	75	5.01	10.9	21.6	1.04–136
f_{anti}	75	0.751	0.746	0.0653	0.440–0.845
Lipid contents(%)	75	0.670	0.691	0.192	0.330–1.17

N: The number of all serum samples in our study. SD: Standard deviation; Range: minimum to maximum.

Pan et al. [26] summairzed the DP body burden (median or mean value) in humans from different countries or regions. It was found that the DP levels (Table 1) in this study were within the range of those of the general population living near the contamination sources, such as e-waste recycling areas [25,27,34], municipal solid waste incinerators [35], and previous DP production areas [36]. The levels were higher than those of the general population [35,37–41]. This was consistent with the fact that e-waste recycling activities were an important source of DP for human exposure [35,37–41]. Moreover, the comparison also suggested that pregnant Chinese women living in this area had a relatively high body burden of DP. In addition, the highest levels of DP were found to be 136 ng g^{-1} lw in this study, but Ben et al. reported that DP levels could be up to 900 ng g^{-1} lw in sera of pregnant women and 89.7 ng g^{-1} lw in cord serum [25], and 590 ng g^{-1} lw in breast milk [25]. In addition, placental transfer of DP in humans had been confirmed and DP concentration in the cord serum was estimated to be 38% of that in the maternal serum [8]. These results demonstrated that the bioaccumulation of DP could reach very high levels in pregnant women and their fetuses, urging the need to assess the health risk of pregnant women and their fetuses/infants exposed to DP. Pregnant women and fetuses are sensitive to chemical exposure, so these elevated levels indicated the need to assess the health risk of pregnant women and their fetuses/infants exposed to DP.

3.2. Trimester-Related Characteristics of DP Concentrations

3.2.1. Trimester-Related Concentrations

The Kolmogorov–Smirnov test showed that the data set for trimester-related DP levels (both ww and lw) did not conform to a normal distribution ($p < 0.000$). The median level of DP in serum was 25.9 pg·g^{-1} ww in the 1st trimester, 31.1 pg·g^{-1} ww in the 2nd trimester, and 33.0 pg·g^{-1} ww in the 3rd trimester of pregnancy, respectively (Figure 1B). After lipid adjustment, the median level was 5.59 ng·g^{-1} in the 1st trimester, 5.01 ng·g^{-1} in the 2nd trimester, and 4.30 ng·g^{-1} in the 3rd trimester, respectively (Figure 1A).

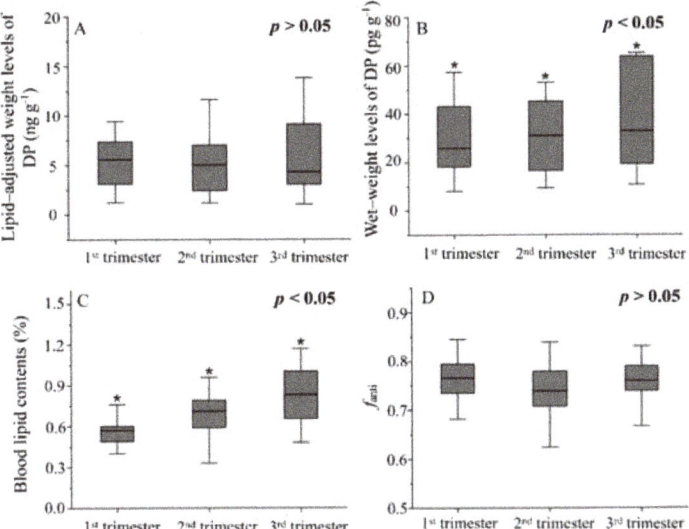

Figure 1. Box plot (minimum, 25% quartile, median, 75% quartile, and maximum; outliers were not shown) of lipid-adjusted levels of DP (**A**), wet-weight levels of DP (**B**), blood lipid contents (**C**), and f_{anti} (**D**) of samples during three trimesters. (* trimester-related levels showed significant differences at $p < 0.05$. The sample sizes for the first, second, and third trimesters were 26, 25, and 24, respectively).

3.2.2. Inter-Trimester Associations

A strong association was found between inter-trimester DP levels, whether wet-weight basis or lipid-adjusted weight basis ($p < 0.01$, $r > 0.675$), as shown in Table 2. For the two isomers, similar results could be found ($p < 0.01$). Age was believed to be an important determinant of the body burden of persistent organic pollutants [22,42,43]. Moreover, day intervals between two sequential samples were also perceived as an important factor affecting the inter-period correlation coefficient [13,16]; however, our results of Partial correlation analysis indicated that the correlations among trimesters still remained significant ($p < 0.01$) (Table 2). Similar, significant inter-trimester correlations were also reported for other POPs in most of the published literature (e.g., PCBs, OCPs, and PFCs), as summarized in Table S3. It seemed to be a general rule that POP levels in sequential maternal blood samples had a significant inter-trimester correlation.

Table 2. Unadjusted/adjusted inter-period correlation of log-transformed trimester-related DP levels.

	Unadjusted			Adjusted Covariate [a]		
	1st–2nd	1st–3rd	2nd–3rd	1st–2nd	1st–3rd	2nd–3rd
Wet-weight (pg g^{-1}) [b]						
syn-DP	0.892	0.785	0.935	0.786	0.675	0.762
anti-DP	0.928	0.862	0.902	0.861	0.791	0.763
∑DPs	0.930	0.859	0.908	0.864	0.831	0.774
Lipid-adjusted weight (ng g^{-1}) [b]						
syn-DP	0.900	0.791	0.907	0.821	0.736	0.772
anti-DP	0.943	0.889	0.889	0.896	0.809	0.698
∑DPs	0.940	0.874	0.895	0.897	0.842	0.723

[a] Correlation coefficients after controlling age and sampling day intervals among samples. [b] All correlations are significant at the 0.01 level (2-tailed).

3.2.3. Variation Patterns

Box plots in Figure 1A,B displayed the variation patterns of DP levels (logarithm transformation) based on lipid-adjusted weight and wet weight, respectively. Based on wet-weight form (Figure 1B), levels of DPs showed a small but significant increase as the pregnancy continued ($p < 0.05$). Specifically, the levels in the 2nd and 3rd trimesters were significantly higher than those in the 1st trimester ($p = 0.009$ and 0.002, respectively, paired-samples t-test), and the levels in the 3rd trimester were also significantly higher than those in the 2nd trimester ($p = 0.040$, paired-samples t-test); however, for lipid-adjusted levels (Figure 1A), the median value was on a downward trend from the 1st to 3rd trimester, but significant differences were not found for log-transformed levels ($p > 0.05$, paired-samples t-test). The results indicated that lipid-adjusted levels of DPs showed a different variation pattern from that of the wet-weight levels in sequential blood samples from pregnant women.

3.2.4. Implications Based on Associations and Variation Patterns

Both inter-trimester associations and variation patterns of DP levels gave two implications about sampling strategies for in utero exposure to DP. The first one was related to the frequency of blood samples taken from the same individual participant, and the other was related to the time window for collecting a sample for different participants.

First, the strong inter-trimester associations of DP levels (both wet weight and lipid-adjusted weight) observed in our study suggested that blood samples taken at one trimester of pregnancy could act as a proxy for the other two trimesters when assessing in utero exposure to DP. It was sufficient to make one single measurement of levels in sera taken within a given time window. Then, the insignificant variation of lipid-adjusted DP levels across trimesters suggested that regardless of the sampling time windows, a single sample measurement based on lipid-adjusted levels could reflect DP exposure throughout gestation; however, for assessment based on wet-weight levels, the sampling time window was required to be within a narrow range because of the significant difference of wet-weight levels by trimesters.

In any case, it should be kept in mind that when a single measurement was expected to reflect DP exposure throughout gestation, the sampling time window based on wet-weight levels should be different from that based on lipid-adjusted weight levels. It was more flexible and practical to use lipid-adjusted DP levels to assess in utero exposure to DP due to the unlimited collection time of blood samples.

3.3. Blood Biochemical Parameters and Effect on Variation of DP Levels

3.3.1. Blood Parameters and Trimester-Related Characteristics

Biochemical parameters of maternal blood provided by the local hospital, including total cholesterol (TC), total triglyceride (TG), high-density lipoprotein (HDL), low-density lipoprotein (LDL), apolipoprotein A (Apo-A), apolipoprotein B (Apo-B), lipoprotein a

[Lp(a)], hemoglobin (Hb), albumin (ALB), and creatinine (Cre) are listed in Table S2. These listed parameters reflected the lipid/protein profiles and blood volume of the pregnant women. The average values of the lipid-related parameters, including TC, TG, Apo-A, Apo-B, and Lp(a), were above or at the top of the reference intervals, while those for Hb, ALB, and Cre, were on the bottom of the reference intervals. The relatively higher lipid-related parameters in pregnant women are believed to be related to the need for more energy supply for mothers and fetuses during pregnancy [14,19,21]. The observed decreases in maternal serum Hb, ALB, and Cre might be associated with pregnancy-related increases in body and blood volume [18,44,45]. Table S2 shows the trimester-related data of these biochemical parameters. TC, TG, and Apo-B significantly increased (paired-sample t-test, $p < 0.05$). In contrast, the non-lipid parameters (e.g., Hb and ALB) showed a significantly decreasing trend (paired-sample t-test, $p < 0.05$). The results supported the occurrence of maternal physiological changes by trimesters, and distinct variation patterns for lipid-related and non-lipid biochemical parameters. Because of the small paired-sample size, data from the 2nd trimester were not included for comparison.

The data of lipid contents conformed to normal distribution. As a whole, the value varied from 0.330% to 1.17%, with 0.691 as the mean value across trimesters (Table 1). The average value (0.691%) was significantly higher ($p < 0.000$, one-sample t-test) than that in the general population (0.5–0.6%) [12]. Then, the average value in the 1st, 2nd, and 3rd trimesters was 0.567%, 0.696%, and 0.820%, respectively. As shown in Figure 1C, trimester-related lipid contents increased gradually and showed significant differences between each other by paired-sample t-test ($p = 0.001$ between the 1st and 2nd trimester, $p < 0.001$ between the 1st and 3rd trimester, and $p = 0.018$ between the 2nd and 3rd trimester, respectively). The increasing tendency of lipid contents during pregnancy was consistent with practical experiences and the reports from other published literature [12].

In short, since the distribution and redistribution of DP in humans might be driven by both blood lipids and proteins [26,27], the trimester-related variation of blood parameters in the circulatory system implied the possible change of DP levels as pregnancy proceeded.

3.3.2. Effect of Blood Biochemistry Parameters on Variation of DP Levels

DP showed certain specific distribution behaviors among human tissues with the preferential accumulation in the blood compared to adipose tissue, placental, and breast milk [8,27], implying the potential strong interaction with blood lipids and non-lipids compositions. Importantly, it had been suspected that both lipid and non-lipid factors in the bloodstream might exert an important impact on the variation of POP levels during pregnancy [18,24]. Our study found that lipid-adjusted DP levels by trimesters were very close to each other (Figure 1A) ($p > 0.05$), while the wet-weight levels increased in parallel with the increase in lipid contents during pregnancy (Figure 1B,C). The stable lipid-adjusted DP levels across pregnancy indicated that lipid content could almost completely correct the differences in inter-trimester DP wet-weight levels; therefore, change in blood lipid contents during pregnancy was the major factor resulting in a change in wet-weight DP levels. Adipose tissue was considered the main sink of DP in humans [26]. The mobilization of store lipids during pregnancy with concomitant redistribution of DP from adipose tissue to the circulatory system might increase serum wet-weight DP concentrations as pregnancy gestation. Hansen et al. [14] found a similar phenomenon in organochlorines (OCs), with wet-weight levels and lipid contents peaking at birth, then this peak disappeared when OC concentrations were adjusted by lipid contents. Hansen et al. [14] believed that the wet-weight OC levels during gestation might be driven by physiological lipids.

In addition, we found a distinct difference in the various patterns of the wet-weight levels between PFCs and PCBs/OCPs, as summarized in Figure 2. The wet-weight serum levels of PFCs usually showed a trimester-related decrease [16–18], while the wet-weight levels of PCBs/OCPs usually showed a trimester-related increase [15,23,24]. It was well known that PFCs were apt to bind to serum albumin [28,29], whereas PCBs/OCPs showed a strong dependence on blood lipids [14]. With the progress of pregnancy, the decreased albumin levels and increase in blood volume might reduce the redistribution of PFCs in the bloodstream, resulting in decreasing wet-weight levels [16]. In contrast, the increasing lipid contents during pregnancy might cause the elevated redistribution of PCBs/OCPs in maternal blood, leading to an increase in serum wet-weight levels [14]. DP showed a similar variation pattern with PCBs/OCPs rather than PFCs, likely ascribing to its more similarity in lipophilicity with PCBs/OCPs. Although the interaction of DP with certain proteins in the bloodstream (e.g., serum lipoproteins and albumin) might play an important role in its distribution in humans [26,27], the lipid-adjusted DP levels by trimesters did not significantly differ from each other, implying that non-lipid factors in blood circulatory system might not dominantly affect the variation patterns of DP levels.

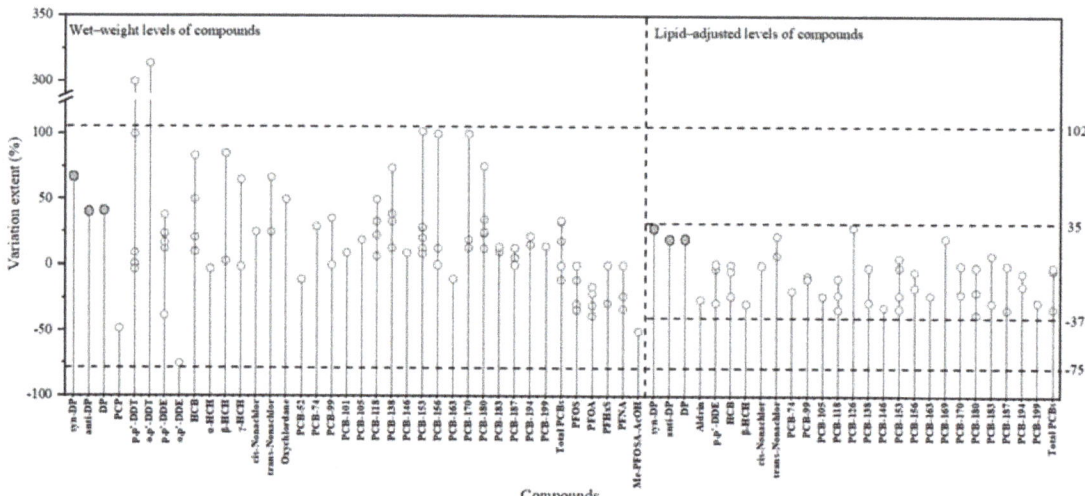

Figure 2. Variation extent (%) of DP in this study and other POPs in published literature. (The grey "○" represents the variation extent of DP in this study, and the blank "○" represents the variation extent of POPs in published literature, detailed information is available in Table S3. The left part of the dotted line represents the levels of DP and POPs based on wet weight, and the right part of the dotted line represents the levels of DP and POPs based on lipid-adjusted weight. The horizontal dotted lines represent the range of POP variation extents based on wet-weight and lipid-adjusted levels.).

In short, the results demonstrated that changes in DP levels observed during pregnancy were associated with variation in blood lipids to a great extent.

3.4. Stereo-Selective Profile of DP during Pregnancy

The stereo-selective profile of two DP isomers was examined by the fraction of anti-DP (anti-DP/$\sum$DP, f_{anti}). In the present study, f_{anti} values were not normally distributed, so that paired samples t-test was used to examine the difference in f_{anti} values between trimesters. As shown in Table 2, the median value of f_{anti} in all sera was 0.751, which was within the data range for the general population [25]. Furthermore, the median value in the 1st, 2nd, and 3rd trimester was 0.758, 0.744, and 0.760, respectively. No significant difference among the three gestational stages was found ($p > 0.05$) (Figure 1D), suggesting that stereo-selective DP bioaccumulation in maternal blood did not occur as a result of pregnancy. Note that the stereoselective bioaccumulation of DP isomers during transport from mother to fetus has been documented [25]. These suggested that both maternal physiological changes during pregnancy and the occurrence of transplacental transport of DP were not enough to influence the status of its stereoselective bioaccumulation in maternal blood.

3.5. Variation Extents in Levels of DP and Other POPs across Gestation

3.5.1. Variation Extent of DP

The variation extent of levels in sequential blood specimens for specific POPs was helpful to outline the fluctuation power of levels across pregnancy. The variation extent of trimester-related DP levels was calculated based on Equations (1) and (2) and the results are displayed in Figure 2. As a result, the median variation extent of wet-weight levels was 30% between the 1st and 2nd trimester, 41% between the 1st and 3rd trimester, and 15% between the 2nd and 3rd trimester, respectively. Based on lipid-adjusted levels, the median value of variation extent was −13%, 20%, and 6%, respectively. Obviously, variation extents based on lipid-adjusted DP levels were less than those based on wet-weight levels, further confirming the effect of blood lipids on DP levels across gestation. In short, it was found that DP levels across trimesters of pregnancy varied by no more than 41% (Figure 2), whether based on wet-weight or lipid-adjusted weight.

3.5.2. Variation Extent of Other POPs

There have been a dozen studies to investigate the variation of POP levels in sequential maternal blood samples across pregnancy [13–18,21–24,37,46–49]. Unfortunately, most published literature focused on the level differences (significant or insignificant) and/or variation patterns (increasing, decreasing, or stable), discussion on the variation extent in POP levels throughout pregnancy was restricted [16,21,23]. To better understand the variation details across pregnancy, we calculated the variation extent of POP levels reported in the published literature according to Equations (1) and (2); however, since only statistical descriptions of POP levels (e.g., mean value, median value, or geometric mean value) were available in the published literature, the statistical values (detailed information available in Table S3), were used to estimate the variation extents. In order to reflect the maximum variation extents of POP levels across pregnancy, the variation extents listed in Figure 2 were calculated by levels from the sequential serum samples with the longest sampling time intervals in the literature. For example, Adetona et al. [23] collected three batches of matched blood samples in the first, second, and third trimesters of pregnancy. The values presented in Figure 2 were data only related to the 1st and 3rd trimesters.

Figure 2 shows that the variation extents based on the lipid-adjusted levels of POPs were negatively limited to 37% and positively limited to 35%, respectively. Variation extents based on wet-weight levels were limited to 75% negatively and 102% positively in all but two studies [48,50]. The largest variation extent of the DDT levels, about 300% as reported by Tasker et al. [48], might be due to the concentrations close to its limitation of detection. In the report of Curley et al. [50], sequential maternal blood samples were taken from only five females, and the levels of both o,p'-DDT and o,p'-DDE, varied widely among individual donors. Nevertheless, Curley et al. only reported the mean value of levels in the same period; the data listed in Figure 2 were calculated by the mean value and possibly contributed to the large variation extents.

Clearly, regardless of wet-weight basis or lipid-weight basis, the variation extent of the POP levels did not exceed 100% for almost all individual compounds reported in the literature.

3.5.3. Implication from Variation Extents

The limited variation extent found in Figure 2 (Table S3) indicated a limited misclassification risk for POP exposure when a single measurement without information on the sampling time window of samples was used. As a result, this observation could inform the appropriate application of sampling strategy to assess in utero exposure to POPs in certain cases. For example, the sampling strategy might be a single measurement without specific sampling time windows for screening purposes. Notably, the reported compounds were limited to PCBs, OCPs, and PFCs in the literature and DP in this study. In addition, the variation extent from the published literature was calculated by statistical values, not by the inter-trimester levels of each participant. Additional studies are needed to insightfully investigate the variation patterns and variation extent of more POPs to generalize this principle.

3.6. Limitations

Although the sampling period covered three trimesters of pregnancy in our study, the days of sampling time intervals between the 1st and 3rd trimester was only 147 (35–196) days (about half of the gestational period). As a result, the relatively short time intervals in this study might restrict our understanding of the full information on the variation in DP levels across the whole pregnancy period (10-month intervals or more). In addition, in terms of the results in Figure 2, it was found that the homologous group of compounds or even the same individual compounds did not subject to uniform variation patterns of levels in different literature [14–18,22–24,37]. These differences might be ascribed to maternal weight gain [20,21], the time since peak exposure (past, recent, and current) [14], as well as baseline concentrations [13,22]. Unfortunately, no information on these factors was available in this study, limiting further discussion of their effect on levels during pregnancy.

4. Conclusions

In summary, a single measurement could be used to assess in utero exposure to DP, whether based on wet-weight levels or lipid-adjusted levels. The accurate evaluation results based on the lipid-adjusted levels could be obtained regardless of the sampling time window of serum samples; however, the assessment results based on wet-weight levels were reliable only when the serum sampling time was limited to a narrow window. Furthermore, the limited variation of period-related levels was found for DP and other POPs, suggesting that it is feasible to use a sample of pregnant women to roughly evaluate in utero exposure risk, regardless of the sampling time window and how the levels were expressed (lw or ww). Stereoselective behaviors of DP isomers in maternal sera did not occur during pregnancy. In conclusion, these results will serve to design an appropriate sampling strategy for assessing in utero exposure to DP. Additional factors need to be considered in future studies, including the time span throughout gestation for sequential blood samples and full information on physiological changes during pregnancy.

Supplementary Materials: The following supporting information can be downloaded at https://www.mdpi.com/article/10.3390/molecules27072242/s1, Table S1: Sampling time intervals between paired samples in different trimesters from the same participant. Table S2: Descriptive statistics for blood biochemical parameters in sequential blood samples. Table S3: Inter-period correlations, variation pattern, and variation extent (%) of POP levels across pregnancy in the present study.

Author Contributions: J.-F.-T.L.: Investigation, Writing—Original draft preparation; X.-H.L.: Writing—Review and editing, Funding acquisition, and Resources; Y.-Y.W.: Formal analysis, Writing—Review and editing; Y.-Y.L.: Writing—Review and editing; Z.-F.Q.: Writing—Review and editing. All authors have read and agreed to the published version of the manuscript.

Funding: This work was funded by the National Natural Science Foundation of China (21976204 and 21777185).

Institutional Review Board Statement: The study was conducted according to the guidelines of the Declaration of Helsinki. The institutional review board of the Research Center for Eco-Environmental Sciences and the First People's Hospital of Wenling approved the study protocol prior to the collection of samples.

Informed Consent Statement: Informed consent was obtained from all subjects involved in the study. Written informed consent has been obtained from the patients to publish this paper.

Data Availability Statement: Not applicable.

Acknowledgments: We gratefully acknowledge the contribution from the volunteer donors and medical staff in this study. This work was funded by the National Natural Science Foundation of China (21976204 and 21777185).

Conflicts of Interest: The authors declare no conflict of interest.

Sample Availability: Not applicable.

References

1. Schuster, J.K.; Harner, T.; Sverko, E. Dechlorane plus in the global atmosphere. *Environ. Sci. Technol. Lett.* **2021**, *8*, 39–45. [CrossRef]
2. Sverko, E.; Tomy, G.T.; Reiner, E.J.; Li, Y.-F.; McCarry, B.E.; Arnot, J.A.; Law, R.J.; Hites, R.A. Dechlorane plus and related compounds in the environment: A review. *Environ. Sci. Technol.* **2011**, *45*, 5088–5098. [CrossRef]
3. Stuer-Lauridsen, F.; Cohr, K.; Andersen, T.T. Health and Environmental Assessment of Alternatives to Deca-BDE in Electrical and Electronic Equipment. Available online: https://www2.mst.dk/udgiv/publications/2007/978-87-7052-351-6/pdf/978-87-7052-352-3.pdf (accessed on 2 January 2022).
4. Zafar, M.I.; Kali, S.; Ali, M.; Riaz, M.A.; Naz, T.; Iqbal, M.M.; Masood, N.; Munawar, K.; Jan, B.; Ahmed, S.; et al. Dechlorane plus as an emerging environmental pollutant in Asia: A review. *Environ. Sci. Pollut. Res.* **2020**, *27*, 42369–42389. [CrossRef] [PubMed]
5. Ren, G.; Yu, Z.; Ma, S.; Li, H.; Peng, P.; Sheng, G.; Fu, J. Determination of dechlorane plus in serum from electronics dismantling workers in South China. *Environ. Sci. Technol.* **2009**, *43*, 9453–9457. [CrossRef]
6. Sühring, R.; Freese, M.; Schneider, M.; Schubert, S.; Pohlmann, J.-D.; Alaee, M.; Wolschke, H.; Hanel, R.; Ebinghaus, R.; Marohn, L. Maternal transfer of emerging brominated and chlorinated flame retardants in European eels. *Sci. Total Environ.* **2015**, *530*, 209–218. [CrossRef] [PubMed]
7. Wu, P.-F.; Yu, L.-L.; Li, L.; Zhang, Y.; Li, X.-H. Maternal transfer of dechloranes and their distribution among tissues in contaminated ducks. *Chemosphere* **2016**, *150*, 514–519. [CrossRef] [PubMed]
8. Ben, Y.-J.; Li, X.-H.; Yang, Y.-L.; Li, L.; Zheng, M.-Y.; Wang, W.-Y.; Xu, X.-B. Placental transfer of dechlorane plus in mother-infant pairs in an e-waste recycling area (Wenling, China). *Environ. Sci. Technol.* **2014**, *48*, 5187–5193. [CrossRef]
9. Chen, X.; Dong, Q.; Chen, Y.; Zhang, Z.; Huang, C.; Zhu, Y.; Zhang, Y. Effects of dechlorane plus exposure on axonal growth, musculature and motor behavior in embryo-larval zebrafish. *Environ. Pollut.* **2017**, *224*, 7–15. [CrossRef] [PubMed]
10. Wu, B.; Liu, S.; Guo, X.; Zhang, Y.; Zhang, X.; Li, M.; Cheng, S. Responses of mouse liver to dechlorane plus exposure by integrative transcriptomic and metabonomic studies. *Environ. Sci. Technol.* **2012**, *46*, 10758–10764. [CrossRef] [PubMed]
11. Fernández-Cruz, T.; Martínez-Carballo, E.; Simal-Gándara, J. Perspective on pre- and post-natal agro-food exposure to persistent organic pollutants and their effects on quality of life. *Environ. Int.* **2017**, *100*, 79–101. [CrossRef]
12. Barr, D.B.; Wang, R.Y.; Needham, L.L. Biologic monitoring of exposure to environmental chemicals throughout the life stages: Requirements and issues for consideration for the National Children's Study. *Environ. Health Perspect.* **2005**, *113*, 1083–1091. [CrossRef] [PubMed]
13. Bloom, M.S.; Louis, G.M.B.; Schisterman, E.F.; Liu, A.; Kostyniak, P.J. Maternal serum polychlorinated biphenyl concentrations across critical windows of human development. *Environ. Health Perspect.* **2007**, *115*, 1320–1324. [CrossRef]
14. Hansen, S.; Nieboer, E.; Odland, J.Ø.; Wilsgaard, T.; Veyhe, A.S.; Sandanger, T.M. Levels of organochlorines and lipids across pregnancy, delivery and postpartum periods in women from Northern Norway. *J. Environ. Monit.* **2010**, *12*, 2128–2137. [CrossRef] [PubMed]
15. Junqué, E.; Garcia, S.; Martínez, M.Á.; Rovira, J.; Schuhmacher, M.; Grimalt, J.O. Changes of organochlorine compound concentrations in maternal serum during pregnancy and comparison to serum cord blood composition. *Environ. Res.* **2020**, *182*, 108994. [CrossRef] [PubMed]
16. Kato, K.; Wong, L.-Y.; Chen, A.; Dunbar, C.; Webster, G.M.; Lanphear, B.P.; Calafat, A.M. Changes in serum concentrations of maternal poly- and perfluoroalkyl substances over the course of pregnancy and predictors of exposure in a multiethnic cohort of Cincinnati, Ohio pregnant women during 2003–2006. *Environ. Sci. Technol.* **2014**, *48*, 9600–9608. [CrossRef] [PubMed]
17. Fisher, M.; Arbuckle, T.E.; Liang, C.L.; LeBlanc, A.; Gaudreau, E.; Foster, W.G.; Haines, D.; Davis, K.; Fraser, W.D. Concentrations of persistent organic pollutants in maternal and cord blood from the maternal-infant research on environmental chemicals (MIREC) cohort study. *Environ. Health* **2016**, *15*, 59. [CrossRef] [PubMed]
18. Glynn, A.; Berger, U.; Bignert, A.; Ullah, S.; Aune, M.; Lignell, S.; Darnerud, P.O. Perfluorinated alkyl acids in blood serum from primiparous women in Sweden: Serial sampling during pregnancy and nursing, and temporal trends 1996–2010. *Environ. Sci. Technol.* **2012**, *46*, 9071–9079. [CrossRef] [PubMed]
19. Chiang, A.N.; Yang, M.L.; Hung, J.H.; Chou, P.; Shyn, S.K.; Ng, H.T. Alterations of serum lipid levels and their biological relevances during and after pregnancy. *Life Sci.* **1995**, *56*, 2367–2375. [CrossRef] [PubMed]
20. Vizcaino, E.; Grimalt, J.O.; Glomstad, B.; Fernández-Somoano, A.; Tardón, A. Gestational weight gain and exposure of newborns to persistent organic pollutants. *Environ. Health Perspect.* **2014**, *122*, 873–879. [CrossRef] [PubMed]
21. Wang, R.Y.; Jain, R.B.; Wolkin, A.F.; Rubin, C.H.; Needham, L.L. Serum concentrations of selected persistent organic pollutants in a sample of pregnant females and changes in their concentrations during gestation. *Environ. Health Perspect.* **2009**, *117*, 1244–1249. [CrossRef]
22. Bloom, M.S.; Buck-Louis, G.M.; Schisterman, E.F.; Kostyniak, P.J.; Vena, J.E. Changes in maternal serum chlorinated pesticide concentrations across critical windows of human reproduction and development. *Environ. Res.* **2009**, *109*, 93–100. [CrossRef]

23. Adetona, O.; Horton, K.; Sjodin, A.; Jones, R.; Hall, D.B.; Aguillar-Villalobos, M.; Cassidy, B.E.; Vena, J.E.; Needham, L.L.; Naeher, L.P. Concentrations of select persistent organic pollutants across pregnancy trimesters in maternal and in cord serum in Trujillo, Peru. *Chemosphere* **2013**, *91*, 1426–1433. [CrossRef]
24. Glynn, A.; Larsdotter, M.; Aune, M.; Darnerud, P.O.; Bjerselius, R.; Bergman, A. Changes in serum concentrations of polychlorinated biphenyls (PCBs), hydroxylated PCB metabolites and pentachlorophenol during pregnancy. *Chemosphere* **2011**, *83*, 144–151. [CrossRef]
25. Ben, Y.-J.; Li, X.-H.; Yang, Y.-L.; Li, L.; Di, J.-P.; Wang, W.-Y.; Zhou, R.-F.; Xiao, K.; Zheng, M.-Y.; Tian, Y.; et al. Dechlorane plus and its dechlorinated analogs from an e-waste recycling center in maternal serum and breast milk of women in Wenling, China. *Environ. Pollut.* **2013**, *173*, 176–181. [CrossRef] [PubMed]
26. Pan, H.-Y.; Li, J.-F.-T.; Li, X.-H.; Yang, Y.-L.; Qin, Z.-F.; Li, J.-B.; Li, Y.-Y. Transfer of dechlorane plus between human breast milk and adipose tissue and comparison with legacy lipophilic compounds. *Environ. Pollut.* **2020**, *265*, 115096. [CrossRef] [PubMed]
27. Yin, J.-F.; Li, J.-F.-T.; Li, X.-H.; Yang, Y.-L.; Qin, Z.-F. Bioaccumulation and transfer characteristics of dechlorane plus in human adipose tissue and blood stream and the underlying mechanisms. *Sci. Total Environ.* **2020**, *700*, 134391. [CrossRef]
28. Bischel, H.N.; MacManus-Spencer, L.A.; Zhang, C.; Luthy, R.G. Strong associations of short-chain perfluoroalkyl acids with serum albumin and investigation of binding mechanisms. *Environ. Toxicol. Chem.* **2011**, *30*, 2423–2430. [CrossRef]
29. Allendorf, F.; Berger, U.; Goss, K.-U.; Ulrich, N. Partition coefficients of four perfluoroalkyl acid alternatives between bovine serum albumin (BSA) and water in comparison to ten classical perfluoroalkyl acids. *Environ. Sci. Process. Impacts* **2019**, *21*, 1852–1863. [CrossRef]
30. Kishi, R.; Nakajima, T.; Goudarzi, H.; Kobayashi, S.; Sasaki, S.; Okada, E.; Miyashita, C.; Itoh, S.; Araki, A.; Ikeno, T.; et al. The association of prenatal exposure to perfluorinated chemicals with maternal essential and long-chain polyunsaturated fatty acids during pregnancy and the birth weight of their offspring: The Hokkaido Study. *Environ. Health Perspect.* **2015**, *123*, 1038–1045. [CrossRef] [PubMed]
31. Soechitram, S.D.; Athanasiadou, M.; Hovander, L.; Bergman, Å.; Sauer, P.J.J. Fetal exposure to PCBs and their hydroxylated metabolites in a Dutch cohort. *Environ. Health Perspect.* **2004**, *112*, 1208–1212. [CrossRef]
32. Zhang, H.; Wang, P.; Li, Y.; Shang, H.; Wang, Y.; Wang, T.; Zhang, Q.; Jiang, G. Assessment on the occupational exposure of manufacturing workers to dechlorane plus through blood and hair analysis. *Environ. Sci. Technol.* **2013**, *47*, 10567–10573. [CrossRef]
33. Izard, J.; Limberger, R.J. Rapid screening method for quantitation of bacterial cell lipids from whole cells. *J. Microbiol. Methods* **2003**, *55*, 411–418. [CrossRef]
34. Yang, Q.; Qiu, X.; Li, R.; Liu, S.; Li, K.; Wang, F.; Zhu, P.; Li, G.; Zhu, T. Exposure to typical persistent organic pollutants from an electronic waste recycling site in Northern China. *Chemosphere* **2013**, *91*, 205–211. [CrossRef] [PubMed]
35. Brasseur, C.; Pirard, C.; Scholl, G.; De Pauw, E.; Viel, J.-F.; Shen, L.; Reiner, E.; Focant, J.-F. Levels of dechloranes and polybrominated diphenyl ethers (PBDEs) in human serum from France. *Environ. Int.* **2014**, *65*, 33–40. [CrossRef]
36. He, S.; Li, M.; Jin, J.; Wang, Y.; Bu, Y.; Xu, M.; Yang, X.; Liu, A. Concentrations and trends of halogenated flame retardants in the pooled serum of residents of Laizhou Bay, China. *Environ. Toxicol. Chem.* **2013**, *32*, 1242–1247. [CrossRef] [PubMed]
37. Fromme, H.; Mosch, C.; Morovitz, M.; Alba-Alejandre, I.; Boehmer, S.; Kiranoglu, M.; Faber, F.; Hannibal, I.; Genzel-Boroviczény, O.; Koletzko, B.; et al. Pre- and postnatal exposure to perfluorinated compounds (PFCs). *Environ. Sci. Technol.* **2010**, *44*, 7123–7129. [CrossRef] [PubMed]
38. Kim, J.-T.; Oh, D.; Choi, S.-D.; Chang, Y.-S. Factors associated with partitioning behavior of persistent organic pollutants in a feto-maternal system: A multiple linear regression approach. *Chemosphere* **2021**, *263*, 128247. [CrossRef] [PubMed]
39. Zhou, S.N.; Siddique, S.; Lavoie, L.; Takser, L.; Abdelouahab, N.; Zhu, J. Hexachloronorbornene-based flame retardants in humans: Levels in maternal serum and milk. *Environ. Int.* **2014**, *66*, 11–17. [CrossRef] [PubMed]
40. Cequier, E.; Marcé-Recasens, R.M.; Becher, G.; Thomsen, C. Comparing human exposure to emerging and legacy flame retardants from the indoor environment and diet with concentrations measured in serum. *Environ. Int.* **2015**, *74*, 54–59. [CrossRef] [PubMed]
41. Cequier, E.; Marcé, R.M.; Becher, G.; Thomsen, C. Determination of emerging halogenated flame retardants and polybrominated diphenyl ethers in serum by gas chromatography mass spectrometry. *J. Chromatogr. A* **2013**, *1310*, 126–132. [CrossRef] [PubMed]
42. Li, W.-L.; Qi, H.; Ma, W.-L.; Liu, L.-Y.; Zhang, Z.; Zhu, N.-Z.; Mohammed, M.O.; Li, Y.-F. Occurrence, behavior and human health risk assessment of dechlorane plus and related compounds in indoor dust of China. *Chemosphere* **2015**, *134*, 166–171. [CrossRef]
43. Liu, X.; Yu, G.; Cao, Z.; Wang, B.; Huang, J.; Deng, S.; Wang, Y.; Shen, H.; Peng, X. Estimation of human exposure to halogenated flame retardants through dermal adsorption by skin wipe. *Chemosphere* **2017**, *168*, 272–278. [CrossRef]
44. Longo, L.D. Maternal blood volume and cardiac output during pregnancy: A hypothesis of endocrinologic control. *Am. J. Physiol.* **1983**, *245*, R720–R729. [CrossRef] [PubMed]
45. Bernstein, I.M.; Ziegler, W.; Badger, G.J. Plasma volume expansion in early pregnancy. *Obstet. Gynecol.* **2001**, *97*, 669–672. [CrossRef] [PubMed]
46. Jarrell, J.; Chan, S.; Hauser, R.; Hu, H. Longitudinal assessment of PCBs and chlorinated pesticides in pregnant women from Western Canada. *Environ. Health* **2005**, *4*, 10. [CrossRef] [PubMed]
47. Longnecker, M.P.; Klebanoff, M.A.; Gladen, B.C.; Berendes, H.W. Serial levels of serum organochlorines during pregnancy and postpartum. *Arch. Environ. Health* **1999**, *54*, 110–114. [CrossRef] [PubMed]

48. Takser, L.; Mergler, D.; Baldwin, M.; Grosbois, S.D.; Smargiassi, A.; Lafond, J. Thyroid hormones in pregnancy in relation to environmental exposure to organochlorine compounds and mercury. *Environ. Health Perspect.* **2005**, *113*, 1039–1045. [CrossRef]
49. Rončević, N.; Pavkov, S.; Galetin-Smith, R.; Vukavić, T.; Vojinović, M.; Djordjevic, M. Serum concentrations of organochlorine compounds during pregnancy and the newborn. *Bull. Environ. Contam. Toxicol.* **1987**, *38*, 117–124. [CrossRef] [PubMed]
50. Curley, A.; Kimbrough, R. Chlorinated hydrocarbon insecticides in plasma and milk of pregnant and lactating women. *Arch. Environ. Health* **1969**, *18*, 156–164. [CrossRef] [PubMed]

Detection of Bisphenol A and Four Analogues in Atmospheric Emissions in Petrochemical Complexes Producing Polypropylene in South America

Joaquín Hernández Fernández [1,*], Yoleima Guerra [2] and Heidi Cano [3]

1. Chemistry Program, Department of Natural and Exact Sciences, San Pablo Campus, University of Cartagena, Cartagena 130015, Colombia
2. Centro de Investigación en Ciencias e Ingeniería, CECOPAT&A, Cartagena 131001, Colombia; yoleima.guerra@cecopat.com
3. Department of Civil and Environment Engineering, Universidad de la Costa, Barranquilla 080002, Colombia; hcano3@cuc.edu.co
* Correspondence: hernandez548@hotmail.com; Tel.: +57-301-5624990

Abstract: Because of its toxicity and impacts on the environment and human health, bisphenol A (BPA) has been controlled in numerous industrialized nations, increasing demand for bisphenol analogues (BP) for its replacement. However, the consequences of these chemicals on the environment and the health of persons exposed to their emissions are still being researched. The emissions from polypropylene manufacturing facilities in Colombia and Brazil were evaluated in this study, and the presence of bisphenol A and four BPs was detected among the gaseous compounds released, with total concentrations of BPs ($\sum$BP) between 92 and 1565 ng g^{-1}. As the melt flow index (MFI) of the polymer rises, so does the quantity of volatiles in its matrix that are eliminated during deodorization, indicating that the MFI and the amount of bisphenol released have a directly proportional connection.

Keywords: bisphenol A; bisphenol analogues; emissions; polypropylene

1. Introduction

Bisphenol A (2,2-bis-(4-hydroxyphenyl)-propane; BPA), its derivatives, and analogues are chemical products that are part of the compounds with two hydroxyphenyl functional groups in their chemical structure [1]. With a production of more than 5 million tons per year, BPA is one of the most produced products among bisphenols (BPs) [2] and is widely used as a raw material in various plastics (powder paints, polycarbonate plastics, epoxy resins, and paper coatings). However, BPA can be considered as one of the most important industrial additives in the production of resins; it is a highly polluting substance mainly generated by effluents from the plastic industry, leachate from plastic waste deposited in landfills, electronic waste, industrial painting, and BPA production [3]. This chemical, together with toxic dyes and nonylphenol, is filtered into the environment due to the natural degradation and decomposition of synthetic plastics, reaching concentrations of BPA of up to 100 mg L^{-1} in water at lethal concentration 50 (LC50) or effective concentration 50 (EC50) of 10 mg L^{-1} in water effluents from plastic production activities [4,5].

The effects of BPA on human health and the environment have been extensively studied, leading to the determination that this substance is responsible for disrupting the normal functions of hormones [6]. These compounds are known as endocrine-disrupting chemicals (EDC), since they are responsible for altering the functioning of estrogenic and androgenic hormones and are commonly found among the contaminants identified in soils and in bodies of water [7]. The negative effects of bisphenol A on the development of the human neuronal, cardiovascular, immune, and metabolic systems have also been reported [8,9], which is why the use of this substance has been limited or prohibited in many countries, and bisphenol analogues have been implemented as a solution to this problem [1,10]. However, these compounds have been considered toxic; several studies have evaluated their effects on human health, and it is considered a possible genotoxin. It has even been evidenced in human blood, urine, and breast milk, evidencing the long exposure they have inadvertently had to this compound [11–13]. Therefore, the identification and quantification of new sources of generation and new derivatives of BPA not reported in the literature are important in order to have a greater inventory of their sources of origin and propose greater regulations for their application. Several studies have investigated the occurrence and profiles of BPs in sediments, indoor dust, paper products (e.g., currency), and human urine [14,15]. Reports of ambient levels of BPs in gaseous emissions from PP-producing plants are scarce, focusing on countries such as the United States, China, Japan, and Korea [1,16].

Colombia and Brazil are countries that have polypropylene production plants which have been previously studied for the detection, identification, and quantification of the concentration of contaminants that are of interest and can affect the health of the surrounding inhabitants [17–24]. This allows us to observe that the development of analytical methods to identify and quantify other types of contaminants, such as bisphenols, for example, is becoming increasingly relevant in order to publicize the levels of these contaminants, which can subsequently be regulated and controlled to guarantee greater sustainability of the regions. It is imperative to assess the contribution of BP emission pathways to the environment in order to delineate the environmental fates, risks, and management of these chemicals.

In this study, gaseous emissions from polypropylene production plants in Colombia and Brazil are evaluated to identify the presence of BP. The samples were taken during the production of polypropylene with different melt flow indexes to determine if this influences the amount of BPs emissions during the removal of volatile compounds in the deodorization column, using analysis techniques such as solid phase extraction (SPE) and high-performance liquid chromatography (HPLC) for the detection of bisphenols.

2. Materials and Methods

2.1. Sampling Sites

The selected polypropylene plants are situated in the industrial zones of Colombia and Brazil. The four key steps of the manufacturing process of PP are: 1. reception, purification, and storage of raw materials; 2. polymerization; 3. addition, extrusion, and pelletizing; and 4. desorber, as shown in Figure 1. The study focuses on the fundamental processes of deodorization. The gaseous samples are collected at the top of the desorber. The desorber works with water vapor ranging from 600 to 1200 kg h^{-1} and steam ranging from 120 to 140 °C.

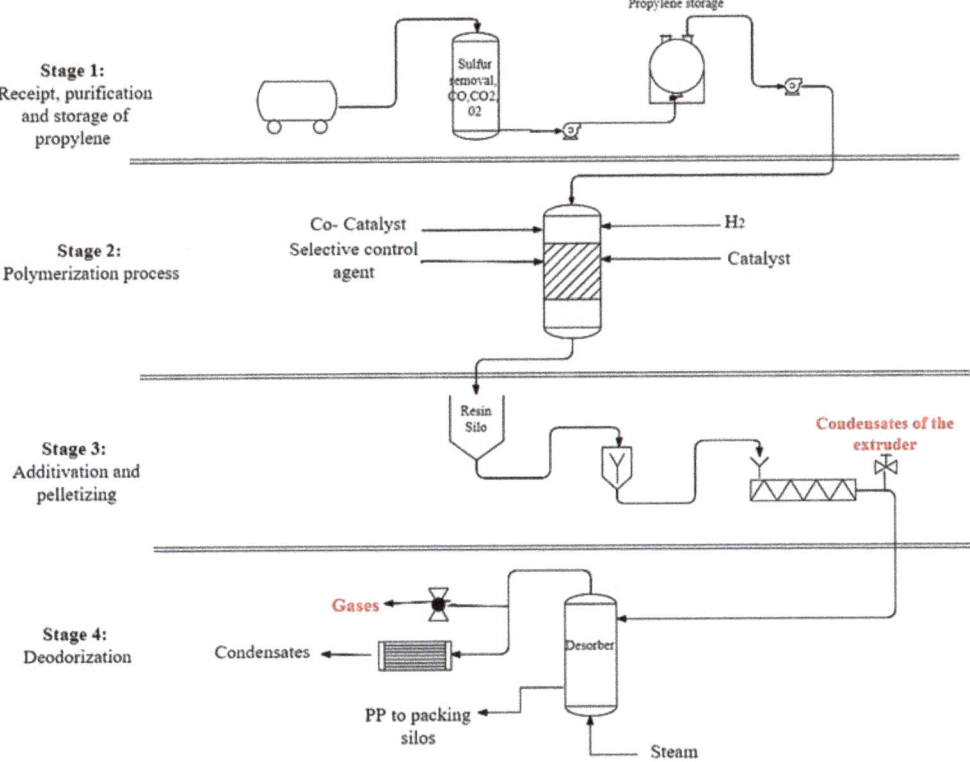

Figure 1. Polypropylene production stages and sampling points.

2.2. Sample Collection

The study was conducted during the manufacturing of three classes of PP with MFIs of 5, 50, and 80. Polypropylene samples with this MFI were used because it was shown in previous studies that the value of this characteristic of PP influences the amount of volatile compounds present in the polymeric matrix, following the methodology explained by J. Hernández (2020) [17].

The desorber produced effluent from the desorption of six grades of PP. Because of the construction of the column, each grade of PP took 4 h to desorb in the desorber. To have a better understanding of the stability of the process, samples were obtained in triplicate every hour. This allowed the variability of the desorber to be evaluated, as well as the influence of time spent in different phases of the process and the migration of the phenols to the condensates.

One meter away from the highest point of the column is where the sample point for the desorber is situated. This point features a nipple and a steel tube in the shape of a flute with a longitude equal to the diameter of the column. This flute features 20 evenly spaced orifices, allowing sampling of more than 95% of the diameter of the column. The end of the flute has two outlets, one of which is attached to a vacuum pump that operates at a constant pressure and the other to a metal cylinder that is coated in sulfurite. With the help of this system, it is possible to collect an isokinetic sample such that each of the 20 orifices receives the same quantity of sample due to equal pressure. A representative isokinetic sample is ensured by the 20 orifices, which allow more than 95% of the laminar flow of gases from the column to enter. The gases are mingled within the flute of each orifice before being later collected into the sulfurite-lined interior of the steel cylinder. For this investigation, a total of 72 gaseous samples—each obtained in triplicate—were used. The desorber allowed each grade of PP to desorb for 4 h.

2.3. Instrumental Analysis

BP structures of interest are shown in Figure 2. For the development of the analysis, the procedure of Sunggyu L. et al. (2015) [25] was followed. An Applied Biosystems API 2000 electrospray triple quadrupole mass spectrometer (ESI-MS/MS; Applied Biosystems, Foster City, CA, USA) and an Agilent 1100 Series HPLC (Agilent Technologies Inc., Santa Clara, CA, USA) outfitted with a binary pump and an autosampler were used to measure the concentrations of BPs in sample extracts. For LC separation, a Javelin guard column (Betasil C18, 20 × 2.1 mm) was attached to an analytical column (Betasil C18, 100 × 2.1 mm; Thermo Electron Corporation, Waltham, MA, USA). The injection has a 10 L volume. With a gradient as follows, the mobile phase consisted of methanol and water flowing at a rate of 0.3 mL min^{-1}: 15 percent methanol for the first two minutes; 15–50 percent methanol for the next five minutes; 50 percent methanol for the next eight minutes; 90–99 percent methanol for the next twenty minutes; and 15 percent methanol for the final thirty minutes. It was performed using the negative ion multiple reaction monitoring (MRM) mode. By injecting several substances into the mass spectrometer using a flow injection technique, the MS/MS settings were tuned. Both the impact gas and the curtain gas were nitrogen. The quantification was performed using an external calibration technique, and it was adjusted for C12-BPA recoveries [25].

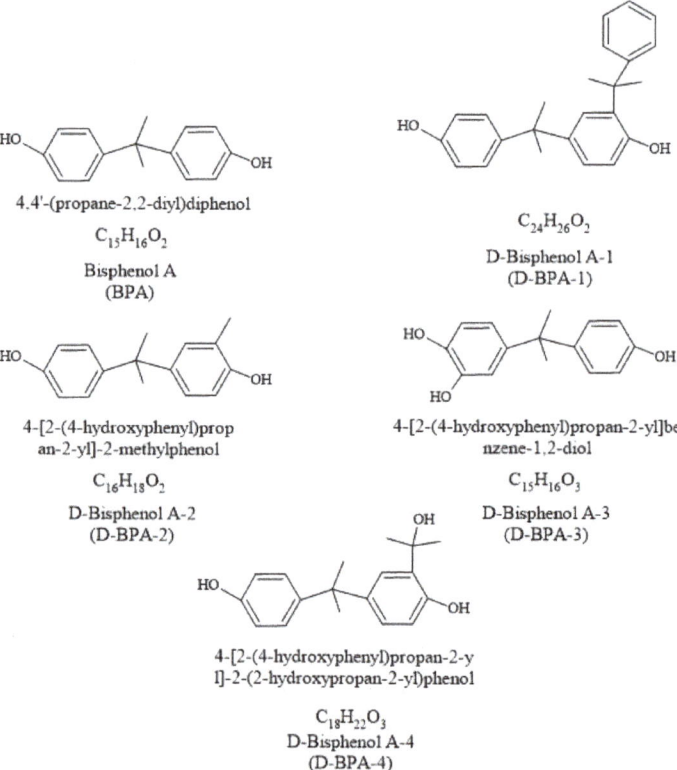

Figure 2. Structures and identifications of the BPs of interest.

2.4. Treatment of Samples Based on SPE

2.4.1. Pretreatment

The work sample was blended and cooled at 25 °C before being filtered through a 0.22 m PTFE Teflon filter to aid in further sample preparation and limit microbial activity.

2.4.2. Preconcentration and Cleaning

Conditioning of Strata X-33 cartridges (6 mL, 500 mg) was conducted at this step using 5 mL of MeOH followed by 5 mL of distilled water. After that, 15 mL of the material was uploaded at a rate of 1 mL per minute. After percolating the entire sample, the cartridges were rinsed with 3 mL of MeOH:H_2O 80:20. The chemicals retained in the solid phase were eluted with 10 mL of ACN. The eluate was evaporated until dry using a nitrogen stream at 5 psi. The finished extract was reconstituted with ACN to a final volume of 1 mL, yielding a 10:1 pre-concentration [17].

2.5. Quality Assurance and Quality Control (QA/QC)

Spiked blanks and matrix-spiked samples were frequently tested to confirm the analytical approach utilized in our investigation. BP recoveries in spiked blanks ($n = 4$) varied from 75.4 to 1.44 percent (mean SD) for D-BPA-1, and 256 to 1.32 percent for D-BPA-2 (30 ng for each compound and 20 ng for C_{12}-BPA). BP recoveries in matrix-spiked samples ($n = 4$) varied from 45.3 to 1.7 percent for D-BPA-3 and 325 to 5.14 percent for D-BPA-4 (30 ng for each individual component and 20 ng for C_{12}-BPA). The lowest admissible calibration standard and a notional sample weight of 0.1 g were used to compute the limit of quantitation (LOQs). The computed LOQs for BPA, D-BPA-1, and D-BPA-2 were 0.40 ng g^{-1}; 0.9 ng g^{-1} for D-BPA-3; and 1.5 ng g^{-1} for D-BPA-4. After every 20 samples, a midway calibration standard was injected to assess for drift in instrumental sensitivity. A pure solvent (methanol) was injected at regular intervals to check for BP carryover between the samples examined. The daily injection of 10 calibration standards at concentrations ranging from 0.01 to 250 ng mL^{-1} confirmed the instrumental calibration, and the linearity of the calibration curve (r) was more than 0.99 for each of the target compounds. BP concentrations are reported as a dry weight (dw).

To perform statistical analysis, concentrations below the LOQ were assigned a value of one-half of the corresponding LOQ rather than a zero value for the computation of the mean and median. To investigate variations in BP concentrations across three distinct PP types, a one-way ANOVA with Tukey's test was used. To analyze differences in BP concentrations throughout the manufacturing of three classes of PP with fluidity indices (MFI) of 5, 50, and 80, a one-way ANOVA with Tukey's test was used [25].

3. Results

Occurrence and Concentrations of BPs in Atmospheric Emissions

BPs concentrations in atmospheric emissions collected from PP production plants in Colombia and Brazil are summarized in Table 1. The sampling points were established with alphanumeric codes that contained the letter of each country and the MFI of interest. For Colombia, the sampling points for BP analogue emissions were EC-MFI5, EC-MFI50, and EC-MFI80, thus guaranteeing the taking of gaseous samples when this plant was producing PP with an MFI of 5, 50, and 80, respectively. For the sampling of emissions at the plant in Brazil, points EB-MFI5, EB-MFI50, and EB-MFI80 were selected. The concentrations of total BPs (BPs; the sum of 5 bisphenol analogues) ranged from 92 to 1565 (average: 628) ng g^{-1} dw. BPA and all its analogues presented detection rates (RD) of 100% in all the samples studied, and this was independent of the type of MFI of PP that was synthesized. The average concentrations of the BPs detected in simple emission gas were in the order of BPA (338.56 ng g^{-1} dw), D-BPA-1 (131.32 ng g^{-1} dw), D-BPA-2 (51.97 ng g^{-1} dw), D-BPA-3 (62.07 ng g^{-1} dw), and D-BPA-4 (42.67 ng g^{-1} dw), indicating that BPA is a by-product of the degradation of PP in this type of industrial process. Between the two sampling points in Colombia and Brazil and under the three different sampling conditions at each plant, the highest concentrations of BPA were found in the industrial emissions from the Brazil plant during the production of PP with an MFI of 80 (mean: 980.59 ng g^{-1} dw), followed by the emissions of this same plant with a production of MFI of 50 (708.11 ng g^{-1} dw) and the emissions of the Colombia plant when it produced PP with an MFI of 80 (670.83 ng g^{-1} dw). Our results suggest that environmental contamination by BPs in South America has a very significant contribution from industrial activities rather than domestic activities and, therefore, corrective measures must be taken in a very short time. The concentrations of BPA, D-BPA-1, and D-BPA-3 in all gas emissions samples from PP production plants in South America ranged from 45 to 947 (average: 338.556) ng g^{-1} dw, from 15 to 542 (average: 131.32) ng g^{-1} dw, and from 16 to 124 (average: 62.07) ng g^{-1} dw, respectively.

Table 1. Concentrations of bisphenol analogues (ng g^{-1} dry weight) in atmospheric emissions from PP production plants in Colombia and Brazil.

Country	Sampling Point	Parameters	BPA	D-BPA-1	D-BPA-2	D-BPA-3	D-BPA-4	$\sum$BPs
Colombia	EC-MFI5	Production MFI 5 (n = 16) Mean Median Range DR	239.71 214 45–546 100	83.29 64 15–341 100	21.81 16.7 0.57–81.6 100	27.29 25.9 15.6–43.5 100	16.97 13.2 4.26–51.6 100	389.07 393.7 92.45–976.75 100
	EC-MFI50	Production MFI 50 (n = 16) Mean Median Range DR	264.88 246 73–576 100	104.06 88 26–421 100	49.14 46.6 1.85–134.5 100	52.68 49.4 29.3–75.3 100	16.97 13.2 4.26–51.6 100	487.72 510.9 160.16–1129.15 100
	EC-MFI80	Production MFI 80 (n = 16) Mean Median Range DR	324.76 345 95–624 100	136.35 99 42–542 100	78.72 76.9 6.5–172.5 100	86.51 91.6 46.7–121.1 100	44.49 34.6 19.4–99.7 100	670.83 698.6 337.6–1383.4 100
Brazil	EB-MFI5	Production MFI 5 (n = 16) Mean Median Range DR	301.01 316 79–615 100	105.53 87 34–369 100	35.09 34.6 2.43–99.3 100	45.66 42.1 31.2–74.3 100	35.99 33.9 9.75–81.3 100	523.28 530 186.75–1123.2 100
	EB-MFI50	Production MFI 50 (n = 16) Mean Median Range DR	377.12 419 99–721 100	151.71 141 49–455 100	51.13 46.5 4.65–112.1 100	70.41 66.5 47.6–99.2 100	57.75 55.6 25.4–98.4 100	708.11 699.8 308.05–1339.7 100
	EB-MFI80	Production MFI 80 (n = 16) Mean Median Range DR	523.88 536 184–947 100	207.00 196 75–523 100	75.95 75.3 10.8–201.5 100	89.90 88.5 66.2–124.3 100	83.86 77.2 46.5–142.5 100	980.59 1002.6 430.5–1565.1 100
		Total (n = 40) Mean Median Range DR	338.56 294 45–947 100	131.32 99 15–542 100	51.97 44.35 0.6–201 100	62.07 64.8 16–124 100	42.67 35.5 4–142 100	626.60 542.95 92–1565 100

In both the Colombian and Brazilian plants, the concentrations of BPA and its analogues varied significantly in the PP production samples, with MFIs of 5, 50, and 80. The fluidity and crystallinity of the material increases as the MFI value increases. The PP of MFI of 80 has shorter chains in its structures, and this makes the content of volatiles in its polymeric matrix higher. Therefore, it would be expected that during the PP deodorization process, additive residues or PP degradation by-products can be more easily removed from the polymer matrix. The results allow us to observe that the means of the total emissions of BP in Colombia were 389.07, 487.72, and 670.83 ng g^{-1} dw, respectively, being directly proportional to the MFI values. In the Brazilian plant, the same relationship of BP with MFI was also observed, except that the concentrations of BPs were higher at 523.28, 708.11, and 980.59 ng g^{-1} dw, respectively.

Concentrations of BPA, D-BPA-1, D-BPA-2, D-BPA3, and D-BPA-4 were 9.5, 19.95, 55.62, 48.19, and 0% higher, respectively, in emissions from the Colombia plant during production of a PP MFI50 compared to the production of a PP MFI5, as seen in Figure 3. Regarding the production of MFI80, the increases were 26.19, 38.91, 72.30, 68.45, and 61.87% higher than the emissions in MFI5. These significant differences could be appreciated when comparing only the variation in the fluidity of the PP. However, when comparing the plant in Colombia with the one in Brazil, we also noticed important differences, and this can be seen for each BPA analogue. In this way, the results in the plant in Brazil were always 20% higher than in the plant in Colombia. For MFI5, the values of BPA, D-BPA-1, D-BPA-2, D-BPA3, and D-BPA-4 were 20.37, 21.07, 37.85, 40.23, and 52.85% higher, respectively, in the Brazilian plant. During the production of MFI50, differences greater than 25% were observed for BPA, D-BPA-1, D-BPA3, and D-BPA-4, while for D-BPA2, the difference was only 3.89%. During the production of MFI80, emissions in Brazil exceeded 34% for BPA, D-BPA-1, and D-BPA-4, while the difference for D-BPA2 and D-BPA3 was only 3.65 and 3.78%, respectively.

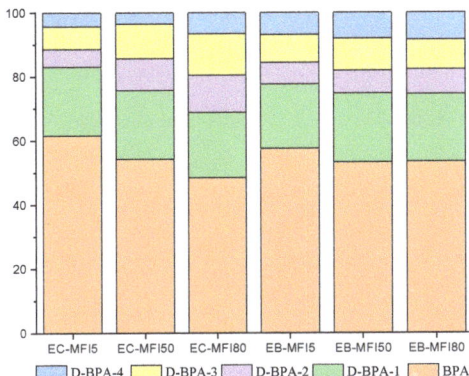

Figure 3. Comparisons of the % of BPA, D-BPA-1, D-BPA-2, D-BPA-3, and D-BPA-4 in Colombia and Brazil.

The values obtained in bisphenol emissions have not been reported before for petrochemical plants. However, when compared with the data obtained in other studies of bisphenol emissions in homes, offices, and urban and residential areas, it is evident that the results obtained are superior to those obtained in Greece, Spain, and Romania [26–28]. In a study carried out by Wang et al. (2015) [26], dust samples were taken inside houses in different countries, including Colombia, which obtained values similar to those reported (500 ng g^{-1}).

The values obtained in this study are worrying. These emissions are made every day of production and are dispersed in the environment, affecting the surrounding population without their knowledge. There are no regulations in these countries on the limit of exposure and emission of BPs allowed to control these compounds. These studies are important because there are no controls on these components. In addition, these companies do not record the BPs concentrations found in this study, which indicates that they may be a product of the decomposition of additives during the production of polypropylene.

4. Conclusions

This is the first industrial-scale investigation on the detection of bisphenols in the polypropylene process. This enables the evaluation of the phases and equipment where bisphenol emissions occur. Furthermore, the detection of compounds similar to bisphenol A makes it possible to evaluate the effects of these bisphenol A substitutes on humans in industrial areas, since the concentrations found and constant exposure to it can cause harmful effects in workers, and protocols to ensure their health are recommended.

Author Contributions: Conceptualization, J.H.F. and H.C.; methodology, J.H.F.; formal analysis, J.H.F.; investigation, Y.G. and H.C.; writing—original draft preparation, Y.G.; supervision, H.C.; project administration, J.H.F. All authors have read and agreed to the published version of the manuscript.

Funding: This research received no external funding.

Institutional Review Board Statement: Not applicable.

Informed Consent Statement: Not applicable.

Data Availability Statement: Not applicable.

Conflicts of Interest: The authors declare no conflict of interest.

References

1. Chen, D.; Kannan, K.; Tan, H.; Zheng, Z.; Feng, Y.-L.; Wu, Y.; Widelka, M. Bisphenol Analogues Other Than BPA: Environmental Occurrence, Human Exposure, and Toxicity—A Review. *Environ. Sci. Technol.* **2016**, *50*, 5438–5453. [CrossRef]
2. Moon, M.K. Concern about the Safety of Bisphenol A Substitutes. *Diabetes Metab. J.* **2019**, *43*, 46–48. [CrossRef] [PubMed]
3. Morin, N.; Arp, H.P.H.; Hale, S.E. Bisphenol A in Solid Waste Materials, Leachate Water, and Air Particles from Norwegian Waste-Handling Facilities: Presence and Partitioning Behavior. *Environ. Sci. Technol.* **2015**, *49*, 7675–7683. [CrossRef]
4. Mihaich, E.M.; Friederich, U.; Caspers, N.; Hall, A.T.; Klecka, G.M.; Dimond, S.S.; Staples, C.A.; Ortego, L.S.; Hentges, S.G. Acute and chronic toxicity testing of bisphenol A with aquatic invertebrates and plants. *Ecotoxicol. Environ. Saf.* **2009**, *72*, 1392–1399. [CrossRef] [PubMed]
5. Belfroid, A.; van Velzen, M.; van der Horst, B.; Vethaak, D. Occurrence of bisphenol A in surface water and uptake in fish: Evaluation of field measurements. *Chemosphere* **2002**, *49*, 97–103. [CrossRef]
6. Gao, H.; Yang, B.-J.; Li, N.; Feng, L.-M.; Shi, X.-Y.; Zhao, W.-H.; Liu, S.-J. Bisphenol A and hormone-associated cancers: Current progress and perspectives. *Medicine* **2015**, *94*, e211. [CrossRef] [PubMed]
7. Catenza, C.J.; Farooq, A.; Shubear, N.S.; Donkor, K.K. A targeted review on fate, occurrence, risk and health implications of bisphenol analogues. *Chemosphere* **2021**, *268*, 129273. [CrossRef]
8. Cimmino, I.; Fiory, F.; Perruolo, G.; Miele, C.; Beguinot, F.; Formisano, P.; Oriente, F. Potential Mechanisms of Bisphenol A (BPA) Contributing to Human Disease. *Int. J. Mol. Sci.* **2020**, *21*, 5761. [CrossRef]
9. Gao, X.; Wang, H.-S. Impact of bisphenol a on the cardiovascular system-epidemiological and experimental evidence and molecular mechanisms. *Int. J. Environ. Res. Public Health* **2014**, *11*, 8399–8413. [CrossRef] [PubMed]
10. Liao, C.; Kannan, K. Concentrations and Profiles of Bisphenol A and Other Bisphenol Analogues in Foodstuffs from the United States and Their Implications for Human Exposure. *J. Agric. Food Chem.* **2013**, *61*, 4655–4662. [CrossRef]
11. Qiu, W.; Zhan, H.; Hu, J.; Zhang, T.; Xu, H.; Wong, M.; Xu, B.; Zheng, C. The occurrence, potential toxicity, and toxicity mechanism of bisphenol S, a substitute of bisphenol A: A critical review of recent progress. *Ecotoxicol. Environ. Saf.* **2019**, *173*, 192–202. [CrossRef] [PubMed]
12. Han, Y.; Fei, Y.; Wang, M.; Xue, Y.; Chen, H.; Liu, Y. Study on the Joint Toxicity of BPZ, BPS, BPC and BPF to Zebrafish. *Molecules* **2021**, *26*, 4180. [CrossRef]
13. Gao, C.; He, H.; Qiu, W.; Zheng, Y.; Chen, Y.; Hu, S.; Zhao, X. Oxidative Stress, Endocrine Disturbance, and Immune Interference in Humans Showed Relationships to Serum Bisphenol Concentrations in a Dense Industrial Area. *Environ. Sci. Technol.* **2021**, *55*, 1953–1963. [CrossRef] [PubMed]
14. Zhang, H.; Quan, Q.; Zhang, M.; Zhang, N.; Zhang, W.; Zhan, M.; Xu, W.; Lu, L.; Fan, J.; Wang, Q. Occurrence of bisphenol A and its alternatives in paired urine and indoor dust from Chinese university students: Implications for human exposure. *Chemosphere* **2020**, *247*, 125987. [CrossRef]
15. Liao, C.; Liu, F.; Kannan, K. Bisphenol S, a New Bisphenol Analogue, in Paper Products and Currency Bills and Its Association with Bisphenol A Residues. *Environ. Sci. Technol.* **2012**, *46*, 6515–6522. [CrossRef]
16. Corrales, J.; Kristofco, L.A.; Steele, W.B.; Yates, B.S.; Breed, C.S.; Williams, E.S.; Brooks, B.W. Global Assessment of Bisphenol A in the Environment: Review and Analysis of Its Occurrence and Bioaccumulation. *Dose-Response Publ. Int. Hormesis Soc.* **2015**, *13*, 15593258–15598308. [CrossRef]

17. Hernández-Fernández, J.; Lopez-Martinez, J.; Barceló, D. Quantification and elimination of substituted synthetic phenols and volatile organic compounds in the wastewater treatment plant during the production of industrial scale polypropylene. *Chemosphere* **2021**, *263*, 128027. [CrossRef] [PubMed]
18. Hernández-Fernández, J. Quantification of arsine and phosphine in industrial atmospheric emissions in Spain and Colombia. Implementation of modified zeolites to reduce the environmental impact of emissions. *Atmospheric Pollut. Res.* **2021**, *12*, 167–176. [CrossRef]
19. Hernández-Fernández, J. Quantification of oxygenates, sulphides, thiols and permanent gases in propylene. A multiple linear regression model to predict the loss of efficiency in polypropylene production on an industrial scale. *J. Chromatogr. A* **2020**, *1628*, 461–478. [CrossRef] [PubMed]
20. Joaquin, H.-F.; Juan, L. Quantification of poisons for Ziegler Natta catalysts and effects on the production of polypropylene by gas chromatographic with simultaneous detection: Pulsed discharge helium ionization, mass spectrometry and flame ionization. *J. Chromatogr. A* **2020**, *1614*, 460736. [CrossRef]
21. Joaquin, H.-F.; Juan, L.-M. Autocatalytic influence of different levels of arsine on the thermal stability and pyrolysis of polypropylene. *J. Anal. Appl. Pyrolysis* **2022**, *161*, 105385. [CrossRef]
22. Hernández-Fernandez, J.; Rodríguez, E. Determination of phenolic antioxidants additives in industrial wastewater from polypropylene production using solid phase extraction with high-performance liquid chromatography. *J. Chromatogr. A* **2019**, *1607*, 460442. [CrossRef]
23. Hernández-Fernández, J.; López-Martínez, J. Experimental study of the auto-catalytic effect of triethylaluminum and TiCl4 residuals at the onset of non-additive polypropylene degradation and their impact on thermo-oxidative degradation and pyrolysis. *J. Anal. Appl. Pyrolysis* **2021**, *155*, 105052. [CrossRef]
24. Hernández-Fernández, J.; Lopez-Martinez, J.; Barceló, D. Development and validation of a methodology for quantifying parts-per-billion levels of arsine and phosphine in nitrogen, hydrogen and liquefied petroleum gas using a variable pressure sampler coupled to gas chromatography-mass spectrometry. *J. Chromatogr. A* **2021**, *1637*, 461833. [CrossRef]
25. Lee, S.; Liao, C.; Song, G.-J.; Ra, K.; Kannan, K.; Moon, H.-B. Emission of bisphenol analogues including bisphenol A and bisphenol F from wastewater treatment plants in Korea. *Chemosphere* **2015**, *119*, 1000–1006. [CrossRef]
26. Wang, W.; Abualnaja, K.O.; Asimakopoulos, A.G.; Covaci, A.; Gevao, B.; Johnson-Restrepo, B.; Kumosani, T.A.; Malarvannan, G.; BinhMinh, T.; Moon, H.-B.; et al. A comparative assessment of human exposure to tetrabromobisphenol A and eight bisphenols including bisphenol A via indoor dust ingestion in twelve countries. *Environ. Int.* **2015**, *83*, 183–191. [CrossRef] [PubMed]
27. Barroso, P.J.; Martín, J.; Santos, J.L.; Aparicio, I.; Alonso, E. Evaluation of the airborne pollution by emerging contaminants using bitter orange (*Citrus aurantium*) tree leaves as biosamplers. *Sci. Total Environ.* **2019**, *677*, 484–492. [CrossRef]
28. Geens, T.; Roosens, L.; Neels, H.; Covaci, A. Assessment of human exposure to Bisphenol-A, Triclosan and Tetrabromobisphenol-A through indoor dust intake in Belgium. *Chemosphere* **2009**, *76*, 755–760. [CrossRef]

Article

Fit-for-Purpose Assessment of QuEChERS LC-MS/MS Methods for Environmental Monitoring of Organotin Compounds in the Bottom Sediments of the Odra River Estuary

Dawid Kucharski [1,*], Robert Stasiuk [2], Przemysław Drzewicz [3], Artur Skowronek [4], Agnieszka Strzelecka [4], Kamila Mianowicz [5] and Joanna Giebułtowicz [1]

1 Department of Bioanalysis and Drugs Analysis, Faculty of Pharmacy, Medical University of Warsaw, Banacha 1, 02-097 Warsaw, Poland; jgiebultowicz@wum.edu.pl
2 Department of Geomicrobiology, Institute of Microbiology, Faculty of Biology, University of Warsaw, Miecznikowa 1, 02-096 Warsaw, Poland; r.stasiuk@biol.uw.edu.pl
3 Polish Geological Institute-National Research Institute, Rakowiecka 4, 00-975 Warsaw, Poland; przemyslaw.drzewicz@pgi.gov.pl
4 Institute of Marine and Environmental Sciences, University of Szczecin, Mickiewicza 16, 70-383 Szczecin, Poland; artur.skowronek@usz.edu.pl (A.S.); agnieszka.strzelecka@usz.edu.pl (A.S.)
5 Interoceanmetal Joint Organization, Cyryla i Metodego 9, 71-541 Szczecin, Poland; k.mianowicz@iom.gov.pl
* Correspondence: dkucharski@wum.edu.pl; Tel.: +48-22-572-0949; Fax: +48-22-572-097

Abstract: Organotin compounds (OTCs) are among the most hazardous substances found in the marine environment and can be determined by either the ISO 23161 method based on extraction with non-polar organic solvents and gas chromatography analysis or by the recently developed QuEChERS method coupled to liquid chromatography-mass spectrometry (LC-MS/MS). To date, the QuEChERS LC/MS and ISO 23161 methods have not been compared in terms of their fit-for-purpose and reliability in the determination of OTCs in bottom sediments. In the case of ISO 23161, due to a large number of interferences gas chromatography-mass spectrometry was not suitable for the determination of OTCs contrary to more selective determination by gas chromatography with an atomic emission detector. Moreover, it has been found that the derivatization of OTCs to volatile compounds, which required prior gas chromatography determination, was strongly affected by the sediments' matrices. As a result, a large amount of reagent was needed for the complete derivatization of the compounds. Contrary to ISO 23161, the QuEChERS LC-MS/MS method did not require the derivatization of OTC and is less prone to interferences. Highly volatile and toxic solvents were not used in the QuEChERS LC-MS/MS method. This makes the method more environmentally friendly according to the principles of green analytical chemistry. QuEChERS LC-MS/MS is suitable for fast and reliable environmental monitoring of OTCs in bottom sediments from the Odra River estuary. However, determination of di- and monobutyltin by the QuEChERS LC-MS/MS method was not possible due to the constraints of the chromatographic system. Hence, further development of this method is needed for monitoring di- and monobutyltin in bottom sediments.

Keywords: bottom sediments; environmental monitoring; Odra River estuary; organotin compounds; trueness; verification

1. Introduction

Organotin compounds (OTCs) belong to the most hazardous substances found in marine environments. The most common OTCs include tributyltin (TBT), triphenyltin (TPhT), monobutyltin (MBT), and dibutyltin (DBT). MBT and DBT are degradation products of TBT [1]. The main sources of TBT and TPhT contamination are antifouling paints that are widely used in the marine industry for the protection of ship hulls and marine structures as well as pesticides used for the preservation of wood, paper, textiles, leather, and plastics. Although the application of OTCs in antifouling coatings has been banned since 2008, the compounds are still present in the environment due to their wide industrial applications (especially in the production of polyurethane foam and silicons) [2,3]. OTCs are introduced into the environment by effluents from wastewater treatment plants. Moreover, the results of many studies indicated the historical contamination of sediments in coastal areas of Poland, especially in the vicinity of harbors and shipyards [4–6].

Due to the lipophilic properties of OTCs (logP $\approx$ 4), routine extraction of organotin compounds from soils, sediments, and sludges is conducted by the use of non-polar organic solvents such as hexane or dichloromethane, frequently with a complexing agent such as tropolone or polycarboxylic acids [7,8]. Recently, to improve extraction efficiency, several sample preparation methods have been developed such as accelerated solvent extraction, microwave extraction, liquid-phase microextraction, solid-phase extraction, and solid-phase microextraction [9]. Those methods allow one to reduce solvent consumption, shorten preparation time, improve recovery, and decrease the limits of detection. The ultrasound-assisted solvent extraction of OTCs from bottom sediments is a well-known ISO standard method [10].

The isolated OTCs are mainly analyzed by gas chromatography (GC) coupled with mass spectrometry (MS), flame photometric detectors, atomic emission detectors (AED), or inductively coupled plasma mass spectrometry [9]. However, OTCs require the derivatization of volatile compounds before their determination by GC-based methods. The most common derivatization procedures include conversion with alkyloborates (e.g., $NaBEt_4$), Grignard reagents, or borohydride species (e.g., sodium borohydride—$NaBH_4$). Regardless of the particular derivatization method, this step is prone to error, tedious, and laborious. Additionally, very toxic and expensive chemical agents are used during the derivatization procedure. As a result, the methods based on derivatization do not follow the rules of green analytical chemistry (GAC). According to GAC rules, the amount of chemicals and samples consumed during an analytical procedure is strongly restricted. Thus, analytical methods without the derivatization step are preferred [11,12]. The recently developed QuEChERS extraction method combined with liquid chromatography coupled with tandem mass spectrometry for the determination of OTCs in bottom sediments meets GAC rules [6,13]. In comparison to the standard method ISO 23161:2018, QuEChERS extraction requires a smaller amount of solvent and chemical agents as well as a lower mass portion of a sample. Extraction of OTCs by QuEChERS also takes less time than other methods.

Although concentrations of OTCs in sediments are generally higher than in the water column [14], the determination of OTCs is very difficult. Bottom sediments contain a large number of compounds, in many cases with high molecular mass that are extracted together with OTCs. As a result, a lower recovery, reliability, and accuracy of the method are frequently observed. Moreover, the derivatization of OTCs to volatile compounds is very ineffective due to the decomposition of the derivatization reagent or its side reactions with co-extracted matrix constituents [15]. Thus, a comprehensive comparison and fit-for-purpose assessment of two analytical methods for the determination of OTCs in bottom sediments is the subject of this study. The QuEChERS extraction method combined with LC-MS/MS is compared with the solvent extraction method combined with GC-MS and GC with atomic emission detector (AED) [10]. Additionally, BCR 646, a reference material, as well as 10 sediments collected from the Odra River estuary area, SW Baltic Sea, were used for testing the comparability and fit-for-purpose assessment of the aforementioned methods.

2. Results and Discussion

The subject of the study was the comparison of the standard method ISO 23161 for the determination of OTCs in bottom sediments with the newly developed QuEChERS LC-MS/MS method. This is a continuation of our earlier studies on the application of the QuEChERS method for OTCs determination [6]. Detailed comparisons of extraction procedures and separation parameters are presented in Tables S1 and S2. In the case of ISO 23161, acid extraction and subsequent derivatization with tetraethylborate are applied in the determination of tin compounds. This method required a larger amount of sediment (at least 1 g d.w.), whereas, in the case of QuEChERS extraction, 125 mg d.w. of the sample amount was enough for the analysis. Analyzing organic tin compounds according to the ISO method requires also a higher amount of reagents. Thus, the cost of the single analysis seems to be higher in comparison to the QuEChERS LC-MS/MS method. Additionally, hexane as the analyte extractant is used in the ISO method. The application of hexane in analytical chemistry is strongly limited by occupational and health safety regulations due to its high neurotoxicity and detrimental effect on lab workers as well as the environment [16]. Contrary to ISO 23161, acetonitrile with 5% formic acid as the organic phase is used in the QuEChERS method. Although acetonitrile is a hazardous reagent, it is generally less toxic than hexane due to its low volatility. According to the Pfizer company's solvent selection guidelines, the application of acetonitrile in analytical procedures is preferable due to lower toxicity and environmental risk in comparison to other solvents [17]. However, the main disadvantage of the ISO method is the derivatization of OTCs to volatile compounds before their determination by GC. Derivatization of OTCs with sodium tetraethylborate (NaBEt$_4$) in the presence of other compounds extracted from complex matrices, such as sediments, requires a large volume of very expensive and toxic reagents. Due to the decomposition of derivatization reagent or the side reactions with other matrix constituents, only part of OTCs may undergo derivatization to volatile compounds [18].

2.1. Fit-for-Purpose of the Methods

The standard method ISO 23161 and QuEChERS LC-MS/MS method were tested based on reference material BCR 646 and sediments collected from the Odra River estuary. In the case of standard method ISO 23161, co-extracted matrix constituents strongly interfered in the determination of OTCs by GC-MS. Application of GC-MS required additional clean-up of the solvent extract before derivatization of OTCs to the volatile compounds. The clean-up step is a source of additional bias in the method. It is also very tedious and time-consuming. The application of a very selective GC-AED method allowed for the precise and accurate determination of OTCs in bottom sediments without the clean-up step (Figure 1B). Application QuEChERS LC-MS/MS method resulted in a sufficient separation. The peaks of TBT and TPhT were identified and integrated easily (Figure 1A). Although MBT and DBT are quantitatively extracted by the QuEChERS method, the compounds were not detected by LC-MS/MS due to the constraints of the chromatographic system. The application of ion trap-mass spectrometry or ICP–MS as a detector, addition of tropolone to the eluent, or application of another stationary phase in the LC separation column may solve the problem [19–21].

Due to the necessity of solvent exchange, the QuEChERS extraction method is not recommended for sample preparation before determination by gas chromatography. Therefore, QuEChERS LC-MS/MS and standard method ISO 23161 combined with GC-AED were taken for further comparison.

2.2. Verification of the Methods

Analytical methods ISO 23161 GC-AED and QuEChERS LC-MS/MS were verified according to Eurachem Guide [22] in terms of limit of detection (LOD), the limit of quantification (LOQ), linearity, accuracy, precision, and recovery (Table 1). The LOQ and LOD values were similar for both methods. In the case of QuEChERS LC-MS/MS, higher LOD values were observed for TPhT (5.0 ng·g^{-1} d.w.) due to the high signal-to-noise ratio. The coefficients of determination (R^2) of the calibration curves were $\geq$0.99. The accuracy did not meet the acceptance criteria (85–115%) for MBT in ISO 23161GC-AED and was in the range of 80–85%. The precision for MBT in ISO 23161-GC-AED did not meet the acceptance criteria ($\leq$15%) and was 27%. For the other OTCs, the precision was lower than 15%. The recovery in most cases was higher than 85%. For the ISO 23161 GC-AED method, MBT recovery was lower than 80%.

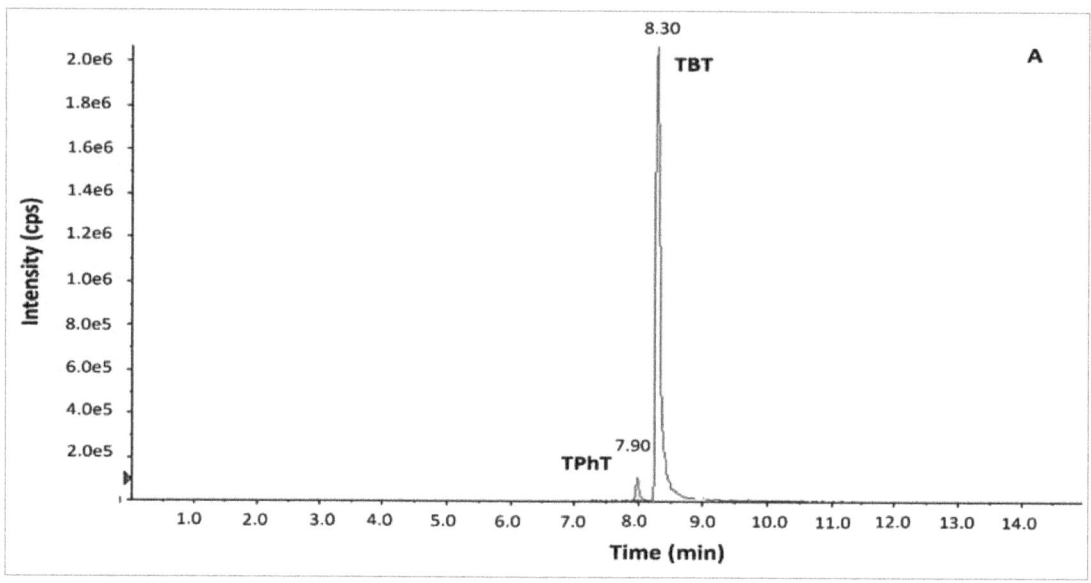
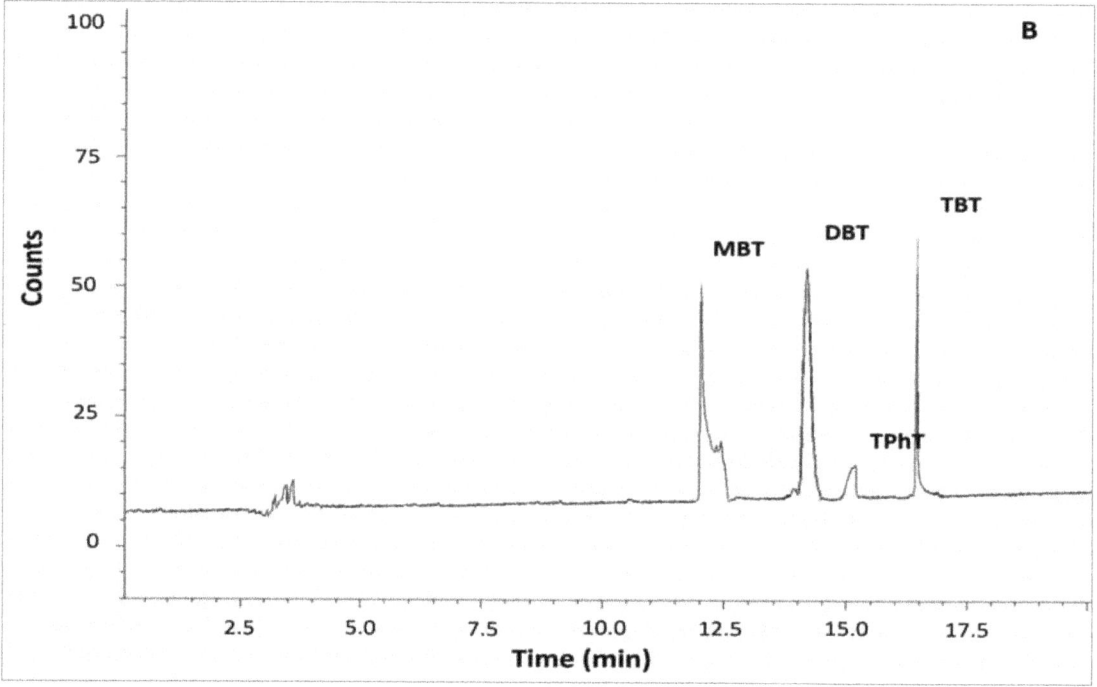

Figure 1. Chromatograms of organotin compounds determined in BCR sediments by (A) QuEChERS coupled to LC-MS/MS (B) ISO 23161 GC-AED method. Determination of DBT and MBT was only possible using the method ISO 23161 GC-AED. Abbreviations: MBT, monobutyltin; DBT, dibutyltin; TBT, tributyltin; TPhT, triphenyltin.

Table 1. Validation parameters of QuEChERS LC-MS/MS and ISO 23161 GC-AED methods for monobutyltin (MBT), dibutyltin (DBT), tributyltin (TBT), and triphenyltin (TPhT).

Parameters	QuEChERS (LC-MS/MS)	ISO 23161 (GC-AED)
Limit of quantification (LOQ)	2.0 ng·g^{-1} d.w. TBT 5.0 ng·g^{-1} d.w. TPhT	1.7 ng·g^{-1} d.w. MBT 2.4 ng·g^{-1} d.w. DBT 2.5 ng·g^{-1} d.w. TBT 3.0 ng·g^{-1} d.w. TPhT
Limit of detection (LOD)	0.6 ng·g^{-1} d.w. TBT 1.5 ng·g^{-1} d.w. TPhT	0.5 ng·g^{-1} d.w. MBT 0.7 ng·g^{-1} d.w. DBT 0.7 ng·g^{-1} d.w. TBT 0.9 ng·g^{-1} d.w. TPhT
Analytical ranges (linearity)	2.0–4000.0 ng·g^{-1} d.w.; R^2_{TBT} = 0.99 5.0–4000.0 ng·g^{-1} d.w.; R^2_{TPhT} = 0.99	1.7–4000.0 ng·g^{-1} d.w; R^2_{MBT} = 0.99 2.4–4000.0 ng·g^{-1} d.w.; R^2_{DBT} = 0.99 2.5–4000.0 ng·g^{-1} d.w.; R^2_{TBT} = 0.99 3.0–4000.0 ng·g^{-1} d.w.; R^2_{TPhT} = 0.99
Accuracy (n = 6) [a]	98.0–105.0%-TBT 91.0–101.0%-TPhT	80.0–85.0%-MBT 88.0–93.0%-DBT 88.0–92.0%-TBT 87.0–105.0%-TPhT
Precision (n = 6) [b]	3.0%-TBT 4.9%-TPhT	27.0%-MBT 15.0%-DBT 6.0%-TBT 2.0%-TPhT
Recovery (n = 6)	93%-TBT 91%-TPhT	79%-MBT 86%-DBT 90%-TBT 91%-TPhT

[a] acceptance criteria: 85–115%; [b] acceptance criteria: ≤15%

The trueness of both methods was also tested. The differences between OTCs concentration in the dry weight of the sample assigned to the certified reference and the mean of concetration determined by the QuEChERS LC-MS/MS method were as follow: B_{TBT} = 30.0 ng·g^{-1} d.w. and B_{TPhT} = 1.1 ng·g^{-1} d.w. The concentration determined by ISO 23161-GC-AED were as follow B_{TBT} = 47.8 ng·g^{-1} d.w., B_{TPhT} = 1.2 ng·g^{-1} d.w., B_{DBT} = 68.8 ng·g^{-1} d.w., and B_{MBT} = 103.9 ng·g^{-1} d.w. The differences between determined and reference concentrations of OTC in the material were within U_Δ (calculating according to Equation (1)). Thus, the determined concentrations were compatible with reference concentrations.

2.3. Analytical Method Comparison Based on the Real Samples of Bottom Sediments

The results of the determination of TBT, TPhT in BCR-646 freshwater sediments, and 10 samples of sediment from the Odra river estuary by the use of QuEChERS-LC-MS/MS and ISO 23161 GC-AED methods are presented in Table 2. Incurred Sample Reanalysis (ISR) tool was used to the assessment of the reproducibility of analytical methods [23]. The results for TBT determination in reference and real samples of sediments are presented in Figure 2. The graph presents individual ISR data points and concentration-dependent trends. If the difference between the concentrations determined by the use of the two methods is within 20% of the mean from at least 67% of the measurements, then the two methods are considered comparable. A difference higher than 20% was observed for three samples in low and medium concentrations. The cumulative %ISR calculated for

11 samples was 72.7% and met acceptance criteria ($\geq$67%) for the comparability of the tested analytical methods.

Table 2. The results of (TBT), triphenyltin (TPhT), dibutyltin (DBT), and monobutyltin (MBT) determination in real sediments samples and reference material of freshwater sediment (BCR 646) by the use QuEChERS LC-MS/MS and ISO GC-AED methods. Certified values of BCR-646: TBT-480 ng·g^{-1} ($\pm$80 ng·g^{-1}), TPhT-29 ng·g^{-1} ($\pm$11 ng·g^{-1}), DBT-770 ng·g^{-1} ($\pm$90 ng·g^{-1}), MBT-610 ng·g^{-1} ($\pm$120 ng·g^{-1}).

Sample	QuEChERS LC-MS/MS [a]		ISO GC-AED			
	TBT, n = 3 ng·g^{-1} d.w	TPhT, n = 3 ng·g^{-1} d.w	TBT, n = 3 ng·g^{-1} d.w	TPhT, n = 3 ng·g^{-1} d.w	DBT, n = 3 ng·g^{-1} d.w	MBT, n = 3 ng·g^{-1} d.w
S1	29.4 $\pm$ 1.1	<5.0	19.0 $\pm$ 1.1	<3.0	<2.4	<1.7
S2	177.2 $\pm$ 11.2	<5.0	137.8 $\pm$ 8.3	<3.0	23.4 $\pm$ 3.5	<1.7
S3	453.2 $\pm$ 7.4	<5.0	589.1 $\pm$ 35.4	<3.0	62.3 $\pm$ 9.4	23.9 $\pm$ 6.5
S4	261.3 $\pm$ 12.6	<5.0	291.3 $\pm$ 17.5	<3.0	522.4 $\pm$ 78.4	177.5 $\pm$ 47.9
S5	165.9 $\pm$ 19.3	<5.0	185.4 $\pm$ 11.1	<3.0	17.3 $\pm$ 2.6	52.3 $\pm$ 14.1
S6	483.4 $\pm$ 11.5	<5.0	573.9 $\pm$ 34.4	<3.0	137.4 $\pm$ 20.6	19.6 $\pm$ 5.3
S7	464.5 $\pm$ 22.7	15.1 $\pm$ 0.3	391.2 $\pm$ 23.5	11.0 $\pm$ 0.2	73.8 $\pm$ 11.1	46.2 $\pm$ 12.5
S8	1667.5 $\pm$ 57.8	21.9 $\pm$ 0.6	1862.5 $\pm$ 111.8	29.9 $\pm$ 0.6	231.6 $\pm$ 34.8	156.7 $\pm$ 42.3
S9	118.1 $\pm$ 4.0	<5.0	97.8 $\pm$ 5.9	<3.0	58.6 $\pm$ 8.8	99.6 $\pm$ 26.9
S10	631.1 $\pm$ 24.6	<5.0	743.4 $\pm$ 44.6	<3.0	<2.4	<1.7
BCR [a]	484.2 $\pm$ 14.4	29.6 $\pm$ 1.4	432 $\pm$ 6.1	27.8 $\pm$ 1.8	698.0 $\pm$ 14.2	483.8 $\pm$ 24.7

[a] the analyses of BCR were replicated six times.

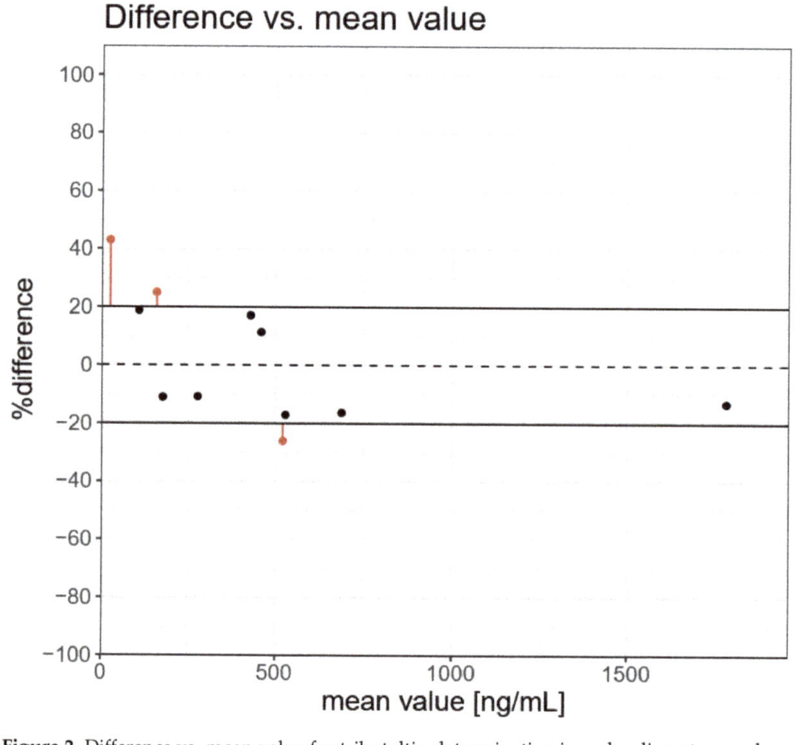

Figure 2. Difference vs. mean value for tributyltin determination in real sediments samples and BCR 464 freshwater sediment. The cumulative %ISR for 11 samples was 72.7% and met acceptance criteria ($\geq$67%) for comparability of the analytical methods tested.

Based on the Wilcoxon test of six replicates of freshwater sediment BCR-464 analyses, it was confirmed that the results of determination of TBT and TPhT by the use of these methods were not statistically different. In the case of TBT determination, the result of the test was $p = 0.0809$ whereas, in the case of TPhT, it was $p = 0.6625$. In real samples of bottom sediments from the Odra River estuary, TBT concentrations in dry weight of sample determined by QuEChERS LC-MS/MS varied from 29.4 to 1667.5 ng·g^{-1}, whereas, in the case of ISO 23161 GC-AED, the determined concentrations varied from 19.0 to 1862.5 ng·g^{-1}. Based on the Wilcoxon signed-rank test, it has been found that the results of TBT determination by the use of these methods were not statistically different ($p = 0.91$). TPhT was found only in the two samples of bottom sediments (samples #7 and #8). The concentration was determined both by the use of QuEChERS LC-MS/MS and ISO 23161 GC-AED (Table 2).

MBT and DBT at dry weight concentration levels above the LOQ were found in samples #7 (19.6 and 177.5 ng·g^{-1}.) and #8 (17.3 and 522.4 ng·g^{-1}), respectively (Table 2). The concentrations of these compounds in the Szczecin Lagoon (sample #3) were close to those reported in 2008 [5]. The highest DBT and MBT concentrations in the dry weight of the sample were 76.0 ng·g^{-1} and 67.0 ng·g^{-1}. In this study, determined DBT and MBT concentrations in the dry weight of the sample were 62.3 and 23.9 ng·g^{-1}, respectively. In 2008, it was reported that TPhT has not been found in sediment samples from the Szczecin Lagoon [5].

In the Szczecin Lagoon, the dry weight concentration of TBT reported in 2008 was 453.2 ng·g^{-1} [5], whereas, in this study, the concentration was 283.1 ng·g^{-1}. The occurrence of OTCs in the Szczecin Lagoon was also reported in 2018 [24]. The highest concentrations were 152.9 ng·g^{-1} for TBT, 66.1 ng·g^{-1} for DBT, and 67.1 ng·g^{-1} for MBT. OTCs concentrations in the sediment collected in the vicinity of the Szczecin harbor (sample #8) were 1667.5 ng·g^{-1} for TBT, 231.6 ng·g^{-1} for DBT, and 156.7 6 ng·g^{-1} MBT. Comparatively, in the sediment collected in Gdynia harbor, the mean dry weight concentrations were 2148.2 ng·g^{-1} for TBT, 751.9 ng·g^{-1} for DBT, and 261.4 ng·g^{-1} for MBT [25]. This indicates that the cleaning of ship surfaces by sandblasting is a source of OTCs pollution. Moreover, OTCs are persistent in bottom sediments. Thus, bottom sediments may be a potential source of environmental pollution by OTCs.

3. Materials and Methods

3.1. Experimental

Pure standards of tributyltin chloride (TBT, 96%), triphenyltin chloride (TPhT, 95%), dibutyltin chloride (DBT, 96%), monobutyltin (MBT, 95%) and internal standard, deuterated tributyltin chloride-d27 (TBT-d27, 96%) were purchased from Sigma Aldrich (St. Louis, MO, USA). HPLC gradient-grade methanol, acetonitrile, and formic acid 98% were purchased from Merck (Darmstadt, Germany). QuEChERS salts: magnesium sulfate, sodium chloride, sodium acetate, and ammonium acetate were purchased from Chempur (Piekary Śląskie, Poland), whereas trisodium citrate from Avantor (Gliwice, Poland). Acetic acid, methanol, hexane, and tetrahydrofurane of GC purity were purchased from Sigma Aldrich (St. Louis, MO, USA). Sodium tetraethylborate purity was 97% from Sigma Aldrich (St. Louis, MO, USA) and disodium citrate was bought from Acros Organics (Morris Plains, NJ, USA). Certificated Reference Material BCR® 646 was purchased from Merck (Darmstadt Germany). Ultrapure water was obtained from a Millipore water purification system (MiliQ, Billerica, MA, USA) equipped with a UV-lamp (resistivity of 18.2 MΩ.cm (at 25 °C) and a TOC value below 5 ppb).

The stock solutions of 1 mg·mL^{-1} of TBT, TPhT, DBT, MBT, and TBT-d27 were prepared in methanol. The working standard solutions were prepared by dilution of the stock solution with an appropriate amount of methanol just before use. All stock solutions were stored at -25 °C.

3.2. Organotin Compounds Extraction

3.2.1. Extraction of Organotin Compounds for LC-MS/MS Analysis (QuEChERS Extraction Procedure)

A dry sample of sediment was weighed (0.125 g) and placed in 2 mL Eppendorf® tubes. As an internal standard, 25 µL of deuterated tributyltin (TBT-d27) was added, followed by 200 µL of MiliQ water. Then, the samples were vigorously shaken by the use of a vortex shaker for 1 min and 250 µL of 5% formic acid in acetonitrile was added. The samples were put in a vessel with ice and 100 mg of ammonium acetate was added. Then they were extracted by ultrasonication for 5 min and shaken for 15 min (1500 rpm). The samples were centrifuged for 5 min (relative centrifuge force was $4472 \times g$) and the supernatant was collected for further analysis.

3.2.2. Extraction of Organotin Compounds for GC Analysis (ISO 23161 Extraction Procedure)

Extraction of OTCs from dry solid samples was carried out according to ISO 23161. Firstly, 3.2 g of lyophilized sediment was ultrasound extracted for 30 min. with 3 mL of the mixture of acetic acid:methanol: water (1:1:1; *v/v/v*). The extraction procedure was repeated with 1.5 mL of the mixture, and liquid phases were combined. Next, the pH was adjusted to 4.5 with acetic acid, and 5 mL of hexane and 10% sodium tetraethylborate in tetrahydrofurane (0.5 mL per gram of solid sample) was added. After that, the organic phase was concentrated to 1 mL.

3.3. Instrumental Methods

3.3.1. LC-MS/MS

Liquid chromatography with tandem mass spectrometry (LC-MS/MS) analysis was performed using Agilent 1260 Infinity (Agilent Technologies, Santa Clara, CA, USA) equipped with a degasser, autosampler, and binary pump, coupled to a Hybrid Triple Quadrupole/Linear Ion trap mass spectrometer (QTRAP® 4000, AB SCIEX, Framingham, MA, USA). The curtain gas, ion source gas 1, ion source gas 2, and collision gas (all high purity nitrogen) were set at 280 kPa, 380 kPa, 410 kPa, and "high" instrument units, respectively. The ion spray voltage and source temperatures were 5500 V and 600 °C, respectively. Kinetex RP-18 column (100 mm, 4.6 mm, particle size 2.6 µm) supplied by Phenomenex (Torrance, CA, USA) was used. The column temperature was 40 °C; the eluent flow rate was 0.5 mL/min. The eluent was prepared from two solutions: A-0.2% formic acid in water and B-0.2% formic acid in acetonitrile. The concentration of solution B in the eluent was 5% for 2 min, after that, the concentration increased to 95% in 7.5 min and for the next 5 min was 95%. The injection volume was 10 µL. The organotin compounds were analyzed in multiple reaction monitoring (MRM) mode. Two ion transitions (precursor ion and fragment ion) were m/z 291 → 179 for TBT, 351 → 196 for TPhT, and 318 → 190 for d-TBT.

3.3.2. GC-MS

The separation of organotin compounds was performed using an Agilent 7890A Series Gas Chromatograph interfaced to an Agilent 5973c Network Mass Selective Detector and an Agilent 7683 Series Injector (Agilent Technologies, Santa Clara, CA, USA). A 5 µL sample was injected in splitless mode (volume relative standard deviation was 0.3%) to an HP-5MS column (30 m × 0.25 mm I.D., 0.25 µm film thickness, Agilent Technologies, Santa Clara, CA, USA) using helium as the carrier gas at 1 mL/min flow. The ion source was maintained at 230 °C; the GC oven was programmed with a temperature gradient starting at 40 °C for 3 min, then increased with a 10 °C/min rate to 220 °C, held for 5 min, after that increased to 20 °C/min rate to 300 °C, and held for 10 min. MS was carried out in the electron-impact mode at an ionizing potential of 70 eV. Mass spectra were recorded in the range of 40–800 mass-to-charge ratio (m/z). Identification of the selected organic compounds was performed with an Agilent Technologies Enhanced ChemStation (G1701EA ver. E.02.00.493) and The Wiley Registry of Mass Spectral Data (version 3.2, Copyright 1988–2000 by Palisade

Corporation with, 8th 213 Edition with Structures, Copyright 2000 by John Wiley and Sons, Inc., Hoboken, NJ, USA) using a 3% cut-off threshold.

3.3.3. GC-AED

Organotin compounds were determined by GC (GC 7890A, Agilent Technologies, USA) coupled with atomic emission detector (AED) model JAS G2350A from Joint Analytical System (Moers, Germany). The samples (5 µL) were manually injected by the use of a 10 µL syringe. The injector was in splitless mode and kept at 280 °C. The organic compounds were separated using an HP-5 column (30 m, 0.32 mm I.D., 0.25 µm particle size, Agilent Technologies, Santa Clara, CA, USA) using helium as the carrier gas (1 mL/min). The temperature gradient was as follows: 40 °C for 3 min, then increased with 10 °C/min rate to 220 °C, held for 5 min, then increased with 20 °C/min rate to 300 °C, and after that, held for 10 min. Helium plasma was used as an excitation source in the AED detector. Detection lines for Sn were 301 and 303 nm. The temperature of the transfer line was 280 °C. The flows of the remaining reaction gases (oxygen, nitrogen, hydrogen) used in the determination of tin were set according to the manufacturer's recommendations. Software from Joint Analytical System, Moers, Germany (version D.02.01 was used for the calculation of OTCs concentrations based on the peak area of the analyte. All determinations were made in triplicate and averaged.

3.4. Verification Procedure

Verification of the analytical methods was performed in terms of limits of quantification (LOQ), limits of detection (LOD), analytical ranges (linearity), accuracy, precision, and recovery. The limit of detection (LOD) was estimated by determining the S/N of the minimum measured concentrations and extrapolating to the S/N value equals 3. Limit of quantification (LOQ) was established as the concentration of OTCs, for which MS signal-to-noise ratio is equal to or greater than 10, with precision below 20% and accuracy ± 20%. Linearity was evaluated based on a relation between peak area ratio (or peak area) and analyte concentration that can be represented as a straight line. A linear regression method was used to test linearity. Calibration curves were prepared in triplicate. The precision and accuracy of the analytical methods were determined based on six repetitions. Recoveries were determined based on six repetitions, comparing peak areas of target analytes extracted from blank samples to the peak areas of standard solution. The certified reference material BCR 646 (European Commission Joint Research Centre Institute for Reference Materials and Measurements, Geel, Belgium) was used to evaluate the accuracy, precision, and trueness of the results. The material consists of a dried and ground freshwater bottom sediment with a particle size (<90 µm), reference concentrations of OTCs with their uncertainty (U_Δ) were as follows: 480 ng·g^{-1} (U_Δ = 80 ng·g^{-1}) for TBT, 770 ng·g^{-1} (U_Δ = 90 ng·g^{-1}) for DBT, 610 ng·g^{-1} for MBT (U_Δ = 120 ng·g^{-1}) and 29 ng·g^{-1} (U_Δ = 11 ng·g^{-1}) for TPhT. Trueness is typically determined in terms of bias (B) by comparison of the obtained results with the reference value. Significance testing of the bias that took into account uncertainty of certificated value was performed. According to ISO Guide 33 [26] (ISO, 2015), the expend uncertainty of the difference between certificated and measured value U_Δ with coverage factor k = 2, corresponding to a level of confidence of 95%, is obtained by Equation (1):

$$U_\Delta = k \cdot \sqrt{u_{ref}^2 + \frac{S_m^2}{n}} \qquad (1)$$

where: u_{ref} is the uncertainty of the reference value taken from the certificate (40 ng·g^{-1} for TBT and 5.5 ng·g^{-1} for TPhT).

s_m is a standard deviation calculated from the measured values.

n is several repeated measurements, (n = 6).

3.5. Statistical Methods

A comparison of the analytical methods was performed by the use of Incurred Sample Reanalysis (ISR) tool developed by Rudzki et al. [23] and the Wilcoxon signed-rank test. For ISR the %difference with fixed acceptance limits was set at −20 and 20% and the %ISR should be at least 67% to confirm the equivalence of the methods. For the Wilcoxon signed-rank test, a p-value of 0.05 or less was considered significant. The statistical analysis of the results was performed with the STATISTICA version 13.1 for Windows (TIBCO Software Inc., Palo Alto, CA, USA).

3.6. Characteristics of Bottom Sediment Samples

Sediment samples were collected from the Odra River estuary by the use of the Van Veen grab sampler, collecting surface sediments up to 20 cm deep. Samples were kept at 4 °C till their delivery to the laboratory. Then, they were frozen at −80 °C, freeze-dried, and stored at −80 °C till analysis. Before analysis, the sediments were grounded in an agate mortar and sieved through a 0.063 mm sieve. Geographical localization of sampling sites and depths of sampling are presented in Table 3.

Table 3. Geographic coordinates, depth of sampling from the water surface, and localization name of sediments collected from the Odra River estuary.

Sample	Geographic Coordinates		Depth of Sampling from the Surface (m)	Localization
	Latitude N	Longitude E		
S1	53°51.209'	014°18.107'	1.5	Karsibór next to Rybaczówka
S2	53°59.570'	014°42.332'	2.1	West Dziwna River
S3	53°41.456'	014°25.297'	4.4	Szczecin Lagoon-Brzózki
S4	53°38.538'	014°35.958'	3.5	Roztoka Odrzańska-near to Stępnicka Bay
S5	53°33.572'	014°34.774'	1.8	Police-Larpia
S6	53°27.621'	014°36.102'	2.2	Sailing Canal, near Święta
S7	53°27.336'	014°35.434'	2.5	West Odra River-Gryfia Shipyard Dock No. 5
S8	53°27.328'	014°35.968'	2.2	Szczecin: opposite Gryfia Island
S9	53°26.300'	014°35.280'	10.3	Elevator "Ewa"
S10	53°23.827'	014°38.205'	3.6	Dąbie Marina

The relevant physicochemical parameters (total organic carbon content (TOC), total nitrogen content (N), acid volatile sulfur content (AVS), total phosphorus content (P), and granulometric composition (% of silt fraction and % of clay fraction)) are presented in Table S3 in Supplementary materials. Elemental analysis (N, P), TOC, and AVS analyzes were performed at the Polish Geological Institute-National Research Institute by the use of analytical methods accredited according to ISO-17025 standard. Granulometric analysis was performed at the University of Szczecin (according to the ISO 13320 standard method).

4. Conclusions

The newly developed QuEChERS LC-MS/MS method for the determination of OTCs in bottom sediments was compared with both GC-MS and GC-AED based standard method ISO-23161. Due to the high number of interferences, the determination of OTCs by GC-MS was very difficult and not recommended for environmental monitoring of bottom sediments. The GC-AED method is more selective and free from interferences. However, the determination of OTCs by QuEChERS LC-MS/MS required less time and it is more convenient for laboratory staff. The method does not require clean-up, derivatization step, and application of very toxic, volatile reagents. Therefore, QuEChERS-LC-MC/MS is more suitable for routine environmental monitoring of OTCs in a large number of bottom sediment samples. The results of OTCs determination in real bottom sediments by QuEChERS-LC-MC/MS and ISO 23161 were compatible. However, the degradation product of TBT, MBT, and DBT could not be detected by LC-MS/MS. The application of other chromatographic settings may likely alleviate the problem of MBT and DBT determination by LC-MS/MS. MBT and DBT were extracted quantitatively by the QuEChERS extraction method. However, the method cannot be applied in the determination of OTC by GC methods due to the necessity of solvent exchange.

Supplementary Materials: The following supporting information can be downloaded at: https://www.mdpi.com/article/10.3390/molecules27154847/s1. Table S1: Parameters of QuEChERS and ISO 23161 extraction procedures; Table S2: Separation and detection parameters of LC-MS/MS, GC-AED, and GC-MS techniques; Table S3: Granulometric composition (% of silt fraction and % of clay fraction) and content in mg·kg-1 of total organic carbon (TOC), total nitrogen content (N), acid volatile sulfur (AVS) and, total phosphorus (P) of sediments collected from the Odra River estuary.

Author Contributions: Formal analysis, D.K., R.S., J.G. and P.D.; resources, D.K., K.M., A.S. (Agnieszka Strzelecka) and A.S. (Artur Skowronek); data curation, D.K., R.S., P.D. and J.G.; methodology, J.G, R.S., D.K. and P.D.; writing–original draft, D.K., J.G. and P.D.; preparation, D.K.; conceptualization, J.G., D.K. and P.D., supervision, J.G. and P.D.; project administration, P.D.; funding acquisition, P.D.; writing—review and editing, J.G., P.D., D.K., K.M., A.S. (Agnieszka Strzelecka) and A.S. (Artur Skowronek). All authors have read and agreed to the published version of the manuscript.

Funding: This research was financially supported by the grant OPUS 11 from the National Science Centre (Poland), entitled "Occurrence of organotin compounds in bottom sediment of the Odra river estuary-environment conditions affecting the presence of the compounds, mobility, their degradation products, and environmental persistence", grant number UMO–2016/21/B/ST10/02391.

Institutional Review Board Statement: Not applicable.

Informed Consent Statement: Not applicable.

Data Availability Statement: Not applicable.

Acknowledgments: D.K. thanks for the doctoral scholarship from Warsaw Medical University and OPUS 11 grant from National Science Centre Poland (grant number UMO–2016/21/B/ST10/02391).

Conflicts of Interest: The authors declare no conflict of interest.

Sample Availability: Samples of the compounds are not available from the authors.

References

1. Olushola Sunday, A.; Alafara, B.A.; Oladele, O.G. Toxicity and speciation analysis of organotin compounds. *Chem. Speciat. Bioavailab.* **2012**, *24*, 216–226. [CrossRef]
2. Concha-Graña, E.; Moscoso-Pérez, C.; Fernández-González, V.; López-Mahía, P.; Gago, J.; León, V.M.; Muniategui-Lorenzo, S. Phthalates, organotin compounds and per-polyfluoroalkyl substances in semiconfined areas of the Spanish coast: Occurrence, sources and risk assessment. *Sci. Total Environ.* **2021**, *780*, 146450. [CrossRef] [PubMed]
3. Chen, Z.; Chen, L.; Chen, C.; Huang, Q.; Wu, L.; Zhang, W. Organotin Contamination in Sediments and Aquatic Organisms from the Yangtze Estuary and Adjacent Marine Environments. *Environ. Eng. Sci.* **2016**, *34*, 227–235. [CrossRef]
4. Radke, B.; Pazikowska-Sapota, G.; Galer-Tatarowicz, K.; Dembska, G. The evaluation of tributyltin concentrations in the sediments from the Port of Gdynia. *Biul. Inst. Mor.* **2018**, *33*, 175–182. [CrossRef]
5. Filipkowska, A.; Kowalewska, G.; Pavoni, B. Organotin compounds in surface sediments of the Southern Baltic coastal zone: A study on the main factors for their accumulation and degradation. *Environ. Sci. Pollut. Res. Int.* **2014**, *21*, 2077–2087. [CrossRef]
6. Kucharski, D.; Drzewicz, P.; Nałęcz-Jawecki, G.; Mianowicz, K.; Skowronek, A.; Giebułtowicz, J. Development and Application of a Novel QuEChERS Method for Monitoring of Tributyltin and Triphenyltin in Bottom Sediments of the Odra River Estuary, North Westernmost Part of Poland. *Molecules* **2020**, *25*, 591. [CrossRef]
7. Smedes, F.; de Jong, A.S.; Davies, I.M. Determination of (mono-, di- and) tributyltin in sediments. Analytical methods. *J. Environ. Monit.* **2000**, *2*, 541–549. [CrossRef]
8. Astruc, A.; Astruc, M.; Pinel, R.; Potin-Gautier, M. Speciation of butyltin compounds by on-line HPLC-ETAA of tropolone complexes in environmental samples. *Appl. Organomet. Chem.* **1992**, *6*, 39–47. [CrossRef]
9. Cole, R.F.; Mills, G.A.; Parker, R.; Bolam, T.; Birchenough, A.; Kröger, S.; Fones, G.R. Trends in the analysis and monitoring of organotins in the aquatic environment. *Trends Environ. Anal. Chem.* **2015**, *8*, 1–11. [CrossRef]

10. *ISO 23161:2018*; Soil Quality—Determination of Selected Organotin Compounds—Gas-Chromatographic Method. International Organization for Standarization: Geneva, Switzerland, 2018.
11. Validated Test Method 8323: Determination of Organotins by Micro-Liquid Chromatography-Electrospray Ion Trap Mass Spectrometry. U. S. Environmental Protection Agency: Washington, DC, USA, 2003. Available online: https://www.epa.gov/sites/default/files/2015-12/documents/8323.pdf (accessed on 14 October 2021).
12. Nowak, P.M.; Wietecha-Posłuszny, R.; Pawliszyn, J. White Analytical Chemistry: An approach to reconcile the principles of Green Analytical Chemistry and functionality. *TrAC Trends Anal. Chem.* **2021**, *138*, 116223. [CrossRef]
13. dos Santos, G.C.; Avellar, A.D.S.; Schwaickhardt, R.D.O.; Bandeira, N.M.G.; Donato, F.F.; Prestes, O.D.; Zanella, R. Effective methods for the determination of triphenyltin residues in surface water and soil samples by high-performance liquid chromatography with tandem mass spectrometry. *Anal. Methods* **2020**, *12*, 2323–2330. [CrossRef]
14. Radke, B.; Namiesnik, J.; Bolalek, J. Pathways and determinations of organometallic compounds in the marine environment with a particular emphasis on organotin. *Oceanol. Hydrobiol. Stud.* **2003**, *32*, 43–71.
15. Parkinson, D.-R.; Dust, J.M. Overview of the current status of sediment chemical analysis: Trends in analytical techniques. *Environ. Rev.* **2010**, *18*, 37–59. [CrossRef]
16. Joshi, D.; Adhikari, N. An Overview on Common Organic Solvents and Their Toxicity. *J. Pharm. Res. Int.* **2019**, *28*, 1–18. [CrossRef]
17. Alfonsi, K.; Colberg, J.; Dunn, P.J.; Fevig, T.; Jennings, S.; Johnson, T.A.; Kleine, H.P.; Knight, C.; Nagy, M.A.; Perry, D.A.; et al. Green chemistry tools to influence a medicinal chemistry and research chemistry based organisation. *Green Chem.* **2008**, *10*, 31–36. [CrossRef]
18. Morabito, R.; Massanisso, P.; Quevauviller, P. Derivatization methods for the determination of organotin compounds in environmental samples. *TrAC Trends Anal. Chem.* **2000**, *19*, 113–119. [CrossRef]
19. Dauchy, X.; Cottier, R.; Batel, A.; Jeannot, R.; Borsier, M.; Astruc, A. Speciation of Butyltin Compounds by High-Performance Liquid Chromatography with Inductively Coupled Plasma Mass Spectrometry Detection. *J. Chromatogr. Sci.* **1993**, *31*, 416–421. [CrossRef]
20. Kadokami, K.; Uehiro, T.; Morita, M.; Fuwa, K. Determination of organotin compounds in water by bonded-phase extraction and high-performance liquid chromatography with long-tube atomic absorption spectrometric detection. *J. Anal. At. Spectrom.* **1988**, *3*, 187–191. [CrossRef]
21. Dauchy, X.; Cottier, R.; Batel, A.; Borsier, M.; Astruc, A.; Astruc, M. Application of butyltin speciation by HPLC/ICP-MS to marine sediments. *Environ. Technol.* **1994**, *15*, 569–576. [CrossRef]
22. Barwick, V.J. The Fitness for Purpose of Analytical Methods—A Laboratory Guide to Method Validation and Related Topics. Available online: www.eurachem.org (accessed on 20 November 2021).
23. Rudzki, P.J.; Biecek, P.; Kaza, M. Comprehensive graphical presentation of data from incurred sample reanalysis. *Bioanalysis* **2017**, *9*, 947–956. [CrossRef]
24. Filipkowska, A.; Kowalewska, G. Butyltins in sediments from the Southern Baltic coastal zone: Is it still a matter of concern, 10 years after implementation of the total ban? *Mar. Pollut. Bull.* **2019**, *146*, 343–348. [CrossRef]
25. Radke, B.; Wasik, A.; Jewell, L.; Pączek, U.; Namiesnik, J. The Speciation of Organotin Compounds in Sediment and Water Samples from the Port of Gdynia. *Soil Sediment Contam. Int. J.* **2013**, *22*, 614–630. [CrossRef]
26. *ISO Guide 33:2015*; Reference Materials-Good Practice in Using Reference Materials. International Organization for Standarization: Geneva, Switzerland, 2015.

MDPI
St. Alban-Anlage 66
4052 Basel
Switzerland
Tel. +41 61 683 77 34
Fax +41 61 302 89 18
www.mdpi.com

Molecules Editorial Office
E-mail: molecules@mdpi.com
www.mdpi.com/journal/molecules

www.ingramcontent.com/pod-product-compliance
Lightning Source LLC
LaVergne TN
LVHW070050120526
838202LV00102B/1978